NASA/TM–2008–214779

HISTORY OF ON-ORBIT SATELLITE FRAGMENTATIONS
14th Edition

Orbital Debris Program Office

Nicholas L. Johnson

Eugene Stansbery
David O. Whitlock
Kira J. Abercromby
Debra Shoots

National Aeronautics and
Space Administration

Lyndon B. Johnson Space Center
Houston, Texas 77058

June 2008

Preface to the Fourteenth Edition

The first edition of the <u>History of On-Orbit Satellite Fragmentations</u> was published by Teledyne Brown Engineering (TBE) in August 1984, under the sponsorship of the NASA Johnson Space Center and with the cooperation of the United States Air Force Space Command and the U.S. Army Ballistic Missile Command. The objective was to bring together information about the 75 satellites, which had at that time experienced noticeable breakups. This update encompasses all known satellite fragmentations. This update is published by the NASA Johnson Space Center, Orbital Debris Program Office with support from the Engineering and Sciences Contract Group.

Since the thirteenth edition (published in May 2004) there have been 21 identified on-orbit breakups and seven anomalous events, for a historical total of 194 fragmentations and 51 anomalous events. This activity has resulted in an approximately 27% increase in the historical cataloged debris count (since 31 December 2003), which includes on-orbit and decayed objects. More significant, an increase of 69% in the on-orbit debris count is observed since the last edition. The reason for these large increases was the intentional destruction of the Fengyun 1C spacecraft on 11 January 2007. If this event had not occurred, the increase in the historical cataloged debris count would have been only 6% (vice 27%) and the increase in on-orbit debris would have been only 8% (vice 69%)

The current authors would like to recognize the substantial contributions of the authors of previous editions of this document. In addition, the assistance of personnel of U.S. Space Command, Air Force Space Command, Naval Network and Space Operations Command (formerly Naval Space Command), and Teledyne Brown Engineering has been vital to the present work.

CONTENTS

ACRONYMS

ADCOM	(USAF) Aerospace Defense Command
AFB	Air Force Base
AFSPC	Air Force Space Command
AFSSS	Air Force Space Surveillance System (formerly NAVSPASUR)
AN/FPS-85	See FPS-85
Asc	Ascending
BMEWS	Ballistic Missile Early Warning System
CIS	Commonwealth of Independent States (see also USSR)
Dsc	Descending
ESA	European Space Agency
ESRO	European Space Research Organization
FPS-85	Phased-array UHF radar at Eglin AFB, Florida
GEO	Geosynchronous Orbit (orbit category)
GEODSS	Ground-based Electro-Optical Deep-space Surveillance System
JSC	Johnson Space Center (NASA)
LEO	Low Earth Orbit, up to 2000 km altitude (orbit category)
NASA	National Aeronautics and Space Administration
NAVSPASUR	Naval Space Surveillance System
NAVSPOC	Naval Space Operations Center
NORAD	North American Aerospace Defense Command
PARCS	Phased-array UHF radar at Cavalier AFB, North Dakota: the Perimeter Acquisition Radar Attack Characterization System
RAE	The Royal Aerospace Establishment
R/B	Rocket Body or Rocket Booster
RORSAT	Radar Ocean Reconnaissance Satellite
SATRAK	PC compatible astrodynamics toolkit
SCC	formerly Space Computational Center (obsolete); now Space Control Center
SOZ	Sistema Obespechanya Zapuska (Proton-K Block DM attitude/ullage motor unit)
SSN	Space Surveillance Network
TBE	Teledyne Brown Engineering
TLE	Two-Line Element Set
USSPACECOM	United States Space Command
USSR/CIS	Union of Soviet Socialist Republics/Commonwealth of Independent States (after 1991)

1.1.1.1.1 SYMBOLS

ΔP	The maximum observed change in the orbital period [min].
ΔI	The maximum observed change in the inclination [ ].

1.0 INTRODUCTION

Since the first serious satellite fragmentation occurred in June 1961 (which instantaneously increased the total Earth satellite population by more than 400%) the issue of space operations within the finite region of space around the Earth has been the subject of increasing interest and concern. The prolific satellite fragmentations of the 1970s and the marked increase in the number of fragmentations in the 1980s served to widen international research into the characteristics and consequences of such events. Continued events in all orbits in later years make definition and historical accounting of those events crucial to future research. Large, manned space stations and the growing number of operational robotic satellites demand a better understanding of the hazards of the dynamic Earth satellite population.

The contribution of satellite fragmentations to the growth of the Earth satellite population is complex and varied. The majority of detectable fragmentation debris has already fallen out of orbit, and the effects of 40% of all fragmentations have completely disappeared. On the other hand, just 10 of more than 4500 space missions flown since 1957 are responsible for 31% of all cataloged artificial Earth satellites presently in orbit (Figure 1.0-1). Moreover, the sources of 8 of these 10 fragmentations were discarded rocket bodies that had operated as designed, but later broke up. The recent fragmentation of Fengyun 1C (1999-025) currently accounts for 17% of all cataloged objects in orbit. Likewise, 8 of these 10 most debris-producing fragmentations occurred more than ten years ago (with one of the two exceptions being intentional). The primary factors affecting the growth of the true Earth satellite population are the international space launch rate, satellite fragmentations, and solar activity. As of 1 August 2007, almost half of the cataloged Earth satellite population was determined to be fragmentation debris, as illustrated in Figure 1.0-2. Also, approximately three out of every four payloads are no longer operational and constitute a separate, but statistically important class of orbital debris.

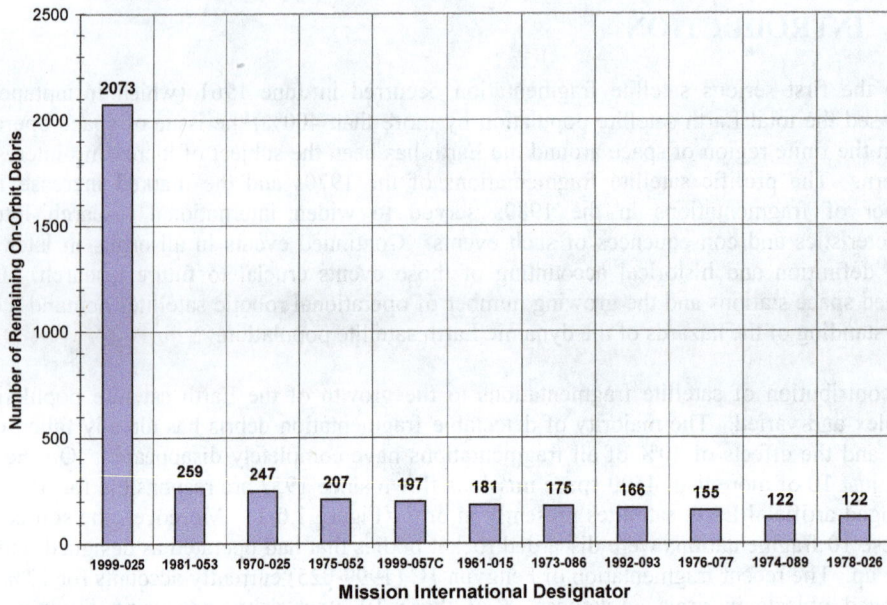

Figure 1.0-1. Magnitude of the 11 largest debris clouds *in orbit* as of 1 August 2007.

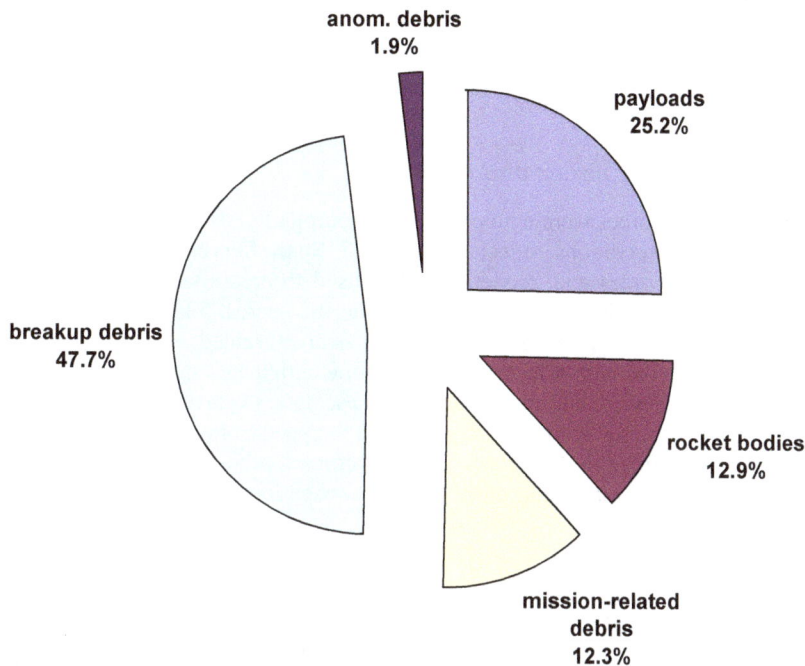

Figure 1.0-2. Relative segments of the cataloged *in-orbit* Earth satellite population.

1.1   Definition of Terms

In this volume, satellite fragmentations are categorized by their assessed nature and, to a lesser degree, by their effect on the near-Earth space environment. A **satellite breakup** is the usually destructive disassociation of an orbital payload, rocket body, or structure, often with a wide range of ejecta velocities. A satellite breakup may be accidental or the result of intentional actions, e.g., due to a propulsion system malfunction or a space weapons test, respectively. An **anomalous event** is the unplanned separation, usually at low velocity, of one or more detectable objects from a satellite, which remains essentially intact. Anomalous events can be caused by material deterioration of items such as thermal blankets, protective shields, or solar panels, or by the impact of small particles. As a general rule, a satellite breakup will produce considerably more debris, both trackable and non-trackable, than an anomalous event. From one perspective, satellite breakups may be viewed as a measure of the effects of man's activity on the environment, while anomalous events may be a measure of the effects of the environment on man-made objects.

Mission-related debris result from the intentional release of objects, usually in small numbers, during normal on-orbit operations. Objects ejected during the deployment, activation, and de-orbit of payloads and during manned operations are examples of mission-related debris. Usually, mission-related debris from a single launch are few in number, but extreme examples occasionally arise, such as the 200 objects from the *Salyut 7* space station or the more than 140

3

objects from the Westford Needles experiment. Although mission-related debris represent a non-trivial portion (approximately 12%) of all satellites in orbit today and, therefore, are a legitimate subject in the study of methods to retard the growth of the Earth satellite population, identification of the thousands of mission-related debris events is beyond the scope of this report.

1.2   Information Sources and Accuracy

A number of data sources were employed in the compilation of this volume. However, nearly all are derived from observations collected by the U.S. Space Surveillance Network (US SSN). Due to the variety of sources and geodetic models used to create satellite orbital element sets, all altitudes cited within this volume are presented to the nearest 5 km, referenced to a mean Earth of radius 6378.145 km. The accuracy of the data presented is not of adequate fidelity for precision analysis, although it is appropriate for the anticipated uses of this text. Complete base element sets are provided, but manipulation of these data, in particular satellite orbit propagation, should be performed only with validated, Air Force Space Command (AFSPC)-derived software, such as the PC compatible SATRAK astrodynamics toolkit. Long-term propagation of these elements is not appropriate regardless of the propagation technique applied and is, therefore, discouraged.

Although all fragmentations are described by the number of debris cataloged and the number of cataloged debris remaining in orbit, these parameters are poor measures of merit and should be used with extreme caution when undertaking comparative analyses. The sensitivity of the SSN and hence the degree to which debris will be detected and cataloged are highly dependent upon satellite altitude and to a lesser degree on satellite inclination. Additionally, historical cataloging practices have changed over the years. Past practices have included cataloging all debris objects associated with a breakup, even if they had already decayed; cataloging almost no pieces from a low altitude breakup when decay of most of the cloud was imminent; and cataloging objects as they were created, regardless of status. These different practices have resulted in an inconsistent historical record.

As a rule of thumb, low altitude, cataloged debris are assessed to be larger than 10 cm in diameter. At higher altitudes objects less than 1 m in diameter may be undetectable. Individual object sensitivities may vary dramatically from this simple generalization. Debris counts for fragmentations occurring in highly elliptical orbits near 63 degrees inclination (Molniya-type) are traditionally low, in part due to stable perigees situated deep in the Southern Hemisphere and often beyond SSN coverage. During a special surveillance session in 1987, as many as 250 uncataloged objects were observed in low inclination, highly elliptical orbits, but reliable tracking and parent identification were not achieved. The disclosure by the Russian Government of the Ekran 2 battery explosion on 25 June 1978 is the first known fragmentation in geostationary orbit. This event was not detected by the SSN, but since the event, four pieces have entered the catalog. Cataloging errors, e.g., identification of an object with the wrong parent satellite, normally are not explicitly noted in this volume since many errors have been or may be corrected.

For fragmentations at very low altitudes, i.e., below 400 km, much of the debris may reenter before detection, identification, and cataloging can be completed. For example, when the debris cloud from Cosmos 1813 passed over a single SSN radar, 846 individual fragments could be discerned. However, the total number of debris officially cataloged only reached 195. Likewise, more than 380 fragments are known to have been injected into Earth orbits (an equal number probably were sent on reentry trajectories) following the USA 19 test, but only 18 debris were entered into the official satellite catalog.

1.3 Environment Overview

To place the debris population component of the orbital environment in context for the reader, it is useful to review the general orbital environment in the near Earth and near geosynchronous regions. Differentiation of the population by source, object type, and orbit type are also included below.

1.3.1 ON-ORBIT SPATIAL DENSITY

The spatial density of resident space objects is a common means of describing the space object environment and is adopted here. Spatial density (objects per unit volume) represents the effective number of spacecraft and other objects as a function of altitude. Effective number, rather than the simple counting of objects, is used because many objects traverse the altitude regions of interest yet contribute little to the local collision hazard, e.g., geosynchronous transfer orbits. Such orbits exhibit an effective contribution to the environment at any given altitude of up to two orders of magnitude less than an object in a circular orbit within this same altitude interval. Thus, circular orbits at or near an orbit of interest normally dominate the hazard environment. The following figure portrays the near Earth (up to 2000 km altitude) environment categorized by intact or debris object types. The densities are subdivided into 10 km altitude intervals and graphed linearly. The epoch of the source data, a US SSN Two-Line Element (TLE) set, is 1 Aug 2007.

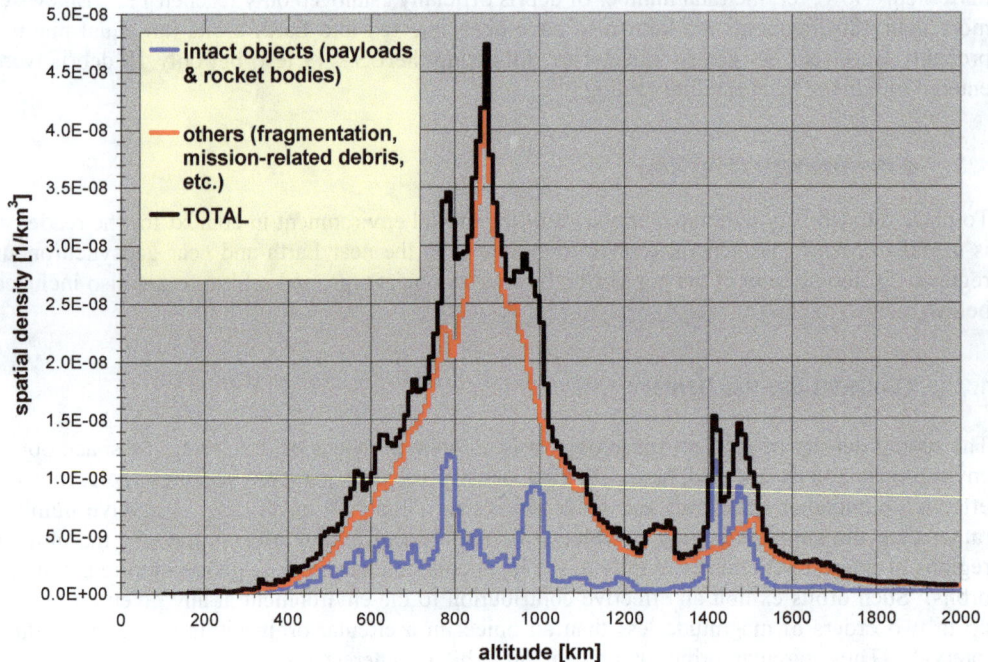

Figure 1.3.1-1. The near Earth (up to 2000 km) altitude population.

It should be noted that some "uncataloged" objects are included in this figure for completeness. These object orbits are reasonably well known, but not yet directly attributed to a specific launch and therefore have not been included in the US SSN catalog. The peak near 890 km is due, in large part, to the recent intentional destruction of the Fengyun 1C derelict weather spacecraft, creating over 2000 cataloged debris and nearly 500 debris still awaiting cataloging. Also clearly visible in this figure are other high-density regions of space. For example, the satellite constellations deployed in LEO in the late 1990s are clearly evident: the IRIDIUM constellation inhabits the altitude region at and about 780 km altitude, while the GLOBALSTAR constellation inhabits the region from 1410-1420 km. Other spacecraft constellations, such as the USSR/CIS communications and navigation constellations, are also visible near 1480 km and 950 km, respectively.

The geosynchronous altitude environment increased in both importance and number of resident space objects over the course of the 1990s and 2000s. Fig. 1.3.1-2 shows the geosynchronous altitude using a logarithmic, vertical spatial density axis and altitude intervals of 25 km. Only objects with an inclination less than 15 degrees were included. Consequently, the spatial density values assume all spacecraft are contained within 15 degrees latitude from the equator. Because high inclination orbits normally do not penetrate this true geosynchronous region, the assumption is appropriate to best categorize the spatial density of this region.

6

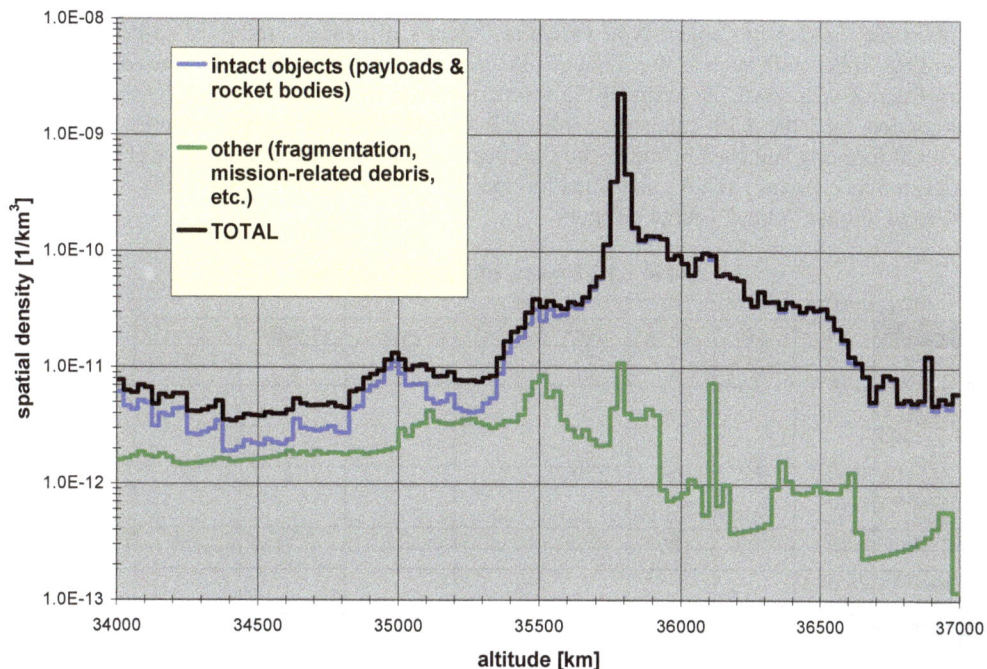

Figure 1.3.1-2. The geosynchronous altitude population.

Because the LEO spatial density chart averages over all inclinations and the GEO spatial density chart averages over inclinations between ±15 degrees, collision rates are not linearly related to the spatial density at any given altitude. Rather, collision rates will vary not only with the spatial density, but also with the inclination-dependent relative velocity. Altitudes dominated by high inclination (70-110°) orbits yield a significantly higher collision rate as compared to those populated by lower inclination orbits. This is because objects in these inclinations can collide at near head-on engagement geometries with objects in complementary inclinations. The exception to this general rule is provided by the commercial constellations in LEO and spacecraft in GEO. The commercial constellations are maintained in precise orbital planes; hence, their expected collision rate would be versus the "background" population only. Therefore, the spikes representing the IRIDIUM and GLOBALSTAR constellations do not present the inordinate collision risk implied by a casual examination. Similarly, the GEO environment is characterized by low collision velocities (< 1.5 km/s) due to the relative motion between controlled and uncontrolled objects.

1.3.2 POPULATION DISTRIBUTION

The distribution of objects by type (e.g., spacecraft, rocket bodies) and source (U.S., the People's Republic of China, etc.) is germane to this discussion since objects are not randomly distributed among these categories. To display this orbital anisotropy, the 1 August 2007 U.S. Satellite Catalog was categorized by these nominal variables. In the following table, most category

7

identities should be obvious to the casual reader; however, several require further identification. For example, among the object type variables, "debris dispensed" refers to so-called "debris dispensing" spacecraft, such as the Soviet/Russian manned orbital stations and the same source's *Romb/Duga-K* spacecraft. In terms of the source variable, spacecraft launched by the USSR are incorporated into the CIS category, while the "ESRO/ESA" category includes only those spacecraft formally launched by either the European Space Research Organization (ESRO) or the European Space Agency (ESA); launches for specific countries, such as Germany or Spain, are distributed into the "other" source category.

Table 1.3.2 Source vs. Type Accounting

on-orbit	US	CIS	France	PRC	India	Japan	ESRO/ESA	Other	totals
payloads	1063	1324	44	61	33	103	36	387	3051
rocket bodies	542	837	97	37	8	35	6	27	1589
debris dispensed	0	0	0	0	0	0	0	0	0
mission related debris	779	507	92	62	1	36	12	5	1494
breakup debris	1666	1524	126	2315	97	2	18	35	5783
anomalous debris	144	82	3	0	0	0	0	0	229
totals	4194	4274	362	2475	139	176	72	454	12146

decayed or beyond Earth orbit	US	CIS	France	PRC	India	Japan	ESRO/ESA	Other	totals
payloads	800	1860	8	50	9	22	18	49	2816
rocket bodies	631	2367	55	68	8	53	5	7	3194
debris dispensed	0	1250	0	0	0	0	0	0	1250
mission related debris	707	4300	123	113	8	81	8	54	5394
breakup debris	2835	3272	474	179	249	22	4	4	7039
anomalous debris	150	5	2	0	0	2	0	0	159
totals	5123	13054	662	410	274	180	35	114	19852
								Grand Total ->	31998

Several salient features are apparent in this table. Debris is dominant among all source variables and the majority of debris (and all other categories of resident space objects) are due to space activities of the U.S., CIS, and PRC. However, individual events from other space-faring nations have also contributed greatly to the local environment in several sun-synchronous orbital regimes. Examples are provided by the 1986 fragmentation of the Ariane SPOT-1/Viking rocket body, the 2000 fragmentation of the Long March 4 CBERS-1/SACI-1 rocket body, and most recently, the intentional destruction of the Fengyun 1C satellite by the People's Republic of China.

A net increase of over 3000 objects on-orbit (almost 4000 total) has been observed since the Thirteenth Edition of this book was published in 2004. Over 2000 of the on-orbit objects were from the intentional destruction of the aforementioned Fengyun 1C spacecraft. Table 1.3.3 shows the net increase or decrease in objects since the Thirteenth Edition. A discouraging feature of Table 1.3.3 is that every object type showed a net increase of on-orbit objects, and only breakup debris from India showed a net decrease of nine objects when evaluating any source and type combination.

Table 1.3.3 Source vs. Type Accounting – Net change since 13th Edition

on-orbit	U.S.	CIS	France	PRC	India	Japan	ESRO/ESA	Other	totals
payloads	82	11	11	23	6	19	3	60	215
rocket bodies	17	22	5	15	2	6	0	13	80
debris dispensed	0	0	0	0	0	0	0	0	0
mission related debris	160	78	5	51	0	17	1	3	315
breakup debris	110	141	12	2065	-9	2	7	35	2363
anomalous debris	38	69	2	0	0	0	0	0	109
totals	407	321	35	2154	-1	44	11	111	3082

decayed or beyond Earth orbit	U.S.	CIS	France	PRC	India	Japan	ESRO/ESA	Other	totals
payloads	22	48	0	7	1	2	3	1	84
rocket bodies	40	67	9	14	1	3	0	3	137
debris dispensed	0	0	0	0	0	0	0	0	0
mission related debris	19	73	11	26	3	17	1	25	175
breakup debris	71	214	5	29	29	21	0	4	373
anomalous debris	15	1	1	0	0	0	0	0	17
totals	167	403	26	76	34	43	4	33	786
								Grand Total ->	3868

2.0   SATELLITE BREAKUPS

This section summarizes the current breakup environment and describes each individual breakup. Each breakup is presented in a two-page format. New classes of breakup types have tended to fuel the background breakup rate, replacing classes of breakups from older on-orbit practices such as the Delta rocket body failures.

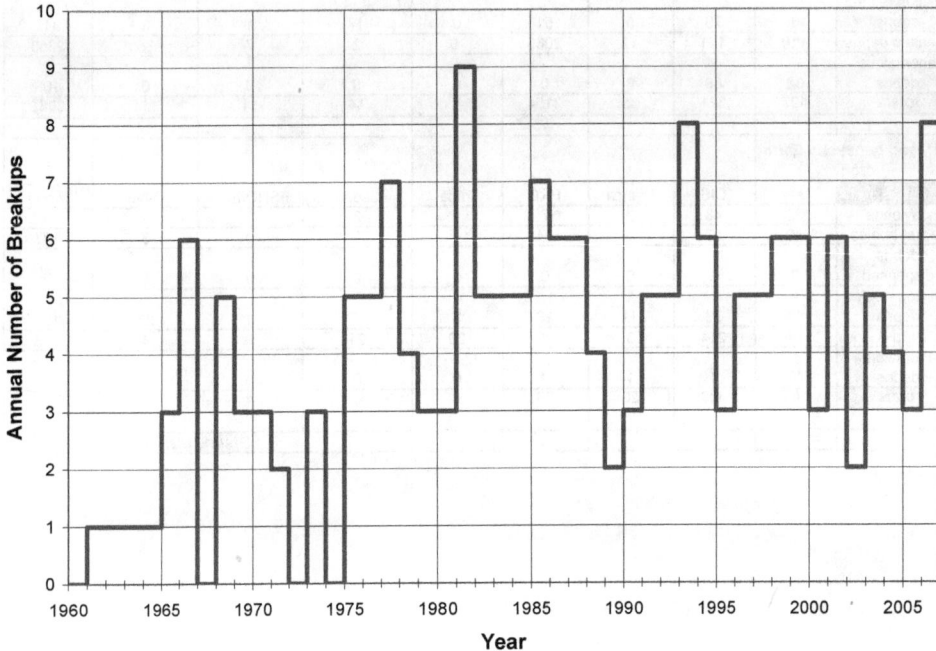

Figure 2.0-1. Number of breakups by year since 1961.

2.1   Background and Status

By far the most important category of man-made, on-orbit objects is satellite breakups, which now account for almost 48% of the total cataloged on-orbit Earth satellite population of 12146 Earth-orbiting objects. Since 1957 a total of 194 satellites are believed to have broken up (Tables 2.1 and 2.2). Breakups due to aerodynamic forces at or near reentry are treated separately from breakups caused by other factors, because aerodynamic breakups occur at the end of the satellite lifetime and, therefore, contribute nothing toward the orbital environment past the very near term. Only a fraction of these breakups are even detected because of the short remaining lifetime of the object and its debris. Twenty-four additional breakups of this aerodynamic nature that have been detected, and these events, are discussed in Chapter 4 and omitted from data included in this chapter.

The primary causes of satellite breakups (Figure 2.1-1) are propulsion-related events and deliberate actions, although the cause for over one in five breakups remains uncertain. This document will continue to carry breakup causes as unknown until a strong case can be made for one of the other cause classifications. Deliberate actions, often associated with activities related to national security, were formerly the most frequently occurring class, although only one such event occurred during the decade from 1997 until the Fengyun 1C event in January 2007. On average, the resulting debris from deliberate actions are short-lived (Figures 2.1-2 and 2.1-3), the exception being Fengyun 1C. Propulsion-related breakups, currently the most frequent class, include catastrophic malfunctions during orbital injection or maneuvers, subsequent explosions based on residual propellants, and failures of active attitude control systems. Breakups of rocket bodies due to propulsion failures are usually more prolific and produce longer-lived debris than the intentional destruction of payloads, often due to the higher altitudes of the malfunctioning rocket bodies rather than the mechanics of the explosive event. Although it may appear obvious that a rocket body breakup should be classified under the "Propulsion" category, rocket body events are carried as "Unknown" until a failure mechanism can be confidently identified for that rocket body design and is associated with a given rocket body event.

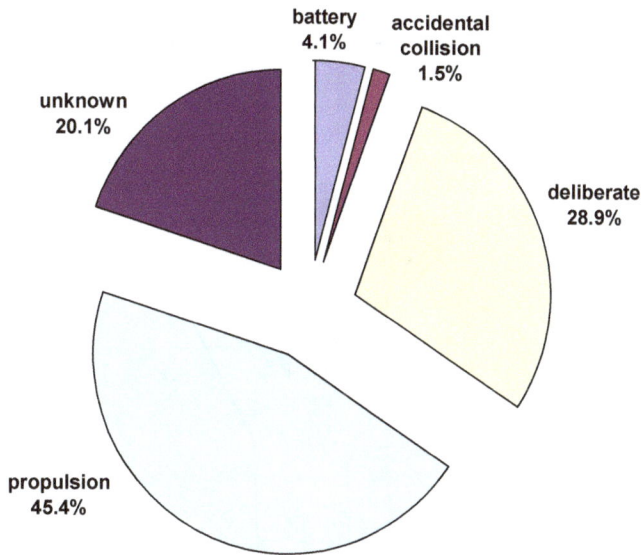

Figure 2.1-1. Causes of known satellite breakups.

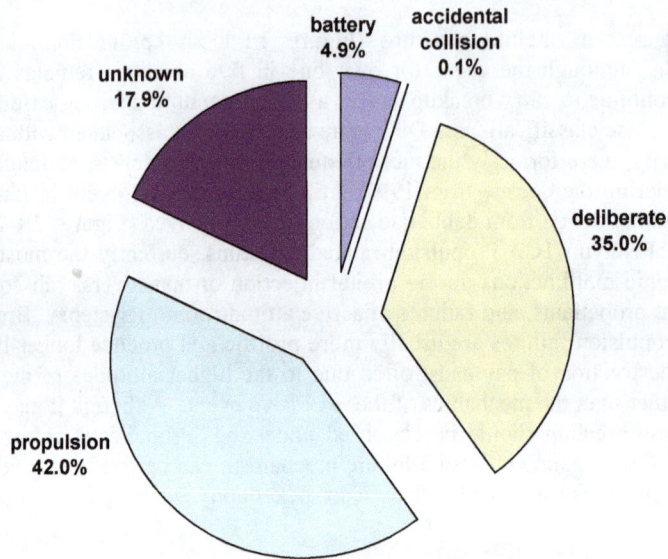

Figure 2.1-2. Proportion of all cataloged satellite breakup debris.

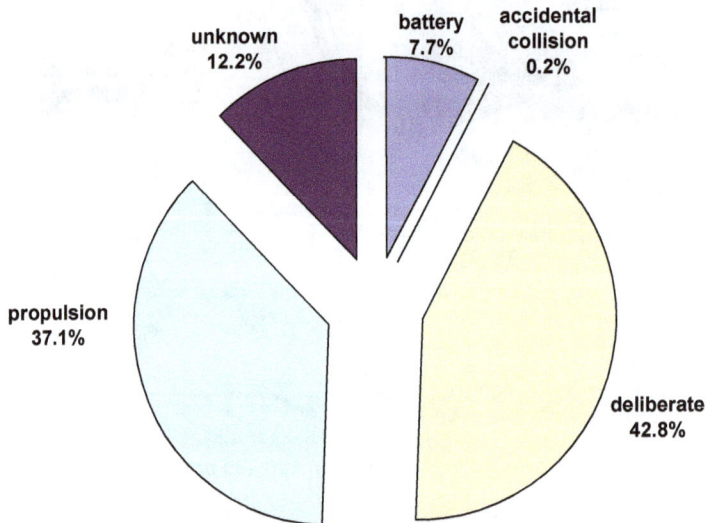

Figure 2.1-3. Proportion of cataloged satellite breakup debris _remaining in orbit_.

The rate of satellite breakups increased noticeably in the 1970s and has continued through the 1990s and into the new millennium at an average pace of approximately five fragmentations per year. There was a period of 11 months from June 2005 to May 2006 in which no known breakups took place. While it would be exciting for this to indicate a trend, there were 13 events in the following 20 months, rendering the previous breakup-free period as nothing more than a statistical anomaly. Increased awareness of potential hazards has resulted in positive actions to mitigate or eliminate many known breakup causes, e.g., Delta second stages, weapons testing, and Cosmos 862-type events. Together, these three programs were responsible for more than one-in-four of all satellite breakups in the decade of the 1980s. The quick response of *Arianespace* and the European Space Agency to the breakup of an Ariane third stage in 1986 is indicative of a desire by most space-faring organizations to operate in near-Earth space responsibly. Today, new series of boosters and satellites have resulted in new breakup sources, such as the fragmentation of a Pegasus HAPS stage in June 1996. Also, the intentional destruction of the Fengyun 1C spacecraft has noticeably increased the percentage of debris from deliberate events as compared to the previous edition of this book.

Figures 2.1-4 and -5 illustrate that a slight majority of the satellite breakup debris total and debris remaining in orbit today have originated from payloads. In previous editions of this book, rocket bodies always had more total and on-orbit debris than payloads (in the case of on-orbit, by a ratio of almost 3:1), but the 2000+ fragments from the Fengyun 1C event have changed the pattern.

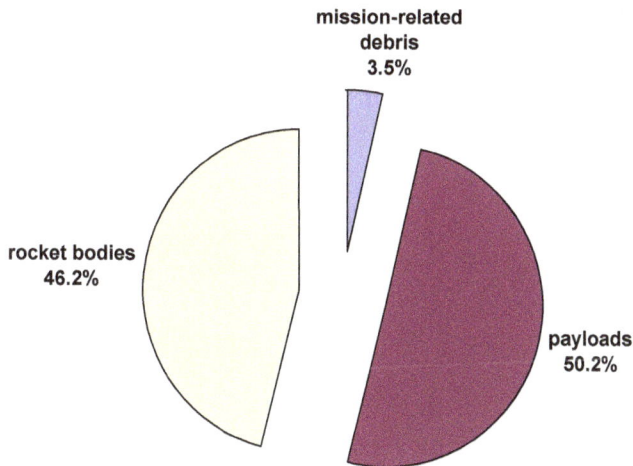

Figure 2.1-4. Sources of all cataloged satellite breakup debris by satellite type.

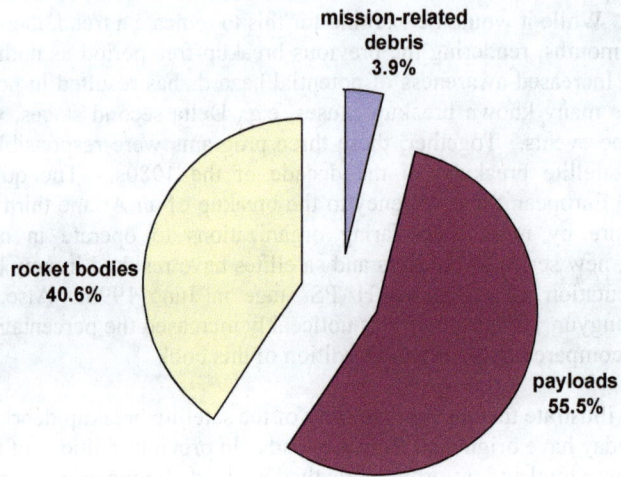

mission-related
debris
3.9%

rocket bodies
40.6%

payloads
55.5%

Figure 2.1-5. Sources of satellite breakup debris in orbit by satellite type.

TABLE 2.1 HISTORY OF SATELLITE BREAKUPS BY LAUNCH DATE

NAME	INTERNATIONAL DESIGNATOR	CATALOG NUMBER	LAUNCH DATE	EVENT DATE	DEBRIS CATALOGED	DEBRIS LEFT	APOGEE (KM)	PERIGEE (KM)	INCLINATION (DEG)	ASSESSED CAUSE	COMMENT
TRANSIT 4A R/B	1961-015C	118	29-Jun-61	29-Jun-61	296	181	995	880	66.8	PROPULSION	ABLESTAR STAGE
SPUTNIK 29	1962-057A	443	24-Oct-62	29-Oct-62	24	0	260	200	65.1	PROPULSION	MOLNIYA FINAL STAGE
ATLAS CENTAUR 2	1963-047A	694	27-Nov-63	27-Nov-63	19	8	1785	475	30.3	PROPULSION	CENTAUR STAGE
COSMOS 50	1964-070A	919	28-Oct-64	5-Nov-64	96	0	220	175	51.2	DELIBERATE	PAYLOAD RECOVERY FAILURE
COSMOS 57	1965-012A	1093	22-Feb-65	22-Feb-65	167	0	425	165	64.8	DELIBERATE	SELF-DESTRUCT
COSMOS 61-63 R/B	1965-020D	1270	15-Mar-65	15-Mar-65	147	18	1825	260	56.1	UNKNOWN	COSMOS SECOND STAGE
OV2-1/LCS 2 R/B	1965-082DM	1822	15-Oct-65	15-Oct-65	473	37	790	710	32.2	PROPULSION	TITAN TRANSTAGE
COSMOS 95	1965-088A	1706	4-Nov-65	15-Jan-66	1	0	520	210	48.4	UNKNOWN	
OPS 3031	1966-012C	2015	15-Feb-66	15-Feb-66	38	0	270	150	96.5	UNKNOWN	INFLATABLE SPHERE
GEMINI 9 ATDA R/B	1966-046B	2188	1-Jun-66	Mid-Jun-66	51	0	275	240	28.8	UNKNOWN	ATLAS CORE STAGE
PAGEOS	1966-056A	2253	24-Jun-66	12-Jul-75	79	2	5170	3200	85.3	UNKNOWN	INFLATABLE SPHERE
				20-Jan-76			5425	2935	85.1	UNKNOWN	
				10-Sep-76						UNKNOWN	
				Mid-Jun-78						UNKNOWN	
				Mid-Sep-84						UNKNOWN	
				Mid-Dec-85						UNKNOWN	
AS-203	1966-059A	2289	5-Jul-66	5-Jul-66	34	0	215	185	32.0	DELIBERATE	SATURN S-IVB STAGE
COSMOS U-1	1966-088A	2437	17-Sep-66	17-Sep-66	53	0	855	140	49.6	DELIBERATE	SELF-DESTRUCT
COSMOS U-2	1966-101A	2536	2-Nov-66	2-Nov-66	41	0	885	145	49.6	DELIBERATE	SELF-DESTRUCT
COSMOS 199	1968-003A	3099	16-Jan-68	24-Jan-68	3	0	355	200	65.6	DELIBERATE	SELF-DESTRUCT
APOLLO 6 R/B (S4B)	1968-025B	3171	4-Apr-68	13-Apr-68	16	0	360	200	32.6	PROPULSION	SATURN S-IVB STAGE
OV2-5 R/B	1968-081E	3432	26-Sep-68	21-Feb-92	4	4	35810	35100	11.9	PROPULSION	TITAN TRANSTAGE
COSMOS 248	1968-090A	3503	19-Oct-68	1-Nov-68	5	0	545	475	62.2	DELIBERATE	DEBRIS IMPACT
COSMOS 249	1968-091A	3504	20-Oct-68	20-Oct-68	108	43	2165	490	62.3	DELIBERATE	SELF-DESTRUCT
COSMOS 252	1968-097A	3530	1-Nov-68	1-Nov-68	139	43	2140	535	62.3	DELIBERATE	SELF-DESTRUCT
METEOR 1-1 R/B	1969-029B	3836	26-Mar-69	28-Mar-69	37	0	850	460	81.2	UNKNOWN	VOSTOK FINAL STAGE
INTELSAT 3 F-5 R/B	1969-064B	4052	26-Jul-69	26-Jul-69	23	1	5445	270	30.4	PROPULSION	TE 364-4 STAGE
OPS 7613 R/B	1969-082AB	4159	30-Sep-69	4-Oct-69	261	73	940	905	70.0	UNKNOWN	AGENA D STAGE
NIMBUS 4 R/B	1970-025C	4367	8-Apr-70	17-Oct-70	373	247	1085	1065	99.9	UNKNOWN	AGENA D STAGE
		4601		23-Jan-85						UNKNOWN	2 ADDITIONAL OBJECTS
		4649		17-Dec-85						UNKNOWN	3 ADDITIONAL OBJECTS
		4610		2-Sep-86						UNKNOWN	2 ADDITIONAL OBJECTS
		4601		23-Dec-91						UNKNOWN	5 ADDITIONAL OBJECTS

15

TABLE 2.1 HISTORY OF SATELLITE BREAKUPS BY LAUNCH DATE (CONT'D)

NAME	INTERNATIONAL DESIGNATOR	CATALOG NUMBER	LAUNCH DATE	EVENT DATE	DEBRIS CATALOGED	DEBRIS LEFT	APOGEE (KM)	PERIGEE (KM)	INCLINATION (DEG)	ASSESSED CAUSE	COMMENT
COSMOS 374	1970-089A	4594	23-Oct-70	23-Oct-70	99	21	2130	530	62.9	DELIBERATE	SELF-DESTRUCT
COSMOS 375	1970-091A	4598	30-Oct-70	30-Oct-70	47	17	2100	525	62.8	DELIBERATE	SELF-DESTRUCT
COSMOS 397	1971-015A	4964	25-Feb-71	25-Feb-71	116	47	2200	575	65.8	DELIBERATE	SELF-DESTRUCT
COSMOS 462	1971-106A	5646	3-Dec-71	3-Dec-71	25	0	1800	230	65.7	DELIBERATE	SELF-DESTRUCT
LANDSAT 1 R/B	1972-058B	6127	23-Jul-72	22-May-75	226	31	910	635	98.3	PROPULSION	DELTA SECOND STAGE
SALYUT 2 R/B	1973-017B	6399	3-Apr-73	3-Apr-73	25	0	245	195	51.5	PROPULSION	PROTON-K SECOND STAGE
COSMOS 554	1973-021A	6432	19-Apr-73	6-May-73	195	0	350	170	72.9	DELIBERATE	SELF-DESTRUCT
NOAA 3 R/B	1973-086B	6921	6-Nov-73	28-Dec-73	197	175	1510	1500	102.1	PROPULSION	DELTA SECOND STAGE
DMSP 5B F5 R/B	1974-015B	7219	16-Mar-74	17-Jan-05	5	5	885	775	99.1	COLLISION, ACCIDENTAL	HIT BY DEBRIS (26207)
NOAA 4 R/B	1974-089D	7532	15-Nov-74	20-Aug-75	146	122	1460	1445	101.7	PROPULSION	DELTA SECOND STAGE
COSMOS 699	1974-103A	7587	24-Dec-74	17-Apr-75	50	0	445	425	65.0	UNKNOWN	COSMOS 699 CLASS
				2-Aug-75			440	415	65.0	UNKNOWN	
LANDSAT 2 R/B	1975-004B	7616	22-Jan-75	9-Feb-76	207	33	915	740	97.8	PROPULSION	DELTA SECOND STAGE
				19-Jun-76			910	745	97.7	PROPULSION	
NIMBUS 6 R/B	1975-052B	7946	12-Jun-75	1-May-91	268	207	1105	1095	99.6	PROPULSION	DELTA SECOND STAGE
COSMOS 758	1975-080A	8191	5-Sep-75	6-Sep-75	76	0	325	175	67.1	DELIBERATE	SELF-DESTRUCT
COSMOS 777	1975-102A	8416	29-Oct-75	25-Jan-76	62	0	440	430	65.0	UNKNOWN	COSMOS 699 CLASS
COSMOS 838	1976-063A	8932	2-Jul-76	17-May-77	40	0	445	415	65.1	UNKNOWN	COSMOS 699 CLASS
COSMOS 839	1976-067A	9011	8-Jul-76	29-Sep-77	69	67	2100	980	65.9	BATTERY	
COSMOS 844	1976-072A	9046	22-Jul-76	25-Jul-76	248	0	355	170	67.1	DELIBERATE	SELF-DESTRUCT
NOAA 5 R/B	1976-077B	9063	29-Jul-76	24-Dec-77	162	155	1520	1505	102.0	PROPULSION	DELTA SECOND STAGE
COSMOS 862	1976-105A	9495	22-Oct-76	15-Mar-77	11	10	39645	765	63.2	DELIBERATE	SELF-DESTRUCT
COSMOS 880	1976-120A	9601	9-Dec-76	27-Nov-78	49	0	620	550	65.8	BATTERY	
COSMOS 884	1976-123A	9614	17-Dec-76	29-Dec-76	2	0	320	170	65.0	DELIBERATE	SELF-DESTRUCT
COSMOS 886	1976-126A	9634	27-Dec-76	27-Dec-76	76	60	2295	595	65.8	DELIBERATE	SELF-DESTRUCT
COSMOS 903	1977-027A	9911	11-Apr-77	8-Jun-78	3	2	39035	1325	63.2	DELIBERATE	SELF-DESTRUCT
COSMOS 917	1977-047A	10059	16-Jun-77	30-Mar-79	8	8	38725	1645	62.9	DELIBERATE	SELF-DESTRUCT
HIMAWARI 1 R/B	1977-065B	10144	14-Jul-77	14-Jul-77	172	63	2025	535	29.0	PROPULSION	DELTA SECOND STAGE
COSMOS 931	1977-068A	10150	20-Jul-77	24-Oct-77	6	5	39665	680	62.9	DELIBERATE	SELF-DESTRUCT
EKRAN 2	1977-092A	10365	20-Sep-77	25-Jun-78	4	4	35800	35785	0.1	BATTERY	
COSMOS 970	1977-121A	10531	21-Dec-77	21-Dec-77	70	65	1140	945	65.8	DELIBERATE	SELF-DESTRUCT
LANDSAT 3 R/B	1978-026C	10704	5-Mar-78	27-Jan-81	210	122	910	900	98.8	PROPULSION	DELTA SECOND STAGE
COSMOS 1030	1978-083A	11015	6-Sep-78	10-Oct-78	7	5	39760	665	62.8	DELIBERATE	SELF-DESTRUCT

TABLE 2.1 HISTORY OF SATELLITE BREAKUPS BY LAUNCH DATE (CONT'D)

NAME	INTERNATIONAL DESIGNATOR	CATALOG NUMBER	LAUNCH DATE	EVENT DATE	DEBRIS CATALOGED	DEBRIS LEFT	APOGEE (KM)	PERIGEE (KM)	INCLINATION (DEG)	ASSESSED CAUSE	COMMENT
NIMBUS 7 R/B	1978-098B	11081	24-Oct-78	26-Dec-81	1	1	955	935	99.3	PROPULSION	DELTA SECOND STAGE
COSMOS 1045 R/B	1978-100D	11087	26-Oct-78	9-May-88	42	37	1705	1685	82.6	PROPULSION	TSYKLON THIRD STAGE
P-78 (SOLWIND)	1979-017A	11278	24-Feb-79	13-Sep-85	285	0	545	515	97.6	DELIBERATE	HYPERVELOCITY IMPACT
COSMOS 1094	1979-033A	11333	18-Apr-79	17-Sep-79	1	0	405	380	65.0	UNKNOWN	COSMOS 699 CLASS
COSMOS 1109	1979-058A	11417	27-Jun-79	Mid-Feb-80	11	11	39425	960	63.3	DELIBERATE	SELF-DESTRUCT
COSMOS 1124	1979-077A	11509	28-Aug-79	9-Sep-79	3	3	39795	570	63.0	DELIBERATE	SELF-DESTRUCT
CAT R/B	1979-104B	11659	24-Dec-79	Apr-80	22	18	33140	180	17.9	PROPULSION	ARIANE 1 FINAL STAGE
COSMOS 1167	1980-021A	11729	14-Mar-80	15-Jul-81	12	0	450	355	65.0	UNKNOWN	COSMOS 699 CLASS
COSMOS 1174	1980-030A	11765	18-Apr-80	18-Apr-80	46	4	1660	380	66.1	DELIBERATE	SELF-DESTRUCT
COSMOS 1191	1980-057A	11871	2-Jul-80	14-May-81	8	8	39255	1110	62.6	DELIBERATE	SELF-DESTRUCT
COSMOS 1217	1980-085A	12032	24-Oct-80	12-Feb-83	7	7	38830	1530	65.2	DELIBERATE	SELF-DESTRUCT
COSMOS 1220	1980-089A	12054	4-Nov-80	20-Jun-82	83	2	885	570	65.0	UNKNOWN	COSMOS 699 CLASS
COSMOS 1247	1981-016A	12303	19-Feb-81	20-Oct-81	7	7	39390	970	63.0	DELIBERATE	SELF-DESTRUCT
COSMOS 1260	1981-028A	12364	20-Mar-81	8-May-82	68	0	750	450	65.0	UNKNOWN	COSMOS 699 CLASS
				10-Aug-82			750	445	65.0	UNKNOWN	
COSMOS 1261	1981-031A	12376	31-Mar-81	Apr/May-81	8	8	39765	610	63.0	DELIBERATE	SELF-DESTRUCT
COSMOS 1275	1981-053A	12504	4-Jun-81	24-Jul-81	310	259	1015	960	83.0	BATTERY	
COSMOS 1278	1981-058A	12547	19-Jun-81	Early-Dec-86	3	0	37690	2665	67.1	DELIBERATE	SELF-DESTRUCT
COSMOS 1285	1981-071A	12627	4-Aug-81	21-Nov-81	18	18	40100	720	63.1	DELIBERATE	SELF-DESTRUCT
COSMOS 1286	1981-072A	12631	4-Aug-81	29-Sep-82	2	0	325	300	65.0	UNKNOWN	COSMOS 699 CLASS
COSMOS 1305 R/B	1981-088F	12827	11-Sep-81	11-Sep-81	8	8	13795	605	62.8	PROPULSION	MOLNIYA FINAL STAGE
COSMOS 1306	1981-089A	12828	14-Sep-81	12-Jul-82	8	0	405	380	64.9	UNKNOWN	COSMOS 699 CLASS
				18-Sep-82	8	0	370	370	64.9	UNKNOWN	
COSMOS 1317	1981-108A	12933	31-Oct-81	Late-Jan-84	4	4	39055	1315	62.8	DELIBERATE	SELF-DESTRUCT
COSMOS 1355	1982-038A	13150	29-Apr-82	8-Aug-83	29	0	395	360	65.1	UNKNOWN	COSMOS 699 CLASS
				1-Feb-84			320	305	65.0	UNKNOWN	
				20-Feb-84			290	270	65.0	UNKNOWN	
COSMOS 1375	1982-055A	13259	6-Jun-82	21-Oct-85	61	58	1000	990	65.8	BATTERY	
COSMOS 1405	1982-088A	13508	4-Sep-82	20-Dec-83	32	0	340	310	65.0	UNKNOWN	COSMOS 699 CLASS
COSMOS 1423 R/B	1982-115E	13696	8-Dec-82	8-Dec-82	29	0	425	235	62.9	PROPULSION	MOLNIYA FINAL STAGE
ASTRON ULLAGE MOTOR	1983-020B	13902	23-Mar-83	3-Sep-84	1	0	1230	220	51.5	PROPULSION	PROTON-K BLOCK DM SOZ
NOAA 8	1983-022A	13923	28-Mar-83	30-Dec-85	8	1	830	805	98.6	BATTERY	
COSMOS 1456	1983-038A	14034	25-Apr-83	13-Aug-83	4	0	39630	730	63.3	DELIBERATE	SELF-DESTRUCT

TABLE 2.1 HISTORY OF SATELLITE BREAKUPS BY LAUNCH DATE (CONT'D)

NAME	INTERNATIONAL DESIGNATOR	CATALOG NUMBER	LAUNCH DATE	EVENT DATE	DEBRIS CATALOGED	DEBRIS LEFT	APOGEE (KM)	PERIGEE (KM)	INCLINATION (DEG)	ASSESSED CAUSE	COMMENT
COSMOS 1461	1983-044A	14064	7-May-83	11-Mar-85	164	5	890	570	65.0	UNKNOWN	COSMOS 699 CLASS
				13-May-85			885	570	65.0	UNKNOWN	
COSMOS 1481	1983-070A	14182	8-Jul-83	9-Jul-83	6	6	39225	625	62.9	DELIBERATE	SELF-DESTRUCT
COSMOS 1484	1983-075A	14207	24-Jul-83	18-Oct-93	49	2	595	550	97.5	UNKNOWN	
COSMOS 1519-1521 ULLAGE MOTOR	1983-127H	14608	29-Dec-83	4-Feb-91	8	3	18805	340	51.9	PROPULSION	PROTON-K BLOCK DM SOZ
PALAPA B2 R/B	1984-011E	14693	3-Feb-84	6-Feb-84	3	0	285	275	28.5	PROPULSION	PAM-D UPPER STAGE
WESTAR 6 R/B	1984-011F	14694	3-Feb-84	3-Feb-84	14	1	310	305	28.5	PROPULSION	PAM-D UPPER STAGE
COSMOS 1588	1984-083A	15167	7-Aug-84	23-Feb-86	45	0	440	410	65.0	UNKNOWN	COSMOS 699 CLASS
COSMOS 1603 ULLAGE MOTOR	1984-106F	15338	28-Sep-84	5-Sep-92	22	1	845	835	66.6	PROPULSION	PROTON-K BLOCK DM SOZ
SPACENET 2/ MARECS B2 R/B	1984-114C	15388	10-Nov-84	20-Nov-84	3	2	35960	325	7.0	PROPULSION	ARIANE 3 FINAL STAGE
COSMOS 1646	1985-030A	15653	18-Apr-85	20-Nov-87	24	0	410	385	65.0	UNKNOWN	COSMOS 699 CLASS
COSMOS 1650-1652 ULLAGE MOTOR	1985-037G	15714	17-May-85	29-Nov-98	4	2	18620	320	52.0	PROPULSION	PROTON-K BLOCK DM SOZ
COSMOS 1654	1985-039A	15734	23-May-85	21-Jun-85	18	0	300	185	64.9	DELIBERATE	SELF-DESTRUCT
COSMOS 1656 ULLAGE MOTOR	1985-042E	15773	30-May-85	5-Jan-88	6	6	860	810	66.6	PROPULSION	PROTON-K BLOCK DM SOZ
COSMOS 1682	1985-082A	16054	19-Sep-85	18-Dec-86	23	0	475	385	65.0	UNKNOWN	COSMOS 699 CLASS
COSMOS 1691	1985-094B	16139	9-Oct-85	22-Nov-85	14	11	1415	1410	82.6	BATTERY	
COSMOS 1703 R/B	1985-108B	16263	22-Nov-85	4-May-06	50	3	640	610	82.5	UNKNOWN	TSYKLON THIRD STAGE
COSMOS 1710-1712 ULLAGE MOTOR	1985-118L	16446	24-Dec-85	29-Dec-91	17	11	18885	655	65.3	PROPULSION	PROTON-K BLOCK DM SOZ
COSMOS 1714 R/B	1985-121F	16439	28-Dec-85	28-Dec-85	2	0	830	165	71.0	PROPULSION	ZENIT SECOND STAGE
SPOT 1/VIKING R/B	1986-019C	16615	22-Feb-86	13-Nov-86	489	30	835	805	98.7	PROPULSION	ARIANE 1 FINAL STAGE
COSMOS 1769	1986-059A	16895	4-Aug-86	21-Sep-87	4	0	445	310	65.0	UNKNOWN	COSMOS 699 CLASS
USA 19	1986-069A	16937	5-Sep-86	5-Sep-86	13	0	745	210	39.1	DELIBERATE	HYPERVELOCITY IMPACT
USA 19 R/B	1986-069B	16938	5-Sep-86	5-Sep-86	5	0	610	220	22.8	DELIBERATE	HYPERVELOCITY IMPACT
COSMOS 1813	1987-004A	17297	15-Jan-87	29-Jan-87	195	0	415	360	72.8	DELIBERATE	SELF-DESTRUCT
COSMOS 1823	1987-020A	17535	20-Feb-87	17-Dec-87	119	44	1525	1480	73.6	BATTERY	
COSMOS 1866	1987-059A	18184	9-Jul-87	26-Jul-87	9	2	255	155	67.1	DELIBERATE	SELF-DESTRUCT
COSMOS 1869	1987-062A	18214	16-Jul-87	27-Nov-97	2	2	635	605	83.0	UNKNOWN	
METEOR 2-16 R/B	1987-068B	18313	18-Aug-87	15-Feb-98	83	18	960	940	82.6	PROPULSION	TSYKLON THIRD STAGE
AUSSAT/ECS R/B	1987-078C	18352	16-Sep-87	Mid-Sep-87	4	1	36515	245	6.9	PROPULSION	ARIANE 3 FINAL STAGE
COSMOS 1883-1885 ULLAGE MOTOR	1987-079G	18374	16-Sep-87	1-Dec-96	14	11	19120	335	64.9	PROPULSION	PROTON-K BLOCK DM SOZ
COSMOS 1883-1885 ULLAGE MOTOR	1987-079H	18375	16-Sep-87	23-Apr-03	39	16	18540	755	65.2	PROPULSION	PROTON-K BLOCK DM SOZ

TABLE 2.1 HISTORY OF SATELLITE BREAKUPS BY LAUNCH DATE (CONT'D)

NAME	INTERNATIONAL DESIGNATOR	CATALOG NUMBER	LAUNCH DATE	EVENT DATE	DEBRIS CATALOGED	DEBRIS LEFT	APOGEE (KM)	PERIGEE (KM)	INCLINATION (DEG)	ASSESSED CAUSE	COMMENT
COSMOS 1906	1987-108A	18713	26-Dec-87	31-Jan-88	37	0	265	245	82.6	DELIBERATE	SELF-DESTRUCT
EKRAN 17 ULLAGE MOTOR	1987-109E	18719	27-Dec-87	22-May-97	1	0	22975	310	46.6	PROPULSION	PROTON-K BLOCK DM SOZ
COSMOS 1916	1988-007A	18823	3-Feb-88	27-Feb-88	1	0	230	150	64.8	DELIBERATE	SELF-DESTRUCT
COSMOS 1934	1988-023A	18985	22-Mar-88	23-Dec-91	3	3	1010	950	83.0	COLLISION, ACCIDENTAL	HIT BY DEBRIS (13475)
INTELSAT 513 R/B	1988-040B	19122	17-May-88	9-Jul-02	5	5	35445	535	7.0	PROPULSION	ARIANE 2 R/B
COSMOS 1970-72 ULLAGE MOTOR	1988-085F	19535	16-Sep-88	4-Aug-03	79	10	18515	720	65.3	PROPULSION	PROTON-K BLOCK DM SOZ
COSMOS 1970-1972 ULLAGE MOTOR	1988-085G	19537	16-Sep-88	9-Mar-99	1	1	18950	300	64.6	PROPULSION	PROTON-K BLOCK DM SOZ
SKYNET 4B/ASTRA 1A R/B	1988-109C	19689	11-Dec-88	17-Feb-98	11	10	35875	435	7.3	PROPULSION	ARIANE 4 H10 FINAL STAGE
COSMOS 1987-1989	1989-001G	19755	10-Jan-89	3-Aug-98	12	7	19055	340	64.9	PROPULSION	PROTON-K BLOCK DM SOZ
COSMOS 1987-1989	1989-001H	19856	10-Jan-89	13-Nov-03	1	1	18740	710	65.4	PROPULSION	PROTON-K BLOCK DM SOZ
GORIZONT 17 ULLAGE MOTOR	1989-004E	19771	26-Jan-89	17-Dec-92	1	0	17575	195	46.7	PROPULSION	PROTON-K BLOCK DM SOZ
INTELSAT 515 R/B	1989-006B	19773	27-Jan-89	1-Jan-01	37	37	35720	510	8.4	PROPULSION	ARIANE 2 R/B
COSMOS 2022-24 ULLAGE MOTOR	1989-039G	20081	31-May-89	10-Jun-06	116	105	18410	655	65.1	PROPULSION	PROTON-K BLOCK DM SOZ
GORIZONT 18 ULLAGE MOTOR	1989-052F	20116	5-Jul-89	12-Jan-93	2	0	36745	260	46.8	PROPULSION	PROTON-K BLOCK DM SOZ
COSMOS 2030	1989-054A	20124	12-Jul-89	28-Jul-89	1	0	215	150	67.1	DELIBERATE	SELF-DESTRUCT
COSMOS 2031	1989-056A	20136	18-Jul-89	31-Aug-89	9	0	365	240	50.5	DELIBERATE	SELF-DESTRUCT
COBE R/B	1989-089B	20323	18-Nov-89	3-Dec-06	26	4	790	685	97.1	UNKNOWN	DELTA SECOND STAGE
COSMOS 2053 R/B	1989-100B	20390	27-Dec-89	18-Apr-99	26	0	485	475	73.5	PROPULSION	TSYKLON THIRD STAGE
COSMOS 2045 ULLAGE MOTOR	1989-101E	20399	27-Dec-89	Jul-92 (?)	14	8	27650	345	47.1	PROPULSION	PROTON-K BLOCK DM SOZ
COSMOS 2079-2081 ULLAGE MOTOR	1990-045G	20631	19-May-90	28-Mar-99	1	1	19065	405	64.8	PROPULSION	PROTON-K BLOCK DM SOZ
FENGYUN 1-2 R/B	1990-081D	20791	3-Sep-90	4-Oct-90	84	67	895	880	98.9	PROPULSION	CZ-4A FINAL STAGE
COSMOS 2101	1990-087A	20828	1-Oct-90	30-Nov-90	4	0	280	195	64.8	DELIBERATE	SELF-DESTRUCT
GORIZONT 22 ULLAGE MOTOR	1990-102E	20957	23-Nov-90	14-Dec-95	2	1	13105	170	46.5	PROPULSION	PROTON-K BLOCK DM SOZ
USA 68	1990-105A	20978	1-Dec-90	1-Dec-90	29	1	850	610	98.9	PROPULSION	TE-M-364-15 UPPER STAGE
COSMOS 2109-11 ULLAGE MOTOR	1990-110G	21012	8-Dec-90	21-Feb-03	1	1	18805	645	65.4	PROPULSION	PROTON-K BLOCK DM SOZ
COSMOS 2109-2111 ULLAGE MOTOR	1990-110H	21013	8-Dec-90	14-Mar-98	2	2	18995	520	65.1	PROPULSION	PROTON-K BLOCK DM SOZ
ITALSAT 1 R/B/ EUTELSAT 2 F2	1991-003C	21057	15-Jan-91	1-May-96	12	9	30930	235	6.7	PROPULSION	ARIANE 4 H10 FINAL STAGE
COSMOS 2125-2132 R/B	1991-009J	21108	12-Feb-91	5-Mar-91	92	92	1725	1460	74.0	PROPULSION	COSMOS SECOND STAGE
COSMOS 2133 ULLAGE MOTOR	1991-010D	21114	12-Feb-91	7-May-94	4	2	21805	225	46.6	PROPULSION	PROTON-K BLOCK DM SOZ

19

TABLE 2.1 HISTORY OF SATELLITE BREAKUPS BY LAUNCH DATE (CONT'D)

NAME	INTERNATIONAL DESIGNATOR	CATALOG NUMBER	LAUNCH DATE	EVENT DATE	DEBRIS CATALOGED	DEBRIS LEFT	APOGEE (KM)	PERIGEE (KM)	INCLINATION (DEG)	ASSESSED CAUSE	COMMENT
ASTRA 1B/MOP 2 R/B	1991-015C	21141	2-Mar-91	27-Apr-94	10	7	17630	205	6.8	PROPULSION	ARIANE 4 H10 FINAL STAGE
COSMOS 2139-41 ULLAGE MOTOR	1991-025G	21226	4-Apr-91	16-Jun-01	1	1	18960	300	64.5	PROPULSION	PROTON-K BLOCK DM SOZ
COSMOS 2157-2162 R/B	1991-068G	21734	28-Sep-91	9-Oct-99	37	37	1485	1410	82.6	PROPULSION	TSYKLON THIRD STAGE
COSMOS 2163	1991-071A	21741	9-Oct-91	6-Dec-91	1	0	260	185	64.8	DELIBERATE	SELF-DESTRUCT
INTELSAT 601 R/B	1991-075B	21766	29-Oct-91	24-Dec-01	12	9	28505	230	7.2	PROPULSION	ARIANE 4 H10 FINAL STAGE
USA 73 (DMSP 5D2 F11)	1991-082A	21798	28-Nov-91	15-Apr-04	79	67	850	830	98.7	UNKNOWN	
TELECOM 2B/ INMARSAT 2 R/B	1992-021C	21941	15-Apr-92	21-Apr-93	14	13	34080	235	4.0	PROPULSION	ARIANE H10+ FINAL STAGE
INSAT 2A/ EUTELSAT 2F4 R/B	1992-041C	22032	9-Jul-92	2-Feb-02	1	1	26550	250	7.0	PROPULSION	ARIANE 4 H10 FINAL STAGE
COSMOS 2204-06 ULLAGE MOTOR	1992-047G	22066	30-Jul-92	10-Jul-04	30	25	18820	415	64.9	PROPULSION	PROTON-K BLOCK DM SOZ
COSMOS 2204-2206 ULLAGE MOTOR	1992-047H	22067	30-Jul-92	8-Nov-94	4	3	19035	480	64.8	PROPULSION	PROTON-K BLOCK DM SOZ
GORIZONT 27 ULLAGE MOTOR	1992-082F	22250	27-Nov-92	14-Jul-01	1	0	5340	145	46.5	PROPULSION	PROTON-K BLOCK DM SOZ
COSMOS 2224 ULLAGE MOTOR	1992-088F	22274	17-Dec-92	~22-Apr-05	1	0	21140	200	46.7	PROPULSION	PROTON-K BLOCK DM SOZ
COSMOS 2225	1992-091A	22280	22-Dec-92	18-Feb-93	6	0	280	225	64.9	DELIBERATE	SELF-DESTRUCT
COSMOS 2227 R/B	1992-093B	22285	25-Dec-92	26-Dec-92	226	166	855	845	71.0	PROPULSION	ZENIT-2 SECOND STAGE
				30-Dec-92			855	845	71.0	PROPULSION	ZENIT-2 SECOND STAGE
COSMOS 2237 R/B	1993-016B	22566	26-Mar-93	28-Mar-93	37	35	850	840	71.0	PROPULSION	ZENIT-2 SECOND STAGE
COSMOS 2238	1993-018A	22585	30-Mar-93	1-Dec-94	1	0	305	210	65.0	UNKNOWN	COSMOS 699 CLASS
COSMOS 2243	1993-028A	22641	27-Apr-93	27-Apr-93	1	0	225	180	70.4	DELIBERATE	SELF-DESTRUCT
COSMOS 2259	1993-045A	22716	14-Jul-93	25-Jul-93	1	0	320	175	67.1	DELIBERATE	SELF-DESTRUCT
COSMOS 2262	1993-057A	22789	7-Sep-93	18-Dec-93	1	0	295	170	64.9	DELIBERATE	SELF-DESTRUCT
GORIZONT 29 ULLAGE MOTOR	1993-072E	22925	18-Nov-93	6-Sep-00	1	0	11215	140	46.7	PROPULSION	PROTON-K BLOCK DM SOZ
CLEMINTINE R/B	1994-004B	22974	25-Jan-94	7-Feb-94	1	0	295	240	67.0	PROPULSION	PEGASUS HAPS
STEP II R/B	1994-029B	23106	19-May-94	3-Jun-96	713	64	820	585	82.0	PROPULSION	PEGASUS HAPS
COSMOS 2282 ULLAGE MOTOR	1994-038F	23174	6-Jul-94	21-Oct-95	2	1	34930	280	47.0	PROPULSION	PROTON-K BLOCK DM SOZ
ELEKTRO ULLAGE MOTOR	1994-069E	23338	31-Oct-94	11-May-95	1	0	35465	155	46.9	PROPULSION	PROTON-K BLOCK DM SOZ
RS-15 R/B	1994-085B	23440	26-Dec-94	26-Dec-94	23	21	2200	1880	64.8	UNKNOWN	ROKOT THIRD STAGE
COSMOS 2313	1995-028A	23596	8-Jun-95	26-Jun-97	13	0	325	210	65.0	UNKNOWN	COSMOS 699 CLASS
CERISE	1995-033B	23606	7-Jul-95	24-Jul-96	2	2	675	665	98.1	COLLISION, ACCIDENTAL	HIT BY DEBRIS (18208)
COSMOS 2316-2318 ULLAGE MOTOR	1995-037K	23631	24-Jul-95	21-Nov-00	1	0	18085	150	64.4	PROPULSION	PROTON-K BLOCK DM SOZ

TABLE 2.1 HISTORY OF SATELLITE BREAKUPS BY LAUNCH DATE (CONT'D)

NAME	INTERNATIONAL DESIGNATOR	CATALOG NUMBER	LAUNCH DATE	EVENT DATE	DEBRIS CATALOGED	DEBRIS LEFT	APOGEE (KM)	PERIGEE (KM)	INCLINATION (DEG)	ASSESSED CAUSE	COMMENT
RADUGA 33 R/B	1996-010D	23797	19-Feb-96	19-Feb-96	2	0	36505	240	48.7	PROPULSION	PROTON-K BLOCK DM
GORIZONT 32 ULLAGE MOTOR	1996-034F	23887	23-May-96	13-Dec-99	1	0	5605	145	46.5	PROPULSION	PROTON-K BLOCK DM SOZ
COSMOS 2343	1997-024A	24805	15-May-97	16-Sep-97	1	0	285	225	65.0	DELIBERATE	SELF-DESTRUCT
KUPON ULLAGE MOTOR	1997-070F	25054	12-Nov-97	14-Feb-07	1	1	14160	260	46.6	PROPULSION	PROTON-K BLOCK DM SOZ
COSMOS 2347	1997-079A	25088	9-Dec-97	22-Nov-99	9	0	410	230	65.0	UNKNOWN	COSMOS 699 CLASS
ASIASAT 3 R/B	1997-086D	25129	24-Dec-97	25-Dec-97	1	0	35995	270	51.0	PROPULSION	PROTON-K BLOCK DM
COMETS R/B	1998-011B	25176	21-Feb-98	21-Feb-98	1	0	1880	245	30.0	PROPULSION	H-II SECOND STAGE
FENGYUN 1-3 (aka FENGYUN 1C)	1999-025A	25730	10-May-99	11-Jan-07	2087	2073	865	845	98.6	DELIBERATE	HYPERVELOCITY IMPACT
CBERS 1	1999-057A	25940	14-Oct-99	18-Feb-07	15	11	780	770	98.2	UNKNOWN	
CBERS-1/SACI-1 R/B	1999-057C	25942	14-Oct-99	11-Mar-00	345	197	745	725	98.5	PROPULSION	CZ-4 FINAL STAGE
COSMOS 2367	1999-072A	26040	26-Dec-99	21-Nov-01	17	0	415	405	65.0	UNKNOWN	COSMOS 699 CLASS
COSMOS 2371 ULLAGE MOTOR	2000-036E	26398	4-Jul-00	~1-Sep-06	1	1	21320	220	46.9	PROPULSION	PROTON-K BLOCK DM SOZ
TES R/B	2001-049D	26960	22-Oct-01	19-Dec-01	346	97	675	550	97.9	PROPULSION	PSLV FINAL STAGE
COSMOS 2383	2001-057A	27053	21-Dec-01	28-Feb-04	14	0	400	220	65.0	UNKNOWN	COSMOS 699 CLASS
COSMOS 2392 ULLAGE MOTOR	2002-037E	27474	25-Jul-02	1-Jun-05	61	7	835	255	63.7	PROPULSION	PROTON-K BLOCK DM SOZ
COSMOS 2392 ULLAGE MOTOR	2002-037F	27475	25-Jul-02	29-Oct-04	1	0	840	235	63.6	PROPULSION	PROTON-K BLOCK DM SOZ
COSMOS 2399	2003-035A	27856	12-Aug-03	9-Dec-03	22	0	250	175	64.9	DELIBERATE	SELF-DESTRUCT
ALOS-1 R/B	2006-002B	28932	24-Jan-06	8-Aug-06	22	1	700	550	98.2	UNKNOWN	H-IIA SECOND STAGE
ARABSAT 4 BRIZ-M R/B	2006-006B	28944	28-Feb-06	19-Feb-07	1	1	14705	495	51.5	PROPULSION	PROTON-K BRIZ-M STAGE
IGS 3A R/B	2006-037B	29394	11-Sep-06	28-Dec-06	1	1	490	430	97.2	UNKNOWN	H-IIA SECOND STAGE
COSMOS 2423	2006-039A	29402	14-Sep-06	17-Nov-06	31	1	285	200	64.9	DELIBERATE	SELF-DESTRUCT
DMSP 5D-3 F17 R/B	2006-050B	29523	4-Nov-06	4-Nov-06	64	61	865	830	98.8	UNKNOWN	DELTA IV SECOND STAGE
BEIDOU 2A	2007-003A	30323	2-Feb-07	2-Feb-07	2	2	41775	195	25.0	UNKNOWN	
				TOTAL	12819	5783					

21

TABLE 2.2 HISTORY OF SATELLITE BREAKUPS BY EVENT DATE

NAME	INTERNATIONAL DESIGNATOR	CATALOG NUMBER	LAUNCH DATE	EVENT DATE	DEBRIS CATALOGED	DEBRIS LEFT	APOGEE (KM)	PERIGEE (KM)	INCLINATION (DEG)	ASSESSED CAUSE	COMMENT
TRANSIT 4A R/B	1961-015C	118	29-Jun-61	29-Jun-61	296	181	995	880	66.8	PROPULSION	ABLESTAR STAGE
SPUTNIK 29	1962-057A	443	24-Oct-62	29-Oct-62	24	0	260	200	65.1	PROPULSION	MOLNIYA FINAL STAGE
ATLAS CENTAUR 2	1963-047A	694	27-Nov-63	27-Nov-63	19	8	1785	475	30.3	PROPULSION	CENTAUR STAGE
COSMOS 50	1964-070A	919	28-Oct-64	5-Nov-64	96	0	220	175	51.2	DELIBERATE	PAYLOAD RECOVERY FAILURE
COSMOS 57	1965-012A	1093	22-Feb-65	22-Feb-65	167	0	425	165	64.8	DELIBERATE	SELF-DESTRUCT
COSMOS 61-63 R/B	1965-020D	1270	15-Mar-65	15-Mar-65	147	18	1825	260	56.1	UNKNOWN	COSMOS SECOND STAGE
OV2-1/LCS 2 R/B	1965-082DM	1822	15-Oct-65	15-Oct-65	473	37	790	710	32.2	PROPULSION	TITAN TRANSTAGE
COSMOS 95	1965-088A	1706	4-Nov-65	15-Jan-66	1	0	520	210	48.4	UNKNOWN	
OPS 3031	1966-012C	2015	15-Feb-66	15-Feb-66	38	0	270	150	96.5	UNKNOWN	INFLATABLE SPHERE
GEMINI 9 ATDA R/B	1966-046B	2188	1-Jun-66	Mid-Jun-66	51	0	275	240	28.8	UNKNOWN	ATLAS CORE STAGE
AS-203	1966-059A	2289	5-Jul-66	5-Jul-66	34	0	215	185	32.0	DELIBERATE	SATURN S-IVB STAGE
COSMOS U-1	1966-088A	2437	17-Sep-66	17-Sep-66	53	0	855	140	49.6	DELIBERATE	SELF-DESTRUCT
COSMOS U-2	1966-101A	2536	2-Nov-66	2-Nov-66	41	0	885	145	49.6	DELIBERATE	SELF-DESTRUCT
COSMOS 199	1968-003A	3099	16-Jan-68	24-Jan-68	3	0	355	200	65.6	DELIBERATE	SELF-DESTRUCT
APOLLO 6 R/B (S4B)	1968-025B	3171	4-Apr-68	13-Apr-68	16	0	360	200	32.6	PROPULSION	SATURN S-IVB STAGE
COSMOS 249	1968-091A	3504	20-Oct-68	20-Oct-68	108	43	2165	490	62.3	DELIBERATE	SELF-DESTRUCT
COSMOS 248	1968-090A	3503	19-Oct-68	1-Nov-68	5	0	545	475	62.2	DELIBERATE	DEBRIS IMPACT
COSMOS 252	1968-097A	3530	1-Nov-68	1-Nov-68	139	43	2140	535	62.3	DELIBERATE	SELF-DESTRUCT
METEOR 1-1 R/B	1969-029B	3836	26-Mar-69	28-Mar-69	37	0	850	460	81.2	UNKNOWN	VOSTOK FINAL STAGE
INTELSAT 3 F-5 R/B	1969-064B	4052	26-Jul-69	26-Jul-69	23	1	5445	270	30.4	PROPULSION	TE 364-4 STAGE
OPS 7613 R/B	1969-082AB	4159	30-Sep-69	4-Oct-69	261	73	940	905	70.0	UNKNOWN	AGENA D STAGE
NIMBUS 4 R/B	1970-025C	4367	8-Apr-70	17-Oct-70	373	247	1085	1065	99.9	UNKNOWN	AGENA D STAGE
		4601		23-Jan-85						UNKNOWN	2 ADDITIONAL OBJECTS
		4649		17-Dec-85						UNKNOWN	3 ADDITIONAL OBJECTS
		4610		2-Sep-86						UNKNOWN	2 ADDITIONAL OBJECTS
		4601		23-Dec-91						UNKNOWN	5 ADDITIONAL OBJECTS
COSMOS 374	1970-089A	4594	23-Oct-70	23-Oct-70	99	21	2130	530	62.9	DELIBERATE	SELF-DESTRUCT
COSMOS 375	1970-091A	4598	30-Oct-70	30-Oct-70	47	17	2100	525	62.8	DELIBERATE	SELF-DESTRUCT
COSMOS 397	1971-015A	4964	25-Feb-71	25-Feb-71	116	47	2200	575	65.8	DELIBERATE	SELF-DESTRUCT
COSMOS 462	1971-106A	5646	3-Dec-71	3-Dec-71	25	0	1800	230	65.7	DELIBERATE	SELF-DESTRUCT
SALYUT 2 R/B	1973-017B	6399	3-Apr-73	3-Apr-73	25	0	245	195	51.5	PROPULSION	PROTON-K SECOND STAGE
COSMOS 554	1973-021A	6432	19-Apr-73	6-May-73	195	0	350	170	72.9	DELIBERATE	SELF-DESTRUCT
NOAA 3 R/B	1973-086B	6921	6-Nov-73	28-Dec-73	197	175	1510	1500	102.1	PROPULSION	DELTA SECOND STAGE

22

TABLE 2.2 HISTORY OF SATELLITE BREAKUPS BY EVENT DATE (CONT'D)

NAME	INTERNATIONAL DESIGNATOR	CATALOG NUMBER	LAUNCH DATE	EVENT DATE	DEBRIS CATALOGED	DEBRIS LEFT	APOGEE (KM)	PERIGEE (KM)	INCLINATION (DEG)	ASSESSED CAUSE	COMMENT
COSMOS 699	1974-103A	7587	24-Dec-74	17-Apr-75	50	0	445	425	65.0	UNKNOWN	COSMOS 699 CLASS
				2-Aug-75			440	415	65.0	UNKNOWN	
LANDSAT 1 R/B	1972-058B	6127	23-Jul-72	22-May-75	226	31	910	635	98.3	PROPULSION	DELTA SECOND STAGE
PAGEOS	1966-056A	2253	24-Jun-66	12-Jul-75	79	2	5170	3200	85.3	UNKNOWN	INFLATABLE SPHERE
				20-Jan-76			5425	2935	85.1	UNKNOWN	
				10-Sep-76						UNKNOWN	
				Mid-Jun-78						UNKNOWN	
				Mid-Sep-84						UNKNOWN	
				Mid-Dec-85						UNKNOWN	
NOAA 4 R/B	1974-089D	7532	15-Nov-74	20-Aug-75	146	122	1460	1445	101.7	PROPULSION	DELTA SECOND STAGE
COSMOS 758	1975-080A	8191	5-Sep-75	6-Sep-75	76	0	325	175	67.1	DELIBERATE	SELF-DESTRUCT
COSMOS 777	1975-102A	8416	29-Oct-75	25-Jan-76	62	0	440	430	65.0	UNKNOWN	COSMOS 699 CLASS
LANDSAT 2 R/B	1975-004B	7616	22-Jan-75	9-Feb-76	207	33	915	740	97.8	PROPULSION	DELTA SECOND STAGE
				19-Jun-76			910	745	97.7	PROPULSION	
COSMOS 844	1976-072A	9046	22-Jul-76	25-Jul-76	248	0	355	170	67.1	DELIBERATE	SELF-DESTRUCT
COSMOS 886	1976-126A	9634	27-Dec-76	27-Dec-76	76	60	2295	595	65.8	DELIBERATE	SELF-DESTRUCT
COSMOS 884	1976-123A	9614	17-Dec-76	29-Dec-76	2	0	320	170	65.0	DELIBERATE	SELF-DESTRUCT
COSMOS 862	1976-105A	9495	22-Oct-76	15-Mar-77	11	10	39645	765	63.2	DELIBERATE	SELF-DESTRUCT
COSMOS 838	1976-063A	8932	2-Jul-76	17-May-77	40	0	445	415	65.1	UNKNOWN	COSMOS 699 CLASS
HIMAWARI 1 R/B	1977-065B	10144	14-Jul-77	14-Jul-77	172	63	2025	535	29.0	PROPULSION	DELTA SECOND STAGE
COSMOS 839	1976-067A	9011	8-Jul-76	29-Sep-77	69	67	2100	980	65.9	BATTERY	
COSMOS 931	1977-068A	10150	20-Jul-77	24-Oct-77	6	5	39665	680	62.9	DELIBERATE	SELF-DESTRUCT
COSMOS 970	1977-121A	10531	21-Dec-77	21-Dec-77	70	65	1140	945	65.8	DELIBERATE	SELF-DESTRUCT
NOAA 5 R/B	1976-077B	9063	29-Jul-76	24-Dec-77	162	155	1520	1505	102.0	PROPULSION	DELTA SECOND STAGE
COSMOS 903	1977-027A	9911	11-Apr-77	8-Jun-78	3	2	39035	1325	63.2	DELIBERATE	SELF-DESTRUCT
EKRAN 2	1977-092A	10365	20-Sep-77	25-Jun-78	4	4	35800	35785	0.1	BATTERY	
COSMOS 1030	1978-083A	11015	6-Sep-78	10-Oct-78	7	5	39760	665	62.8	DELIBERATE	SELF-DESTRUCT
COSMOS 880	1976-120A	9601	9-Dec-76	27-Nov-78	49	0	620	550	65.8	BATTERY	
COSMOS 917	1977-047A	10059	16-Jun-77	30-Mar-79	8	8	38725	1645	62.9	DELIBERATE	SELF-DESTRUCT
COSMOS 1124	1979-077A	11509	28-Aug-79	9-Sep-79	3	3	39795	570	63.0	DELIBERATE	SELF-DESTRUCT
COSMOS 1094	1979-033A	11333	18-Apr-79	17-Sep-79	1	0	405	380	65.0	UNKNOWN	COSMOS 699 CLASS
COSMOS 1109	1979-058A	11417	27-Jun-79	Mid-Feb-80	11	11	39425	960	63.3	DELIBERATE	SELF-DESTRUCT
CAT R/B	1979-104B	11659	24-Dec-79	Apr-80	22	18	33140	180	17.9	PROPULSION	ARIANE 1 FINAL STAGE
COSMOS 1174	1980-030A	11765	18-Apr-80	18-Apr-80	46	4	1660	380	66.1	DELIBERATE	SELF-DESTRUCT

TABLE 2.2 HISTORY OF SATELLITE BREAKUPS BY EVENT DATE (CONT'D)

NAME	INTERNATIONAL DESIGNATOR	CATALOG NUMBER	LAUNCH DATE	EVENT DATE	DEBRIS CATALOGED	DEBRIS LEFT	APOGEE (KM)	PERIGEE (KM)	INCLINATION (DEG)	ASSESSED CAUSE	COMMENT
LANDSAT 3 R/B	1978-026C	10704	5-Mar-78	27-Jan-81	210	122	910	900	98.8	PROPULSION	DELTA SECOND STAGE
COSMOS 1261	1981-031A	12376	31-Mar-81	Apr/May-81	8	8	39765	610	63.0	DELIBERATE	SELF-DESTRUCT
COSMOS 1191	1980-057A	11871	2-Jul-80	14-May-81	8	8	39255	1110	62.6	DELIBERATE	SELF-DESTRUCT
COSMOS 1167	1980-021A	11729	14-Mar-80	15-Jul-81	12	0	450	355	65.0	UNKNOWN	COSMOS 699 CLASS
COSMOS 1275	1981-053A	12504	4-Jun-81	24-Jul-81	310	259	1015	960	83.0	BATTERY	
COSMOS 1305 R/B	1981-088F	12827	11-Sep-81	11-Sep-81	8	8	13795	605	62.8	PROPULSION	MOLNIYA FINAL STAGE
COSMOS 1247	1981-016A	12303	19-Feb-81	20-Oct-81	7	7	39390	970	63.0	DELIBERATE	SELF-DESTRUCT
COSMOS 1285	1981-071A	12627	4-Aug-81	21-Nov-81	18	18	40100	720	63.1	DELIBERATE	SELF-DESTRUCT
NIMBUS 7 R/B	1978-098B	11081	24-Oct-78	26-Dec-81	1	1	955	935	99.3	PROPULSION	DELTA SECOND STAGE
COSMOS 1260	1981-028A	12364	20-Mar-81	8-May-82	68	0	750	450	65.0	UNKNOWN	COSMOS 699 CLASS
				10-Aug-82			750	445	65.0	UNKNOWN	
COSMOS 1220	1980-089A	12054	4-Nov-80	20-Jun-82	83	2	885	570	65.0	UNKNOWN	COSMOS 699 CLASS
COSMOS 1306	1981-089A	12828	14-Sep-81	12-Jul-82	8	0	405	380	64.9	UNKNOWN	COSMOS 699 CLASS
				18-Sep-82			370	370	64.9	UNKNOWN	
COSMOS 1286	1981-072A	12631	4-Aug-81	29-Sep-82	2	0	325	300	65.0	UNKNOWN	COSMOS 699 CLASS
COSMOS 1423 R/B	1982-115E	13696	8-Dec-82	8-Dec-82	29	0	425	235	62.9	PROPULSION	MOLNIYA FINAL STAGE
COSMOS 1217	1980-085A	12032	24-Oct-80	12-Feb-83	7	7	38830	1530	65.2	DELIBERATE	SELF-DESTRUCT
COSMOS 1481	1983-070A	14182	8-Jul-83	9-Jul-83	6	6	39225	625	62.9	DELIBERATE	SELF-DESTRUCT
COSMOS 1355	1982-038A	13150	29-Apr-82	8-Aug-83	29	0	395	360	65.1	UNKNOWN	COSMOS 699 CLASS
				1-Feb-84			320	305	65.0	UNKNOWN	
				20-Feb-84			290	270	65.0	UNKNOWN	
COSMOS 1456	1983-038A	14034	25-Apr-83	13-Aug-83	4	0	39630	730	63.3	DELIBERATE	SELF-DESTRUCT
COSMOS 1405	1982-088A	13508	4-Sep-82	20-Dec-83	32	0	340	310	65.0	UNKNOWN	COSMOS 699 CLASS
COSMOS 1317	1981-108A	12933	31-Oct-81	Late-Jan-84	4	4	39055	1315	62.8	DELIBERATE	SELF-DESTRUCT
WESTAR 6 R/B	1984-011F	14694	3-Feb-84	3-Feb-84	14	1	310	305	28.5	PROPULSION	PAM-D UPPER STAGE
PALAPA B2 R/B	1984-011E	14693	3-Feb-84	6-Feb-84	3	0	285	275	28.5	PROPULSION	PAM-D UPPER STAGE
ASTRON ULLAGE MOTOR	1983-020B	13902	23-Mar-83	3-Sep-84	1	0	1230	220	51.5	PROPULSION	PROTON-K BLOCK DM SOZ
SPACENET 2/ MARECS B2 R/B	1984-114C	15388	10-Nov-84	20-Nov-84	3	2	35960	325	7.0	PROPULSION	ARIANE 3 FINAL STAGE
COSMOS 1461	1983-044A	14064	7-May-83	11-Mar-85	164	5	890	570	65.0	UNKNOWN	COSMOS 699 CLASS
				13-May-85			885	570	65.0	UNKNOWN	
COSMOS 1654	1985-039A	15734	23-May-85	21-Jun-85	18	0	300	185	64.9	DELIBERATE	SELF-DESTRUCT
P-78 (SOLWIND)	1979-017A	11278	24-Feb-79	13-Sep-85	285	0	545	515	97.6	DELIBERATE	HYPERVELOCITY IMPACT
COSMOS 1375	1982-055A	13259	6-Jun-82	21-Oct-85	61	58	1000	990	65.8	BATTERY	

TABLE 2.2 HISTORY OF SATELLITE BREAKUPS BY EVENT DATE (CONT'D)

NAME	INTERNATIONAL DESIGNATOR	CATALOG NUMBER	LAUNCH DATE	EVENT DATE	DEBRIS CATALOGED	DEBRIS LEFT	APOGEE (KM)	PERIGEE (KM)	INCLINATION (DEG)	ASSESSED CAUSE	COMMENT
COSMOS 1691	1985-094B	16139	9-Oct-85	22-Nov-85	14	11	1415	1410	82.6	BATTERY	
COSMOS 1714 R/B	1985-121F	16439	28-Dec-85	28-Dec-85	2	0	830	165	71.0	PROPULSION	ZENIT SECOND STAGE
NOAA 8	1983-022A	13923	28-Mar-83	30-Dec-85	8	1	830	805	98.6	BATTERY	
COSMOS 1588	1984-083A	15167	7-Aug-84	23-Feb-86	45	0	440	410	65.0	UNKNOWN	COSMOS 699 CLASS
USA 19	1986-069A	16937	5-Sep-86	5-Sep-86	13	0	745	210	39.1	DELIBERATE	HYPERVELOCITY IMPACT
USA 19 R/B	1986-069B	16938	5-Sep-86	5-Sep-86	5	0	610	220	22.8	DELIBERATE	HYPERVELOCITY IMPACT
SPOT 1/VIKING R/B	1986-019C	16615	22-Feb-86	13-Nov-86	489	30	835	805	98.7	PROPULSION	ARIANE 1 FINAL STAGE
COSMOS 1278	1981-058A	12547	19-Jun-81	Early-Dec-86	3	0	37690	2665	67.1	DELIBERATE	SELF-DESTRUCT
COSMOS 1682	1985-082A	16054	19-Sep-85	18-Dec-86	23	0	475	385	65.0	UNKNOWN	COSMOS 699 CLASS
COSMOS 1813	1987-004A	17297	15-Jan-87	29-Jan-87	195	0	415	360	72.8	DELIBERATE	SELF-DESTRUCT
COSMOS 1866	1987-059A	18184	9-Jul-87	26-Jul-87	9	0	255	155	67.1	DELIBERATE	SELF-DESTRUCT
AUSSAT/ECS R/B	1987-078C	18352	16-Sep-87	Mid-Sep-87	4	1	36515	245	6.9	PROPULSION	ARIANE 3 FINAL STAGE
COSMOS 1769	1986-059A	16895	4-Aug-86	21-Sep-87	4	0	445	310	65.0	UNKNOWN	COSMOS 699 CLASS
COSMOS 1646	1985-030A	15653	18-Apr-85	20-Nov-87	24	0	410	385	65.0	UNKNOWN	COSMOS 699 CLASS
COSMOS 1823	1987-020A	17535	20-Feb-87	17-Dec-87	119	44	1525	1480	73.6	BATTERY	
COSMOS 1656 ULLAGE MOTOR	1985-042E	15773	30-May-85	5-Jan-88	6	6	860	810	66.6	PROPULSION	PROTON-K BLOCK DM SOZ
COSMOS 1906	1987-108A	18713	26-Dec-87	31-Jan-88	37	0	265	245	82.6	DELIBERATE	SELF-DESTRUCT
COSMOS 1916	1988-007A	18823	3-Feb-88	27-Feb-88	1	0	230	150	64.8	DELIBERATE	SELF-DESTRUCT
COSMOS 1045 R/B	1978-100D	11087	26-Oct-78	9-May-88	42	37	1705	1685	82.6	PROPULSION	TSYKLON THIRD STAGE
COSMOS 2030	1989-054A	20124	12-Jul-89	28-Jul-89	1	0	215	150	67.1	DELIBERATE	SELF-DESTRUCT
COSMOS 2031	1989-056A	20136	18-Jul-89	31-Aug-89	9	0	365	240	50.5	DELIBERATE	SELF-DESTRUCT
FENGYUN 1-2 R/B	1990-081D	20791	3-Sep-90	4-Oct-90	84	67	895	880	98.9	PROPULSION	CZ-4A FINAL STAGE
COSMOS 2101	1990-087A	20828	1-Oct-90	30-Nov-90	4	0	280	195	64.8	DELIBERATE	SELF-DESTRUCT
USA 68	1990-105A	20978	1-Dec-90	1-Dec-90	29	1	850	610	98.9	PROPULSION	TE-M-364-15 UPPER STAGE
COSMOS 1519-1521 ULLAGE MOTOR	1983-127H	14608	29-Dec-83	4-Feb-91	8	3	18805	340	51.9	PROPULSION	PROTON-K BLOCK DM SOZ
COSMOS 2125-2132 R/B	1991-009J	21108	12-Feb-91	5-Mar-91	92	92	1725	1460	74.0	PROPULSION	COSMOS SECOND STAGE
NIMBUS 6 R/B	1975-052B	7946	12-Jun-75	1-May-91	268	207	1105	1095	99.6	PROPULSION	DELTA SECOND STAGE
COSMOS 2163	1991-071A	21741	9-Oct-91	6-Dec-91	1	0	260	185	64.8	DELIBERATE	SELF-DESTRUCT
COSMOS 1934	1988-023A	18985	22-Mar-88	23-Dec-91	3	3	1010	950	83.0	COLLISION, ACCIDENTAL	HIT BY DEBRIS (13475)
COSMOS 1710-1712 ULLAGE MOTOR	1985-118L	16446	24-Dec-85	29-Dec-91	17	11	18885	655	65.3	PROPULSION	PROTON-K BLOCK DM SOZ
OV2-5 R/B	1968-081E	3432	26-Sep-68	21-Feb-92	4	4	35810	35100	11.9	PROPULSION	TITAN TRANSTAGE
COSMOS 2045 ULLAGE MOTOR	1989-101E	20399	27-Dec-89	Jul-92 (?)	14	8	27650	345	47.1	PROPULSION	PROTON-K BLOCK DM SOZ

TABLE 2.2 HISTORY OF SATELLITE BREAKUPS BY EVENT DATE (CONT'D)

NAME	INTERNATIONAL DESIGNATOR	CATALOG NUMBER	LAUNCH DATE	EVENT DATE	DEBRIS CATALOGED	DEBRIS LEFT	APOGEE (KM)	PERIGEE (KM)	INCLINATION (DEG)	ASSESSED CAUSE	COMMENT
COSMOS 1603 ULLAGE MOTOR	1984-106F	15338	28-Sep-84	5-Sep-92	22	1	845	835	66.6	PROPULSION	PROTON-K BLOCK DM SOZ
GORIZONT 17 ULLAGE MOTOR	1989-004E	19771	26-Jan-89	17-Dec-92	1	0	17575	195	46.7	PROPULSION	PROTON-K BLOCK DM SOZ
COSMOS 2227 R/B	1992-093B	22285	25-Dec-92	26-Dec-92	226	166	855	845	71.0	PROPULSION	ZENIT-2 SECOND STAGE
				30-Dec-92			855	845	71.0		
GORIZONT 18 ULLAGE MOTOR	1989-052F	20116	5-Jul-89	12-Jan-93	2	0	36745	260	46.8	PROPULSION	PROTON-K BLOCK DM SOZ
COSMOS 2225	1992-091A	22280	22-Dec-92	18-Feb-93	6	0	280	225	64.9	DELIBERATE	SELF-DESTRUCT
COSMOS 2237 R/B	1993-016B	22566	26-Mar-93	28-Mar-93	37	35	850	840	71.0	PROPULSION	ZENIT-2 SECOND STAGE
TELECOM 2B/ INMARSAT 2 R/B	1992-021C	21941	15-Apr-92	21-Apr-93	14	13	34080	235	4.0	PROPULSION	ARIANE H10+ FINAL STAGE
COSMOS 2243	1993-028A	22641	27-Apr-93	27-Apr-93	1	0	225	180	70.4	DELIBERATE	SELF-DESTRUCT
COSMOS 2259	1993-045A	22716	14-Jul-93	25-Jul-93	1	0	320	175	67.1	DELIBERATE	SELF-DESTRUCT
COSMOS 1484	1983-075A	14207	24-Jul-83	18-Oct-93	49	2	595	550	97.5	UNKNOWN	SELF-DESTRUCT
COSMOS 2262	1993-057A	22789	7-Sep-93	18-Dec-93	1	0	295	170	64.9	DELIBERATE	SELF-DESTRUCT
CLEMENTINE R/B	1994-004B	22974	25-Jan-94	7-Feb-94	1	0	295	240	67.0	PROPULSION	
ASTRA 1B/MOP 2 R/B	1991-015C	21141	2-Mar-91	27-Apr-94	10	7	17630	205	6.8	PROPULSION	ARIANE 4 H10 FINAL STAGE
COSMOS 2133 ULLAGE MOTOR	1991-010D	21114	12-Feb-91	7-May-94	4	2	21805	225	46.6	PROPULSION	PROTON-K BLOCK DM SOZ
COSMOS 2204-2206 ULLAGE MOTOR	1992-047H	22067	30-Jul-92	8-Nov-94	4	3	19035	480	64.8	PROPULSION	PROTON-K BLOCK DM SOZ
COSMOS 2238	1993-018A	22585	30-Mar-93	1-Dec-94	1	0	305	210	65.0	UNKNOWN	COSMOS 699 CLASS
RS-15 R/B	1994-085B	23440	26-Dec-94	26-Dec-94	23	21	2200	1880	64.8	UNKNOWN	ROKOT THIRD STAGE
ELEKTRO ULLAGE MOTOR	1994-069E	23338	31-Oct-94	11-May-95	1	0	35465	155	46.9	PROPULSION	PROTON-K BLOCK DM SOZ
COSMOS 2282 ULLAGE MOTOR	1994-038F	23174	6-Jul-94	21-Oct-95	2	1	34930	280	47.0	PROPULSION	PROTON-K BLOCK DM SOZ
GORIZONT 22 ULLAGE MOTOR	1990-102E	20957	23-Nov-90	14-Dec-95	2	1	13105	170	46.5	PROPULSION	PROTON-K BLOCK DM SOZ
RADUGA 33 R/B	1996-010D	23797	19-Feb-96	19-Feb-96	2	0	36505	240	48.7	PROPULSION	PROTON-K BLOCK DM
ITALSAT 1 R/B/ EUTELSAT 2 F2	1991-003C	21057	15-Jan-91	1-May-96	12	9	30930	235	6.7	PROPULSION	ARIANE 4 H10 FINAL STAGE
STEP II R/B	1994-029B	23106	19-May-94	3-Jun-96	713	64	820	585	82.0	PROPULSION	PEGASUS HAPS
CERISE	1995-033B	23606	7-Jul-95	24-Jul-96	2	2	675	665	98.1	COLLISION, ACCIDENTAL	HIT BY DEBRIS (18208)
COSMOS 1883-1885 ULLAGE MOTOR	1987-079G	18374	16-Sep-87	1-Dec-96	14	11	19120	335	64.9	PROPULSION	PROTON-K BLOCK DM SOZ
EKRAN 17 ULLAGE MOTOR	1987-109E	18719	27-Dec-87	22-May-97	1	0	22975	310	46.6	PROPULSION	PROTON-K BLOCK DM SOZ
COSMOS 2313	1995-028A	23596	8-Jun-95	26-Jun-97	13	0	325	210	65.0	UNKNOWN	COSMOS 699 CLASS
COSMOS 2343	1997-024A	24805	15-May-97	16-Sep-97	1	0	285	225	65.0	DELIBERATE	SELF-DESTRUCT

TABLE 2.2 HISTORY OF SATELLITE BREAKUPS BY EVENT DATE (CONT'D)

NAME	INTERNATIONAL DESIGNATOR	CATALOG NUMBER	LAUNCH DATE	EVENT DATE	DEBRIS CATALOGED	DEBRIS LEFT	APOGEE (KM)	PERIGEE (KM)	INCLINATION (DEG)	ASSESSED CAUSE	COMMENT
COSMOS 1869	1987-062A	18214	16-Jul-87	27-Nov-97	2	2	635	605	83.0	UNKNOWN	
ASIASAT 3 R/B	1997-086D	25129	24-Dec-97	25-Dec-97	1	0	35995	270	51.0	PROPULSION	PROTON-K BLOCK DM
METEOR 2-16 R/B	1987-068B	18313	18-Aug-87	15-Feb-98	83	18	960	940	82.6	PROPULSION	TSYKLON THIRD STAGE
SKYNET 4B/ASTRA 1A R/B	1988-109C	19689	11-Dec-88	17-Feb-98	11	10	35875	435	7.3	PROPULSION	ARIANE 4 H10 FINAL STAGE
COMETS R/B	1998-011B	25176	21-Feb-98	21-Feb-98	1	0	1880	245	30.0	PROPULSION	H-II SECOND STAGE
COSMOS 2109-2111 ULLAGE MOTOR	1990-110H	21013	8-Dec-90	14-Mar-98	2	2	18995	520	65.1	PROPULSION	PROTON-K BLOCK DM SOZ
COSMOS 1987-1989	1989-001G	19755	10-Jan-89	3-Aug-98	12	7	19055	340	64.9	PROPULSION	PROTON-K BLOCK DM SOZ
COSMOS 1650-1652 ULLAGE MOTOR	1985-037G	15714	17-May-85	29-Nov-98	4	2	18620	320	52.0	PROPULSION	PROTON-K BLOCK DM SOZ
COSMOS 1970-1972 ULLAGE MOTOR	1988-085G	19537	16-Sep-88	9-Mar-99	1	1	18950	300	64.6	PROPULSION	PROTON-K BLOCK DM SOZ
COSMOS 2079-2081 ULLAGE MOTOR	1990-045G	20631	19-May-90	28-Mar-99	1	1	19065	405	64.8	PROPULSION	PROTON-K BLOCK DM SOZ
COSMOS 2053 R/B	1989-100B	20390	27-Dec-89	18-Apr-99	26	0	485	475	73.5	PROPULSION	TSYKLON THIRD STAGE
COSMOS 2157-2162 R/B	1991-068G	21734	28-Sep-91	9-Oct-99	37	37	1485	1410	82.6	PROPULSION	TSYKLON THIRD STAGE
COSMOS 2347	1997-079A	25088	9-Dec-97	22-Nov-99	9	0	410	230	65.0	UNKNOWN	COSMOS 699 CLASS
GORIZONT 32 ULLAGE MOTOR	1996-034F	23887	23-May-96	13-Dec-99	1	0	5605	145	46.5	PROPULSION	PROTON-K BLOCK DM SOZ
CBERS-1/SACI-1 R/B	1999-057C	25942	14-Oct-99	11-Mar-00	345	197	745	725	98.5	PROPULSION	CZ-4 FINAL STAGE
GORIZONT 29 ULLAGE MOTOR	1993-072E	22925	18-Nov-93	6-Sep-00	1	0	11215	140	46.7	PROPULSION	PROTON-K BLOCK DM SOZ
COSMOS 2316-2318 ULLAGE MOTOR	1995-037K	23631	24-Jul-95	21-Nov-00	1	0	18085	150	64.4	PROPULSION	PROTON-K BLOCK DM SOZ
INTELSAT 515 R/B	1989-006B	19773	27-Jan-89	1-Jan-01	37	37	35720	510	8.4	PROPULSION	ARIANE 2 R/B
COSMOS 2139-41 ULLAGE MOTOR	1991-025G	21226	4-Apr-91	16-Jun-01	1	1	18960	300	64.5	PROPULSION	PROTON-K BLOCK DM SOZ
GORIZONT 27 ULLAGE MOTOR	1992-082F	22250	27-Nov-92	14-Jul-01	1	0	5340	145	46.5	PROPULSION	PROTON-K BLOCK DM SOZ
COSMOS 2367	1999-072A	26040	26-Dec-99	21-Nov-01	17	0	415	405	65.0	UNKNOWN	COSMOS 699 CLASS
TES R/B	2001-049D	26960	22-Oct-01	19-Dec-01	346	97	675	550	97.9	PROPULSION	PSLV FINAL STAGE
INTELSAT 601 R/B	1991-075B	21766	29-Oct-91	24-Dec-01	12	9	28505	230	7.2	PROPULSION	ARIANE 4 H10 FINAL STAGE
INSAT 2A/ EUTELSAT 2F4 R/B	1992-041C	22032	9-Jul-92	2-Feb-02	1	1	26550	250	7.0	PROPULSION	ARIANE 4 H10 FINAL STAGE
INTELSAT 513 R/B	1988-040B	19122	17-May-88	9-Jul-02	5	5	35445	535	7.0	PROPULSION	ARIANE 2 R/B
COSMOS 2109-11 ULLAGE MOTOR	1990-110G	21012	8-Dec-90	21-Feb-03	1	1	18805	645	65.4	PROPULSION	PROTON-K BLOCK DM SOZ
COSMOS 1883-1885	1987-079H	18375	16-Sep-87	23-Apr-03	39	16	18540	755	65.2	PROPULSION	PROTON-K BLOCK DM SOZ
COSMOS 1970-72 ULLAGE MOTOR	1988-085F	19535	16-Sep-88	4-Aug-03	79	10	18515	720	65.3	PROPULSION	PROTON-K BLOCK DM SOZ
COSMOS 1987-1989 ULLAGE MOTOR	1989-001H	19856	10-Jan-89	13-Nov-03	1	1	18740	710	65.4	PROPULSION	PROTON-K BLOCK DM SOZ

27

TABLE 2.2 HISTORY OF SATELLITE BREAKUPS BY EVENT DATE (CONT'D)

NAME	INTERNATIONAL DESIGNATOR	CATALOG NUMBER	LAUNCH DATE	EVENT DATE	DEBRIS CATALOGED	DEBRIS LEFT	APOGEE (KM)	PERIGEE (KM)	INCLINATION (DEG)	ASSESSED CAUSE	COMMENT
COSMOS 2399	2003-035A	27856	12-Aug-03	9-Dec-03	22	0	250	175	64.9	DELIBERATE	SELF-DESTRUCT
COSMOS 2383	2001-057A	27053	21-Dec-01	28-Feb-04	14	0	400	220	65.0	UNKNOWN	COSMOS 699 CLASS
USA 73 (DMSP 5D2 F11)	1991-082A	21798	28-Nov-91	15-Apr-04	79	67	850	830	98.7	UNKNOWN	
COSMOS 2204-06 ULLAGE MOTOR	1992-047G	22066	30-Jul-92	10-Jul-04	30	25	18820	415	64.9	PROPULSION	PROTON-K BLOCK DM SOZ
COSMOS 2392 ULLAGE MOTOR	2002-037F	27475	25-Jul-02	29-Oct-04	1	0	840	235	63.6	PROPULSION	PROTON-K BLOCK DM SOZ
DMSP 5B F5 R/B	1974-015B	7219	16-Mar-74	17-Jan-05	5	5	885	775	99.1	COLLISION, ACCIDENTAL	HIT BY DEBRIS (26207)
COSMOS 2224 ULLAGE MOTOR	1992-088F	22274	17-Dec-92	~22-Apr-05	1	0	21140	200	46.7	PROPULSION	PROTON-K BLOCK DM SOZ
COSMOS 2392 ULLAGE MOTOR	2002-037E	27474	25-Jul-02	1-Jun-05	61	7	835	255	63.7	PROPULSION	PROTON-K BLOCK DM SOZ
COSMOS 1703 R/B	1985-108B	16263	22-Nov-85	4-May-06	50	3	640	610	82.5	UNKNOWN	TSYKLON THIRD STAGE
COSMOS 2022-24 ULLAGE MOTOR	1989-039G	20081	31-May-89	10-Jun-06	116	105	18410	655	65.1	PROPULSION	PROTON-K BLOCK DM SOZ
ALOS-1 R/B	2006-002B	28932	24-Jan-06	8-Aug-06	22	1	700	550	98.2	UNKNOWN	H-IIA SECOND STAGE
COSMOS 2371 ULLAGE MOTOR	2000-036E	26398	4-Jul-00	~1-Sep-06	1	1	21320	220	46.9	PROPULSION	PROTON-K BLOCK DM SOZ
DMSP 5D-3 F17 R/B	2006-050B	29523	4-Nov-06	4-Nov-06	64	61	865	830	98.8	UNKNOWN	DELTA IV SECOND STAGE
COSMOS 2423	2006-039A	29402	14-Sep-06	17-Nov-06	31	1	285	200	64.9	DELIBERATE	SELF-DESTRUCT
COBE R/B	1989-089B	20323	18-Nov-89	3-Dec-06	26	4	790	685	97.1	UNKNOWN	DELTA SECOND STAGE
IGS 3A R/B	2006-037B	29394	11-Sep-06	28-Dec-06	1	1	490	430	97.2	UNKNOWN	H-IIA SECOND STAGE
FENGYUN 1-3 (aka FENGYUN 1C)	1999-025A	25730	10-May-99	11-Jan-07	2087	2073	865	845	98.6	DELIBERATE	HYPERVELOCITY IMPACT
BEIDOU 2A	2007-003A	30323	2-Feb-07	2-Feb-07	2	2	41775	195	25.0	UNKNOWN	
KUPON ULLAGE MOTOR	1997-070F	25054	12-Nov-97	14-Feb-07	1	1	14160	260	46.6	PROPULSION	PROTON-K BLOCK DM SOZ
CBERS 1	1999-057A	25940	14-Oct-99	18-Feb-07	15	11	780	770	98.2	UNKNOWN	
ARABSAT 4 BRIZ-M R/B	2006-006B	28944	28-Feb-06	19-Feb-07	1	1	14705	495	51.5	PROPULSION	PROTON-K BRIZ-M STAGE
				TOTAL	12819	5783					

2.2 Identified Satellite Breakups

The remainder of this section devotes two pages to each identified satellite breakup. Each satellite is listed by common name, international designator, and satellite number. The satellite is then described in terms of type, ownership, launch date, and physical characteristics. The third grouping defines the breakup event by time, location, altitude, and assessed cause. In almost all cases, the calculated time of the event has been determined by the US SSN. Next, the last available element set for the satellite prior to the breakup is provided.

Contents of the pre- or post-event elements are described in Table 2.2-1. The epoch time's format consists of the last two digits of a year (YY) followed by a fractional day of year (DDD.DDDDDDDD). Three propagation scheme drag coefficients are available in a TLE, which form the basis of the orbital element data presented in this subsection. Though not all TLEs possess data for all three, they are described here for completeness; these are denoted as drag coefficients peculiar to the US SSN SGP, SGP4, and SGP8 orbit propagators. The data items $\dot{n}/2$ (pronounced "n dot over two") and $\ddot{n}/6$ (pronounced "n double dot over six") refer to the first and second order time derivatives of the mean motion n and represent phenomenological series expansion coefficient fits to the observed change in mean motion. The SGP4 propagator is the accepted standard for orbit propagation.

Table 2.1-1. TLE Numerical Data, as incorporated into this section's "Pre-/Post-Event Elements" for all fragmentation events.

DATA ITEM	FORMAT/UNITS
Epoch time	YYDDD.DDDDDDDD
$\dot{n}/2$ (SGP) <u>or</u> B (SGP8)	[rev/day^2] <u>or</u> [m^2/kg]
$\ddot{n}/6$ (SGP)	[rev/day^3]
B* (SGP4)	[1/Earth radii]
Eccentricity e	[-]
Inclination i	[°]
Right ascension of ascending node Ω	[°]
Argument of perigee ω	[°]
Mean anomaly M	[°]
Mean motion n	[rev/day]

If the breakup occurred soon after launch or after a maneuver and before an element set could be generated, the most appropriate post-event element set is given. The maximum observed changes in the orbital period (ΔP) and inclination (ΔI), referenced to the parent's pre-event element set, are then summarized. The reader is reminded that for a given event, the magnitudes of the resultant ΔP and ΔI are a function of the satellite's latitude and altitude. Comparisons of these values from one event to another cannot be made directly. Additionally, inclination changes measure only one portion of the fragmentation orbital plane change. Changes in Right Ascension also occur in most events and can account for some plane change fragmentation energy.

A general summary of the event, actions leading to the event, debris cataloging progress, and evaluations of the event are collected under the Comments heading. Documents that relate directly to the subject breakup or to breakups of satellites of this type are then listed. Gabbard diagrams of the early debris cloud prior to the effects of perturbations, if the data were available, are reconstructed. These diagrams often include uncataloged as well as cataloged debris data. When used correctly, Gabbard diagrams can provide important insights into the features of the fragmentation.

SATELLITE DATA

TYPE:	Ablestar Stage
OWNER:	US
LAUNCH DATE:	29.18 Jun 1961
DRY MASS (KG):	625
MAIN BODY:	Flared cylinder; 1.6 m diameter by 4.8 m length
MAJOR APPENDAGES:	None
ATTITUDE CONTROL:	None at time of the event
ENERGY SOURCES:	On-board propellants, range safety device

EVENT DATA

DATE:	29 Jun 1961	LOCATION:	28N, 254E (dsc)
TIME:	0608 GMT	ASSESSED CAUSE:	Propulsion
ALTITUDE:	990 km		

POST-EVENT ELEMENTS

EPOCH:	61187.36647288	MEAN ANOMALY:	72.1786
RIGHT ASCENSION:	79.1120	MEAN MOTION:	13.86864257
INCLINATION:	66.8199	MEAN MOTION DOT/2:	.0
ECCENTRICITY:	.0078181	MEAN MOTION DOT DOT/6:	.0
ARG. OF PERIGEE:	288.2398	BSTAR:	.0

DEBRIS CLOUD DATA

MAXIMUM ΔP:	15.5 min
MAXIMUM ΔI:	1.3 deg

COMMENTS

This is the first known satellite fragmentation. The Ablestar stage performed two main burns and a small payload separation retro burn to successfully deploy three payloads (Transit 4A, Injun, and Solrad 3), although the Injun and Solrad 3 satellites did not separate from one another as planned. The event occurred approximately 77 minutes after orbital insertion and was photographically imaged by the Organ Pass, NM, Baker-Nunn camera system. Fragmentation coincided with cessation of the 378 MHz beacon on the Ablestar stage at 0608:10 GMT. At the time of the event, 100 kg of hypergolic propellants remained on board. This was the first time an Ablestar stage did not vent the fuel tank during payload separation. After a thorough investigation, fuel venting was recommended for future missions. No reliable elements are available prior to the event. Elements above are for one of the payloads with parameters believed to be very similar to those for the Ablestar at the time of the event.

REFERENCE DOCUMENTS

Transit 4-A Ablestar Vehicle Fragmentation Study (Preliminary), Report TOR-930 (2102)-6, Flight Test Planning and Evaluation Department, Transit Program Office, USAF Systems Command, Inglewood, 28 August 1961.

Description, Operation and Performance of Ablestar Stage AJ10-104S, S/N 008 (Transit 4-A), T.W. Fehr and J.K. Stark, Report No. 2102, Spacecraft Division, Aerojet-General Corporation, Azusa, October 1961.

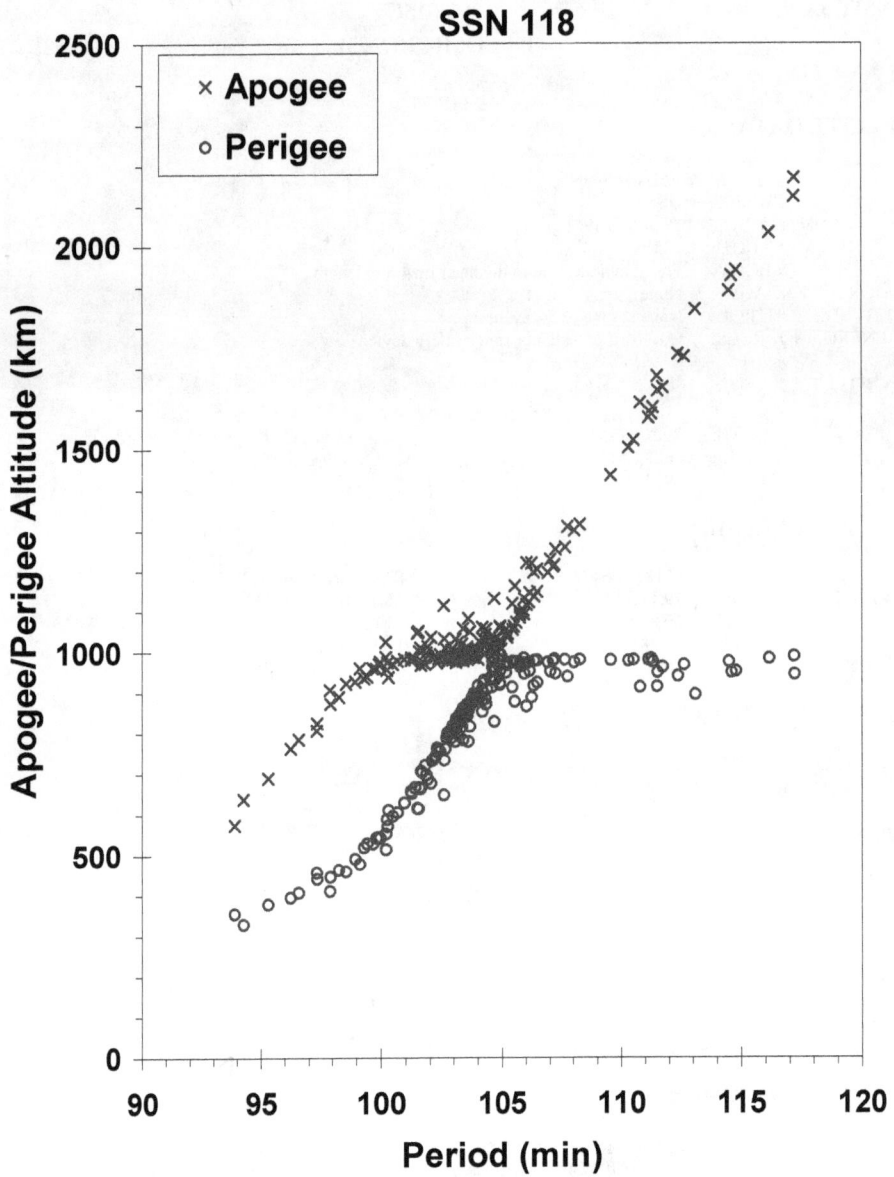

SSN 118

× Apogee
○ Perigee

Apogee/Perigee Altitude (km)

Period (min)

Transit 4A R/B debris cloud of 201 cataloged fragments in May 1964 as reconstructed from the US SSN database.

1962-057A
(1962-BETA IOTA 1)

SATELLITE DATA

TYPE:	Payload and R/B(s) (?)
OWNER:	CIS
LAUNCH DATE:	24.75 Oct 1962
DRY MASS (KG):	1500
MAIN BODY:	Cylinder; 2.6 m diameter by 7.15 m length
MAJOR APPENDAGES:	None
ATTITUDE CONTROL:	Unknown at time of event
ENERGY SOURCES:	On-board propellants

EVENT DATA

DATE:	29 Oct 1962	LOCATION:	Unknown
TIME:	Unknown	ASSESSED CAUSE:	Propulsion
ALTITUDE:	~200 km		

PRE-EVENT ELEMENTS

EPOCH:	62297.80327270	MEAN ANOMALY:	229.0409
RIGHT ASCENSION:	336.4972	MEAN MOTION:	16.15589719
INCLINATION:	65.1128	MEAN MOTION DOT/2:	.01124103
ECCENTRICITY:	.0044520	MEAN MOTION DOT DOT/6:	.0
ARG. OF PERIGEE:	92.2650	BSTAR:	.0

DEBRIS CLOUD DATA

MAXIMUM ΔP:	Unknown
MAXIMUM ΔI:	0.6 deg

COMMENTS

Sputnik 29 (also known as Sputnik 22) was not acknowledged at launch by the USSR and was probably a Mars probe that failed to leave Earth orbit. This was apparently the fourth orbital failure of the Molniya third stage since 25 August 1962. No Molniya orbital (3rd) stage nor final (4th) stage was cataloged after launch. Possible that orbital and final stages never separated. Sputnik 29 was officially decayed 29 October 1962 but no debris were cataloged before 11 November. Consequently, ΔP cannot be calculated. Source of the fragmentation was probably the fully-fueled Molniya final stage.

REFERENCE DOCUMENT

History of Soviet/Russian Satellite Fragmentations-A Joint U.S.-Russian Investigation, N. L. Johnson et al, Kaman Sciences Corporation, October 1995.

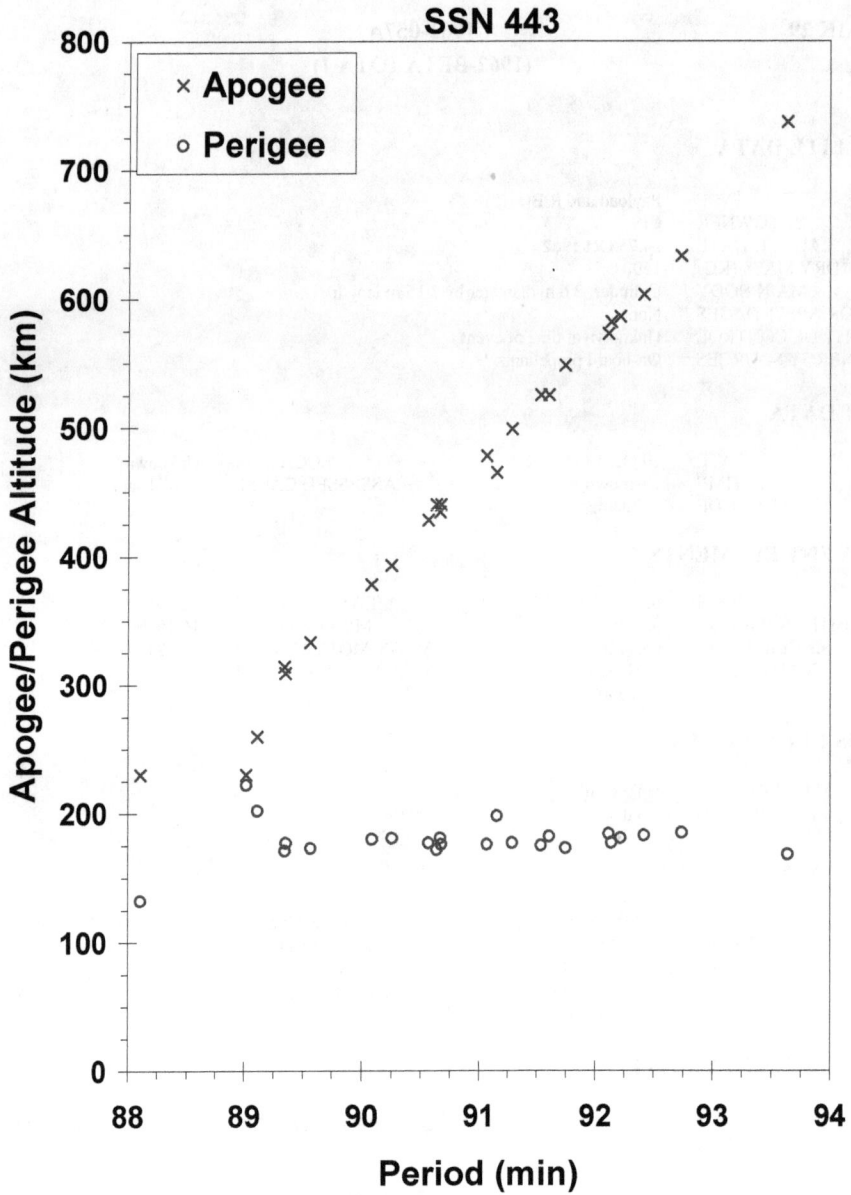

Sputnik 29 debris cloud of 24 fragments cataloged by mid-December 1962 as reconstructed from the US SSN database.

SATELLITE DATA

TYPE:	Centaur Stage
OWNER:	US
LAUNCH DATE:	27.79 Nov 1963
DRY MASS (KG):	4600
MAIN BODY:	Cylinder; 3 m diameter by 9 m length
MAJOR APPENDAGES:	None
ATTITUDE CONTROL:	Unknown at time of the event
ENERGY SOURCES:	Unknown

EVENT DATA

DATE:	27 Nov 1963	LOCATION:	Unknown
TIME:	Unknown	ASSESSED CAUSE:	Propulsion
ALTITUDE:	Unknown		

POST-EVENT ELEMENTS

EPOCH:	63336.85832214	MEAN ANOMALY:	213.1623
RIGHT ASCENSION:	135.1828	MEAN MOTION:	13.34437775
INCLINATION:	30.3440	MEAN MOTION DOT/2:	.00003262
ECCENTRICITY:	.0869282	MEAN MOTION DOT DOT/6:	.0
ARG. OF PERIGEE:	151.8246	BSTAR:	.0

DEBRIS CLOUD DATA

MAXIMUM ΔP:	0.9 min
MAXIMUM ΔI:	0.4 deg

COMMENTS

First Centaur stage to reach Earth orbit. No payload was carried. After orbital insertion, residual liquid hydrogen vaporized, resulting in an increase in tank pressurization. Venting via an aft tube then induced a pin-wheel tumble that reached 48 rpm a little more than 1 hour after launch. At the beginning of the third orbit insulation blankets around the Centaur stage were thrown off. Subsequent Centaur missions were not subject to this phenomenon that was caused by the unique configuration of Atlas Centaur 2. First six fragments were cataloged within 1 week of launch. Centaur stage retains large radar cross-section, while all debris are substantially smaller.

REFERENCE DOCUMENT

Supplementary Information on AC-2 Post-Injection Flight Events, W.S. Hicks, Memorandum BXN63-521, 27 December 1963.

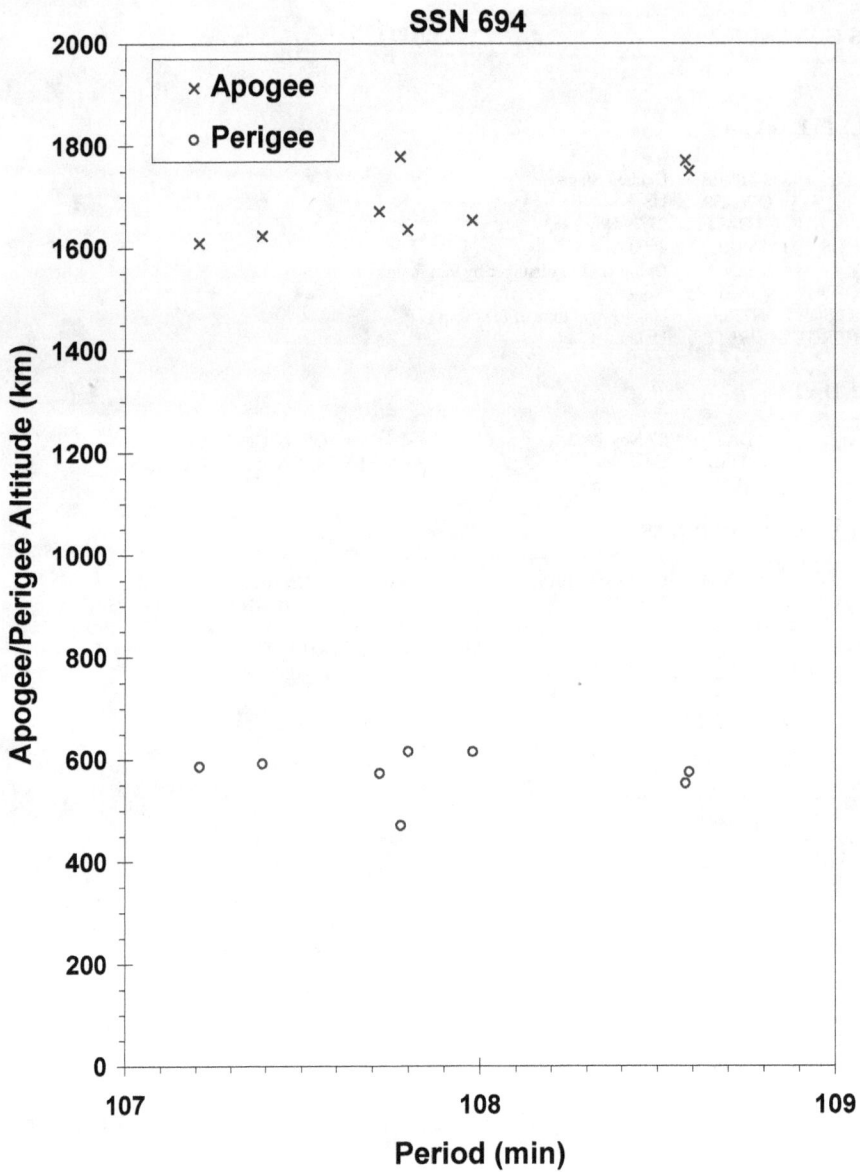

Atlas Centaur 2 debris cloud of 8 fragments 5 months after the event as reconstructed from the US SSN database.

34

SATELLITE DATA

TYPE:	Payload
OWNER:	CIS
LAUNCH DATE:	28.45 Oct 1964
DRY MASS (KG):	4750
MAIN BODY:	Sphere-cylinder; 2.4 m diameter by 4.3 m length
MAJOR APPENDAGES:	None
ATTITUDE CONTROL:	Active, 3-axis
ENERGY SOURCES:	On-board propellants, 10 kg TNT explosive charge

EVENT DATA

DATE:	5 Nov 1964	LOCATION:	Unknown
TIME:	Unknown	ASSESSED CAUSE:	Deliberate
ALTITUDE:	~200 km		

PRE-EVENT ELEMENTS

EPOCH:	64303.72916435	MEAN ANOMALY:	46.7488
RIGHT ASCENSION:	198.5952	MEAN MOTION:	16.23335350
INCLINATION:	51.2318	MEAN MOTION DOT/2:	.00269057
ECCENTRICITY:	.0034483	MEAN MOTION DOT DOT/6:	.0
ARG. OF PERIGEE:	312.9624	BSTAR:	.0

DEBRIS CLOUD DATA

MAXIMUM ΔP:	Unknown
MAXIMUM ΔI:	Unknown

COMMENTS

Spacecraft was destroyed after a malfunction prevented reentry and landing in the Soviet Union. Event occurred on the anticipated day of recovery. All debris were cataloged without elements. A probable fragment from this event reentered on 12 November 1964, landing in Malawi. See cited reference below.

REFERENCE DOCUMENTS

The Examination of a Sample of Space Debris, P.H.H. Bishop and K.F. Rogers, Technical Report 65165, Royal Aircraft Establishment, Farnborough Hants, August 1965.

History of Soviet/Russian Satellite Fragmentations-A Joint U.S.-Russian Investigation, N. L. Johnson et al, Kaman Sciences Corporation, October 1995.

Insufficient data to construct a Gabbard diagram.

SATELLITE DATA

TYPE:	Payload
OWNER:	CIS
LAUNCH DATE:	22.32 Feb 1965
DRY MASS (KG):	5500
MAIN BODY:	Sphere-cylinder; 2.4 m diameter by 6 m length
MAJOR APPENDAGES:	None
ATTITUDE CONTROL:	Active, 3-axis
ENERGY SOURCES:	On-board propellants, 10 kg TNT explosive charge

EVENT DATA

DATE:	22 Feb 1965	LOCATION:	64N, 80E (asc)
TIME:	0957 GMT	ASSESSED CAUSE:	Deliberate
ALTITUDE:	380 km		

POST-EVENT ELEMENTS

EPOCH:	65056.64509999	MEAN ANOMALY:	293.2095
RIGHT ASCENSION:	288.1532	MEAN MOTION:	15.92461677
INCLINATION:	64.7411	MEAN MOTION DOT/2:	.01501524
ECCENTRICITY:	.0182240	MEAN MOTION DOT DOT/6:	.0048063
ARG. OF PERIGEE:	68.7266	BSTAR:	.0

DEBRIS CLOUD DATA

MAXIMUM ΔP:	4.4 min
MAXIMUM ΔI:	0.9 deg

COMMENTS

Cosmos 57 was an unmanned precursor for the manned Voskhod 2 mission that took place in March 1965. Spacecraft fragmented a little more than 2 hours after launch when operational ground instructions were misinterpreted by the on-board command system and the self-destruct system was activated. No elements available for Cosmos 57, but the rocket body elements are provided above. The Royal Aircraft Establishment published the following parameters for Cosmos 57 for 22.4 February: 165 km by 427 km, 64.74 degree inclination, 64 degree argument of perigee. A total of 35 debris were cataloged without elements. Event may have occurred a little later than the time calculated above.

REFERENCE DOCUMENTS

The 1093 Breakup, D.J. Watson, BMEWS-ADC Systems Engineering Memorandum BSM-1000-16, 16 June 1965.

"To Save Man: A Conversation with the General Designer of Life-Support and Rescue Systems, Hero of Socialist Labor G.I. Severin", Pravda, Moscow, 26 June 1989, p. 4.

"Pages From a Diary: He Soared Freely Above the Earth", Sovetskaya Rossiya, Moscow, 17 March 1990, p. 6.

History of Soviet/Russian Satellite Fragmentations-A Joint U.S.-Russian Investigation, N. L. Johnson et al, Kaman Sciences Corporation, October 1995.

"The Kamanin Diaries 1964-1966", B. Hendrickx, Journal of the Interplanetary Society, Vol. 51, 1998, pp. 421-422.

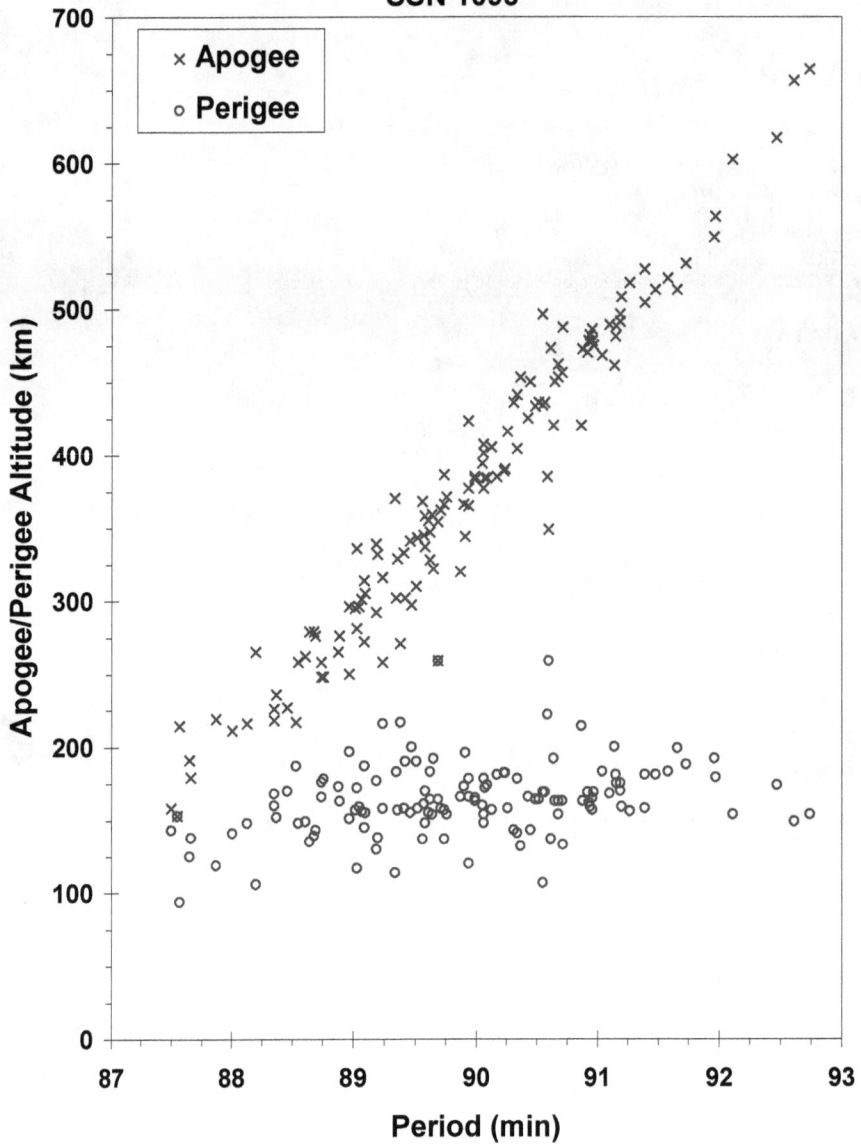

SSN 1093

Cosmos 57 debris cloud of 133 fragments cataloged within 1 month of the event as reconstructed from the US SSN database.

SATELLITE DATA

TYPE:	Cosmos Second Stage
OWNER:	CIS
LAUNCH DATE:	15.46 Mar 1965
DRY MASS (KG):	1600
MAIN BODY:	Cylinder; 2.4 m diameter by 5 m length
MAJOR APPENDAGES:	None
ATTITUDE CONTROL:	None at time of the event.
ENERGY SOURCES:	Unknown

EVENT DATA

DATE:	15 Mar 1965	LOCATION:	51S, 162E (dsc)
TIME:	1714 GMT	ASSESSED CAUSE:	Unknown
ALTITUDE:	1640 km		

POST-EVENT ELEMENTS

EPOCH:	65074.89183830	MEAN ANOMALY:	265.7165
RIGHT ASCENSION:	357.3218	MEAN MOTION:	13.57884745
INCLINATION:	56.0538	MEAN MOTION DOT/2:	.00231832
ECCENTRICITY:	.1056119	MEAN MOTION DOT DOT/6:	.0
ARG. OF PERIGEE:	106.1560	BSTAR:	.0

DEBRIS CLOUD DATA

MAXIMUM	ΔP:	10.3 min
MAXIMUM	ΔI:	0.4 deg

COMMENTS

This is the first confirmed case of the fragmentation of the Cosmos 3 (SL-8 or C-1) second stage. This was the third mission to deploy three payloads and was a repeat of the Cosmos 54-56 mission 3 weeks earlier. The event occurred a little more than 6 hours after the successful deployment of the three payloads. Elements above are the first developed for the rocket body and are about 4 hours after the event. Official debris cataloging did not begin for 6 weeks.

REFERENCE DOCUMENTS

"Fragmentations of Asteroids and Artificial Satellites in Orbit", W. Wiesel, Icarus, Vol. 34, 1978, pp. 99-116.

History of Soviet/Russian Satellite Fragmentations-A Joint U.S.-Russian Investigation, N. L. Johnson et al, Kaman Sciences Corporation, October 1995.

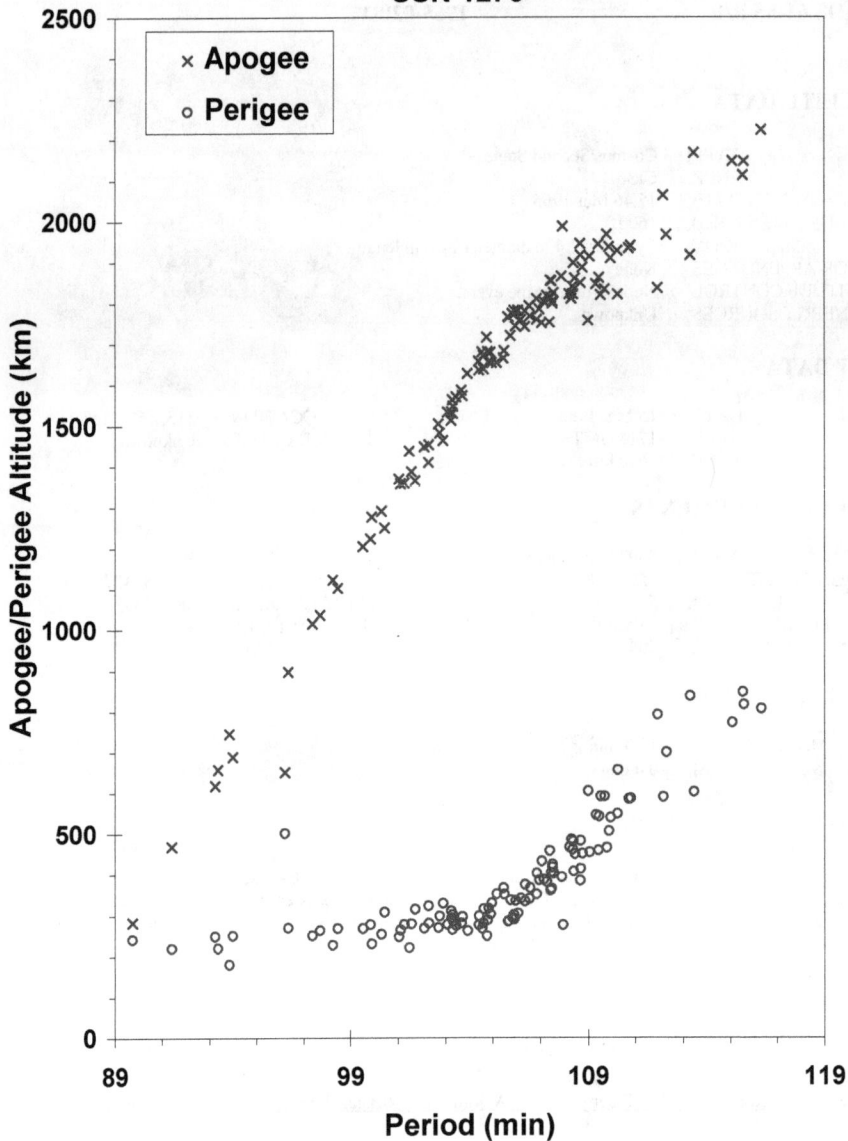

SSN 1270

× Apogee
○ Perigee

Cosmos 61-63 R/B debris cloud of 113 fragments 8 months after the event as reconstructed from the US SSN database.

SATELLITE DATA

TYPE:	Payload
OWNER:	USSR
LAUNCH DATE:	4.23 Nov 1965
DRY MASS (KG):	400
MAIN BODY:	Ellipsoid; 1.2 m diameter by 1.8 m length
MAJOR APPENDAGES:	Unknown
ATTITUDE CONTROL:	Unknown
ENERGY SOURCES:	Unknown

EVENT DATA

DATE:	15 Jan 1966	LOCATION:	Unknown
TIME:	Unknown	ASSESSED CAUSE:	Unknown
ALTITUDE:	Unknown		

PRE-EVENT ELEMENTS

EPOCH:	66009.5	MEAN ANOMALY:	Unknown
RIGHT ASCENSION:	Unknown	MEAN MOTION:	16.09757275
INCLINATION:	48.39	MEAN MOTION DOT/2:	Unknown
ECCENTRICITY:	0.009282	MEAN MOTION DOT DOT/6:	Unknown
ARG. OF PERIGEE:	77	BSTAR:	Unknown

DEBRIS CLOUD DATA

MAXIMUM ΔP:	Unknown
MAXIMUM ΔI:	Unknown

COMMENTS

Pre-event elements were taken from RAE Table of Earth Satellites. Cosmos 95 was placed into a low Earth orbit on 4 November 1965. Within 2 weeks nearly two dozen debris had been detected and were later cataloged. However, the nature of the debris, i.e. breakup versus operational, was not determined. The last of these debris decayed naturally by 6 January 1966. Russian records indicate that a breakup may have occurred on 15 January 1966, 3 days before the 400 kg spacecraft itself reentered. No other information on this event has been discovered, and no debris remains in orbit.

REFERENCE DOCUMENT

History of Soviet/Russian Satellite Fragmentations-A Joint U.S.-Russian Investigation, N. L. Johnson et al, Kaman Sciences Corporation, October 1995.

Insufficient data to construct a Gabbard diagram.

SATELLITE DATA

TYPE:	Titan 3C-4 Transtage
OWNER:	US
LAUNCH DATE:	15.72 Oct 1965
DRY MASS (KG):	2500
MAIN BODY:	Cylinder; 3 m diameter by 6 m length
MAJOR APPENDAGES:	None
ATTITUDE CONTROL:	Active, 3-axis
ENERGY SOURCES:	On-board propellants

EVENT DATA

DATE:	15 Oct 1965	LOCATION:	22S, 108E (asc)
TIME:	1820 GMT	ASSESSED CAUSE:	Propulsion
ALTITUDE:	740 km		

POST-EVENT ELEMENTS

EPOCH:	65361.23126396	MEAN ANOMALY:	237.1066
RIGHT ASCENSION:	21.5316	MEAN MOTION:	14.54928550
INCLINATION:	32.1697	MEAN MOTION DOT/2:	.00000268
ECCENTRICITY:	.0072678	MEAN MOTION DOT DOT/6:	.071801
ARG. OF PERIGEE:	123.6068	BSTAR:	.0

DEBRIS CLOUD DATA

MAXIMUM ΔP:	4.1 min
MAXIMUM ΔI:	1.4 deg

COMMENTS

This was the second test of the Titan 3C-4 Transtage with AJ10-138 engine using hypergolic propellants. Event occurred one-half revolution after launch following second ignition that may have been accompanied with vehicle tumbling. LCS 2 payload was to have been deployed at 735 km circular while OV2-1 was to have been released later in an orbit of 735 km by about 7400 km. Transtage also malfunctioned on next mission in December 1965. The main remnant of the rocket body was recently identified as Satellite No. 1822 (1965-082DM). Previous editions of this book had identified the main rocket body remnant as Satellite No. 1640 (1965-082B).

REFERENCE DOCUMENT

TRW Space Log, Winter 1965-66, Vol. 5, No. 4, T.L. Branigan, ed., TRW Systems, Redondo Beach, 1966, pp. 15-17.

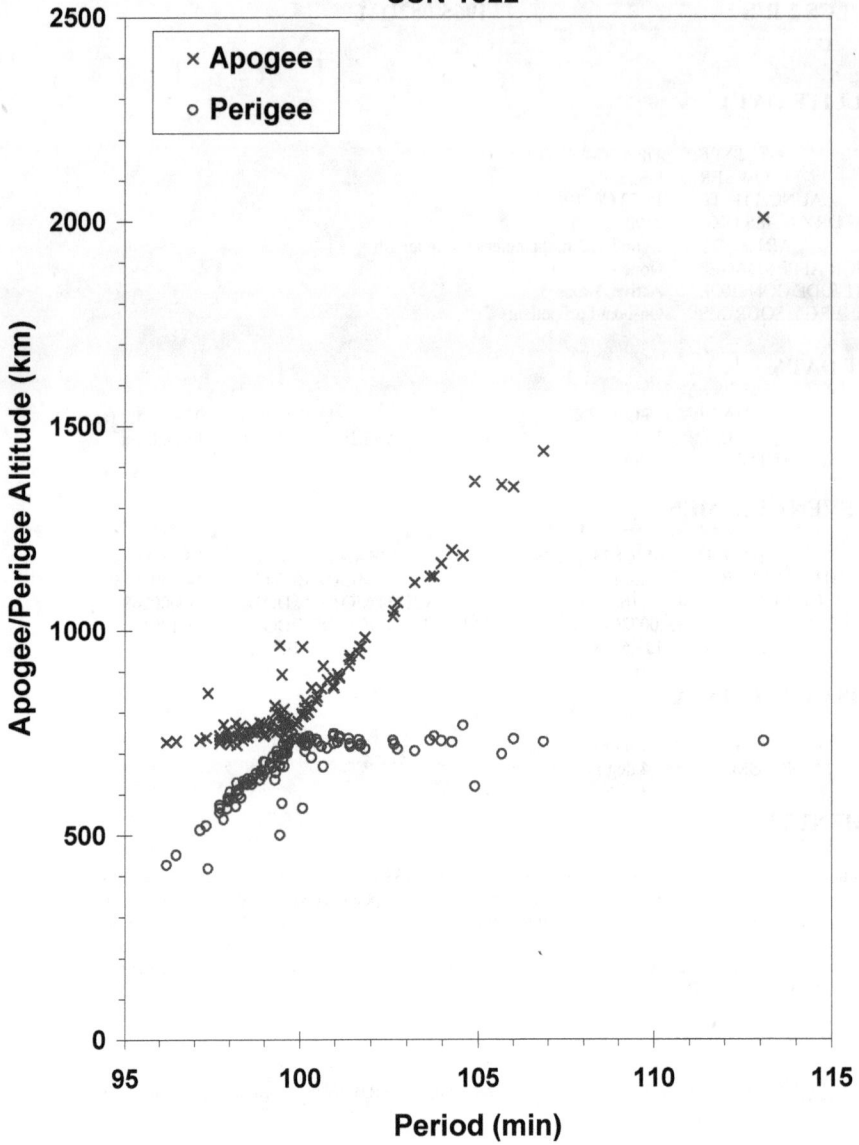

SSN 1822

OV2-1/LCS 2 R/B debris cloud of 103 cataloged fragments 6 weeks after the event as reconstructed from the US SSN database.

SATELLITE DATA

TYPE:	Payload
OWNER:	US
LAUNCH DATE:	15.85 Feb 1966
DRY MASS (KG):	4
MAIN BODY:	Sphere; 0.3 m diameter
MAJOR APPENDAGES:	None
ATTITUDE CONTROL:	None
ENERGY SOURCES:	Unknown

EVENT DATA

DATE:	15 Feb 1966	LOCATION:	Unknown
TIME:	Unknown	ASSESSED CAUSE:	Unknown
ALTITUDE:	~200 km		

POST-EVENT ELEMENTS

EPOCH:	66047.01671304	MEAN ANOMALY:	234.6777
RIGHT ASCENSION:	148.6481	MEAN MOTION:	16.20030654
INCLINATION:	96.5380	MEAN MOTION DOT/2:	.01298049
ECCENTRICITY:	.0108362	MEAN MOTION DOT DOT/6:	.0053719
ARG. OF PERIGEE:	126.3670	BSTAR:	.0

DEBRIS CLOUD DATA

MAXIMUM ΔP:	Unknown
MAXIMUM ΔI:	0.6 deg

COMMENTS

OPS 3031 was an inflated sphere also known as Bluebell 2. It was deployed from satellite 2012, which was an Agena D stage carrying a separate payload. Elements above are for satellite 2012. Debris cataloging began 19 February after many debris had already decayed. Consequently, ΔP cannot be calculated. OPS 3031 and all debris decayed within 1 week of launch.

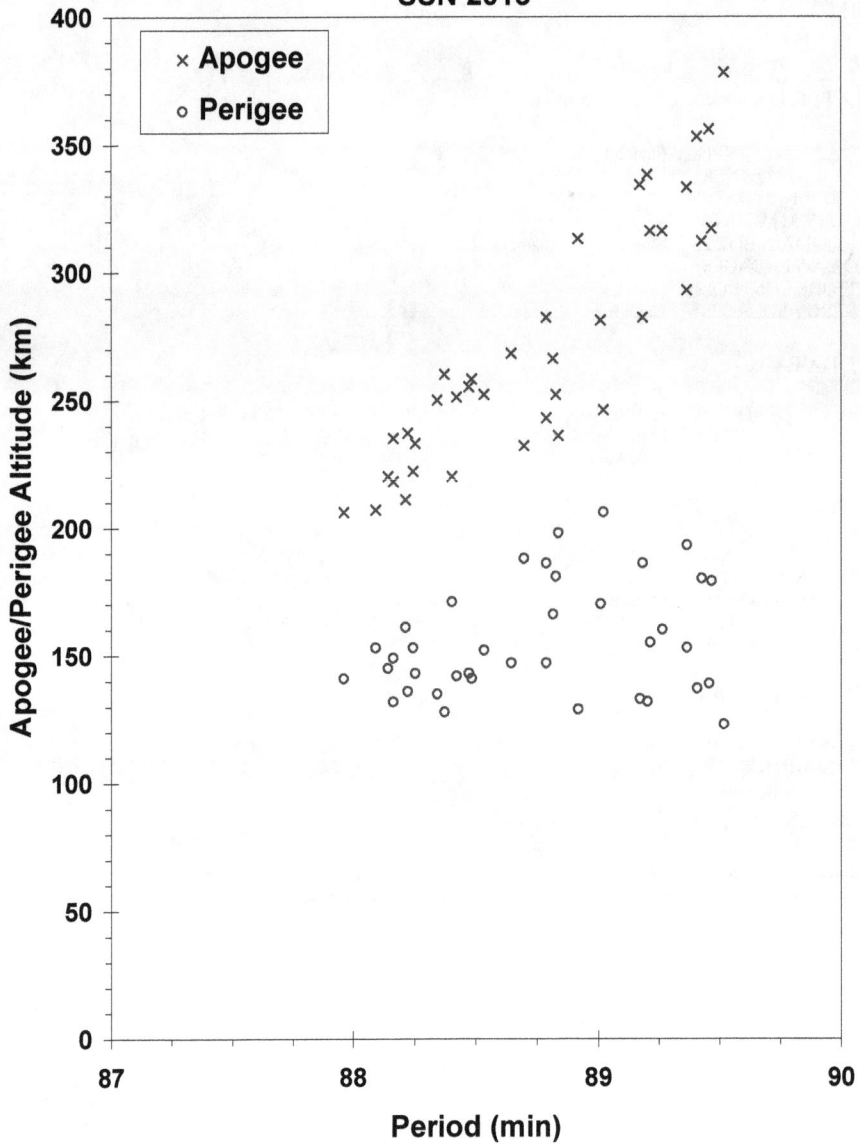

SSN 2015

OPS 3031 debris cloud of 38 fragments as initially cataloged by the US SSN during February 1966.

SATELLITE DATA

TYPE:	Atlas Core Stage
OWNER:	US
LAUNCH DATE:	1.63 Jun 1966
DRY MASS (KG):	3400
MAIN BODY:	Cylinder; 3 m diameter by 20 m length
MAJOR APPENDAGES:	None
ATTITUDE CONTROL:	None at time of the event.
ENERGY SOURCES:	Unknown

EVENT DATA

DATE:	Mid-Jun 1966	LOCATION:	Unknown
TIME:	Unknown	ASSESSED CAUSE:	Unknown
ALTITUDE	~250 km		

PRE-EVENT ELEMENTS

EPOCH:	66164.96883397	MEAN ANOMALY:	224.9775
RIGHT ASCENSION:	223.9064	MEAN MOTION:	16.05545399
INCLINATION:	28.7968	MEAN MOTION DOT/2:	.00654808
ECCENTRICITY:	.0025152	MEAN MOTION DOT DOT/6:	.0010778
ARG. OF PERIGEE:	135.2510	BSTAR:	.0

DEBRIS CLOUD DATA

MAXIMUM ΔP:	5.5 min
MAXIMUM ΔI:	1.5 deg

COMMENTS

This stage successfully deployed the Augmented Target Docking Adapter (ATDA) for the Gemini 9 mission. The elements above are the last available for the rocket body. Debris cataloging began on 21 June. Debris decay dates ranged from 21 June to 4 July with the rocket body officially decaying on 22 June. A review of NASA archives for this mission revealed no documented anomaly with the Atlas booster. Discussions in 1989 with General Dynamics personnel involved in the mission (Mr. Phil Genser of General Dynamics, San Diego) also failed to uncover any knowledge of the event. Pressure relief valves should have relieved pressurization increases, particularly in the oxygen tank. Possible failure of the oxygen relief valve could not be ruled out.

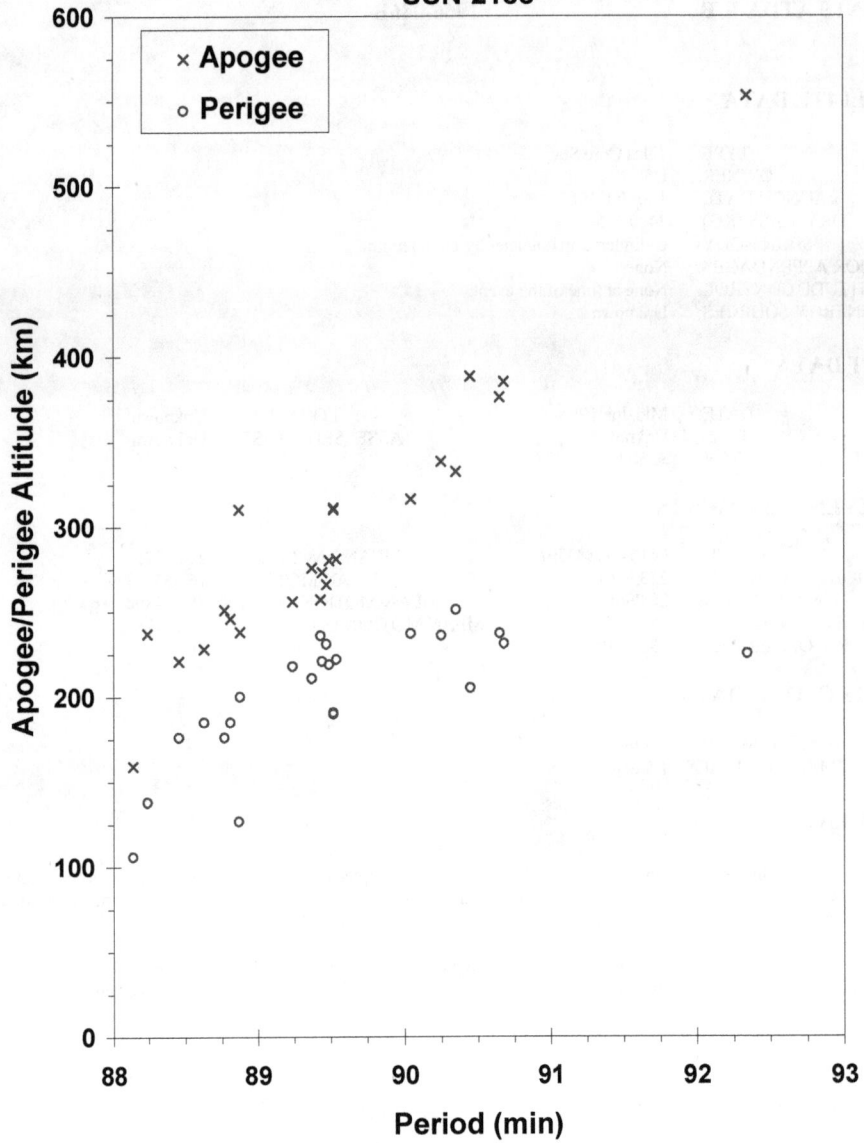

SSN 2188

Gemini 9 ATDA R/B debris cloud of 24 fragments cataloged between 21 and 24 June as reconstructed from the US SSN database.

SATELLITE DATA

TYPE:	Payload
OWNER:	US
LAUNCH DATE:	24.01 Jun 1966
DRY MASS (KG):	55
MAIN BODY:	Sphere; 30 m diameter
MAJOR APPENDAGES:	None
ATTITUDE CONTROL:	None
ENERGY SOURCES:	None

EVENT DATA (1)

DATE:	12 Jul 1975	LOCATION:	67N, 135E (dsc)
TIME:	2248 GMT	ASSESSED CAUSE:	Unknown
ALTITUDE:	5145 km		

PRE-EVENT ELEMENTS (1)

EPOCH:	75192.78059719	MEAN ANOMALY:	67.9594
RIGHT ASCENSION:	238.7429	MEAN MOTION:	7.99684492
INCLINATION:	85.2811	MEAN MOTION DOT/2:	.00001217
ECCENTRICITY:	.0931904	MEAN MOTION DOT DOT/6:	.0
ARG. OF PERIGEE:	281.8264	BSTAR:	.77087

EVENT DATA (2)

DATE:	20 Jan 1976	LOCATION:	Unknown
TIME:	Unknown	ASSESSED CAUSE:	Unknown
ALTITUDE:	Unknown		

PRE-EVENT ELEMENTS (2)

EPOCH:	76019.86486339	MEAN ANOMALY:	305.5539
RIGHT ASCENSION:	209.8639	MEAN MOTION:	8.00368182
INCLINATION:	85.0720	MEAN MOTION DOT/2:	.0
ECCENTRICITY:	.1179567	MEAN MOTION DOT DOT/6:	.0
ARG. OF PERIGEE:	66.4633	BSTAR:	.0

DEBRIS CLOUD DATA

MAXIMUM ΔP:	0.1 min*
MAXIMUM ΔI:	0.7 deg*

*Based on 1st event data

COMMENTS

PAGEOS (Passive Geodetic Earth-Orbiting Satellite) was an inflated balloon made of thin Mylar with an aluminum coating. The first fragmentation event occurred 9 years after launch and resulted in 11 new cataloged objects. The second event was detected by D.G. King-Hele of the RAE, and NAVSPASUR confirmed 44 additional fragments. By August 1976 no additional debris had been cataloged but 19 objects were being tracked in orbits with mean motions near 8 and eccentricities between 0.16 and 0.34. Due to the character of PAGEOS and its subsequent debris, natural perturbations had little effect on orbital period but strongly increased eccentricity by simultaneously lowering perigee and raising apogee. About 10 September 1976 one of the 19 unofficial objects is believed to have broken up into perhaps more than 250 new pieces, none of which were cataloged prior to reentry. Eighteen objects were later cataloged during 7-8 October 1976. On the first anniversary of the second fragmentation (20 Jan

1977), 45 fragments were cataloged without elements and immediately decayed administratively. Additional fragmentations are suspected to have taken place in June 1978, September 1984, and December 1985. Historically, radar tracking of PAGEOS debris has been extremely difficult and cross-tagging frequent. Cause for the second and subsequent events may be material deterioration under environmental stress. A suspected PAGEOS fragment, SSN 5994, which was cataloged as a Westford Needles object, fragmented on 8 September 1995 and again on 14 September 1995 with 12 associated objects.

REFERENCE DOCUMENT

Spacetrack System Data Related to Some Non-Routine Events Through May 1981, J.R. Gabbard, Technical Memorandum 81-6, DCS/Plans, Hdqtrs NORAD/ADCOM, Colorado Springs, 30 June 1981.

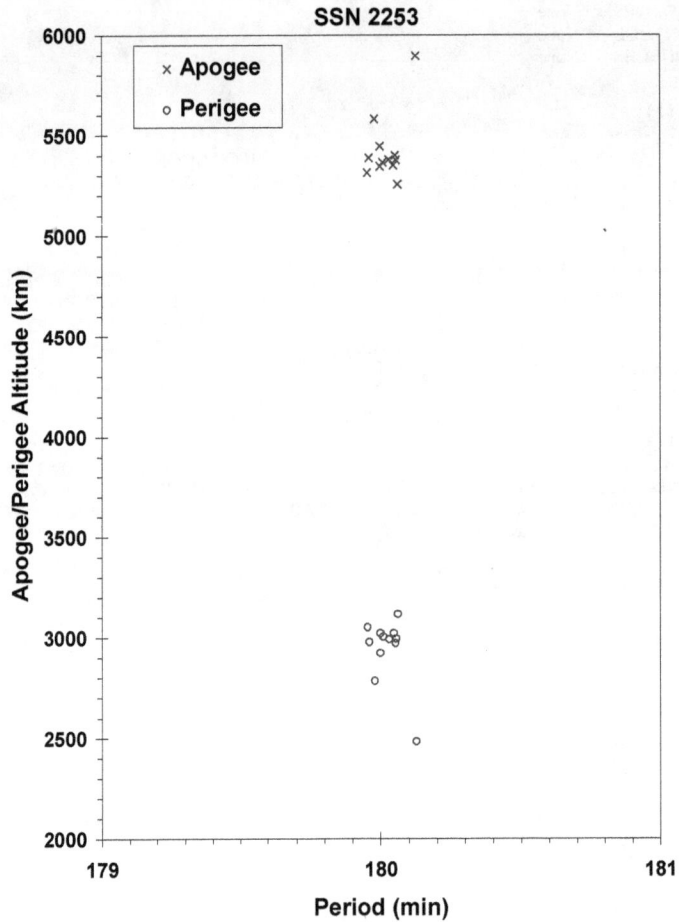

PAGEOS debris cloud of 12 fragments 5 weeks after the first event as reconstructed from the US SSN database.

SATELLITE DATA

TYPE:	Saturn SIVB Stage
OWNER:	US
LAUNCH DATE:	5.62 Jul 1966
DRY MASS (KG):	26,600
MAIN BODY:	Cylinder; 6.6 m diameter by 28.3 m length
MAJOR APPENDAGES:	None
ATTITUDE CONTROL:	Active, 3-axis
ENERGY SOURCES:	Attitude control and pressurization systems

EVENT DATA

DATE:	5 Jul 1966	LOCATION:	20N, 277E (dsc)
TIME:	2111 GMT	ASSESSED CAUSE:	Deliberate
ALTITUDE:	205 km		

PRE-EVENT ELEMENTS

EPOCH:	66186.73481847	MEAN ANOMALY:	353.9219
RIGHT ASCENSION:	5.5870	MEAN MOTION:	16.27379993
INCLINATION:	31.9810	MEAN MOTION DOT/2:	.03796193
ECCENTRICITY:	.0022272	MEAN MOTION DOT DOT/6:	.17429
ARG. OF PERIGEE:	6.1632	BSTAR:	.0

DEBRIS CLOUD DATA

MAXIMUM ΔP:	3.5 min
MAXIMUM ΔI:	1.4 deg

COMMENTS

This was the second flight of the SIVB stage. After orbital insertion, the vehicle was intentionally subjected to dynamic integrity tests, including high gravity loadings during attitude control maneuvers and high pressure tests. The vehicle finally broke up after exceeding structural design limits with a propellant tank bulkhead differential pressure in excess of 23.7 N/cm^2. The fragmentation occurred early on the fifth revolution. Elements for the first fragments were not cataloged until 8 July.

REFERENCE DOCUMENT

Saturn AS-203 Evaluation Bulletin, No. 2, R-AERO-F-142-66, J.P. Lindberg, NASA Marshall Space Flight Center, Alabama, 21 July 1966.

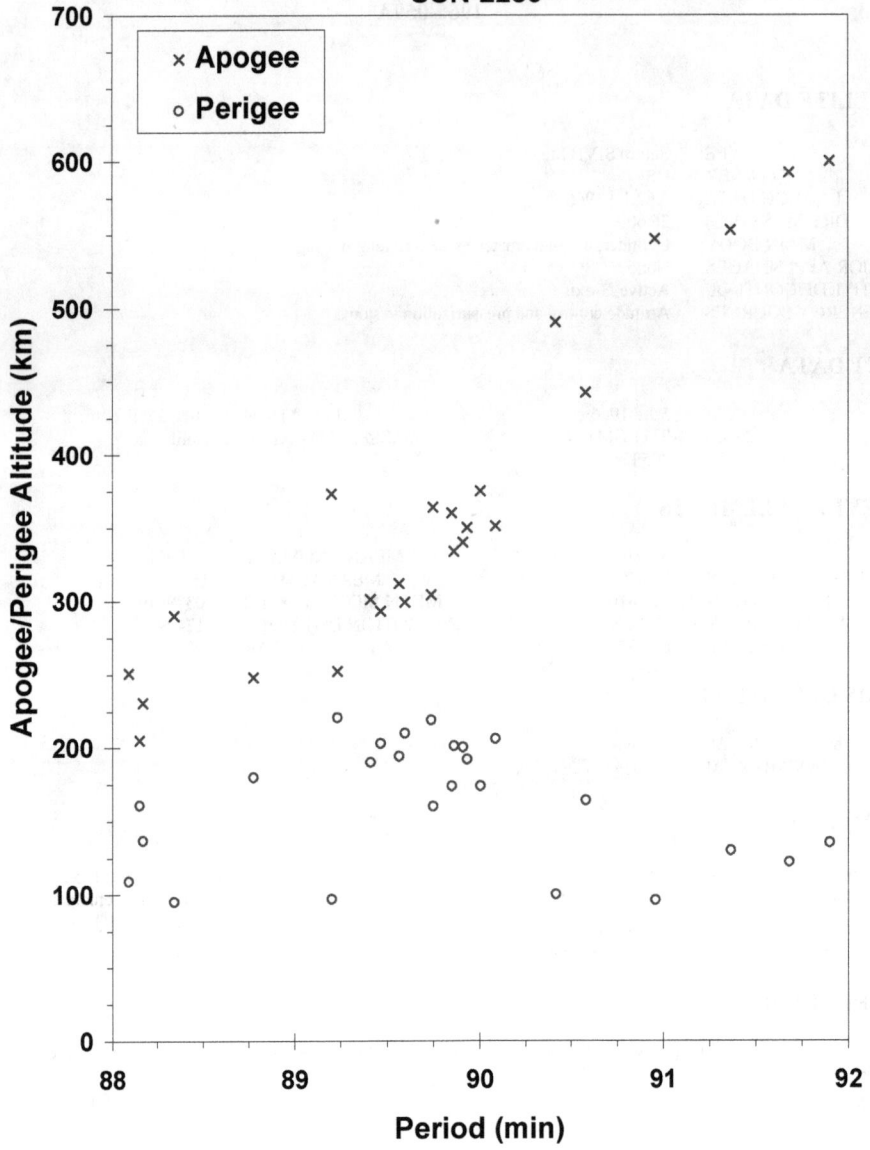

SSN 2289

Apogee/Perigee Altitude (km) vs Period (min)

AS-203 debris cloud of 25 fragments using orbits developed within 1 week of the event as reconstructed from the US SSN database.

SATELLITE DATA

TYPE:	Unknown
OWNER:	CIS
LAUNCH DATE:	17.94 Sep 1966
DRY MASS (KG):	Unknown
MAIN BODY:	Cone-cylinder; 1.5 m diameter by 6 m length
MAJOR APPENDAGES:	None
ATTITUDE CONTROL:	Unknown
ENERGY SOURCES:	Explosive device

EVENT DATA

DATE:	17 Sep 1966	LOCATION:	Unknown
TIME:	Unknown	ASSESSED CAUSE:	Deliberate
ALTITUDE:	~300 km		

POST-EVENT ELEMENTS

EPOCH:	66261.0	MEAN ANOMALY:	283
RIGHT ASCENSION:	338	MEAN MOTION:	14.879
INCLINATION:	49.63	MEAN MOTION DOT/2:	.0
ECCENTRICITY:	.063	MEAN MOTION DOT DOT/6:	.0
ARG. OF PERIGEE:	83	BSTAR:	.0

DEBRIS CLOUD DATA

MAXIMUM ΔP:	Unknown
MAXIMUM ΔI:	Unknown

COMMENTS

This was the first of two missions of this type flown in 1966 and not acknowledged by the USSR. The identity of the parent orbit is uncertain. Satellite 2437 was the first cataloged fragment. The above elements are taken or derived from the RAE Table of Earth Satellites. The debris distribution is consistent with a fragmentation near 300 km. Failure of the payload led to immediate activation of the self-destruct system.

REFERENCE DOCUMENT

History of Soviet/Russian Satellite Fragmentations-A Joint U.S.-Russian Investigation, N. L. Johnson et al, Kaman Sciences Corporation, October 1995.

SSN 2437

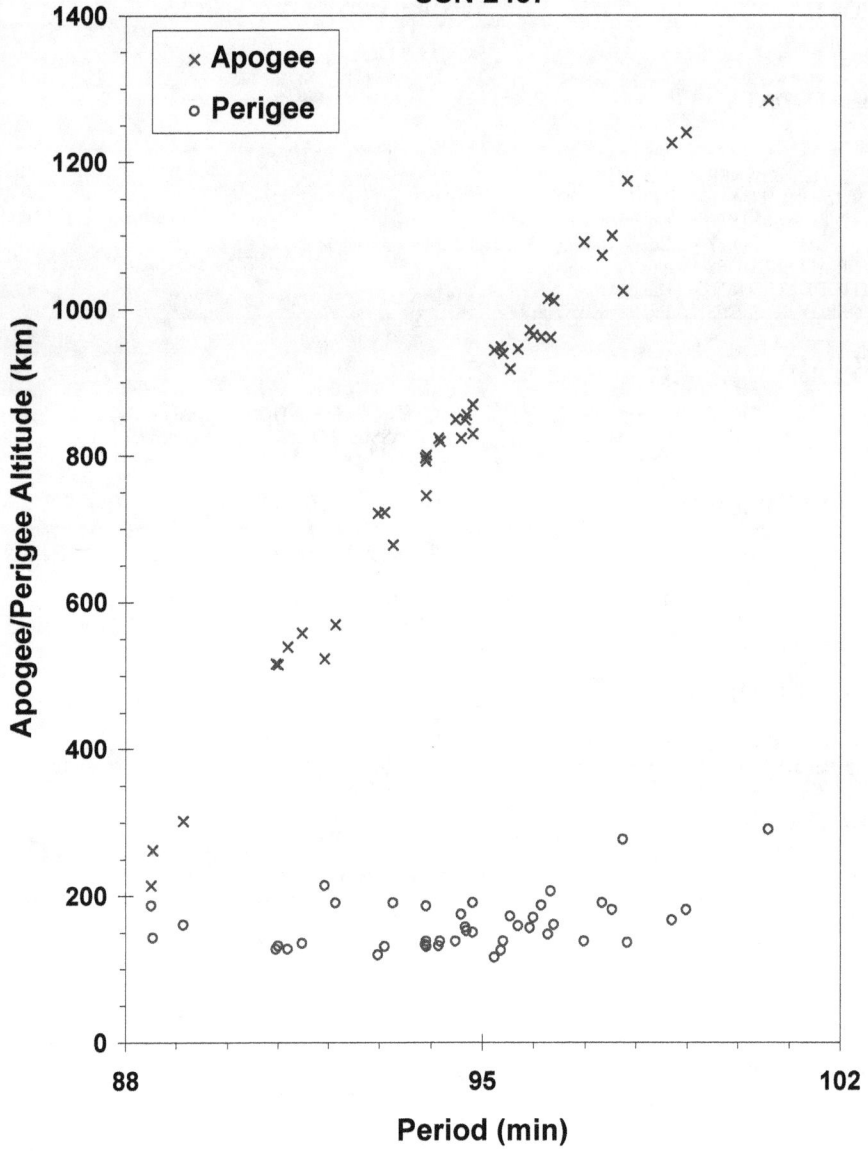

Cosmos U-1 debris cloud of 43 fragments cataloged by 5 October 1966 as reconstructed from the US SSN database.

COSMOS U-2   1966-101A   2536

SATELLITE DATA

                    TYPE: Unknown
                   OWNER: CIS
            LAUNCH DATE: 2.03 Nov 1966
          DRY MASS (KG):   Unknown
              MAIN BODY: Cone-cylinder; 1.5 m diameter by 6 m length
        MAJOR APPENDAGES: None
        ATTITUDE CONTROL: Unknown
          ENERGY SOURCES: Explosive device

EVENT DATA

              DATE: 2 Nov 1966   LOCATION: Unknown
              TIME: Unknown   ASSESSED CAUSE: Deliberate
          ALTITUDE: ~225 km

POST-EVENT ELEMENTS

              EPOCH: 66309.99121234   MEAN ANOMALY: 265.7893
    RIGHT ASCENSION: 35.2944   MEAN MOTION: 15.17033022
        INCLINATION: 49.5617   MEAN MOTION DOT/2:   .01866914
       ECCENTRICITY: .05339049   MEAN MOTION DOT DOT/6:   .0043309
     ARG. OF PERIGEE: 100.3324   BSTAR: .0

DEBRIS CLOUD DATA

        MAXIMUM ΔP: Unknown
        MAXIMUM ΔI: Unknown

COMMENTS

This was the second mission of this type flown in 1966 and not acknowledged by the USSR. No elements were cataloged until 3 days after the launch. The identity of the parent orbit is uncertain. Satellite 2536 was the first object cataloged and was near the center of the debris cloud. The debris distribution is consistent with a fragmentation near 225 km. Failure of the payload led to immediate activation of the self-destruct system.

REFERENCE DOCUMENT

History of Soviet/Russian Satellite Fragmentations-A Joint U.S.-Russian Investigation, N. L. Johnson et al, Kaman Sciences Corporation, October 1995.

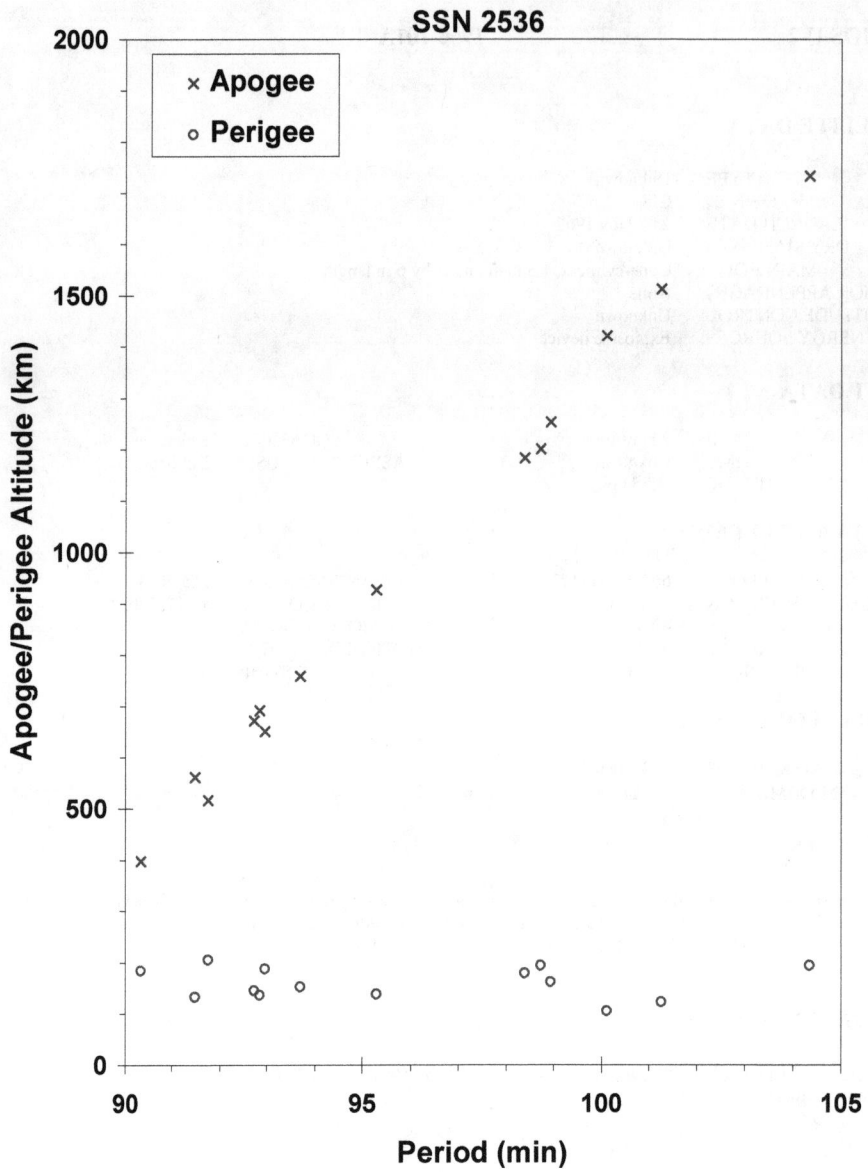

SSN 2536

Cosmos U-2 debris cloud composed of 14 different orbits as developed by the US SSN within 1 week of the event.

56

SATELLITE DATA

TYPE:	Payload
OWNER:	CIS
LAUNCH DATE:	16.50 Jan 1968
DRY MASS (KG):	5500
MAIN BODY:	Sphere-cylinder; 2.4 m diameter by 6.0 m length
MAJOR APPENDAGES:	None
ATTITUDE CONTROL:	Active, 3-axis
ENERGY SOURCES:	On-board propellants, 10 kg TNT explosive charge

EVENT DATA

DATE:	24 Jan 1968	LOCATION:	Unknown
TIME:	Unknown	ASSESSED CAUSE:	Deliberate
ALTITUDE:	Unknown		

PRE-EVENT ELEMENTS

EPOCH:	68024. 25242706	MEAN ANOMALY:	305.4920
RIGHT ASCENSION:	247.4278	MEAN MOTION:	15.98596524
INCLINATION:	65.6289	MEAN MOTION DOT/2:	0.00196964
ECCENTRICITY:	0.0118074	MEAN MOTION DOT DOT/6:	0
ARG. OF PERIGEE:	55.7254	BSTAR:	0

DEBRIS CLOUD DATA

MAXIMUM ΔP:	Unknown
MAXIMUM ΔI:	Unknown

COMMENTS

Spacecraft was destroyed after a malfunction prevented reentry and landing in the Soviet Union.

REFERENCE DOCUMENT

History of Soviet/Russian Satellite Fragmentations-A Joint U.S.-Russian Investigation, N. L. Johnson et al, Kaman Sciences Corporation, October 1995.

Insufficient data to construct a Gabbard diagram.

SATELLITE DATA

TYPE:	Saturn SIVB Stage
OWNER:	US
LAUNCH DATE:	4.50 Apr 1968
DRY MASS (KG):	30,000
MAIN BODY:	Cylinder; 6.6 m diameter by 30 m length (?)
MAJOR APPENDAGES:	None
ATTITUDE CONTROL:	None at time of the event.
ENERGY SOURCES:	On-board propellants

EVENT DATA

DATE:	13 Apr 1968	LOCATION:	32N, 245E (asc)
TIME:	1054 GMT	ASSESSED CAUSE:	Propulsion
ALTITUDE:	330 km		

PRE-EVENT ELEMENTS

EPOCH:	68103.56521409	MEAN ANOMALY:	151.0074
RIGHT ASCENSION:	177.3270	MEAN MOTION:	15.97292993
INCLINATION:	32.5869	MEAN MOTION DOT/2:	.00302835
ECCENTRICITY:	.0120930	MEAN MOTION DOT DOT/6:	.0
ARG. OF PERIGEE:	208.3921	BSTAR:	.0

DEBRIS CLOUD DATA

MAXIMUM ΔP: 0.7 min
MAXIMUM ΔI: 0.1 deg

COMMENTS

This Saturn SIVB Stage was fitted with an 11,800 kg mock Lunar Module (LM). The SIVB stage was programmed for a second firing to place the Apollo 6 vehicle into a more eccentric orbit, but the restart did not occur. The Apollo 6 payload was separated, leaving the SIVB stage and the LM in a low Earth orbit. Vaporization and venting of residual liquid oxygen induced a tumble to the SIVB stage that reached 30 rpm by 13 April. On this date the axial loads on the LM attach strap fittings and support struts were exceeded, resulting in separation of the LM from the SIVB along with numerous debris. Five fragments were cataloged without elements.

REFERENCE DOCUMENT

Apollo 6 Mission Anomaly Report No. 6, Unexpected Structural Indications During Launch Phase (Review Copy), MSC-PT-R-68-22, prepared by Apollo 6 Mission Evaluation Team, Marshall Space Flight Center, Alabama, and Manned Spacecraft Center, Texas, 1968.

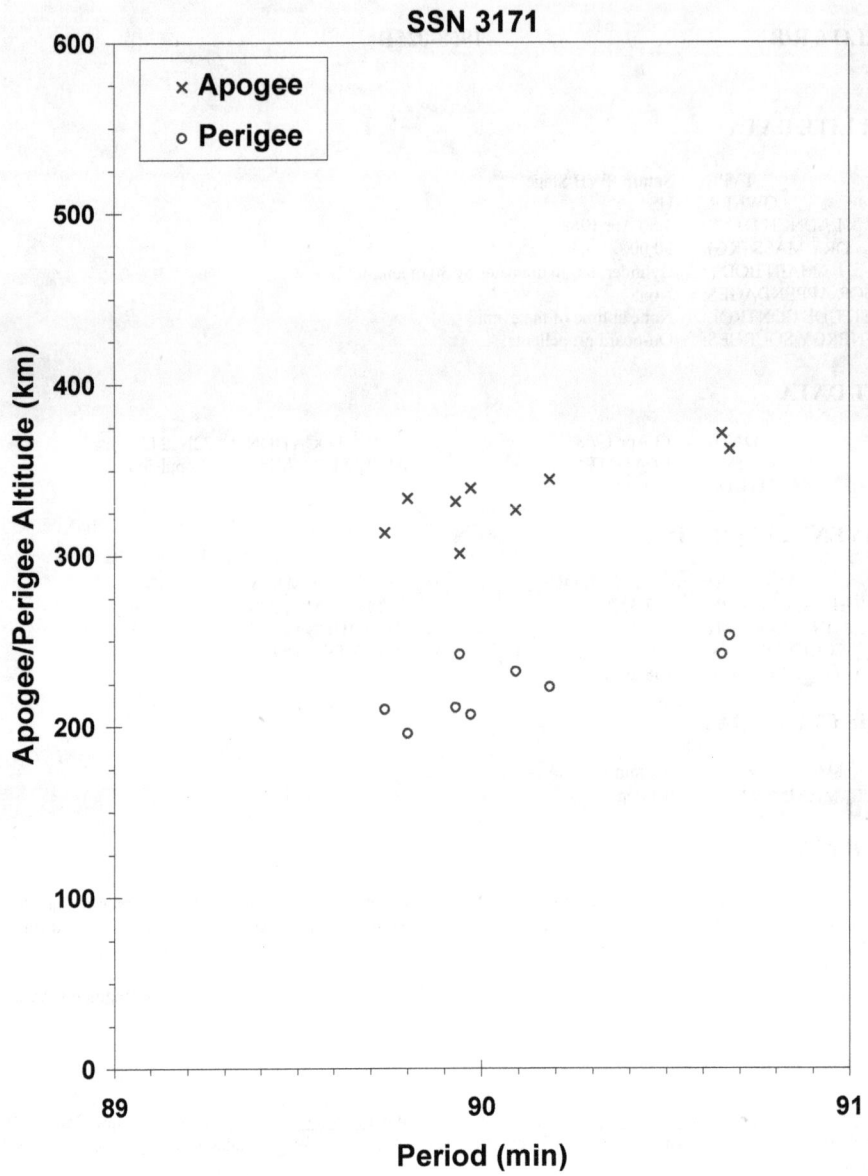

SSN 3171

Apogee/Perigee Altitude (km) vs Period (min)

× Apogee
○ Perigee

Apollo 6 R/B debris cloud of 9 fragments 4 days after the event as reconstructed from the US SSN database.

SATELLITE DATA

TYPE:	Titan 3C Transtage
OWNER:	US
LAUNCH DATE:	26.32 Sep 1968
DRY MASS (KG):	2500
MAIN BODY:	Cylinder; 3 m diameter by 6 m length
MAJOR APPENDAGES:	None
ATTITUDE CONTROL:	Active, 3-axis
ENERGY SOURCES:	On-board propellants

EVENT DATA

DATE:	21 Feb 1992	LOCATION:	Unknown (~ 197E)
TIME:	0931 GMT	ASSESSED CAUSE:	Propulsion
ALTITUDE:	~ 35600		

PRE-EVENT ELEMENTS

EPOCH:	92043.23217642	MEAN ANOMALY:	284.5600
RIGHT ASCENSION:	21.8025	MEAN MOTION:	1.01459126
INCLINATION:	11.9035	MEAN MOTION DOT/2:	.00000174
ECCENTRICITY:	.0084771	MEAN MOTION DOT DOT/6:	.0
ARG. OF PERIGEE:	76.2786	BSTAR:	.0

DEBRIS CLOUD DATA

MAXIMUM ΔP:	Unknown
MAXIMUM ΔI:	Unknown

COMMENTS

This was the second major fragmentation of a Titan 3C Transtage (the first was 1965-082DM). This transtage released ERS-28 (also known as OV5-2) in highly eccentric transfer orbit, then released LES-6 and ERS-21 (also known as OV5-4) in synchronous orbit, before slightly decelerating and releasing OV2-5 into a slightly lower orbit. This rocket body successfully completed its mission and remained on-orbit for 281 months before fragmenting. Mr. Bob Brock, operating the Maui GEODSS sensor, observed this transtage as it fragmented, liberating a reported 20 objects.

REFERENCE DOCUMENTS

TRW Space Log, Winter 1968-69 edition, Vol. 8, No. 4, H. T. Seaborn, ed., TRW Systems Group, Redondo Beach, pp. 32-35.

"Debris in Geosynchronous Orbits", A.F. Pensa et al, Space Forum, special issue, 1st International Workshop on Space Debris, Moscow, October 1995.

Insufficient data to construct a Gabbard Diagram

SATELLITE DATA

TYPE:	Payload
OWNER:	CIS
LAUNCH DATE:	19.18 Oct 1968
DRY MASS (KG):	1400
MAIN BODY:	Irregular; 1.8 m by 4.2 m
MAJOR APPENDAGES:	Unknown
ATTITUDE CONTROL:	Active, 3-axis
ENERGY SOURCES:	On-board propellants

EVENT DATA

DATE:	1 Nov 1968	LOCATION:	55N, 104E (dsc)
TIME:	0412 GMT	ASSESSED CAUSE:	Deliberate
ALTITUDE:	540 km		

PRE-EVENT ELEMENTS

EPOCH:	68304.83833772	MEAN ANOMALY:	61.1261
RIGHT ASCENSION:	82.2502	MEAN MOTION:	15.19330723
INCLINATION:	62.2495	MEAN MOTION DOT/2:	.00016932
ECCENTRICITY:	.0050333	MEAN MOTION DOT DOT/6:	.0
ARG. OF PERIGEE:	298.4670	BSTAR:	.0

DEBRIS CLOUD DATA

MAXIMUM ΔP:	Unknown
MAXIMUM ΔI:	0.1 deg

COMMENTS

Cosmos 248 was the target of rendezvous for the Cosmos 249 and Cosmos 252 tests. Calculations suggest the few fragments detected from Cosmos 248 were released within 10 minutes of the Cosmos 252 event that took place in the vicinity of Cosmos 248. The four observed fragments were not cataloged until 4-6 weeks after the event, preventing an accurate assessment of the event due to drag effects. It is possible that the Cosmos 248 event occurred immediately after the rendezvous and was a direct result of interaction with Cosmos 252 debris.

REFERENCE DOCUMENTS

"Artificial Satellite Break-Ups (Part 2): Soviet Anti-Satellite Program", N.L. Johnson, Journal of the British Interplanetary Society, August 1983, pp. 357-362.

History of Soviet/Russian Satellite Fragmentations-A Joint U.S.-Russian Investigation, N. L. Johnson et al, Kaman Sciences Corporation, October 1995.

Insufficient data to construct a Gabbard diagram.

SATELLITE DATA

TYPE:	Payload
OWNER:	CIS
LAUNCH DATE:	20.17 Oct 1968
DRY MASS (KG):	1400
MAIN BODY:	Irregular; 1.8 m by 4.2 m
MAJOR APPENDAGES:	None
ATTITUDE CONTROL:	Active, 3-axis
ENERGY SOURCES:	On-board propellants, explosive charge

EVENT DATA

DATE:	20 Oct 1968	LOCATION:	57S, 181E (asc)
TIME:	1427 GMT	ASSESSED CAUSE:	Deliberate
ALTITUDE:	1995 km		

POST-EVENT ELEMENTS

EPOCH:	68294.85197372	MEAN ANOMALY:	295.3555
RIGHT ASCENSION:	118.4255	MEAN MOTION:	12.83515528
INCLINATION:	62.3313	MEAN MOTION DOT/2:	.0
ECCENTRICITY:	.1088260	MEAN MOTION DOT DOT/6:	.0
ARG. OF PERIGEE:	76.6147	BSTAR:	.0

DEBRIS CLOUD DATA

MAXIMUM ΔP:	3.9 min
MAXIMUM ΔI:	0.4 deg

COMMENTS

Cosmos 249 was the first of a class of maneuverable spacecraft flown to rendezvous within 4 hours with another Cosmos satellite. In 9 of 20 such missions, orbital debris clouds were created by the active spacecraft, and in one case a passive (target) spacecraft also spawned a few fragments. Fragmentations occurred either in the vicinity of the passive satellite or a few hours after the rendezvous. In the case of Cosmos 249, the spacecraft was launched on a two-revolution rendezvous with Cosmos 248. After a close approach, Cosmos 249 continued on before its warhead was intentionally fired. The elements above are the first available for the final orbit. Some debris from Cosmos 249 and Cosmos 252 have been cross-tagged.

REFERENCE DOCUMENTS

"Artificial Satellite Break-Ups (Part 2): Soviet Anti-Satellite Program", N.L. Johnson, Journal of the British Interplanetary Society, August 1983, pp. 357-362.

History of Soviet/Russian Satellite Fragmentations-A Joint U.S.-Russian Investigation, N. L. Johnson et al, Kaman Sciences Corporation, October 1995.

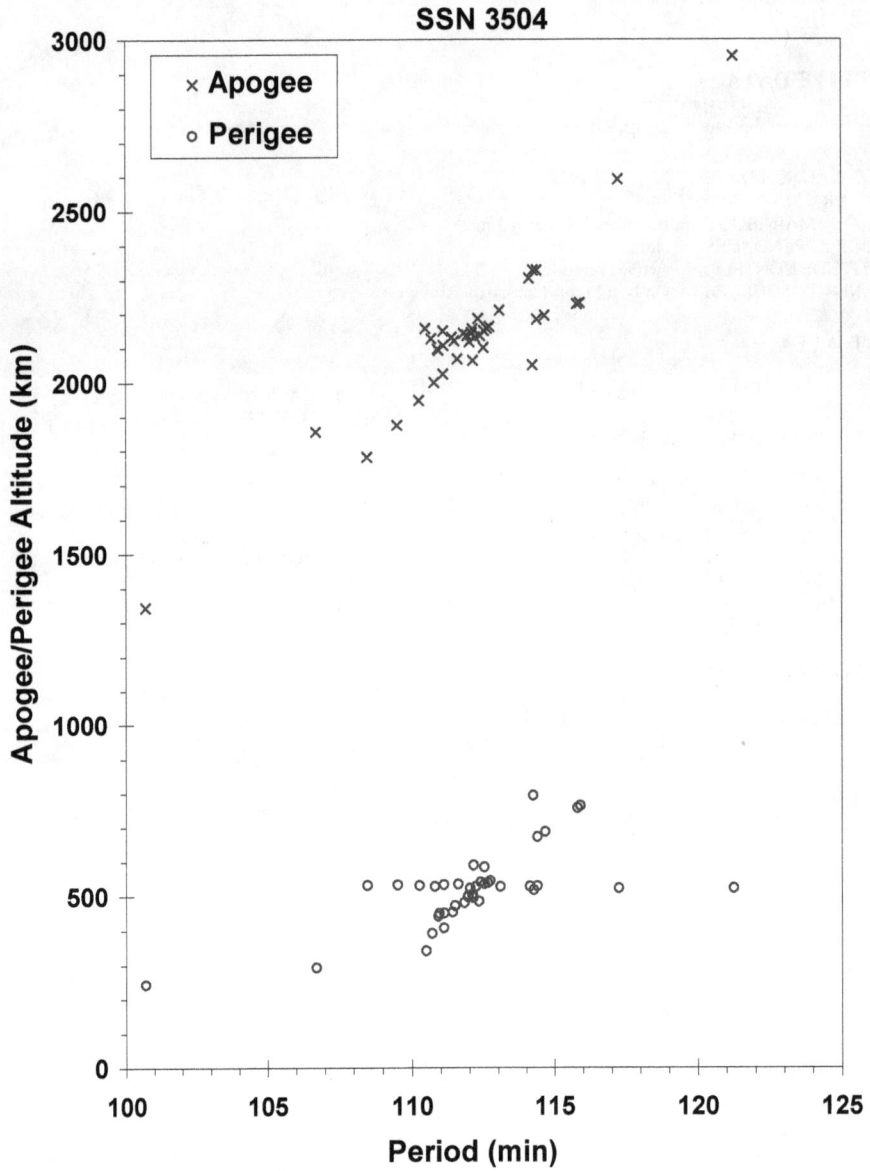

SSN 3504

Cosmos 249 cataloged debris cloud of 43 fragments 4 months after the event as reconstructed from the US SSN database. Cross-tagging with Cosmos 252 debris is evident.

SATELLITE DATA

TYPE:	Payload
OWNER:	CIS
LAUNCH DATE:	1.02 Nov 1968
DRY MASS (KG):	1400
MAIN BODY:	Irregular; 1.8 m by 4.2 m
MAJOR APPENDAGES:	None
ATTITUDE CONTROL:	Active, 3-axis
ENERGY SOURCES:	On-board propellants, explosive charge

EVENT DATA

DATE:	1 Nov 1968	LOCATION:	58N, 34E (asc)
TIME:	0402 GMT	ASSESSED CAUSE:	Deliberate
ALTITUDE:	535 km		

POST-EVENT ELEMENTS

EPOCH:	68306.70122094	MEAN ANOMALY:	297.5777
RIGHT ASCENSION:	76.5565	MEAN MOTION:	12.81276799
INCLINATION:	62.3351	MEAN MOTION DOT/2:	.00811969
ECCENTRICITY:	.1040368	MEAN MOTION DOT DOT/6:	.0
ARG. OF PERIGEE:	73.6953	BSTAR:	.0

DEBRIS CLOUD DATA

MAXIMUM ΔP:	8.7 min
MAXIMUM ΔI:	0.5 deg

COMMENTS

Cosmos 252 was launched on a two-revolution rendezvous with Cosmos 248. The fragmentation occurred in the vicinity of Cosmos 248. Cosmos 252 was part of the test series begun with Cosmos 249. Elements above are for the orbit of the spacecraft after final maneuver, which took place immediately before fragmentation. Some debris from Cosmos 249 and Cosmos 252 have been cross-tagged.

REFERENCE DOCUMENTS

"Artificial Satellite Break-Ups (Part 2): Soviet Anti-Satellite Program", N.L. Johnson, Journal of the British Interplanetary Society, August 1983, pp. 357-362.

History of Soviet/Russian Satellite Fragmentations-A Joint U.S.-Russian Investigation, N. L. Johnson et al, Kaman Sciences Corporation, October 1995.

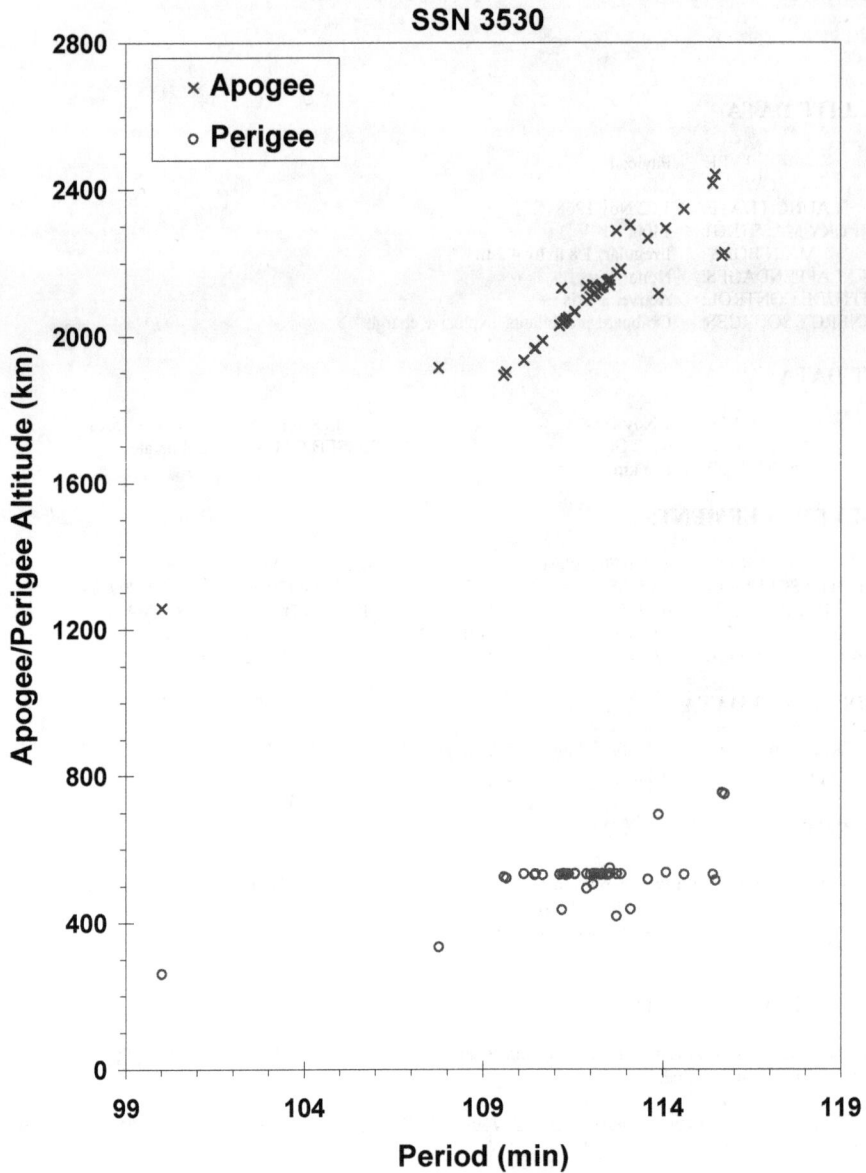

SSN 3530

Cosmos 252 cataloged debris cloud of 43 fragments 4 months after the event as reconstructed from the US SSN database. Cross-tagging with the Cosmos 249 cloud is evident.

SATELLITE DATA

TYPE:	Vostok Second Stage
OWNER:	CIS
LAUNCH DATE:	26.52 Mar 1969
DRY MASS (KG):	1440
MAIN BODY:	Cylinder; 2.6 m diameter by 3.8 m length
MAJOR APPENDAGES:	None
ATTITUDE CONTROL:	None at time of the event.
ENERGY SOURCES:	Unknown

EVENT DATA

DATE:	28 Mar 1969	LOCATION:	59N, 91E (dsc)
TIME:	1845 GMT	ASSESSED CAUSE:	Unknown
ALTITUDE:	555 km		

PRE-EVENT ELEMENTS

EPOCH:	69087.21308063	MEAN ANOMALY:	175.1148
RIGHT ASCENSION:	33.3926	MEAN MOTION:	14.71400174
INCLINATION:	81.1687	MEAN MOTION DOT/2:	.0
ECCENTRICITY:	.0276787	MEAN MOTION DOT DOT/6:	.0
ARG. OF PERIGEE:	184.7318	BSTAR:	.0

DEBRIS CLOUD DATA

MAXIMUM ΔP:	2.4 min
MAXIMUM ΔI:	0.5 deg

COMMENTS

The vehicle successfully deployed the Meteor 1-1 payload into the desired orbit. An object believed to be the rocket body was found on 27 March in an orbit (1) of 565 km by 755 km, similar to earlier missions of the Vostok second stage. Early on 28 March an object was found in an orbit (2) of 460 km by 850 km with elements as indicated above. Analysis indicates that a transition from orbit (1) to orbit (2) was possible during the latter part of 27 March. Debris analysis clearly indicates that the orbit of the parent satellite had to be similar to orbit (2). Radar cross-section data supports the belief that the post-event object in the center of the debris cloud is the rocket body. No object was found in orbit (1) after the event.

REFERENCE DOCUMENT

History of the Soviet/Russian Satellite Fragmentations-A Joint U.S.-Russian Investigation, N. L. Johnson et al, Kaman Sciences Corporation, October 1995.

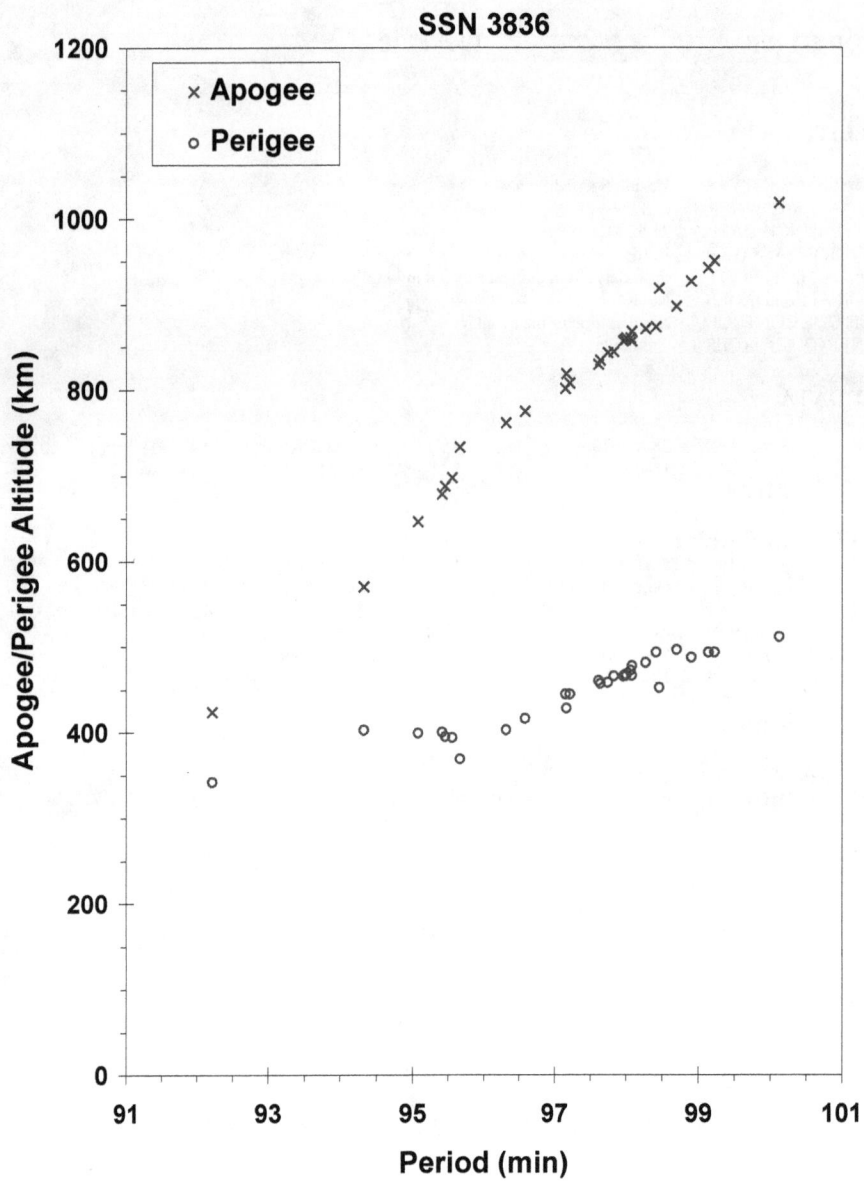

SSN 3836

- × Apogee
- ○ Perigee

Apogee/Perigee Altitude (km)

Period (min)

Meteor 1-1 R/B debris cloud of 31 fragments 2 months after the event as reconstructed from the US SSN database.

SATELLITE DATA

TYPE:	TE 364-4 (STAR 37E motor)
OWNER:	US
LAUNCH DATE:	26.09 Jul 1969
DRY MASS (KG):	1100 (70 without solid propellants)
MAIN BODY:	Sphere-nozzle; 1.0 m by 1.8 m
MAJOR APPENDAGES:	None
ATTITUDE CONTROL:	Active, 3-axis
ENERGY SOURCES:	On-board propellants

EVENT DATA

DATE:	26 Jul 1969	LOCATION:	0N, 333E (dsc)
TIME:	0228 GMT	ASSESSED CAUSE:	Propulsion
ALTITUDE:	270 km		

POST-EVENT ELEMENTS

EPOCH:	69208.17261261	MEAN ANOMALY:	166.4542
RIGHT ASCENSION:	130.0186	MEAN MOTION:	9.78100102
INCLINATION:	30.3692	MEAN MOTION DOT/2:	.00000270
ECCENTRICITY:	.2800849	MEAN MOTION DOT DOT/6:	.0
ARG. OF PERIGEE:	187.9970	BSTAR:	.0

DEBRIS CLOUD DATA

MAXIMUM ΔP:	Unknown
MAXIMUM ΔI:	1.2 deg

COMMENTS

This solid-propellant upper stage failed soon after ignition, following a normal launch. The cause of the failure is assessed to be a possible rupture of the motor casing or nozzle. See similar failures of two PAM-D upper stages in 1984. Elements above are first developed for the rocket body about 1 day after the event. Rocket body may later have been cross-tagged with satellite 4053. Validity of debris identification and cataloging after 1969 is suspect.

REFERENCE DOCUMENT

TRW Space Log, Winter 1969-70 edition, Vol. 9, No. 4, W.A. Donop, ed., TRW Systems Group, Redondo Beach, pp. 34-36.

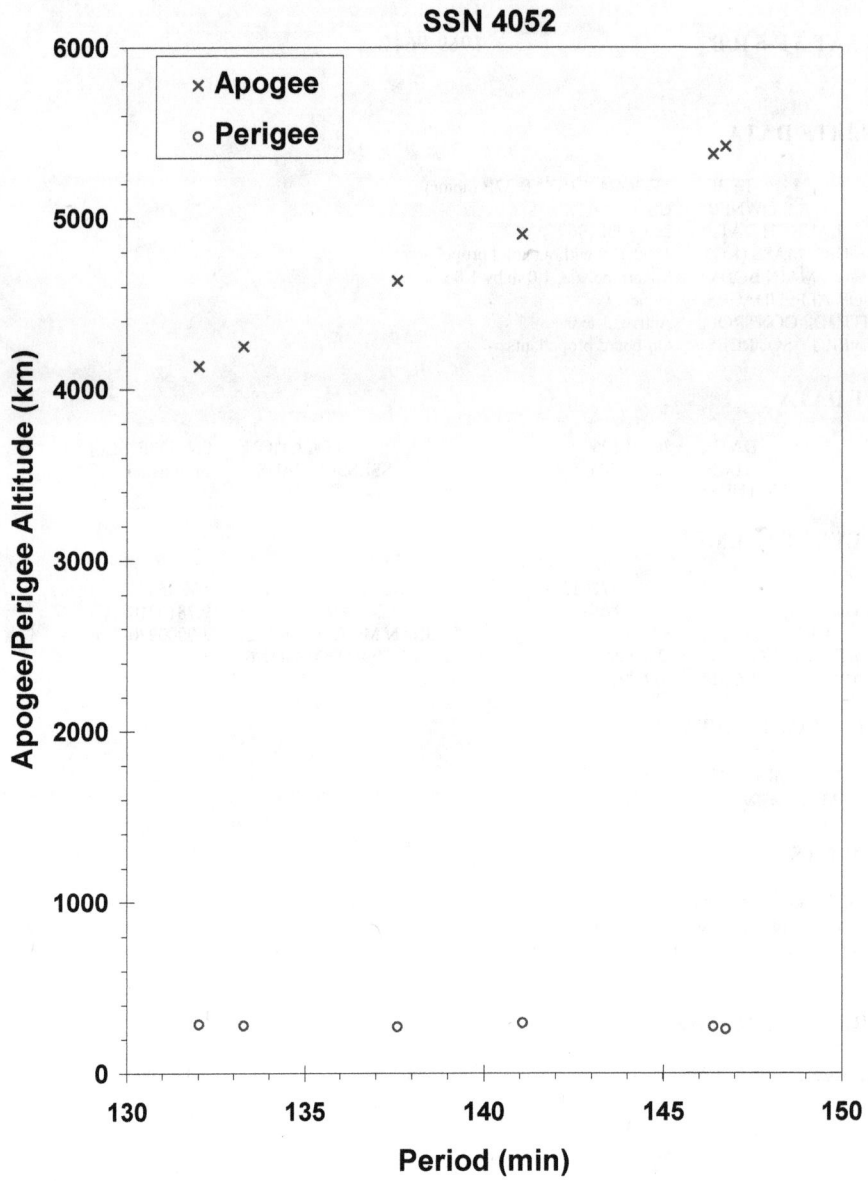

SSN 4052

Intelsat 3 F-5 R/B debris cloud of 6 fragments 10 days after the event as reconstructed from the US SSN database.

SATELLITE DATA

TYPE:	Agena D Stage
OWNER:	US
LAUNCH DATE:	30.57 Sep 1969
DRY MASS (KG):	600
MAIN BODY:	Cylinder; 1.5 m diameter by 7.1 m length
MAJOR APPENDAGES:	None
ATTITUDE CONTROL:	None at time of the event.
ENERGY SOURCES:	Unknown

EVENT DATA

DATE:	4 Oct 1969	LOCATION:	54N, 178E (dsc)
TIME:	1553 GMT	ASSESSED CAUSE:	Unknown
ALTITUDE:	920 km		

POST-EVENT ELEMENTS

EPOCH:	69295.54249482	MEAN ANOMALY:	274.0514
RIGHT ASCENSION:	243.5157	MEAN MOTION:	13.68701087
INCLINATION:	69.9611	MEAN MOTION DOT/2:	.00000064
ECCENTRICITY:	.0117819	MEAN MOTION DOT DOT/6:	.0
ARG. OF PERIGEE:	87.4011	BSTAR:	.0

DEBRIS CLOUD DATA

MAXIMUM ΔP:	3.1 min
MAXIMUM ΔI:	1.0 deg

COMMENTS

This was the first of two Agena D stages to fragment in a span of only 12 months. The vehicle delivered ten payloads to an orbit of about 905 km by 940 km. Four days later, before the rocket body had been cataloged, a large fragmentation occurred. What appeared to be the largest piece of the rocket body was found in the orbit described by the elements above almost 3 weeks after the event. See 1967-53 as a reference to an earlier mission of this type. Both missions were sponsored by DOD and public information is limited.

REFERENCE DOCUMENT

"Fragmentations of Asteroids and Artificial Satellites in Orbit", W. Wiesel, <u>Icarus</u>, Vol. 34, 1978, pp. 99-116.

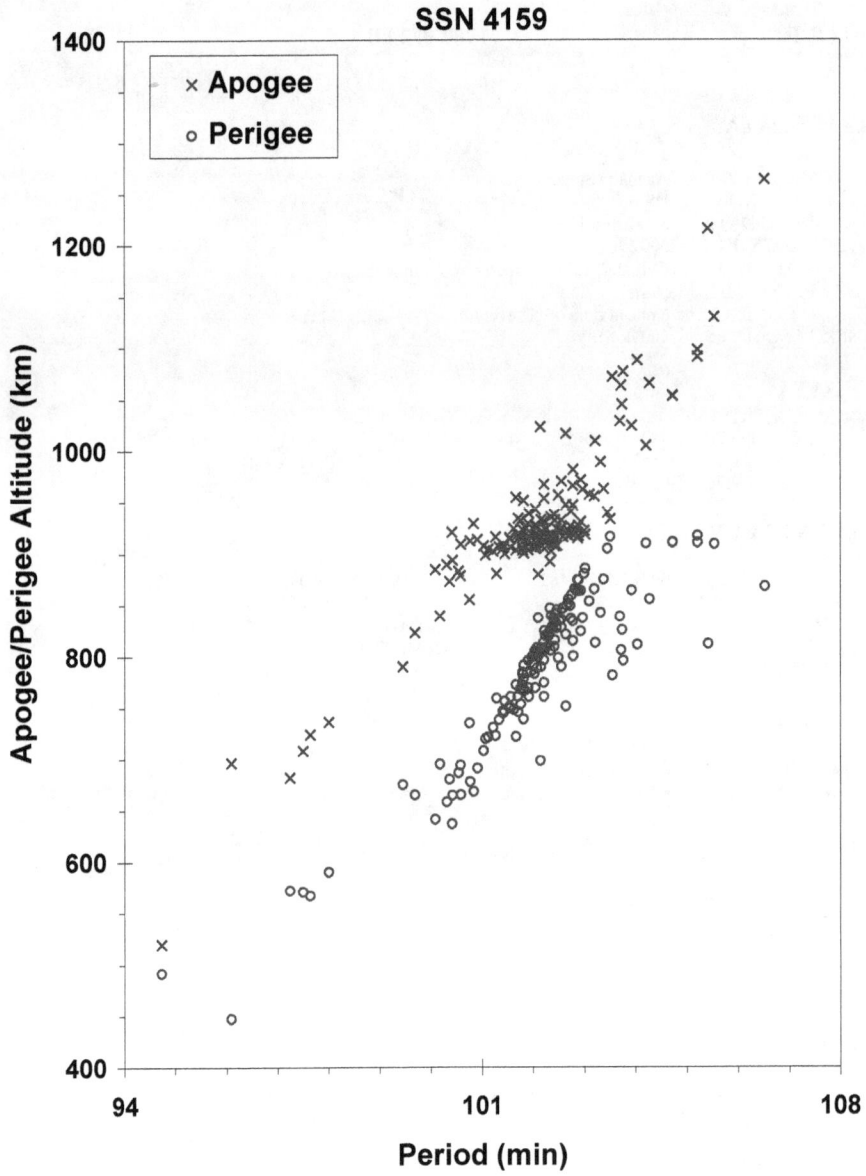

SSN 4159

OPS 7613 R/B debris cloud (excluding 10 payloads) of 152 fragments 8 months after the event. The largest fragment was found in an eccentric orbit with an orbital period of more than 105 min and is presumed to be the rocket body remnant.

SATELLITE DATA

TYPE:	Agena D Stage
OWNER:	US
LAUNCH DATE:	8.35 Apr 1970
DRY MASS (KG):	600
MAIN BODY:	Cylinder; 1.5 m diameter by 7.1 m length
MAJOR APPENDAGES:	None
ATTITUDE CONTROL:	None at time of the event.
ENERGY SOURCES:	Unknown

EVENT DATA

DATE:	17 Oct 1970	LOCATION:	50S, 142E (asc)
TIME:	0317 GMT	ASSESSED CAUSE:	Unknown
ALTITUDE:	1075 km		

PRE-EVENT ELEMENTS

EPOCH:	70289.33183878	MEAN ANOMALY:	141.3434
RIGHT ASCENSION:	203.5235	MEAN MOTION:	13.49254887
INCLINATION:	99.8780	MEAN MOTION DOT/2:	.0
ECCENTRICITY:	.0016616	MEAN MOTION DOT DOT/6:	.0
ARG. OF PERIGEE:	218.6463	BSTAR:	.0

DEBRIS CLOUD DATA

MAXIMUM ΔP:	14.2 min
MAXIMUM ΔI:	0.8 deg

COMMENTS

This was the second Agena D stage to fragment in a span of only 12 months. The event occurred 6 months after the successful deployment of the Nimbus 4 payload. Twice in 1985, again in 1986, once in 1991, and twice in 1995 Nimbus 4 R/B debris spawned a few additional fragments, accounting for an additional 16 new debris objects between the 6 sub-events.

REFERENCE DOCUMENTS

"Fragmentations of Asteroids and Artificial Satellites in Orbit", W. Wiesel, Icarus, Vol. 34, 1978, pp. 99-116.

"Analysis of the Nimbus 4 Rocket Body Breakup and Subsequent Debris Anomalies", N.L. Johnson, Kaman Sciences Corporation, February 1992.

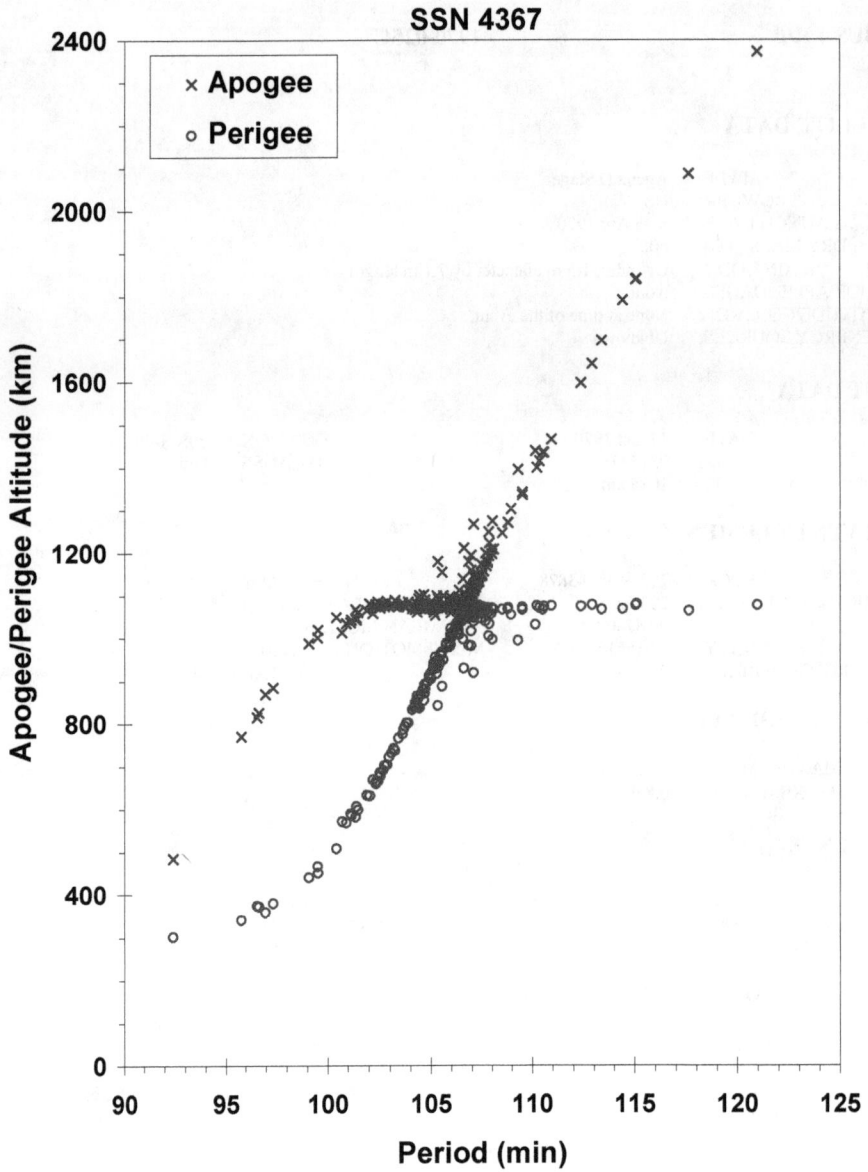

SSN 4367

Nimbus 4 R/B debris cloud of 246 fragments 8 months after the event as reconstructed from the US SSN database. Some lower period fragments already exhibit the effects of natural decay.

SATELLITE DATA

TYPE:	Payload
OWNER:	CIS
LAUNCH DATE:	23.18 Oct 1970
DRY MASS (KG):	1400
MAIN BODY:	Irregular; 1.8 m by 4.2 m
MAJOR APPENDAGES:	None
ATTITUDE CONTROL:	Active, 3-axis
ENERGY SOURCES:	On-board propellants, explosive charge

EVENT DATA

DATE:	23 Oct 1970	LOCATION:	22S, 217E (asc)
TIME:	1513 GMT	ASSESSED CAUSE:	Deliberate
ALTITUDE:	1195 km		

PRE-EVENT ELEMENTS

EPOCH:	70296.40542099	MEAN ANOMALY:	309.5623
RIGHT ASCENSION:	129.1049	MEAN MOTION:	12.82808179
INCLINATION:	62.9380	MEAN MOTION DOT/2:	.00019973
ECCENTRICITY:	.1039489	MEAN MOTION DOT DOT/6:	.0
ARG. OF PERIGEE:	60.4933	BSTAR:	.0

DEBRIS CLOUD DATA

MAXIMUM ΔP:	Unknown
MAXIMUM ΔI:	Unknown

COMMENTS

Cosmos 374 was launched on a two-revolution rendezvous with Cosmos 373. After a close approach, Cosmos 374 continued on before its warhead was intentionally fired. Cosmos 374 was part of test series begun with Cosmos 249. Considerable cross-cataloging of Cosmos 374 and Cosmos 375 debris occurred; therefore, ΔP and ΔI are not calculated.

REFERENCE DOCUMENTS

"Artificial Satellite Break-Ups (Part 2): Soviet Anti-Satellite Program", N.L. Johnson, Journal of the British Interplanetary Society, August 1983, pp. 357-362.

History of Soviet/Russian Satellite Fragmentations-A Joint U.S.-Russian Investigation, N. L. Johnson et al, Kaman Sciences Corporation, October 1995.

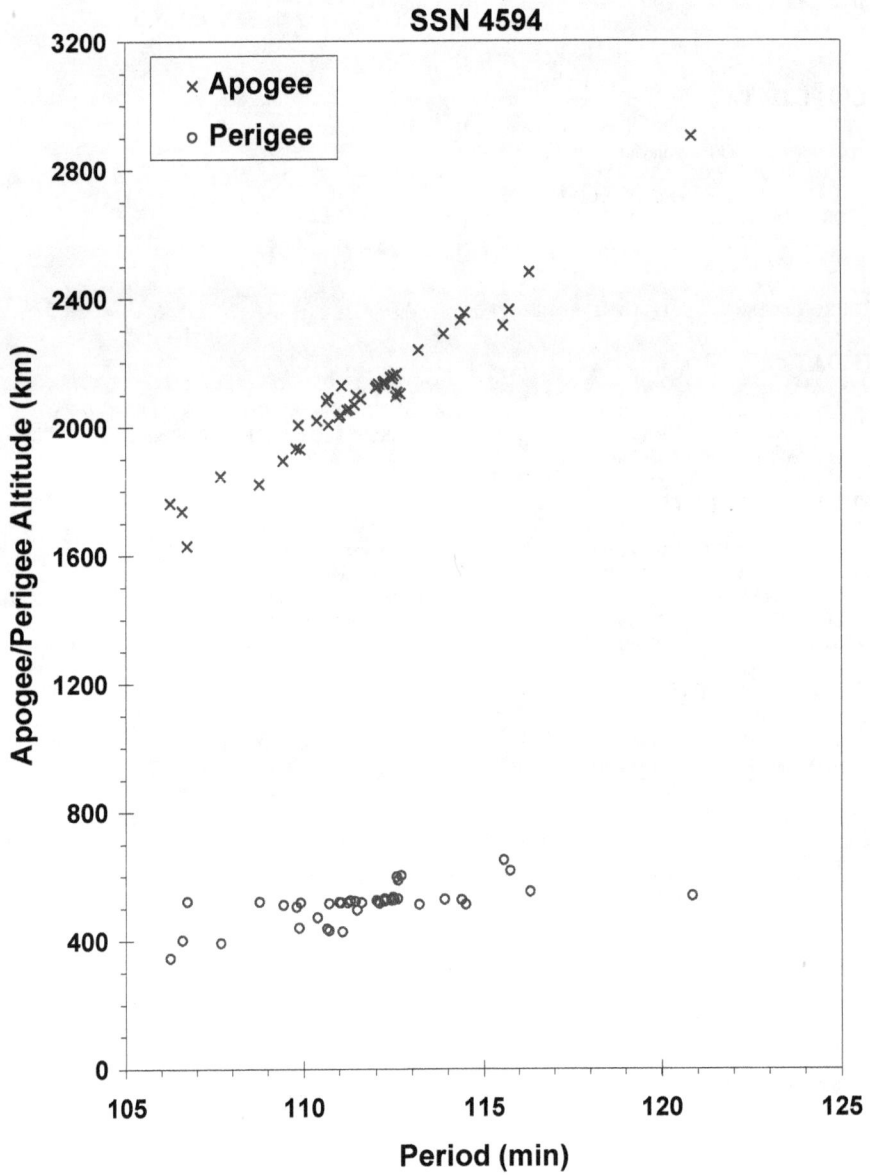

SSN 4594

Cosmos 374 official debris cloud of 43 fragments 5 months after the event as reconstructed from the US SSN database. All fragments were cataloged after the Cosmos 375 fragmentation, and some contamination exists.

SATELLITE DATA

TYPE:	Payload
OWNER:	CIS
LAUNCH DATE:	30.09 Oct 1970
DRY MASS (KG):	1400
MAIN BODY:	Irregular; 1.8 m by 4.2 m
MAJOR APPENDAGES:	None
ATTITUDE CONTROL:	Active, 3-axis
ENERGY SOURCES:	On-board propellants, explosive charge

EVENT DATA

DATE:	30 Oct 1970	LOCATION:	54N, 23E (asc)
TIME:	0600 GMT	ASSESSED CAUSE:	Deliberate
ALTITUDE:	535 km		

POST-EVENT ELEMENTS

EPOCH:	70306.81102869	MEAN ANOMALY:	313.3102
RIGHT ASCENSION:	96.4080	MEAN MOTION:	12.87482205
INCLINATION:	62.8057	MEAN MOTION DOT/2:	.00009999
ECCENTRICITY:	.1022289	MEAN MOTION DOT DOT/6:	.0
ARG. OF PERIGEE:	56.0864	BSTAR:	.0

DEBRIS CLOUD DATA

MAXIMUM ΔP:	Unknown
MAXIMUM ΔI:	Unknown

COMMENTS

Cosmos 375 was launched on a two-revolution rendezvous with Cosmos 373. The fragmentation occurred in the vicinity of Cosmos 373. Cosmos 375 was part of test series begun with Cosmos 249. Elements above are first reliable ones for orbit after final maneuver that took place immediately before fragmentation. Considerable cross-cataloging of Cosmos 374 and Cosmos 375 debris occurred; therefore, ΔP and ΔI are not calculated.

REFERENCE DOCUMENTS

"Artificial Satellite Break-Ups (Part 2): Soviet Anti-Satellite Program", N.L. Johnson, Journal of the British Interplanetary Society, August 1983, pp. 357-362.

History of Soviet/Russian Satellite Fragmentations-A Joint U.S.-Russian Investigation, N. L. Johnson et al, Kaman Sciences Corporation, October 1995.

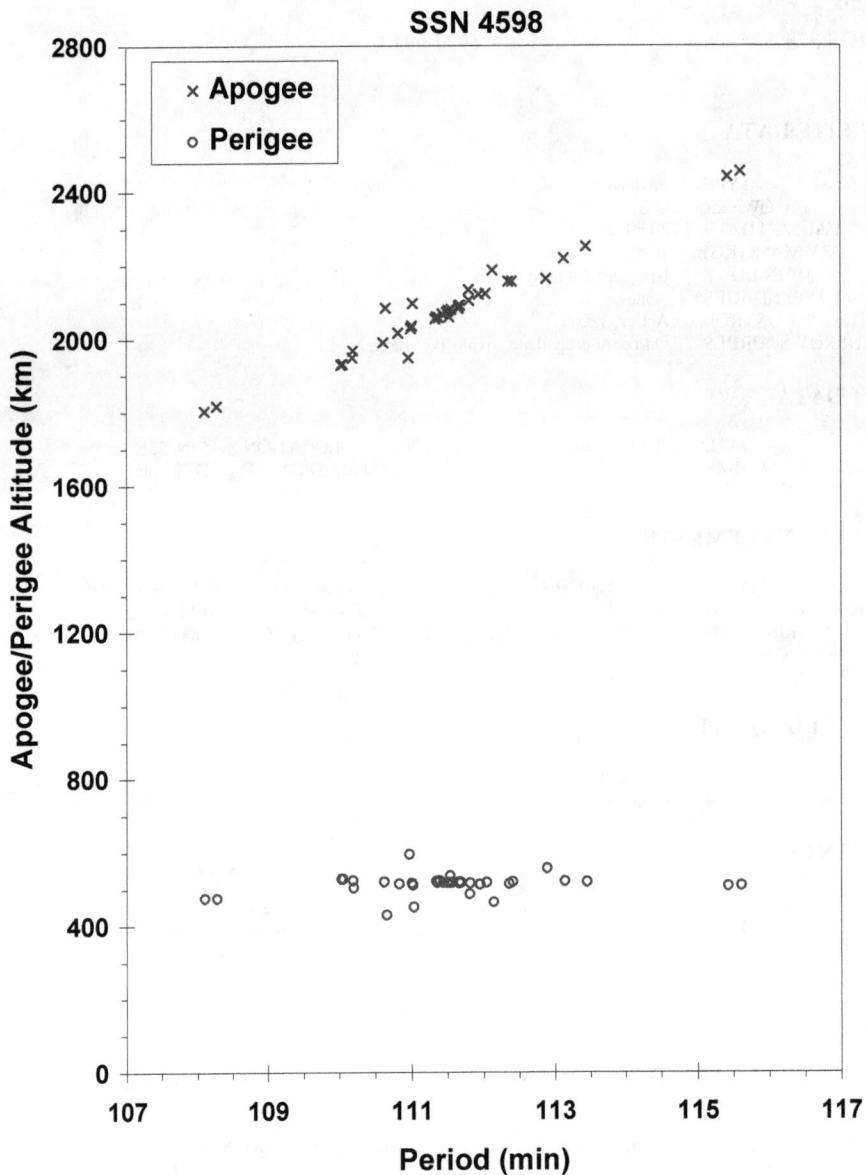

SSN 4598

Cosmos 375 debris cloud of 38 fragments about 4 months after the event as reconstructed from the US SSN database. Some contamination exists with Cosmos 374 debris.

SATELLITE DATA

TYPE:	Payload
OWNER:	CIS
LAUNCH DATE:	25.47 Feb 1971
DRY MASS (KG):	1400
MAIN BODY:	Irregular; 1.8 m by 4.2 m
MAJOR APPENDAGES:	None
ATTITUDE CONTROL:	Active, 3-axis
ENERGY SOURCES:	On-board propellants, explosive charge

EVENT DATA

DATE:	25 Feb 1971	LOCATION:	54N, 21E (asc)
TIME:	1431 GMT	ASSESSED CAUSE:	Deliberate
ALTITUDE:	585 km		

POST-EVENT ELEMENTS

EPOCH:	71057.77590281	MEAN ANOMALY:	318.5528
RIGHT ASCENSION:	352.8670	MEAN MOTION:	12.68709606
INCLINATION:	65.7618	MEAN MOTION DOT/2:	.00013192
ECCENTRICITY:	.1046189	MEAN MOTION DOT DOT/6:	.0
ARG. OF PERIGEE:	50.3064	BSTAR:	.0

DEBRIS CLOUD DATA

MAXIMUM ΔP:	2.8 min
MAXIMUM ΔI:	1.2 deg

COMMENTS

Cosmos 397 was launched on a two-revolution rendezvous with Cosmos 394. The fragmentation occurred in the vicinity of Cosmos 394. Cosmos 397 was part of the test series begun with Cosmos 249. Elements above are first available for orbit after final maneuver that took place immediately before fragmentation.

REFERENCE DOCUMENTS

"Artificial Satellite Break-Ups (Part 2): Soviet Anti-Satellite Program", N.L. Johnson, Journal of the British Interplanetary Society, August 1983, pp. 357-362.

History of Soviet/Russian Satellite Fragmentations-A Joint U.S.-Russian Investigation, N. L. Johnson et al, Kaman Sciences Corporation, October 1995.

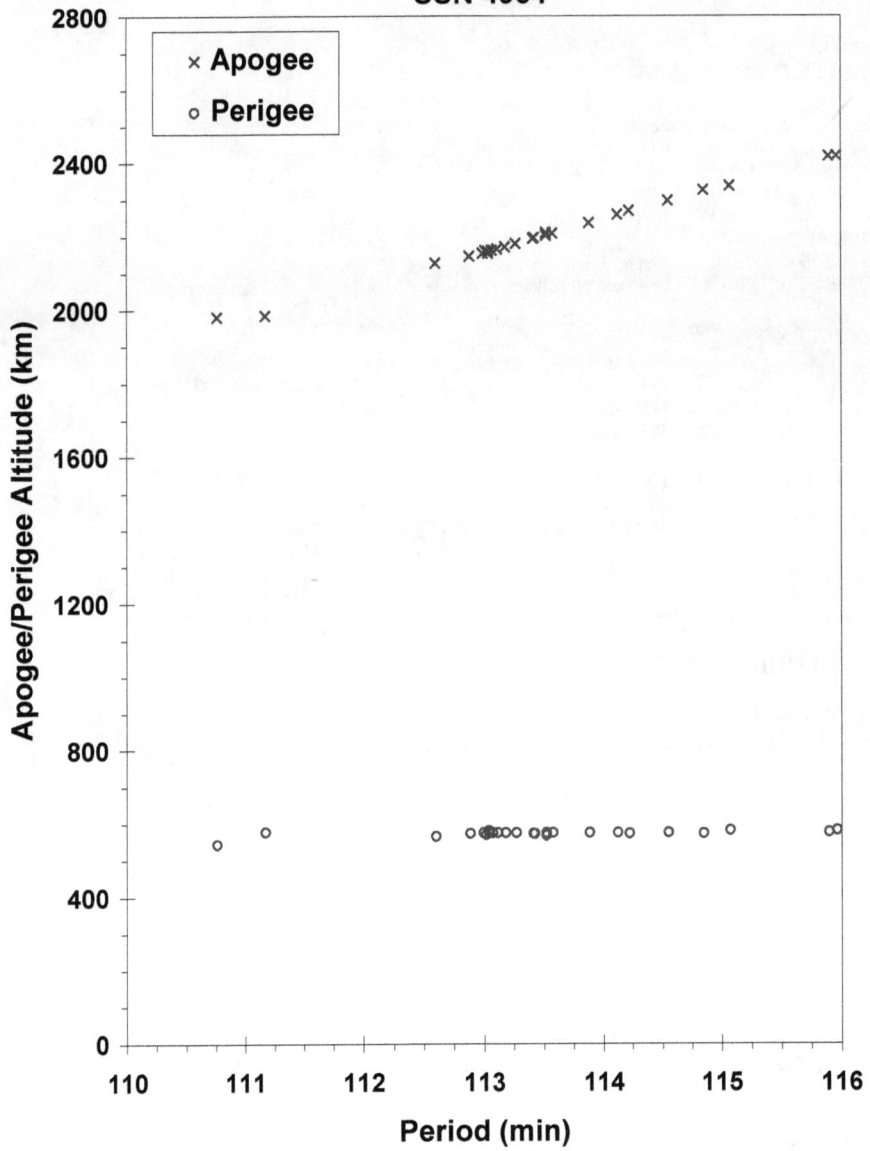

SSN 4964

Cosmos 397 cataloged debris cloud of 26 fragments about 7 weeks after the event as reconstructed from the US SSN database.

SATELLITE DATA

TYPE:	Payload
OWNER:	CIS
LAUNCH DATE:	3.55 Dec 1971
DRY MASS (KG):	1400
MAIN BODY:	Irregular; 1.8 m by 4.2 m
MAJOR APPENDAGES:	None
ATTITUDE CONTROL:	Active, 3-axis
ENERGY SOURCES:	On-board propellants, explosive charge

EVENT DATA

DATE:	3 Dec 1971	LOCATION:	51N, 7E (asc)
TIME:	1651 GMT	ASSESSED CAUSE:	Deliberate
ALTITUDE:	230 km		

POST-EVENT ELEMENTS

EPOCH:	71339.01001769	MEAN ANOMALY:	316.0762
RIGHT ASCENSION:	294.0999	MEAN MOTION:	13.65823046
INCLINATION:	65.7483	MEAN MOTION DOT/2:	.00001349
ECCENTRICITY:	.1062360	MEAN MOTION DOT DOT/6:	.0
ARG. OF PERIGEE:	53.3215	BSTAR:	.0

DEBRIS CLOUD DATA

MAXIMUM ΔP:	3.6 min
MAXIMUM ΔI:	0.7 deg

COMMENTS

Cosmos 462 was launched on a two-revolution rendezvous with Cosmos 459. The fragmentation occurred in the vicinity of Cosmos 459. Cosmos 462 was part of test series begun with Cosmos 249. Elements above are first available for orbit after final maneuver that took place immediately before fragmentation.

REFERENCE DOCUMENTS

"Artificial Satellite Break-Ups (Part 2): Soviet Anti-Satellite Program", N.L. Johnson, Journal of the British Interplanetary Society, August 1983, pp. 357-362.

History of Soviet/Russian Satellite Fragmentations-A Joint U.S.-Russian Investigation, N. L. Johnson et al, Kaman Sciences Corporation, October 1995.

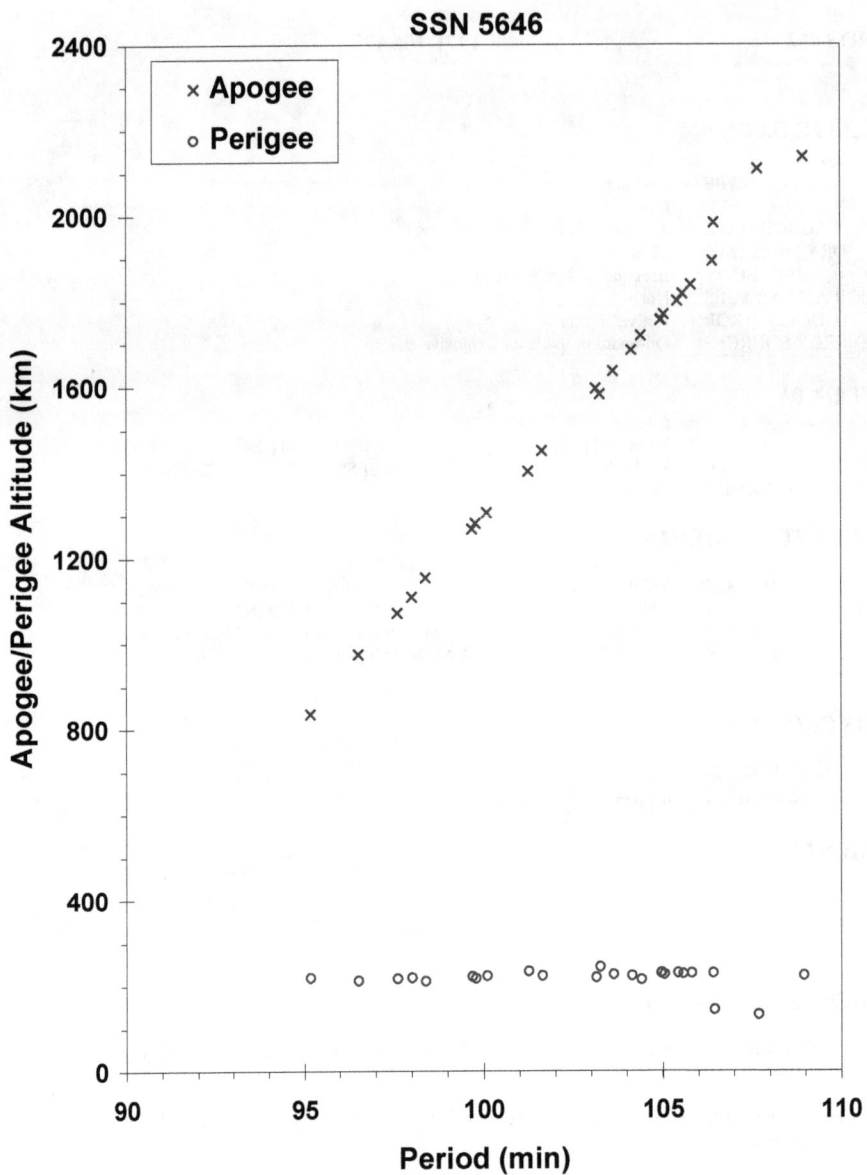

SSN 5646

Cosmos 462 debris cloud of 25 cataloged fragments within 1 week of the event as reconstructed from the US SSN database.

SATELLITE DATA

TYPE:	Delta Second Stage (900)
OWNER:	US
LAUNCH DATE:	23.75 Jul 1972
DRY MASS (KG):	800
MAIN BODY:	Cylinder-nozzle; 1.4 m diameter by 6.3 m length
MAJOR APPENDAGES:	None
ATTITUDE CONTROL:	None at time of the event.
ENERGY SOURCES:	On-board propellants, range safety device

EVENT DATA

DATE:	22 May 1975	LOCATION:	34S, 46E (asc)
TIME:	1827 GMT	ASSESSED CAUSE:	Propulsion
ALTITUDE:	730 km		

PRE-EVENT ELEMENTS

EPOCH:	75142.56642671	MEAN ANOMALY:	323.2981
RIGHT ASCENSION:	196.3353	MEAN MOTION:	14.36209995
INCLINATION:	98.3439	MEAN MOTION DOT/2:	.00000060
ECCENTRICITY:	.0193108	MEAN MOTION DOT DOT/6:	.0
ARG. OF PERIGEE:	38.1650	BSTAR:	.000027579

DEBRIS CLOUD DATA

MAXIMUM ΔP:	9.3 min
MAXIMUM ΔI:	1.0 deg

COMMENTS

This was the second Delta Second Stage to experience a severe fragmentation. The event occurred 34 months after the successful deployment of the Landsat 1 payload. Cause of the explosion is assessed to be related to the nearly 150 kg of residual propellants and characteristics of the sun-synchronous orbit.

REFERENCE DOCUMENTS

Dynamics of Satellite Disintegration, R. Dasenbrock, B. Kaufman, and W. Heard, NRL Report 7954, Naval Research Laboratory, Washington, 30 January 1976.

"Fragmentations of Asteroids and Artificial Satellites in Orbit", W. Wiesel, Icarus, Vol. 34, 1978, pp. 99-116.

Explosion of Satellite 10704 and other Delta Second Stage Rockets, J.R. Gabbard, Technical Memorandum 81-5, DCS Plans, Hdqtrs NORAD/ADCOM, Colorado Springs, May 1981.

Investigation of Delta Second Stage On-Orbit Explosions, C.S. Gumpel, Report MDC-H0047, McDonnell Douglas Astronautics Company - West, Huntington Beach, April 1982.

A Later Look at Delta Second Stage On-Orbit Explosions, J.R. Gabbard, Technical Report CS85-BMDSC-00-24, Teledyne Brown Engineering, Colorado Springs, March 1985.

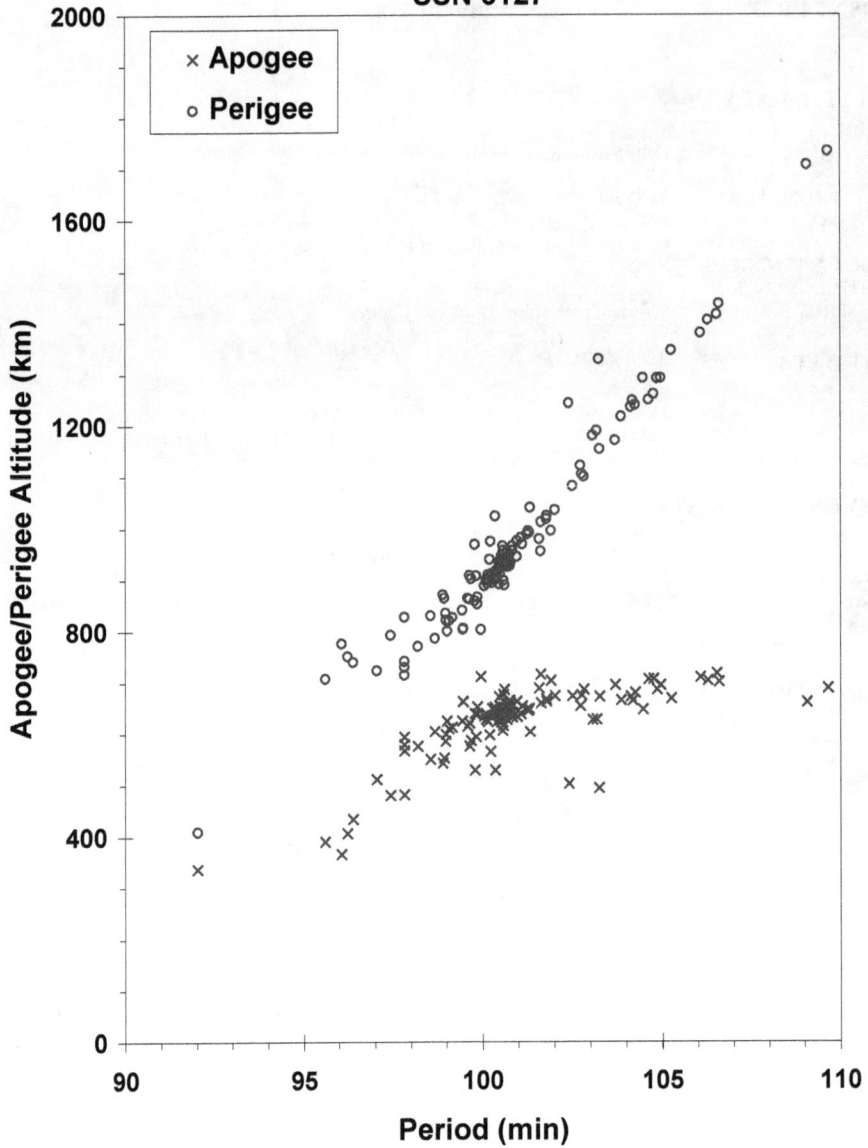

Landsat 1 R/B debris cloud of 133 fragments 4 months after the event as reconstructed from the US SSN database.

SALYUT 2 R/B   1973-017B   6399

SATELLITE DATA

TYPE:	Proton Third Stage
OWNER:	CIS
LAUNCH DATE:	3.38 Apr 1973
DRY MASS (KG):	4000
MAIN BODY:	Cylinder; 4.0 m diameter by 12.0 m length
MAJOR APPENDAGES:	None
ATTITUDE CONTROL:	None at time of the event.
ENERGY SOURCES:	On-board propellants

EVENT DATA

DATE:	3 Apr 1973	LOCATION:	45N, 290E (dsc)
TIME:	2236 GMT	ASSESSED CAUSE:	Propulsion
ALTITUDE:	225 km		

PRE-EVENT ELEMENTS

EPOCH:	73093.61404736	MEAN ANOMALY:	357.9254
RIGHT ASCENSION:	334.5652	MEAN MOTION:	16.20127597
INCLINATION:	51.4798	MEAN MOTION DOT/2:	.00508885
ECCENTRICITY:	.0037670	MEAN MOTION DOT DOT/6:	.0
ARG. OF PERIGEE:	2.1878	BSTAR:	.0

DEBRIS CLOUD DATA

MAXIMUM ΔP:	3.8 min
MAXIMUM ΔI:	0.5 deg

COMMENTS

This is the only known fragmentation of the Proton third stage. The event occurred less than 14 hours after reaching orbit. The event was apparently unrelated to the later payload malfunction. NAVSPASUR counted at least 95 objects shortly after the event, but most reentered before being officially cataloged. Information uncovered by Mr. Nicholas Johnson during an information exchange with Russian officials in the Spring of 1993 revealed that residual propellants resulted in an over-pressurization of the rocket body, causing this fragmentation. After this event, the Russians reported that the Proton third stage has been vented to avoid future events of this nature.

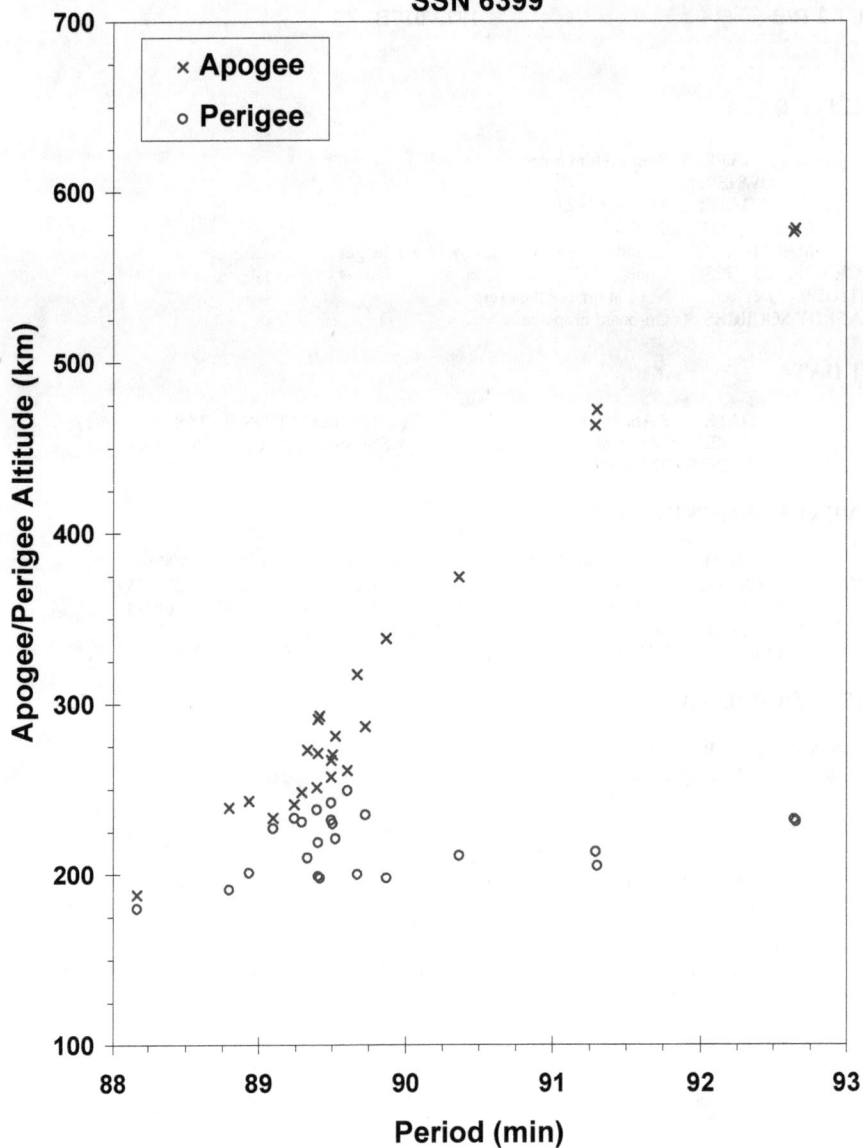

SSN 6399

Salyut 2 R/B debris cloud of 25 fragments as reconstructed from the US SSN database. Most elements were developed within 2 days of the event.

SATELLITE DATA

TYPE:	Payload
OWNER:	CIS
LAUNCH DATE:	19.38 Apr 1973
DRY MASS (KG):	6300
MAIN BODY:	Sphere-cylinder; 2.8 m diameter by 6.5 m length
MAJOR APPENDAGES:	None
ATTITUDE CONTROL:	Active, 3-axis
ENERGY SOURCES:	On-board propellants, explosive charge

EVENT DATA

DATE:	6 May 1973	LOCATION:	71S, 215E (asc)
TIME:	0724 GMT	ASSESSED CAUSE:	Deliberate
ALTITUDE:	310 km		

PRE-EVENT ELEMENTS

EPOCH:	73125.63953480	MEAN ANOMALY:	337.7411
RIGHT ASCENSION:	305.5573	MEAN MOTION:	16.05578988
INCLINATION:	72.8514	MEAN MOTION DOT/2:	.00433078
ECCENTRICITY:	.0137599	MEAN MOTION DOT DOT/6:	.00010923
ARG. OF PERIGEE:	22.9846	BSTAR:	.0

DEBRIS CLOUD DATA

MAXIMUM ΔP:	6.0 min
MAXIMUM ΔI:	1.3 deg

COMMENTS

Spacecraft was destroyed after a malfunction prevented controlled reentry and landing in the Soviet Union. A total of 88 fragments were cataloged without elements.

REFERENCE DOCUMENT

History of Soviet/Russian Satellite Fragmentations-A Joint U.S.-Russian Investigation, N. L. Johnson et al, Kaman Sciences Corporation, October 1995.

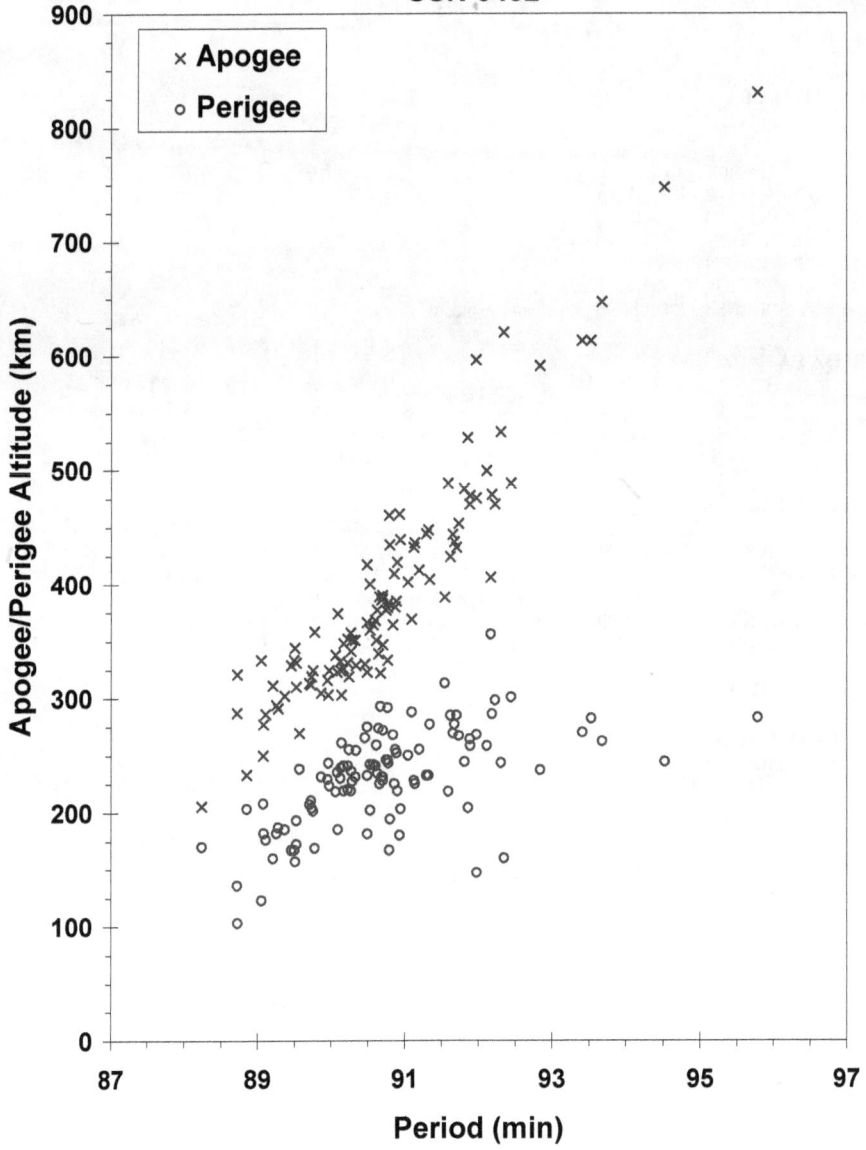

SSN 6432

Apogee/Perigee Altitude (km) vs Period (min)

× Apogee
○ Perigee

Cosmos 554 debris cloud of 107 fragments using initial elements as developed over several weeks. Some decay effects are present. Source is the US SSN database.

SATELLITE DATA

TYPE:	Delta Second Stage (300)
OWNER:	US
LAUNCH DATE:	6.71 Nov 1973
DRY MASS (KG):	840
MAIN BODY:	Cylinder-nozzle; 1.4 m diameter by 6.3 m length
MAJOR APPENDAGES:	None
ATTITUDE CONTROL:	None at time of the event.
ENERGY SOURCES:	On-board propellants, range safety device

EVENT DATA

DATE:	28 Dec 1973	LOCATION:	37S, 181E (asc)
TIME:	0904 GMT	ASSESSED CAUSE:	Propulsion
ALTITUDE:	1515 km		

PRE-EVENT ELEMENTS

EPOCH:	73359.56303028	MEAN ANOMALY:	202.2816
RIGHT ASCENSION:	41.7242	MEAN MOTION:	12.40088347
INCLINATION:	102.0500	MEAN MOTION DOT/2:	.00000577
ECCENTRICITY:	.0005689	MEAN MOTION DOT DOT/6:	.000000056523
ARG. OF PERIGEE:	157.8450	BSTAR:	.0

DEBRIS CLOUD DATA

MAXIMUM ΔP:	10.4 min
MAXIMUM ΔI:	1.4 deg

COMMENTS

This was the first of seven Delta Second Stages to experience severe fragmentations between 1973 and 1981. Six of the seven stages were left in mid-morning, sun-synchronous orbits with residual propellants. Fragmentations occurred from 2-35 months after launch. The seventh stage exploded within hours of launch on a geosynchronous mission. The assessed cause in all cases is a propellant-induced explosion. Depletion burns to remove residual propellants were initiated in 1981, and no vented Delta Second Stages have fragmented since. In the case of the NOAA 3 R/B, fragmentation took place nearly 2 months after successful deployment of the NOAA 3 payload. Approximately 130 kg of propellants were left on board.

REFERENCE DOCUMENTS

Dynamics of Satellite Disintegration, R. Dasenbrock, B. Kaufman, and W. Heard, NRL Report 7954, Naval Research Laboratory, Washington, 30 January 1976.

Explosion of Satellite 10704 and other Delta Second Stage Rockets, J.R. Gabbard, Technical Memorandum 81-5, DCS Plans, Hdqtrs NORAD/ADCOM, Colorado Springs, May 1981.

Investigation of Delta Second Stage On-Orbit Explosions, C.S. Gumpel, Report MDC-H0047, McDonnell Douglas Astronautics Company - West, Huntington Beach, April 1982.

A Later Look at Delta Second Stage On-Orbit Explosions, J.R. Gabbard, Technical Report CS85-BMDSC-00-24, Teledyne Brown Engineering, Colorado Springs, March 1985.

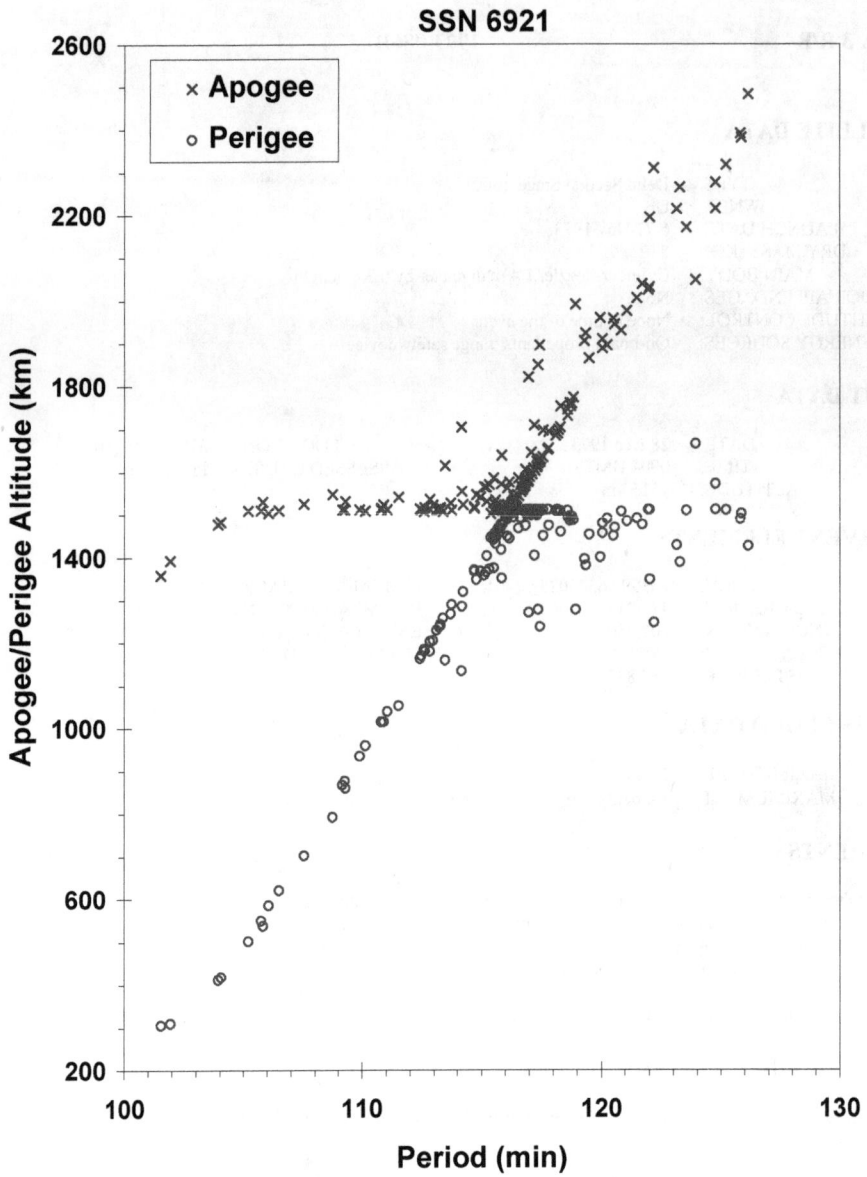

NOAA 3 R/B debris cloud of 160 fragments 4 months after the event as reconstructed from the US SSN database.

SATELLITE DATA

TYPE:	Rocket Body
OWNER:	US
LAUNCH DATE:	16.33 Mar 1974
DRY MASS (KG):	50
MAIN BODY:	Box; 1 m long by 1 m wide by 1 m high
MAJOR APPENDAGES:	None
ATTITUDE CONTROL:	None at time of event
ENERGY SOURCES:	None at time of event

EVENT DATA

DATE:	17 Jan 2005	LOCATION:	80.6S, 306.8E
TIME:	0214 GMT	ASSESSED CAUSE:	Accidental Collision
ALTITUDE:	885 km		

PRE-EVENT ELEMENTS

EPOCH:	05016.54972523	MEAN ANOMALY:	256.1717
RIGHT ASCENSION:	350.2846	MEAN MOTION:	14.24162249
INCLINATION:	99.0928	MEAN MOTION DOT/2:	0.00000028
ECCENTRICITY:	.0066248	MEAN MOTION DOT DOT/6:	0.0
ARG. OF PERIGEE:	104.6813	BSTAR:	0.000031607

DEBRIS CLOUD DATA

MAXIMUM ΔP:	Unknown
MAXIMUM ΔI:	Unknown

COMMENTS

The THOR 2A upper stage collided with a piece of fragmentation debris (1999-57CV, SSN# 26207) from the March 2000 explosion of the third stage of a Chinese CZ-4 launch vehicle (1999-57C, SSN# 25942). This was the third historical collision of cataloged objects; the collision occurred over Antarctica as both object were near the southernmost point in their respective retrograde orbits. The relative velocity of the collision was just under 6 km/s. 1999-57CV is believed to be relatively small, with a radar cross-section of 600 cm^2. The collision produced less than 10 cataloged debris. Because the upper stage had been in orbit for over 30 years, it is believed there was no on-board propellant remaining at the time of the event.

REFERENCE DOCUMENT

"Accidental Collisions of Cataloged Satellites Identified", The Orbital Debris Quarterly News, NASA JSC, April 2005. Available online at http://www.orbitaldebris.jsc nasa.gov/newsletter/pdfs/ODQNv9i2.pdf.

Insufficient data to construct a Gabbard diagram.

SATELLITE DATA

TYPE:	Delta Second Stage (2310)
OWNER:	US
LAUNCH DATE:	15.72 Nov 1974
DRY MASS (KG):	840
MAIN BODY:	Cylinder-nozzle; 1.4 m diameter by 5.8 m length
MAJOR APPENDAGES:	Mini-skirt; 2.4m by 0.3 m
ATTITUDE CONTROL:	None at time of the event.
ENERGY SOURCES:	On-board propellants, range safety device

EVENT DATA

DATE:	20 Aug 1975	LOCATION:	52S, 278E (dsc)
TIME:	1307 GMT	ASSESSED CAUSE:	Propulsion
ALTITUDE:	1465 km		

PRE-EVENT ELEMENTS

EPOCH:	75231.53619619	MEAN ANOMALY:	309.0001
RIGHT ASCENSION:	277.2201	MEAN MOTION:	12.52826370
INCLINATION:	101.6940	MEAN MOTION DOT/2:	.00000083
ECCENTRICITY:	.0009694	MEAN MOTION DOT DOT/6:	.0
ARG. OF PERIGEE:	51.1891	BSTAR:	.0

DEBRIS CLOUD DATA

MAXIMUM ΔP:	15.7 min
MAXIMUM ΔI:	1.8 deg

COMMENTS

This was the third Delta Second Stage to experience a severe fragmentation. The event occurred 9 months after the successful deployment of the NOAA 4 payload. Cause of the explosion is assessed to be related to the estimated more than 200 kg of residual propellants and characteristics of the sun-synchronous orbit. A fragment from this event (satellite number 8138) may have generated six or more additional pieces in September 1981.

REFERENCE DOCUMENTS

Explosion of Satellite 10704 and other Delta Second Stage Rockets, J.R. Gabbard, Technical Memorandum 81-5, DCS Plans, Hdqtrs NORAD/ADCOM, Colorado Springs, May 1981.

Investigation of Delta Second Stage On-Orbit Explosions, C.S. Gumpel, Report MDC-H0047, McDonnell Douglas Astronautics Company - West, Huntington Beach, April 1982.

A Later Look at Delta Second Stage On-Orbit Explosions, J.R. Gabbard, Technical Report CS85-BMDSC-00-24, Teledyne Brown Engineering, Colorado Springs, March 1985.

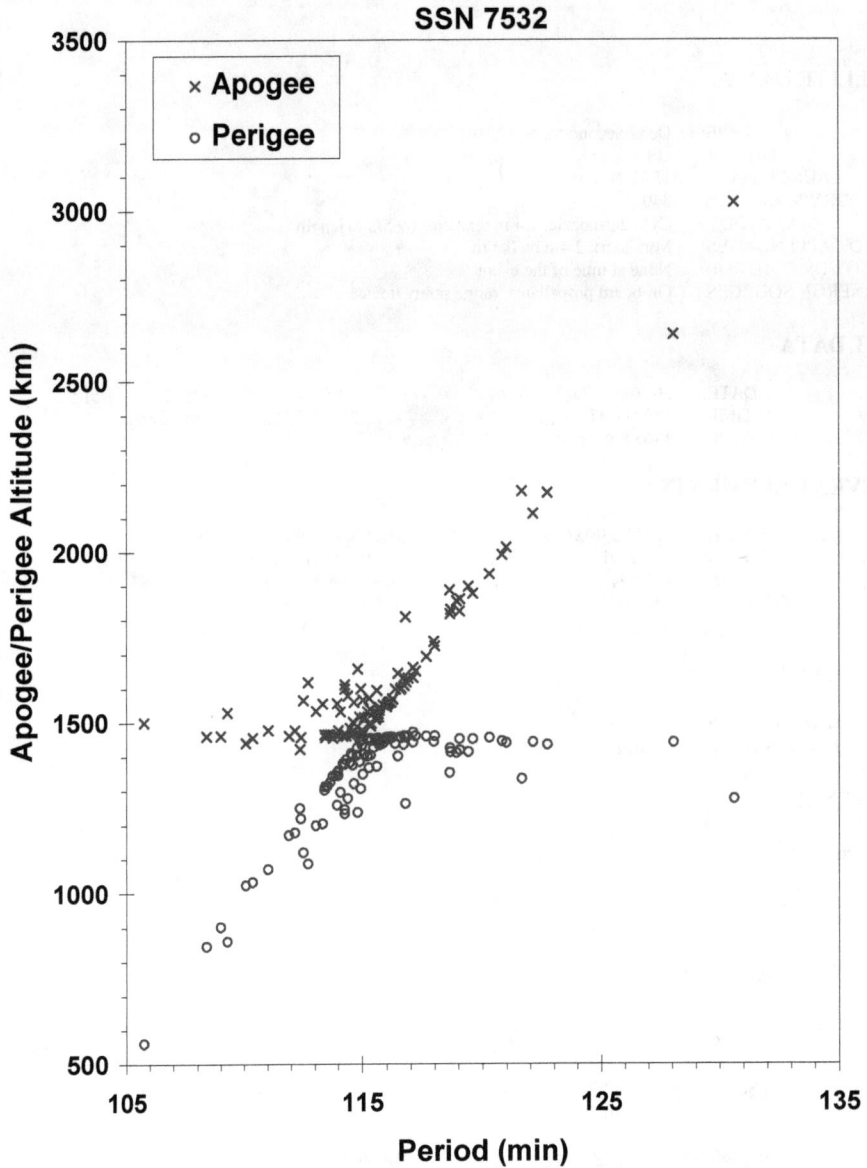

SSN 7532

NOAA 4 R/B debris cloud of 101 fragments 6 months after the event as reconstructed from the US SSN database.

SATELLITE DATA

TYPE:	Payload
OWNER:	CIS
LAUNCH DATE:	24.46 Dec 1974
DRY MASS (KG):	3000
MAIN BODY:	Cylinder; 1.3 m diameter by 17 m length
MAJOR APPENDAGES:	Solar panels
ATTITUDE CONTROL:	Active, 3-axis
ENERGY SOURCES:	On-board propellants, explosive charge (?)

EVENT DATA (1)

DATE:	17 Apr 1975	LOCATION:	01N, 278E (dsc)
TIME:	2148 GMT	ASSESSED CAUSE:	Unknown
ALTITUDE:	440 km		

PRE-EVENT ELEMENTS (1)

EPOCH:	75107.81173798	MEAN ANOMALY:	71.8460
RIGHT ASCENSION:	271.0743	MEAN MOTION:	15.44155646
INCLINATION:	65.0355	MEAN MOTION DOT/2:	.00007106
ECCENTRICITY:	.0014224	MEAN MOTION DOT DOT/6:	.0
ARG. OF PERIGEE:	288.1084	BSTAR:	.0

EVENT DATA (2)

DATE:	2 Aug 1975	LOCATION:	02S, 258E (dsc)
TIME:	1623 GMT	ASSESSED CAUSE:	Unknown
ALTITUDE:	435 km		

PRE-EVENT ELEMENTS (2)

EPOCH:	75214.45597981	MEAN ANOMALY:	68.4232
RIGHT ASCENSION:	274.3453	MEAN MOTION:	15.46205523
INCLINATION:	65.0458	MEAN MOTION DOT/2:	.00001715
ECCENTRICITY:	.0020980	MEAN MOTION DOT DOT/6:	.0
ARG. OF PERIGEE:	291.4623	BSTAR:	.0

DEBRIS CLOUD DATA

MAXIMUM ΔP:	3.5 min*
MAXIMUM ΔI:	0.9 deg*

*Based on NRL analysis

COMMENTS

Cosmos 699 was the first of a new type spacecraft. Many members of this class have experienced breakups. Beginning in 1988 old spacecraft have been commanded to lower perigee at end of life, resulting in an accelerated natural decay with fewer fragmentations. For several spacecraft, two distinct events have been detected and observational data suggest that the spacecraft remain essentially intact after each event. In most cases, breakups occur after spacecraft has ceased orbit maintenance and entered natural decay. Debris are sometimes highly unidirectional. In the case of Cosmos 699, the spacecraft had been in a regime of natural decay for 1 month at the time of the event.

REFERENCE DOCUMENTS

An Analysis of the Breakup of Satellite 1974-103A (Cosmos 699), W. B. Heard, NRL Report 7991, Naval Research Laboratory, Washington, 23 April 1976.

"Artificial Satellite Break-Ups (Part 1): Soviet Ocean Surveillance Satellites", N. L. Johnson, Journal of the British Interplanetary Society, February 1983, pp. 51-58.

History of the Soviet/Russian Satellite Fragmentations-A Joint U.S.-Russian Investigation, N. L. Johnson et al, Kaman Sciences Corporation, October 1995.

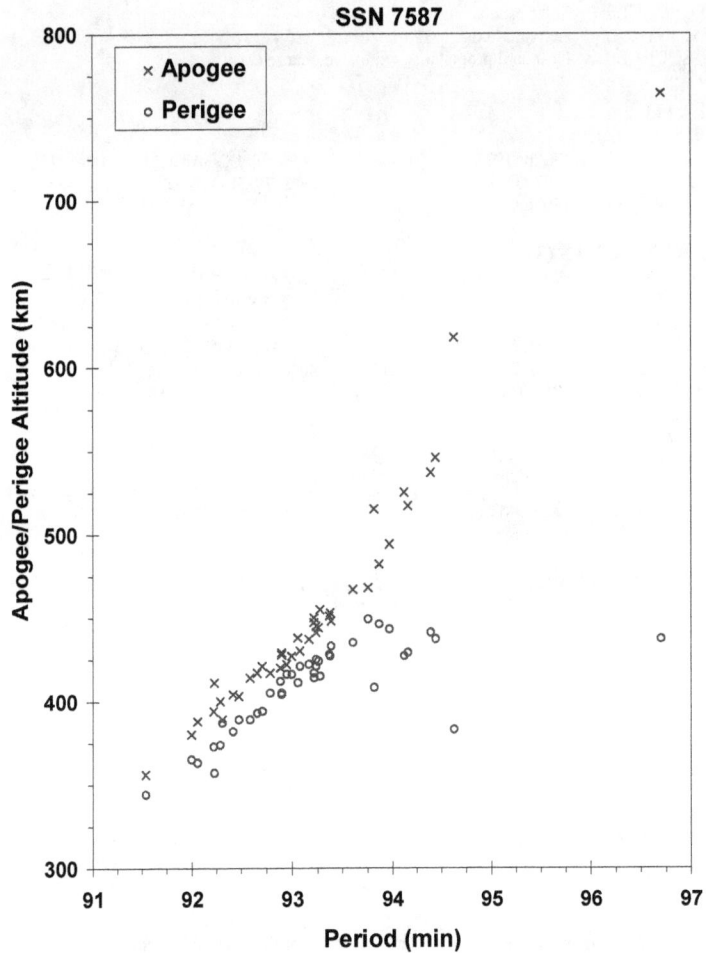

Cosmos 699 debris cloud of 41 fragments after the first breakup event as reconstructed from radar observations following the first breakup event.

SATELLITE DATA

TYPE:	Delta Second Stage (2910)
OWNER:	US
LAUNCH DATE:	22.75 Jan 1975
DRY MASS (KG):	840
MAIN BODY:	Cylinder-nozzle; 1.4 m diameter by 5.8 m length
MAJOR APPENDAGES:	Mini-skirt; 2.4 m by 0.2 m
ATTITUDE CONTROL:	None at time of the event.
ENERGY SOURCES:	On-board propellants, range safety device

EVENT DATA (1)

DATE:	9 Feb 1976	LOCATION:	Unknown
TIME:	Unknown	ASSESSED CAUSE:	Propulsion
ALTITUDE:	Unknown		

PRE-EVENT ELEMENTS (1)

EPOCH:	76040.08509016	MEAN ANOMALY:	189.3492
RIGHT ASCENSION:	60.2329	MEAN MOTION:	14.19373945
INCLINATION:	97.7751	MEAN MOTION DOT/2:	.0
ECCENTRICITY:	.0120730	MEAN MOTION DOT DOT/6:	.0
ARG. OF PERIGEE:	170.9843	BSTAR:	.0

EVENT DATA (2)

DATE:	19 Jun 1976	LOCATION:	7N, 344E (dsc)
TIME:	0659 GMT	ASSESSED CAUSE:	Propulsion
ALTITUDE:	750 km		

PRE-EVENT ELEMENTS (2)

EPOCH:	76170.97576375	MEAN ANOMALY:	217.2433
RIGHT ASCENSION:	175.3897	MEAN MOTION:	14.19574919
INCLINATION:	97.7497	MEAN MOTION DOT/2:	.0
ECCENTRICITY:	.0115288	MEAN MOTION DOT DOT/6:	.0
ARG. OF PERIGEE:	143.6594	BSTAR:	.0

DEBRIS CLOUD DATA

MAXIMUM ΔP:	5.6 min
MAXIMUM ΔI:	2.3 deg

COMMENTS

This was the fourth Delta Second Stage to experience a severe fragmentation. The first event occurred almost 13 months after the successful deployment of the Landsat 2 payload. Only 14 fragments were cataloged after the first event and all possessed orbital period changes of less than 0.6 min. Four months later a much larger fragmentation occurred. The cause of the second event is assessed to be related to the estimated 150 kg of residual propellants on board and characteristics of the sun-synchronous orbit.

REFERENCE DOCUMENTS

Explosion of Satellite 10704 and other Delta Second Stage Rockets, J.R. Gabbard, Technical Memorandum 81-5, DCS Plans, Hdqtrs NORAD/ADCOM, Colorado Springs, May 1981.

Investigation of Delta Second Stage On-Orbit Explosions, C.S. Gumpel, Report MDC-H0047, McDonnell Douglas Astronautics Company - West, Huntington Beach, April 1982.

A Later Look at Delta Second Stage On-Orbit Explosions, J.R. Gabbard, Technical Report CS85-BMDSC-00-24, Teledyne Brown Engineering, Colorado Springs, March 1985.

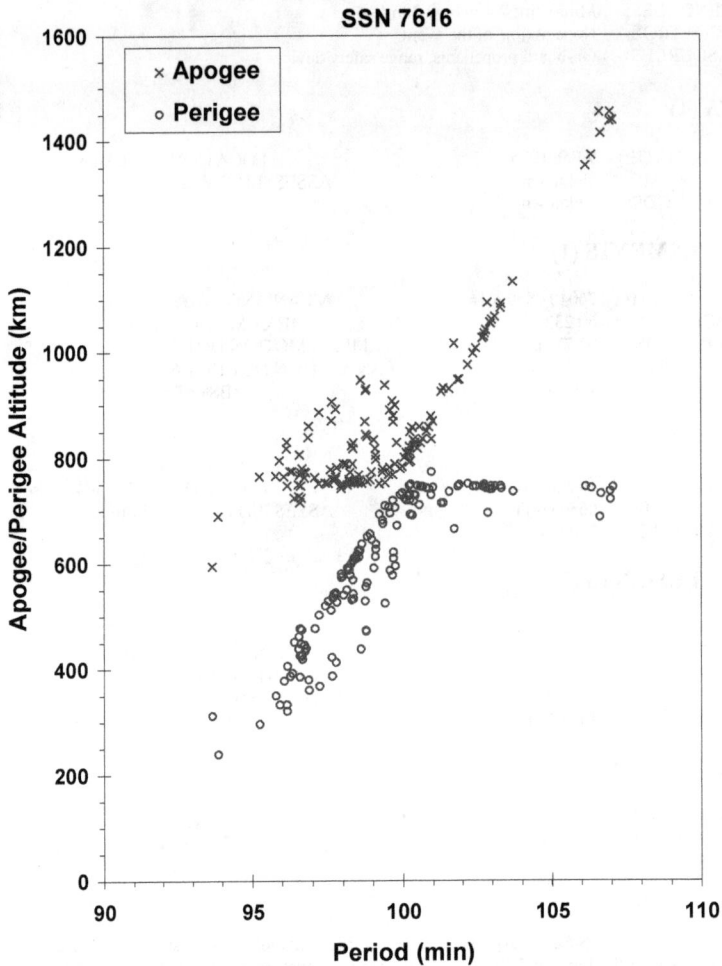

Landsat 2 R/B debris cloud of 147 fragments about 6 weeks after the second event as reconstructed from the US SSN database.

SATELLITE DATA

TYPE:	Delta Second Stage (2910)
OWNER:	US
LAUNCH DATE:	12.34 Jun 1975
DRY MASS (KG):	840
MAIN BODY:	Cylinder-nozzle; 1.4 m diameter by 5.8 m length
MAJOR APPENDAGES:	Mini-skirt; 2.4 m by 0.3 m
ATTITUDE CONTROL:	None at time of the event.
ENERGY SOURCES:	On-board propellants, range safety device

EVENT DATA

DATE:	1 May 1991	LOCATION:	66N, 322E (asc)
TIME:	0856 GMT	ASSESSED CAUSE:	Propulsion
ALTITUDE:	1090 km		

PRE-EVENT ELEMENTS

EPOCH:	91112.56709963	MEAN ANOMALY:	211.7525
RIGHT ASCENSION:	329.2109	MEAN MOTION:	13.43007146
INCLINATION:	99.5801	MEAN MOTION DOT/2:	.00000050
ECCENTRICITY:	.0006217	MEAN MOTION DOT DOT/6:	0.0
ARG. OF PERIGEE:	148.3989	BSTAR:	.0055458

DEBRIS CLOUD DATA

MAXIMUM ΔP:	27.4 min*
MAXIMUM ΔI:	2.4 min*

*Based on uncataloged debris data

COMMENTS

This was the eighth Delta Second Stage to experience a severe fragmentation. The event occurred nearly 191 months after the successful deployment of the Nimbus 6 payload. Cause of the explosion is assessed to be related to the estimated 245 kg of residual propellants on board and characteristics of the sun-synchronous orbit.

REFERENCE DOCUMENTS

The Fragmentation of the Nimbus 6 Rocket Body, D. J. Nauer and N. L. Johnson, Technical Report CS91-TR-JSC-017, Teledyne Brown Engineering, Colorado Springs, Colorado, November 1991.

Nimbus 6 Delta Upper Stage Rocket Body Breakup Report, E. L. Jenkins and H. V. Reynolds, Naval Space Surveillance Center, Dahlgren, Virginia, 1991.

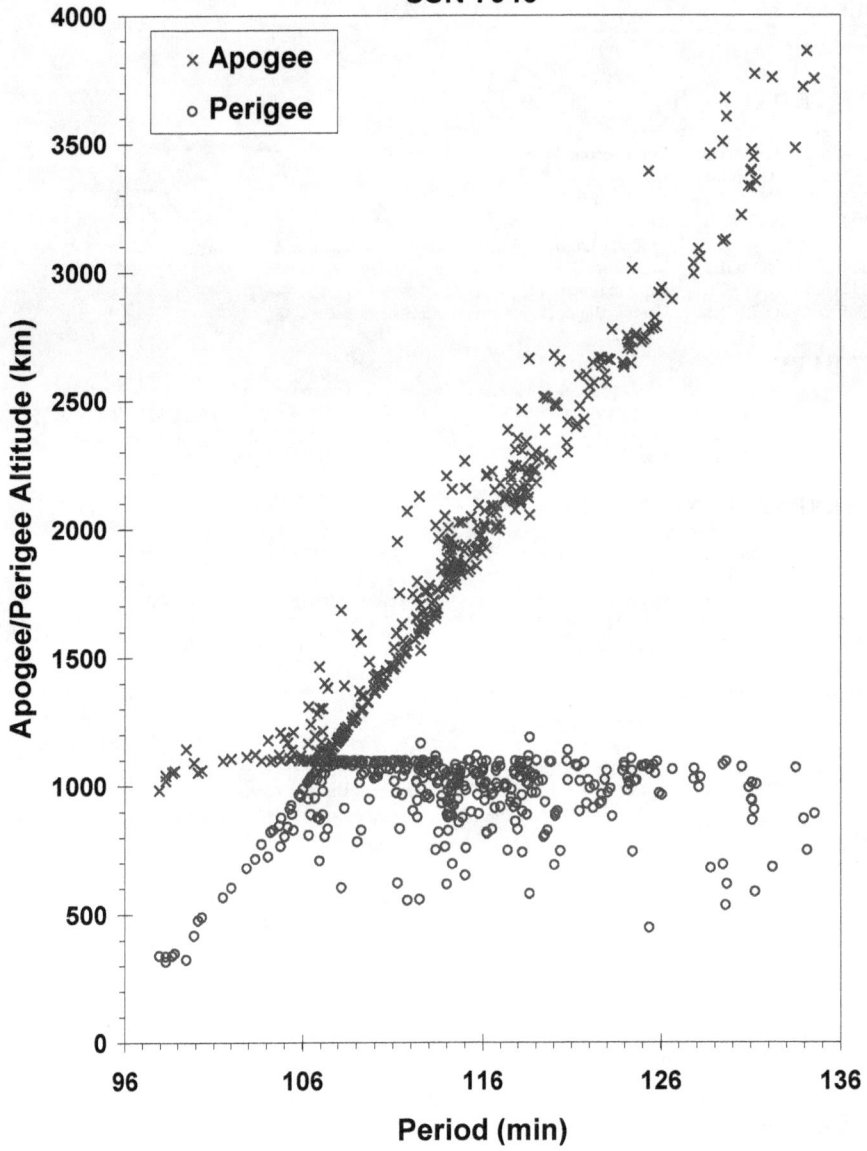

SSN 7946

Nimbus 6 R/B debris cloud of 386 identified fragments within 1 week after the event as reconstructed from Naval Space Surveillance System database. This diagram is taken from the first cited reference.

SATELLITE DATA

TYPE:	Payload
OWNER:	CIS
LAUNCH DATE:	5.62 Sep 1975
DRY MASS (KG):	5700
MAIN BODY:	Cone-cylinder; 2.7 m diameter by 6.3 m length
MAJOR APPENDAGES:	Solar panels
ATTITUDE CONTROL:	Active, 3-axis
ENERGY SOURCES:	On-board propellants, explosive charge

EVENT DATA

DATE:	6 Sep 1975	LOCATION:	32N, 293E (asc)
TIME:	1906 GMT	ASSESSED CAUSE:	Deliberate
ALTITUDE:	185 km		

PRE-EVENT ELEMENTS

EPOCH:	75249.72782895	MEAN ANOMALY:	294.2107
RIGHT ASCENSION:	189.2795	MEAN MOTION:	16.09422927
INCLINATION:	67.1445	MEAN MOTION DOT/2:	.00430774
ECCENTRICITY:	.0113994	MEAN MOTION DOT DOT/6:	.0
ARG. OF PERIGEE:	67.1020	BSTAR:	.0

DEBRIS CLOUD DATA

MAXIMUM ΔP:	Unknown
MAXIMUM ΔI:	Unknown

COMMENTS

Spacecraft was apparently destroyed after a malfunction prevented controlled reentry and landing in the Soviet Union. Most debris reentered before being officially cataloged. All but three official fragments were cataloged without elements.

REFERENCE DOCUMENT

History of Soviet/Russian Satellite Fragmentations-A Joint U.S.-Russian Investigation, N. L. Johnson et al, Kaman Sciences Corporation, October 1995.

Insufficient data to construct a Gabbard diagram.

SATELLITE DATA

TYPE:	Payload
OWNER:	CIS
LAUNCH DATE:	29.46 Oct 1975
DRY MASS (KG):	3000
MAIN BODY:	Cylinder; 1.3 m diameter by 17 m length
MAJOR APPENDAGES:	Solar panels
ATTITUDE CONTROL:	Active, 3-axis
ENERGY SOURCES:	On-board propellants, explosive charge (?)

EVENT DATA

DATE:	25 Jan 1976	LOCATION:	53N, 7E (asc)
TIME:	1400 GMT	ASSESSED CAUSE:	Unknown
ALTITUDE:	440 km		

PRE-EVENT ELEMENTS

EPOCH:	76025.37753295	MEAN ANOMALY:	88.9272
RIGHT ASCENSION:	303.6319	MEAN MOTION:	15.43461781
INCLINATION:	65.0177	MEAN MOTION DOT/2:	.00000373
ECCENTRICITY:	.0009065	MEAN MOTION DOT DOT/6:	.0
ARG. OF PERIGEE:	271.0782	BSTAR:	.0

DEBRIS CLOUD DATA

MAXIMUM ΔP:	1.6 min
MAXIMUM ΔI:	0.4 deg

COMMENTS

Cosmos 777 was the second spacecraft of the Cosmos 699-type to experience a fragmentation. It is the only one to breakup before terminating its precise orbit maintenance pattern and entering a regime of natural decay. A second event may have occurred about 90 minutes after the event cited above.

REFERENCE DOCUMENTS

"Artificial Satellite Break-Ups (Part 1): Soviet Ocean Surveillance Satellites", N. L. Johnson, Journal of the British Interplanetary Society, February 1983, pp. 51-58.

History of Soviet/Russian Satellite Fragmentations-A Joint U.S.-Russian Investigation, N. L. Johnson et al, Kaman Sciences Corporation, October 1995.

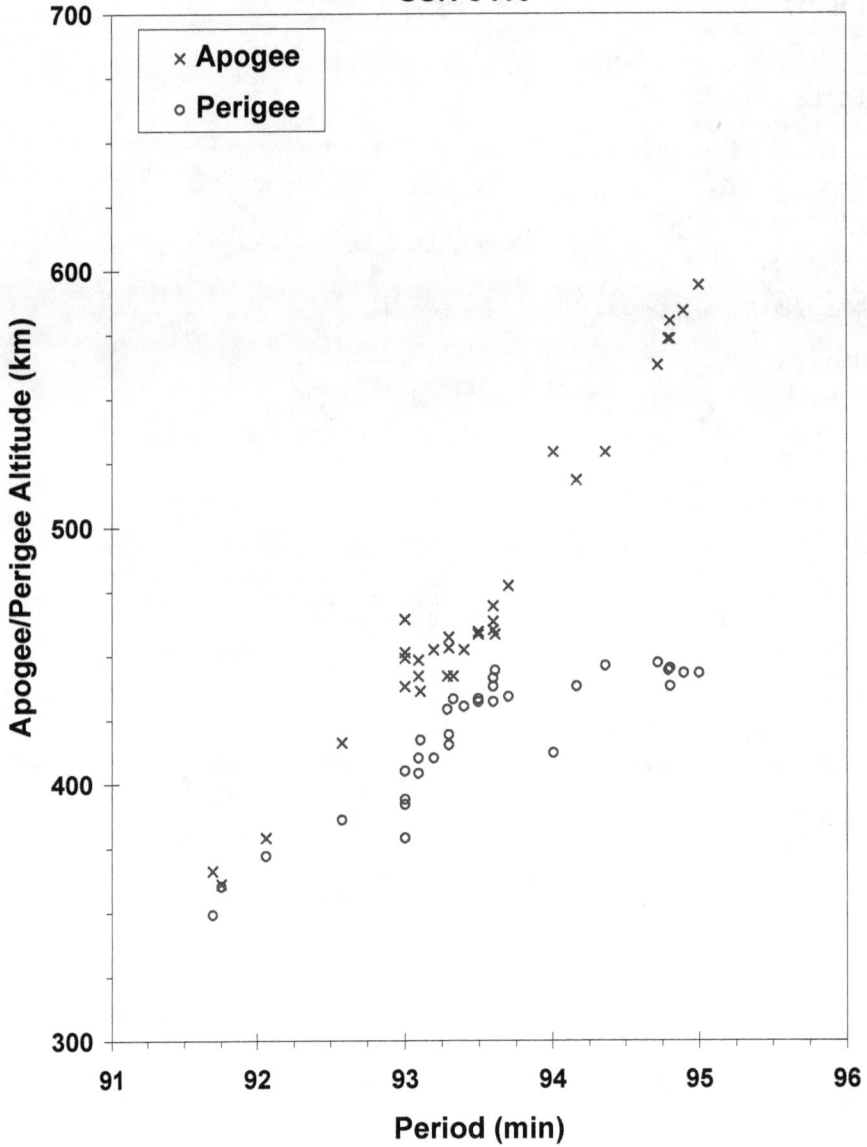

SSN 8416

Cosmos 777 debris cloud of 35 fragments about 10 days after the event as reconstructed from the US SSN database. Some drag effects are already evident.

COSMOS 838   **1976-063A**   **8932**

SATELLITE DATA

TYPE:	Payload
OWNER:	CIS
LAUNCH DATE:	2.44 Jul 1976
DRY MASS (KG):	3000
MAIN BODY:	Cylinder; 1.3 m diameter by 17 m length
MAJOR APPENDAGES:	Solar panels
ATTITUDE CONTROL:	Active, 3-axis
ENERGY SOURCES:	On-board propellants, explosive charge (?)

EVENT DATA

DATE:	17 May 1977	LOCATION:	9S, 284E (dsc)
TIME:	1018 GMT	ASSESSED CAUSE:	Unknown
ALTITUDE:	430 km		

PRE-EVENT ELEMENTS

EPOCH:	77136.94211102	MEAN ANOMALY:	73.5502
RIGHT ASCENSION:	131.3837	MEAN MOTION:	15.45822335
INCLINATION:	65.0556	MEAN MOTION DOT/2:	.00007521
ECCENTRICITY:	.0021270	MEAN MOTION DOT DOT/6:	.0
ARG. OF PERIGEE:	286.3253	BSTAR:	.0

DEBRIS CLOUD DATA

MAXIMUM ΔP:	8.0 min*
MAXIMUM ΔI:	1.1 deg*

*Based on uncataloged debris data

COMMENTS

Cosmos 838 was the third spacecraft of the Cosmos 699-type to experience a fragmentation. Spacecraft had been in a regime of natural decay for 6 months prior to the event. Many debris reentered before being officially cataloged.

REFERENCE DOCUMENTS

"Artificial Satellite Break-Ups (Part 1): Soviet Ocean Surveillance Satellites", N. L. Johnson, Journal of the British Interplanetary Society, February 1983, pp. 51-58.

History of Soviet/Russian Satellite Fragmentations-A Joint U.S.-Russian Investigation, N. L. Johnson et al, Kaman Sciences Corporation, October 1995.

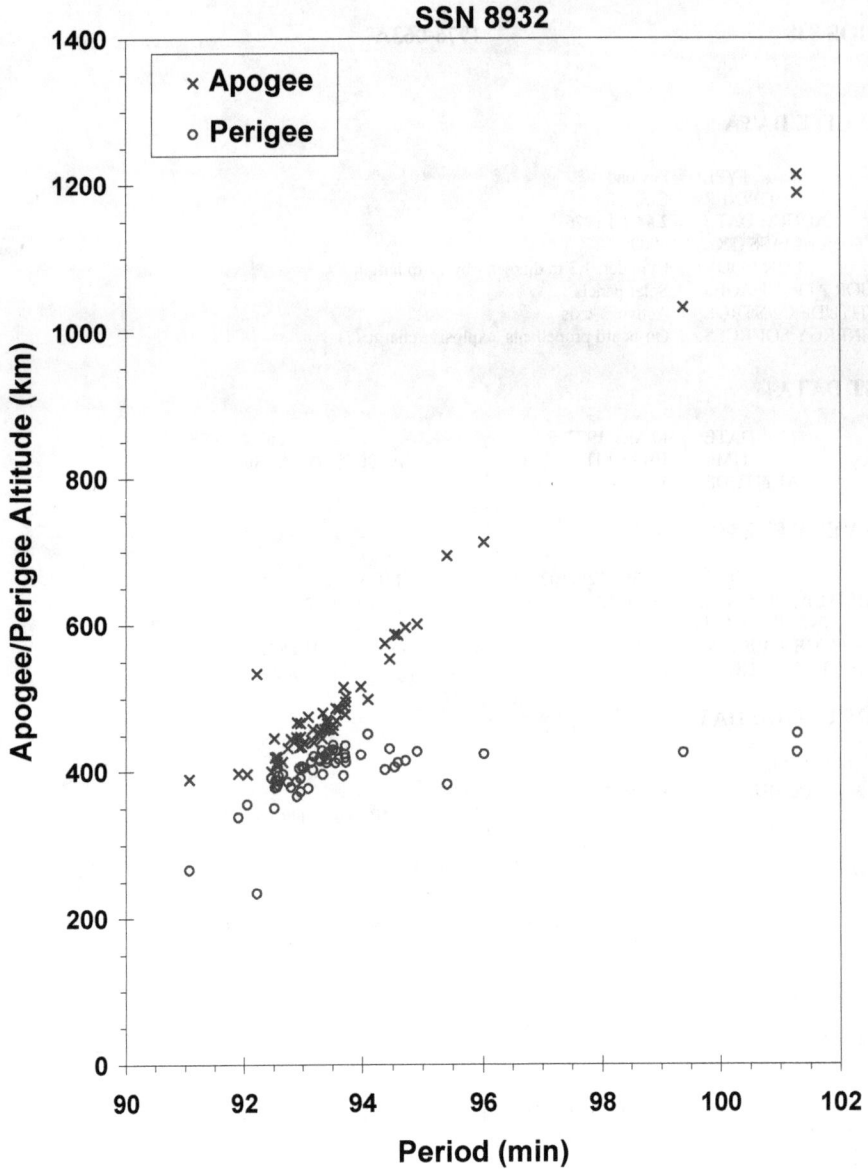

SSN 8932

Cosmos 838 debris cloud of 59 fragments about 1 week after the event as reconstructed from the US SSN database.

SATELLITE DATA

TYPE:	Payload
OWNER:	CIS
LAUNCH DATE:	8.88 Jul 1976
DRY MASS (KG):	650
MAIN BODY:	Polyhedron; 1.4 m by 1.4 m
MAJOR APPENDAGES:	Solar panels, gravity-gradient boom (?)
ATTITUDE CONTROL:	Gravity gradient (?)
ENERGY SOURCES:	Battery

EVENT DATA

DATE:	29 Sep 1977	LOCATION:	33S, 162E (dsc)
TIME:	0717 GMT	ASSESSED CAUSE:	Battery
ALTITUDE:	1910 km		

PRE-EVENT ELEMENTS

EPOCH:	77270.46732078	MEAN ANOMALY:	7.6996
RIGHT ASCENSION:	85.9347	MEAN MOTION:	12.32137908
INCLINATION:	65.8538	MEAN MOTION DOT/2:	.00000367
ECCENTRICITY:	.0706585	MEAN MOTION DOT DOT/6:	.0
ARG. OF PERIGEE:	351.1444	BSTAR:	.0

DEBRIS CLOUD DATA

MAXIMUM ΔP:	2.7 min
MAXIMUM ΔI:	0.3 deg

COMMENTS

Cosmos 839 was the first of three satellites of the same class to experience unexplained fragmentations. These satellites are used in conjunction with the Cosmos 249-type spacecraft, which are deliberately fragmented; but the cause of the Cosmos 839-type events appears to be unrelated since they occur more than 1 year after tests with Cosmos 249-type spacecraft. In the case of Cosmos 839, 14 months elapsed between its test with a Cosmos 249-type spacecraft and its fragmentation. Russian officials have determined that battery malfunctions were the causes of these events.

REFERENCE DOCUMENTS

"Artificial Satellite Break-Ups (Part 2): Soviet Anti-Satellite Program", N.L. Johnson, Journal of the British Interplanetary Society, August 1983, pp. 357-362.

History of Soviet/Russian Satellite Fragmentations-A Joint U.S.-Russian Investigation, N. L. Johnson et al, Kaman Sciences Corporation, October 1995.

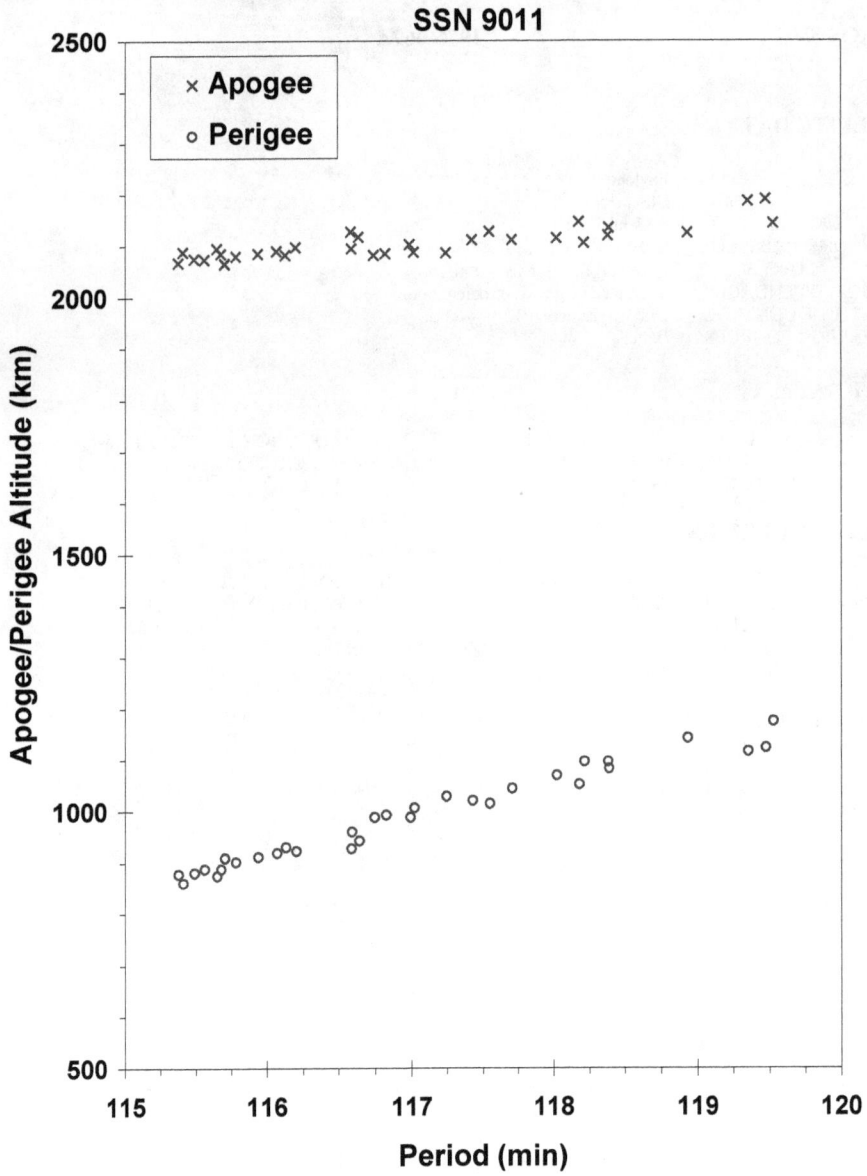

SSN 9011

Cosmos 839 debris cloud of 33 fragments about 5 weeks after the event as reconstructed
from the US SSN database.

SATELLITE DATA

TYPE:	Payload
OWNER:	CIS
LAUNCH DATE:	22.66 Jul 1976
DRY MASS (KG):	5700
MAIN BODY:	Cone-cylinder; 2.7 m diameter by 6.3 m length
MAJOR APPENDAGES:	Solar panels
ATTITUDE CONTROL:	Active, 3-axis
ENERGY SOURCES:	On-board propellants, explosive charge (?)

EVENT DATA

DATE:	25 Jul 1976	LOCATION:	49N, 100E (dsc)
TIME:	1718 GMT	ASSESSED CAUSE:	Deliberate
ALTITUDE:	210 km		

PRE-EVENT ELEMENTS

EPOCH:	76207.45032150	MEAN ANOMALY:	291.2246
RIGHT ASCENSION:	152.6930	MEAN MOTION:	16.04433196
INCLINATION:	67.1467	MEAN MOTION DOT/2:	.00313532
ECCENTRICITY:	.0136374	MEAN MOTION DOT DOT/6:	.0
ARG. OF PERIGEE:	70.3553	BSTAR:	.0

DEBRIS CLOUD DATA

MAXIMUM ΔP:	Unknown
MAXIMUM ΔI:	Unknown

COMMENTS

Spacecraft was apparently destroyed after a malfunction prevented controlled reentry and landing in the Soviet Union. No elements were cataloged on any of the official debris. Most fragments reentered rapidly.

REFERENCE DOCUMENT

History of Soviet/Russian Satellite Fragmentations-A Joint U.S.-Russian Investigation, N. L. Johnson et al, Kaman Sciences Corporation, October 1995.

Insufficient data to construct a Gabbard diagram.

SATELLITE DATA

TYPE:	Delta Second Stage (2310)
OWNER:	US
LAUNCH DATE:	29.71 Jul 1976
DRY MASS (KG):	840
MAIN BODY:	Cylinder-nozzle; 1.4 m diameter by 5.8 m length
MAJOR APPENDAGES:	Mini-skirt; 2.4 m by 0.3 m
ATTITUDE CONTROL:	None at time of the event.
ENERGY SOURCES:	On-board propellants, range safety device

EVENT DATA

DATE:	24 Dec 1977	LOCATION:	40S, 146E (asc)
TIME:	1133 GMT	ASSESSED CAUSE:	Propulsion
ALTITUDE:	1510 km		

PRE-EVENT ELEMENTS

EPOCH:	77354.53228225	MEAN ANOMALY:	330.8663
RIGHT ASCENSION:	38.5560	MEAN MOTION:	12.38394892
INCLINATION:	102.0192	MEAN MOTION DOT/2:	.0
ECCENTRICITY:	.0010085	MEAN MOTION DOT DOT/6:	.0
ARG. OF PERIGEE:	29.2920	BSTAR:	.0

DEBRIS CLOUD DATA

MAXIMUM ΔP:	12.5 min
MAXIMUM ΔI:	3.0 deg

COMMENTS

This was the sixth Delta Second Stage to experience a severe fragmentation. The event occurred 17 months after the successful deployment of the NOAA 5 payload. Cause of the explosion is assessed to be related to the estimated 250 kg of residual propellants on board and characteristics of the sun-synchronous orbit.

REFERENCE DOCUMENTS

Explosion of Satellite 10704 and other Delta Second Stage Rockets, J.R. Gabbard, Technical Memorandum 81-5, DCS Plans, Hdqtrs NORAD/ADCOM, Colorado Springs, May 1981.

Investigation of Delta Second Stage On-Orbit Explosions, C.S. Gumpel, Report MDC-H0047, McDonnell Douglas Astronautics Company - West, Huntington Beach, April 1982.

A Later Look at Delta Second Stage On-Orbit Explosions, J.R. Gabbard, Technical Report CS85-BMDSC-00-24, Teledyne Brown Engineering, Colorado Springs, March 1985.

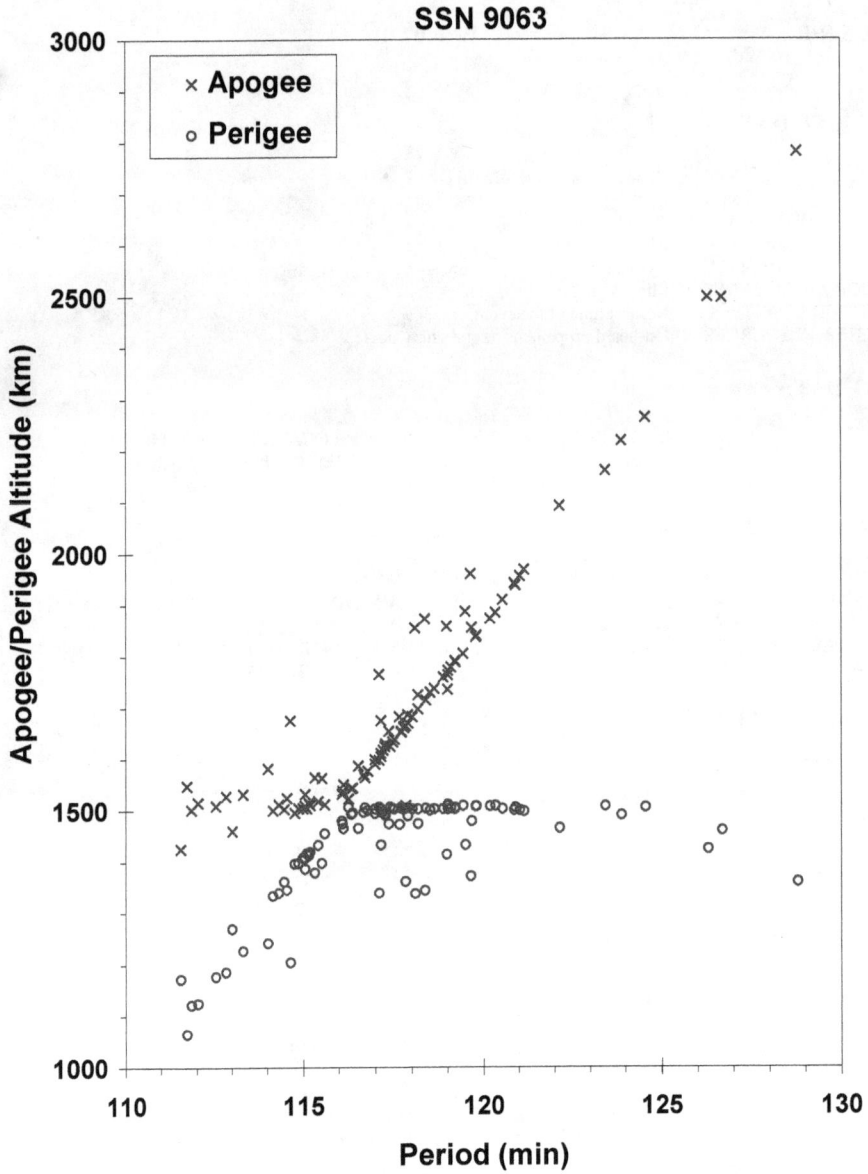

SSN 9063

NOAA 5 R/B debris cloud of 98 fragments about 4 months after the event as reconstructed from the US SSN database.

SATELLITE DATA

TYPE:	Payload
OWNER:	CIS
LAUNCH DATE:	22.38 Oct 1976
DRY MASS (KG):	1250
MAIN BODY:	Cylinder; 1.7 m diameter by 2 m length
MAJOR APPENDAGES:	Solar panels
ATTITUDE CONTROL:	Active, 3-axis
ENERGY SOURCES:	On-board propellants, explosive device

EVENT DATA

DATE:	15 Mar 1977	LOCATION:	39N, 114E (asc)
TIME:	1256 GMT	ASSESSED CAUSE:	Deliberate
ALTITUDE:	5375 km		

PRE-EVENT ELEMENTS

EPOCH:	77066.03986408	MEAN ANOMALY:	4.4196
RIGHT ASCENSION:	98.8078	MEAN MOTION:	2.00311741
INCLINATION:	63.1553	MEAN MOTION DOT/2:	.0
ECCENTRICITY:	.7312859	MEAN MOTION DOT DOT/6:	.0
ARG. OF PERIGEE:	318.6653	BSTAR:	.0

DEBRIS CLOUD DATA

MAXIMUM ΔP:	5.7 min
MAXIMUM ΔI:	0.4 deg

COMMENTS

Cosmos 862 was the first of a new class of operational satellites in highly elliptical, semi-synchronous orbits that experienced a total of 16 fragmentations during the period 1977-1986. Due to the nature of these orbits, which result in high altitudes over the Northern Hemisphere where most surveillance sensors are located, debris detection and tracking is extremely difficult. Only the largest fragments can be seen. Cosmos 862 spacecraft were equipped with self-destruct packages in the event that spacecraft control was lost; this was the cause of breakups until the explosives were removed after Cosmos 1481.

REFERENCE DOCUMENT

History of Soviet/Russian Satellite Fragmentations-A Joint U.S.-Russian Investigation, N. L. Johnson et al, Kaman Sciences Corporation, October 1995.

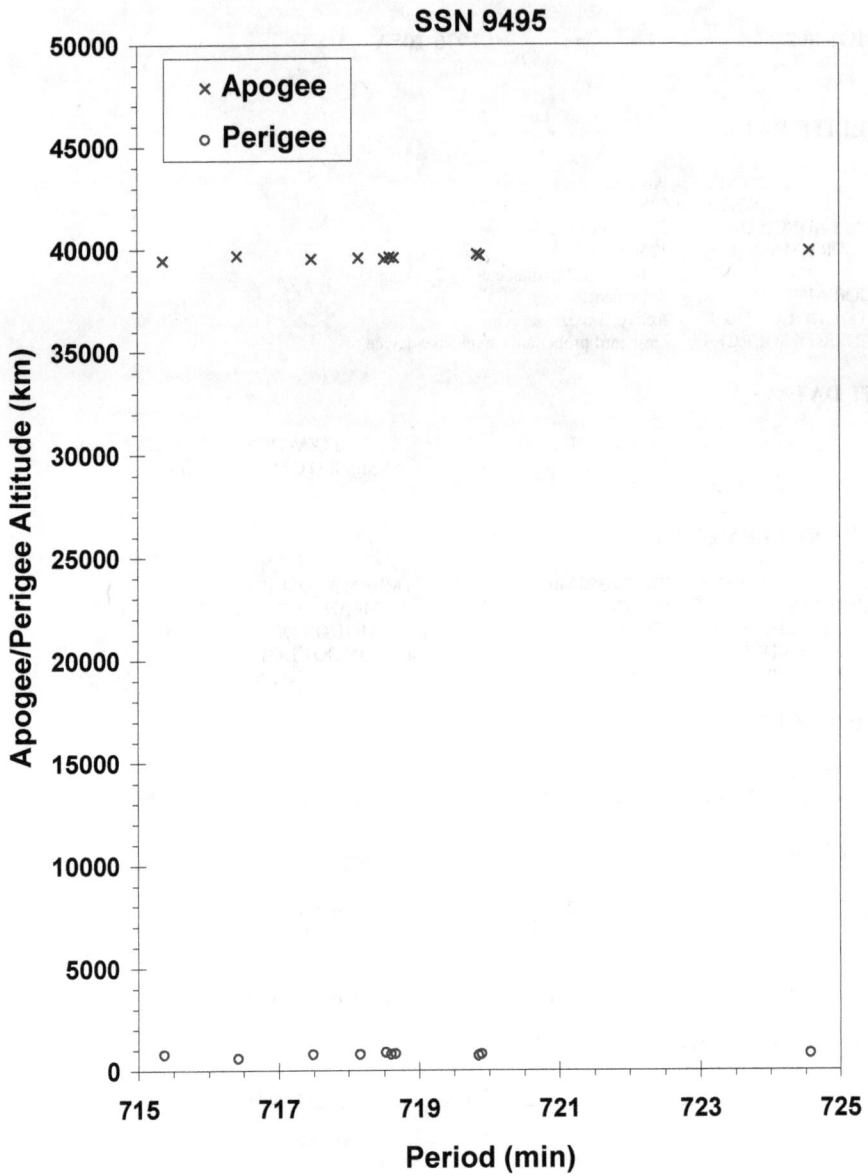

SSN 9495

Cosmos 862 debris cloud of 10 cataloged fragments 2 weeks after the event as reconstructed from the US SSN database.

SATELLITE DATA

TYPE:	Payload
OWNER:	CIS
LAUNCH DATE:	9.84 Dec 1976
DRY MASS (KG):	650
MAIN BODY:	Polyhedron; 1.4 m by 1.4 m
MAJOR APPENDAGES:	Solar panels, gravity-gradient boom (?)
ATTITUDE CONTROL:	Gravity gradient (?)
ENERGY SOURCES:	Battery

EVENT DATA

DATE:	27 Nov 1978	LOCATION:	65S, 306E (dsc)
TIME:	1703 GMT	ASSESSED CAUSE:	Battery
ALTITUDE:	560 km		

PRE-EVENT ELEMENTS

EPOCH:	78331.59395829	MEAN ANOMALY:	55.5772
RIGHT ASCENSION:	11.0317	MEAN MOTION:	14.93841919
INCLINATION:	65.8440	MEAN MOTION DOT/2:	.00000004
ECCENTRICITY:	.0050108	MEAN MOTION DOT DOT/6:	.0
ARG. OF PERIGEE:	304.0553	BSTAR:	.0

DEBRIS CLOUD DATA

MAXIMUM ΔP:	1.3 min*
MAXIMUM ΔI:	0.0 deg*

*Based on uncataloged debris data

COMMENTS

Cosmos 880 was the second spacecraft of the Cosmos 839-type to experience a fragmentation. Although these satellites are used in conjunction with the Cosmos 249-type spacecraft that are deliberately fragmented, the cause of the Cosmos 839-type events appears to be unrelated. In the case of Cosmos 880, 23 months elapsed since its test with a Cosmos 249-type spacecraft. Russian officials have determined that battery malfunctions were the cause of these events.

REFERENCE DOCUMENTS

"Artificial Satellite Break-Ups (Part 2): Soviet Anti-Satellite Program", N.L. Johnson, Journal of the British Interplanetary Society, August 1983, pp. 357-362.

History of Soviet/Russian Satellite Fragmentations-A Joint U.S.-Russian Investigation, N. L. Johnson et al, Kaman Sciences Corporation, October 1995.

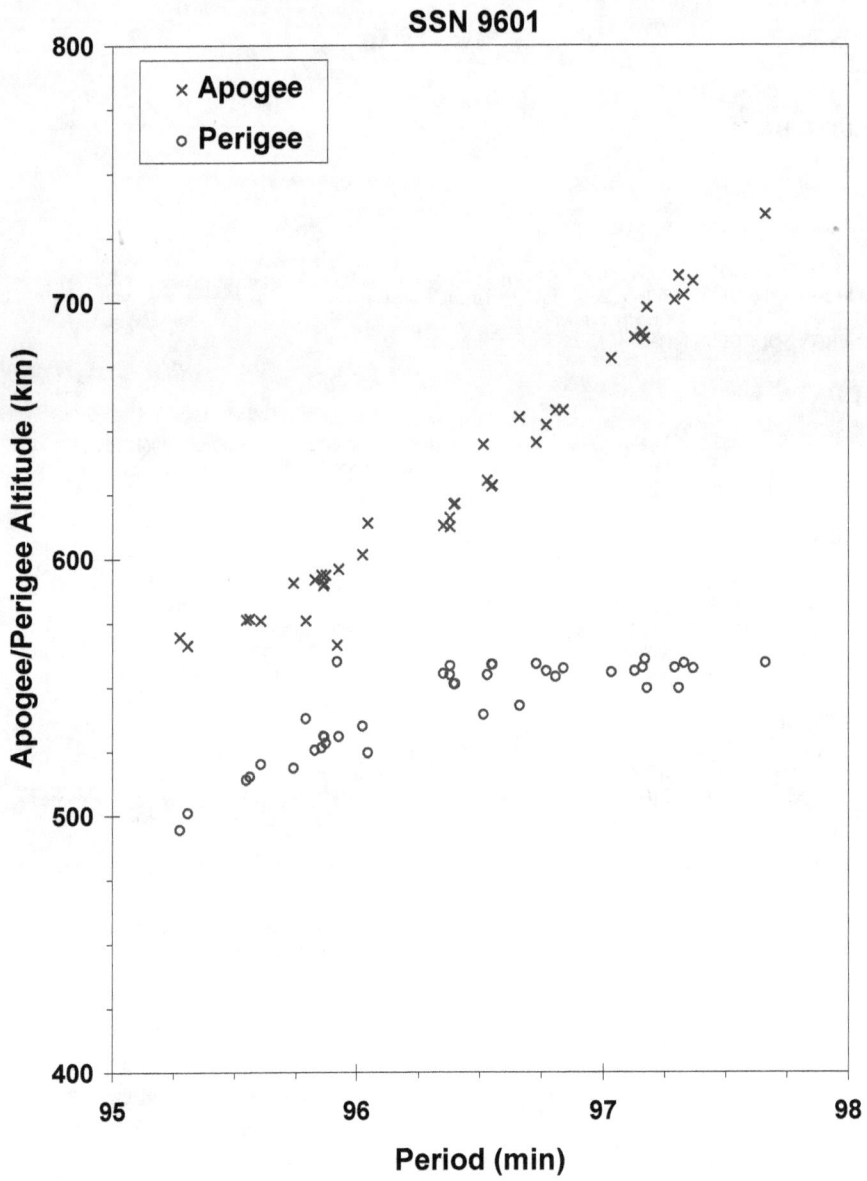

SSN 9601

Cosmos 880 debris cloud of 40 fragments 2 days after the event as reconstructed from the US SSN database.

COSMOS 884   **1976-123A**   **9614**

SATELLITE DATA

TYPE:	Payload
OWNER:	USSR
LAUNCH DATE:	17.40 Dec 1976
DRY MASS (KG):	6300
MAIN BODY:	Sphere-cylinder; 2.4 m diameter by 6.5 m length
MAJOR APPENDAGES:	None
ATTITUDE CONTROL:	Active, 3-axis
ENERGY SOURCES:	On-board propellants, explosive charge

EVENT DATA

DATE:	29 Dec 1976	LOCATION:	Unknown
TIME:	Unknown	ASSESSED CAUSE:	Deliberate
ALTITUDE:	Unknown		

PRE-EVENT ELEMENTS

EPOCH:	76362.45360574	MEAN ANOMALY:	302.3648
RIGHT ASCENSION:	227.6719	MEAN MOTION:	16.11011505
INCLINATION:	65.0214	MEAN MOTION DOT/2:	0.00147448
ECCENTRICITY:	0.0113306	MEAN MOTION DOT DOT/6:	0
ARG. OF PERIGEE:	58.8529	BSTAR:	0

DEBRIS CLOUD DATA

MAXIMUM ΔP:	Unknown
MAXIMUM ΔI:	Unknown

COMMENTS

Spacecraft was destroyed after a malfunction prevented reentry and landing in the Soviet Union. Event identified by Russian officials during investigation cited below.

REFERENCE DOCUMENT

History of Soviet/Russian Satellite Fragmentations-A Joint U.S.-Russian Investigation, N. L. Johnson et al, Kaman Sciences Corporation, October 1995.

Insufficient data to construct a Gabbard diagram.

SATELLITE DATA

TYPE:	Payload
OWNER:	CIS
LAUNCH DATE:	27.53 Dec 1976
DRY MASS (KG):	1400
MAIN BODY:	Irregular; 1.8 m by 4.2 m
MAJOR APPENDAGES:	None
ATTITUDE CONTROL:	Active, 3-axis
ENERGY SOURCES:	On-board propellants, explosive charge

EVENT DATA

DATE:	27 Dec 1976	LOCATION:	65S, 210E (asc)
TIME:	1840 GMT	ASSESSED CAUSE:	Deliberate
ALTITUDE:	2090 km		

POST-EVENT ELEMENTS

EPOCH:	76362.79720829	MEAN ANOMALY:	313.0540
RIGHT ASCENSION:	306.5669	MEAN MOTION:	12.54457816
INCLINATION:	65.8434	MEAN MOTION DOT/2:	.00004000
ECCENTRICITY:	.1087102	MEAN MOTION DOT DOT/6:	.0
ARG. OF PERIGEE:	57.0236	BSTAR:	.0

DEBRIS CLOUD DATA

MAXIMUM ΔP:	4.3 min
MAXIMUM ΔI:	0.2 deg

COMMENTS

Cosmos 886 was launched on a two-revolution rendezvous with Cosmos 880. After a close approach, Cosmos 886 continued on before its warhead was intentionally fired. Cosmos 886 was part of test series begun with Cosmos 249. The elements above are the first available after the final maneuver of Cosmos 886 but represent the revolution immediately after the event.

REFERENCE DOCUMENTS

"Artificial Satellite Break-Ups (Part 2): Soviet Anti-Satellite Program", N.L. Johnson, Journal of the British Interplanetary Society, August 1983, pp. 357-362.

History of Soviet/Russian Satellite Fragmentations-A Joint U.S.-Russian Investigation, N. L. Johnson et al, Kaman Sciences Corporation, October 1995.

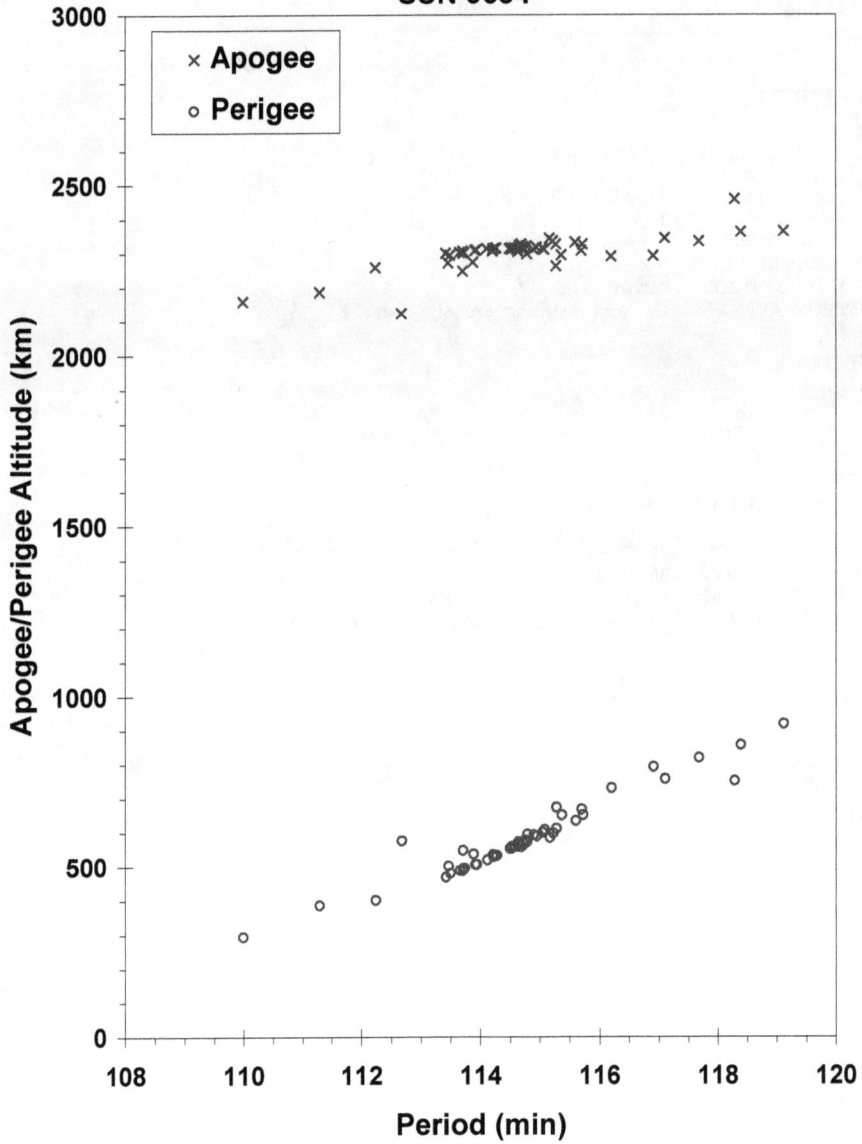

SSN 9634

Cosmos 886 debris cloud of 53 fragments 5 months after the event as reconstructed from the US SSN database.

SATELLITE DATA

TYPE:	Payload
OWNER:	CIS
LAUNCH DATE:	11.07 Apr 1977
DRY MASS (KG):	1250
MAIN BODY:	Cylinder; 1.7 m diameter by 2 m length
MAJOR APPENDAGES:	Solar panels
ATTITUDE CONTROL:	Active, 3-axis
ENERGY SOURCES:	On-board propellants, explosive charge

EVENT DATA

DATE:	8 Jun 1978	LOCATION:	Unknown
TIME:	Unknown	ASSESSED CAUSE:	Deliberate
ALTITUDE:	Unknown		

PRE-EVENT ELEMENTS

EPOCH:	78156.86414074	MEAN ANOMALY:	5.0496
RIGHT ASCENSION:	115.5660	MEAN MOTION:	2.00599850
INCLINATION:	63.1514	MEAN MOTION DOT/2:	.0
ECCENTRICITY:	.7100107	MEAN MOTION DOT DOT/6:	.0
ARG. OF PERIGEE:	319.7397	BSTAR:	.0

DEBRIS CLOUD DATA

MAXIMUM ΔP:	2.6 min*
MAXIMUM ΔI:	0.5 deg*

*See Comments

COMMENTS

Cosmos 903 was another spacecraft of the Cosmos 862-type to experience a fragmentation. One new fragment was cataloged within a week of the event. The ΔP and ΔI values above are based on the lower period (717.5 min) orbit of Cosmos 903 after the event.

REFERENCE DOCUMENT

History of Soviet/Russian Satellite Fragmentations-A Joint U.S.-Russian Investigation, N. L. Johnson et al, Kaman Sciences Corporation, October 1995.

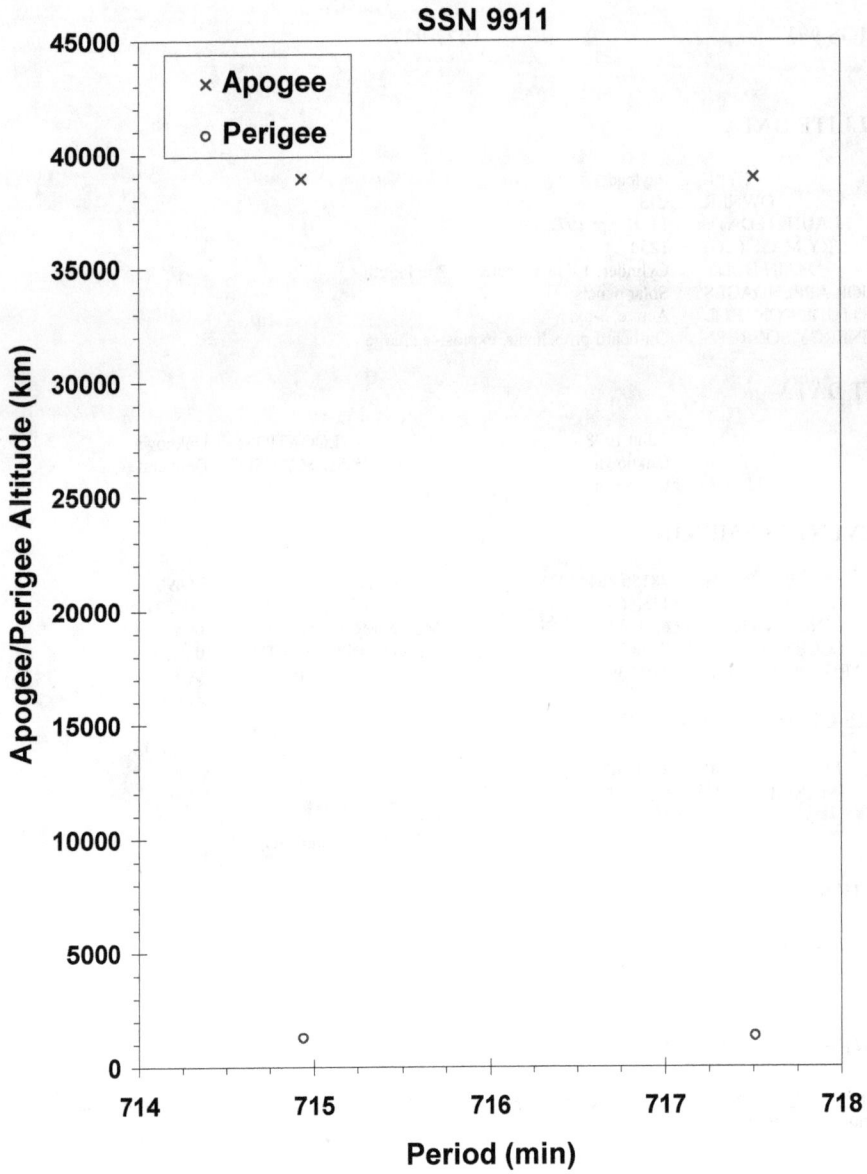

SSN 9911

Apogee/Perigee Altitude (km) vs Period (min)

x Apogee
o Perigee

Cosmos 903 and a single piece of debris 3 weeks after the event as reconstructed from the US SSN database.

SATELLITE DATA

TYPE:	Payload
OWNER:	CIS
LAUNCH DATE:	16.08 Jun 1977
DRY MASS (KG):	1250
MAIN BODY:	Irregular; 1.7 m by 2 m
MAJOR APPENDAGES:	Solar panels
ATTITUDE CONTROL:	Active, 3-axis
ENERGY SOURCES:	On-board propellants, explosive charge

EVENT DATA

DATE:	30 March 1979	LOCATION:	63S, 0E (dsc)
TIME:	1545 GMT	ASSESSED CAUSE:	Deliberate
ALTITUDE:	3280 km		

PRE-EVENT ELEMENTS

EPOCH:	79089.17562851	MEAN ANOMALY:	5.2297
RIGHT ASCENSION:	156.1576	MEAN MOTION:	2.00553521
INCLINATION:	62.9498	MEAN MOTION DOT/2:	.0
ECCENTRICITY:	.6980052	MEAN MOTION DOT DOT/6:	.0
ARG. OF PERIGEE:	322.3289	BSTAR:	.0

DEBRIS CLOUD DATA

MAXIMUM ΔP:	22.6 min*
MAXIMUM ΔI:	0.6 deg*

*Based on uncataloged debris data

COMMENTS

Cosmos 917 was another spacecraft of the Cosmos 862-type to experience a fragmentation.

REFERENCE DOCUMENT

History of Soviet/Russian Satellite Fragmentations-A Joint U.S.-Russian Investigation, N. L. Johnson et al, Kaman Sciences Corporation, October 1995.

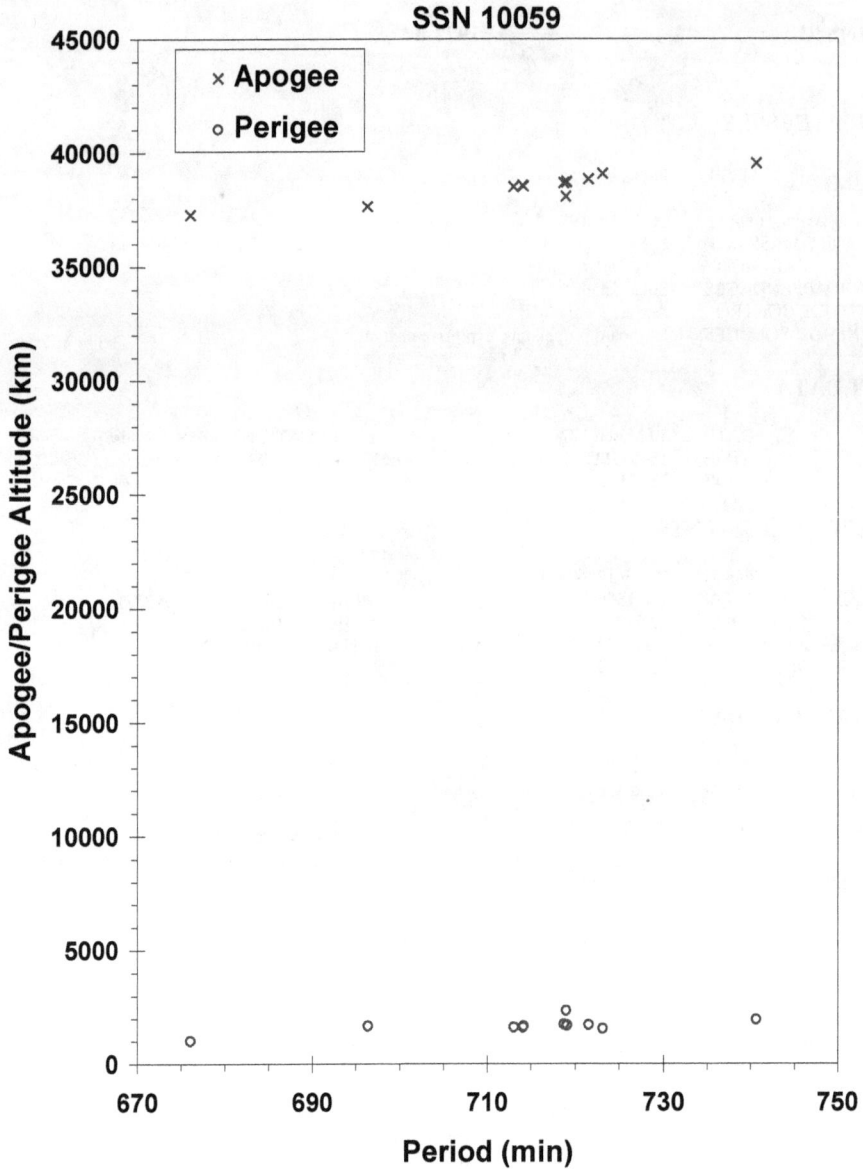

Cosmos 917 debris cloud of 12 fragments about 3 weeks after the event as reconstructed from the US SSN database.

SATELLITE DATA

TYPE:	Delta Second Stage (2914)
OWNER:	US
LAUNCH DATE:	14.44 Jul 1977
DRY MASS (KG):	900
MAIN BODY:	Cylinder-nozzle; 1.4 m diameter by 5.8 m length
MAJOR APPENDAGES:	Mini-skirt; 2.4 m by 0.3 m
ATTITUDE CONTROL:	None at time of the event.
ENERGY SOURCES:	On-board propellants, range safety device

EVENT DATA

DATE:	14 Jul 1977	LOCATION:	14N, 249E (dsc)
TIME:	1612 GMT	ASSESSED CAUSE:	Propulsion
ALTITUDE:	1450 km		

POST-EVENT ELEMENTS

EPOCH:	77197.57445278	MEAN ANOMALY:	303.2693
RIGHT ASCENSION:	262.0317	MEAN MOTION:	12.95114397
INCLINATION:	29.0493	MEAN MOTION DOT/2:	.00007335
ECCENTRICITY:	.0973469	MEAN MOTION DOT DOT/6:	.0
ARG. OF PERIGEE:	66.7255	BSTAR:	.0

DEBRIS CLOUD DATA

MAXIMUM ΔP:	9.7 min*
MAXIMUM ΔI:	3.0 deg*

*Based on uncataloged debris data

COMMENTS

This was the fifth Delta Second Stage to experience a severe fragmentation. It is also the only one that was not in a sun-synchronous orbit, which had performed a depletion burn, and which fragmented on the day of launch. This rocket body did perform its mission successfully, carrying the third stage and the payload into a low Earth orbit. The energy for the breakup is assessed to have been the 40 kg of propellants (mainly oxidizer) remaining after the depletion burn. The elements above are the first available after the depletion burn although also after the event.

REFERENCE DOCUMENTS

Explosion of Satellite 10704 and other Delta Second Stage Rockets, J.R. Gabbard, Technical Memorandum 81-5, DCS Plans, Hdqtrs NORAD/ADCOM, Colorado Springs, May 1981.

Investigation of Delta Second Stage On-Orbit Explosions, C.S. Gumpel, Report MDC-H0047, McDonnell Douglas Astronautics Company - West, Huntington Beach, April 1982.

A Later Look at Delta Second Stage On-Orbit Explosions, J.R. Gabbard, Technical Report CS85-BMDSC-00-24, Teledyne Brown Engineering, Colorado Springs, March 1985.

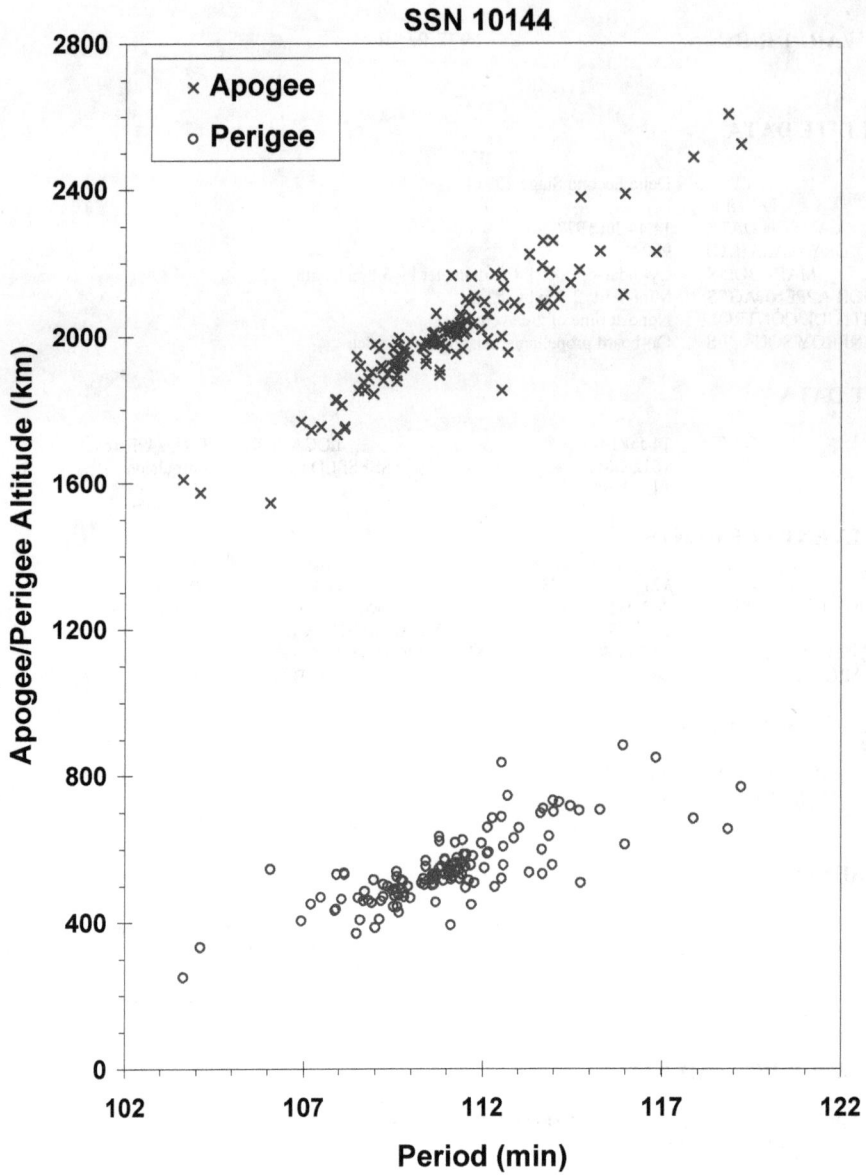

Himawari 1 R/B debris cloud of 132 fragments 5 months after the event as reconstructed from the US SSN database.

SATELLITE DATA

TYPE:	Payload
OWNER:	CIS
LAUNCH DATE:	20.20 Jul 1977
DRY MASS (KG):	1250
MAIN BODY:	Cylinder; 1.7 m diameter by 2 m length
MAJOR APPENDAGES:	Solar panels
ATTITUDE CONTROL:	Active, 3-axis
ENERGY SOURCES:	On-board propellants, explosive charge

EVENT DATA

DATE:	24 Oct 1977	LOCATION:	Unknown
TIME:	Unknown	ASSESSED CAUSE:	Deliberate
ALTITUDE:	Unknown		

PRE-EVENT ELEMENTS

EPOCH:	77289.02131186	MEAN ANOMALY:	4.2624
RIGHT ASCENSION:	305.6648	MEAN MOTION:	2.00651833
INCLINATION:	62.9440	MEAN MOTION DOT/2:	.0
ECCENTRICITY:	.7341055	MEAN MOTION DOT DOT/6:	.0
ARG. OF PERIGEE:	318.8771	BSTAR:	.0

DEBRIS CLOUD DATA

MAXIMUM ΔP:	5.3 min*
MAXIMUM ΔI:	0.7 deg*

*Based on uncataloged debris data

COMMENTS

Cosmos 931 was another spacecraft of the Cosmos 862-type to experience a fragmentation. Debris were not officially cataloged until 4 years after the event.

REFERENCE DOCUMENT

History of Soviet/Russian Satellite Fragmentations-A Joint U.S.-Russian Investigation, N. L. Johnson et al, Kaman Sciences Corporation, October 1995.

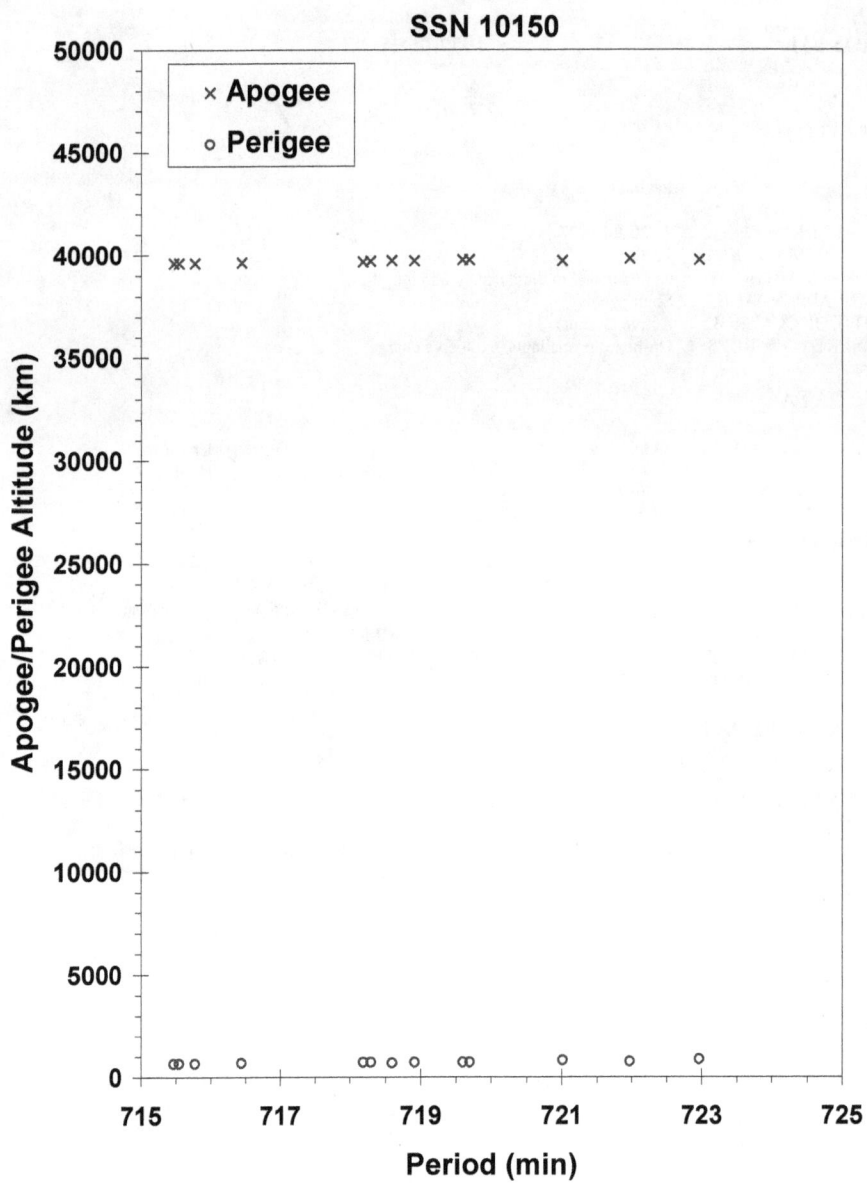

SSN 10150

Cosmos 931 debris cloud of 13 fragments 2 weeks after the event as reconstructed from the US SSN database.

SATELLITE DATA

TYPE:	Payload
OWNER:	CIS
LAUNCH DATE:	20.73 Sept 1977
DRY MASS (KG):	1750
MAIN BODY:	Cone; 2 m by 4 m
MAJOR APPENDAGES:	Plate + 2 solar panels
ATTITUDE CONTROL:	Active, 3-axis
ENERGY SOURCES:	On-board propellants, Battery

EVENT DATA

DATE:	23 Jun 1978	LOCATION:	0.0N, 98.7E
TIME:	Unknown	ASSESSED CAUSE:	Battery
ALTITUDE:	35790 km		

PRE-EVENT ELEMENTS

EPOCH:	88166.03647595	MEAN ANOMALY:	78.3897
RIGHT ASCENSION:	78.3897	MEAN MOTION:	1.00252588
INCLINATION:	0.1137	MEAN MOTION DOT/2:	.0
ECCENTRICITY	.0001436	MEAN MOTION DOT DOT/6:	.0
ARG. OF PERIGEE:	325.2771	BSTAR:	.0

DEBRIS CLOUD DATA

MAXIMUM ΔP:	Unknown
MAXIMUM ΔI:	Unknown

COMMENTS

This event was revealed by Russian officials in an orbital debris meeting in February 1992 in Moscow. This is the first known geostationary orbit fragmentation and was not detected by the Space Surveillance Network (SSN). Russian photographs originally linked to the breakup were later determined to have been misidentified.

REFERENCE DOCUMENTS

History of Soviet/Russian Satellite Fragmentations-A Joint U.S.-Russian Investigation, N. L. Johnson et al, Kaman Sciences Corporation, October 1995.

"Breakup in Review-Two GEO Breakups", Orbital Debris Monitor, April 1992, p 35-36.

Insufficient data to construct a Gabbard diagram.

SATELLITE DATA

TYPE:	Payload
OWNER:	CIS
LAUNCH DATE:	21.44 Dec 1977
DRY MASS (KG):	1400
MAIN BODY:	Cylinder; 2 m diameter by 4 m length
MAJOR APPENDAGES:	None
ATTITUDE CONTROL:	Active, 3-axis
ENERGY SOURCES:	On-board propellants, explosive charge

EVENT DATA

DATE:	21 Dec 1977	LOCATION:	38S, 274E (asc)
TIME:	1710 GMT	ASSESSED CAUSE:	Deliberate
ALTITUDE:	1135 km		

PRE-EVENT ELEMENTS

EPOCH:	77355.65049149	MEAN ANOMALY:	245.5638
RIGHT ASCENSION:	282.1792	MEAN MOTION:	13.58084598
INCLINATION:	65.8467	MEAN MOTION DOT/2:	.00023007
ECCENTRICITY:	.0129854	MEAN MOTION DOT DOT/6:	.0
ARG. OF PERIGEE:	116.3098	BSTAR:	.0

DEBRIS CLOUD DATA

MAXIMUM ΔP:	4.7 min
MAXIMUM ΔI:	1.1 deg

COMMENTS

Cosmos 970 was launched on a two-revolution rendezvous with Cosmos 967. After a close approach, Cosmos 970 continued on before its warhead was intentionally fired. Cosmos 970 was part of test series begun with Cosmos 249.

REFERENCE DOCUMENTS

"Artificial Satellite Break-Ups (Part 2): Soviet Anti-Satellite Program", N. L. Johnson, Journal of the British Interplanetary Society, August 1983, p. 357-362.

History of Soviet/Russian Satellite Fragmentations-A Joint U.S.-Russian Investigation, N. L. Johnson et al, Kaman Sciences Corporation, October 1995.

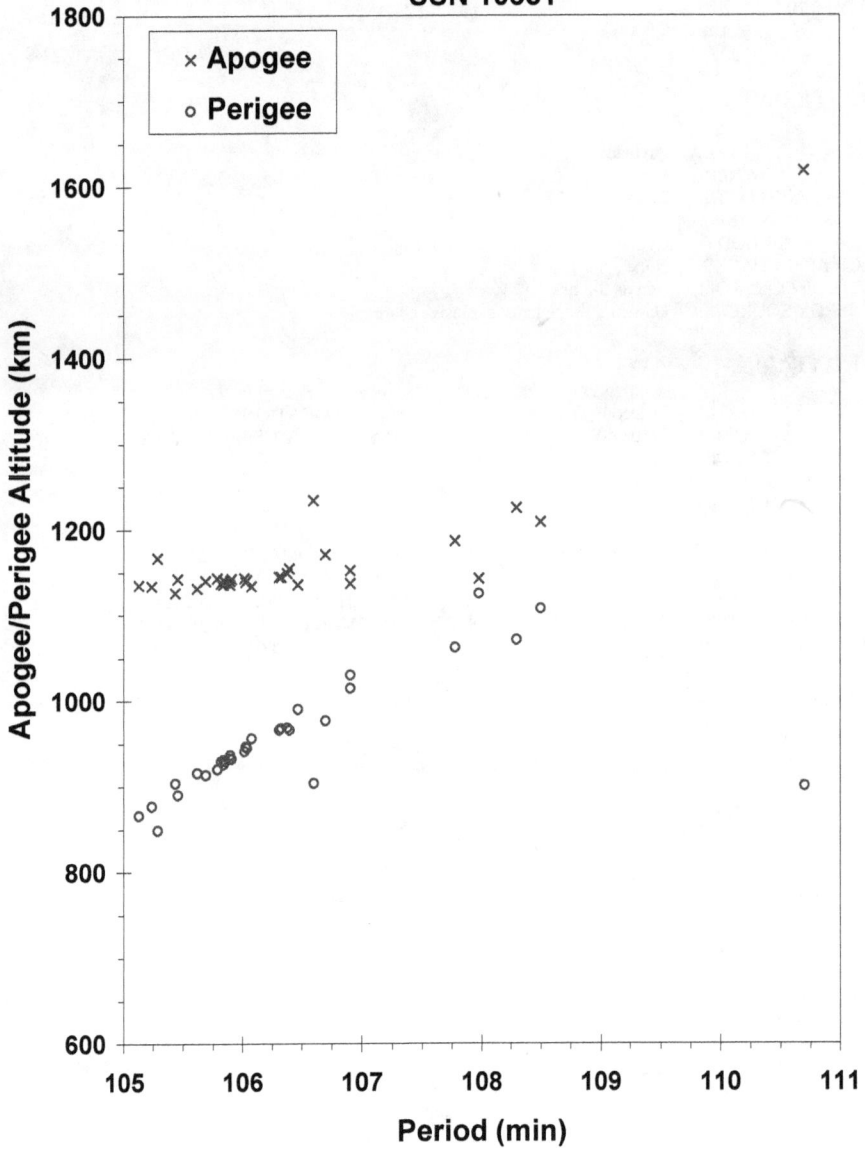

Cosmos 970 debris cloud of 34 fragments about 5 months after the event as reconstructed from the US SSN database.

LANDSAT 3 R/B   **1978-026C**   **10704**

SATELLITE DATA

TYPE:	Delta Second Stage (2910)
OWNER:	US
LAUNCH DATE:	5.75 Mar 1978
DRY MASS (KG):	900
MAIN BODY:	Cylinder-nozzle; 1.4 m diameter by 5.8 m length
MAJOR APPENDAGES:	Mini-skirt; 2.4 m by 0.3 m
ATTITUDE CONTROL:	None at time of the event.
ENERGY SOURCES:	On-board propellants, range safety device

EVENT DATA

DATE:	27 Jan 1981	LOCATION:	80S, 301E (asc)
TIME:	0432 GMT	ASSESSED CAUSE:	Propulsion
ALTITUDE:	910 km		

PRE-EVENT ELEMENTS

EPOCH:	81026.99107090	MEAN ANOMALY:	147.0549
RIGHT ASCENSION:	68.7927	MEAN MOTION:	13.96108433
INCLINATION:	98.8485	MEAN MOTION DOT/2:	.00000434
ECCENTRICITY:	.0006255	MEAN MOTION DOT DOT/6:	.0
ARG. OF PERIGEE:	212.9842	BSTAR:	.00032708

DEBRIS CLOUD DATA

MAXIMUM ΔP:	9.1 min
MAXIMUM ΔI:	0.5 deg

COMMENTS

This was the seventh Delta Second Stage to experience a severe fragmentation. The event occurred nearly 35 months after the successful deployment of the Landsat 3 payload. Cause of the explosion is assessed to be related to the estimated 100 kg of residual propellants on board and characteristics of the sun-synchronous orbit.

REFERENCE DOCUMENTS

Explosion of Satellite 10704 and other Delta Second Stage Rockets, J.R. Gabbard, Technical Memorandum 81-5, DCS Plans, Hdqtrs NORAD/ADCOM, Colorado Springs, May 1981.

Analysis of PARCS Recorded Data on the Breakup of Satellite 10704 on 27 January 1981, S.F. Hoffman and P.P. Shinkunas, Technical Report MSB82-ADC-0138, Teledyne Brown Engineering, Huntsville, February 1982.

Investigation of Delta Second Stage On-Orbit Explosions, C.S. Gumpel, Report MDC-H0047, McDonnell Douglas Astronautics Company - West, Huntington Beach, April 1982.

A Later Look at Delta Second Stage On-Orbit Explosions, J.R. Gabbard, Technical Report CS85-BMDSC-00-24, Teledyne Brown Engineering, Colorado Springs, March 1985.

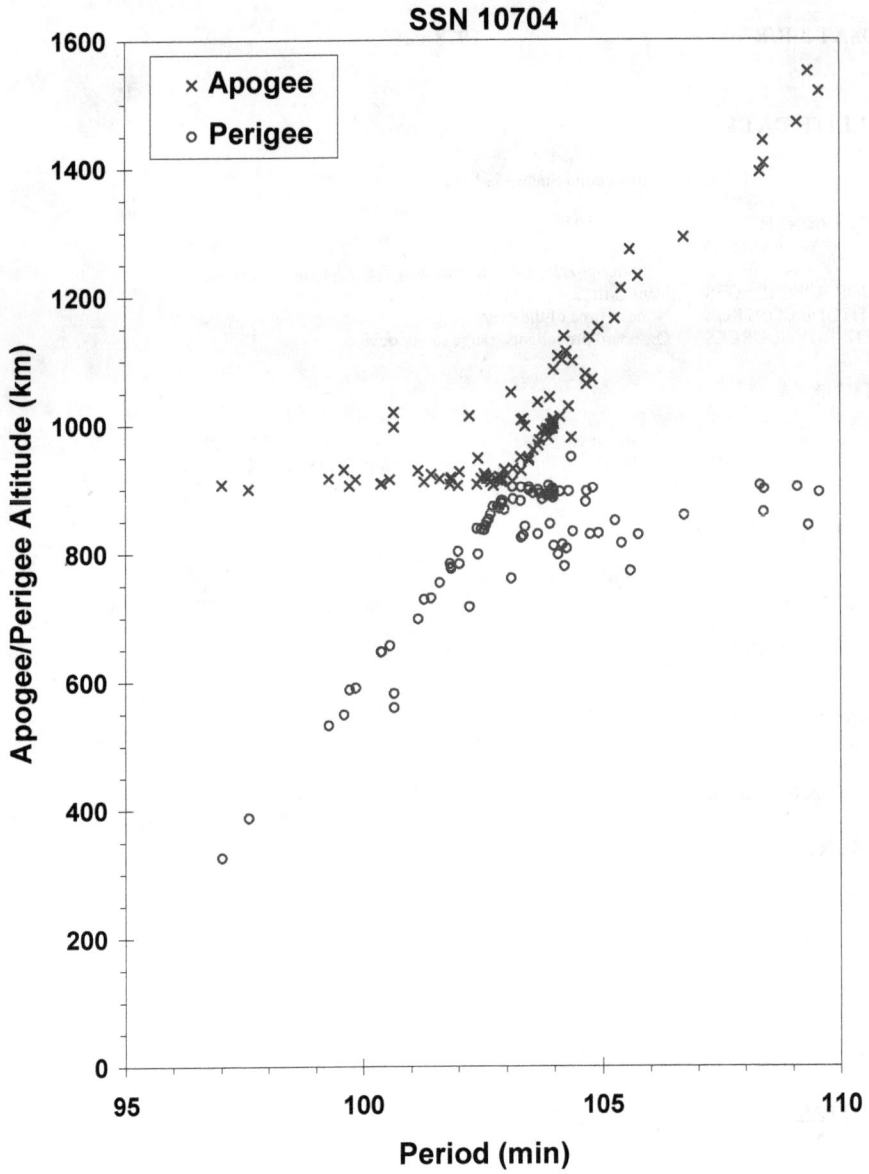

Landsat 3 R/B debris cloud of 90 identified fragments 4 days after the event as reconstructed from the US SSN database.

SATELLITE DATA

TYPE:	Payload
OWNER:	CIS
LAUNCH DATE:	6.13 Sep 1978
DRY MASS (KG):	1250
MAIN BODY:	Cylinder; 1.7 m diameter by 2 m length
MAJOR APPENDAGES:	Solar panels (?)
ATTITUDE CONTROL:	Active, 3-axis
ENERGY SOURCES:	On-board propellants, explosive charge

EVENT DATA

DATE:	10 Oct 1978	LOCATION:	Unknown
TIME:	Unknown	ASSESSED CAUSE:	Deliberate
ALTITUDE:	Unknown		

PRE-EVENT ELEMENTS

EPOCH:	78277.19859350	MEAN ANOMALY:	4.9827
RIGHT ASCENSION:	336.7676	MEAN MOTION:	2.00213289
INCLINATION:	62.8388	MEAN MOTION DOT/2:	.0
ECCENTRICITY:	.7350882	MEAN MOTION DOT DOT/6:	.0
ARG. OF PERIGEE:	318.4262	BSTAR:	.0

DEBRIS CLOUD DATA

MAXIMUM ΔP:	Unknown
MAXIMUM ΔI:	Unknown

COMMENTS

Cosmos 1030 was another spacecraft of the Cosmos 862-type to experience a fragmentation. After entering a Molniya-type transfer orbit on 6 September, Cosmos 1030 maneuvered about 14 September to enter an operational orbit. Elements on the first identifiable fragment did not appear until a year after the event. Official cataloging of debris did not begin until 3 years after the event.

REFERENCE DOCUMENT

History of Soviet/Russian Satellite Fragmentations-A Joint U.S.-Russian Investigation, N. L. Johnson et al, Kaman Sciences Corporation, October 1995.

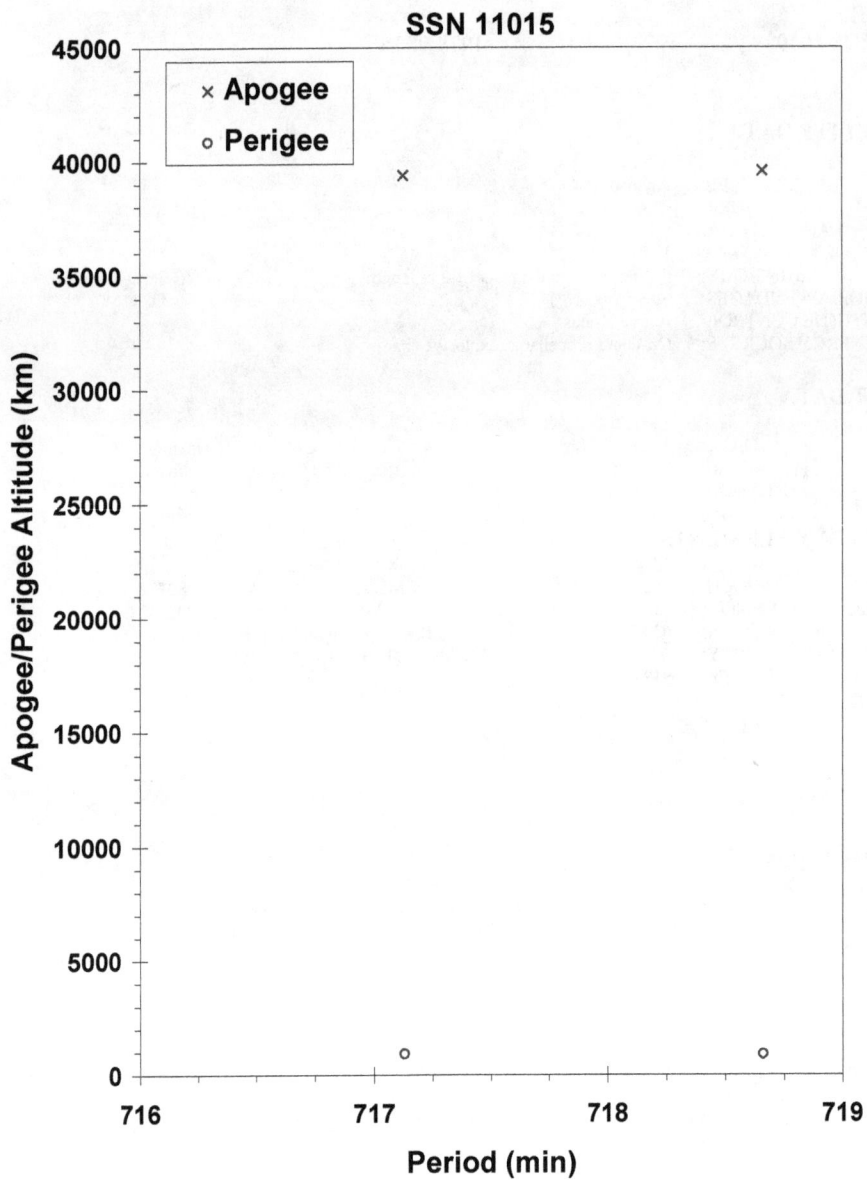

SSN 11015

Cosmos 1030 and a single debris fragment 1 year after the event as reconstructed from the US SSN database.

NIMBUS 7 R/B   **1978-098B**   **11081**

SATELLITE DATA

TYPE:	Delta Second Stage (2910)
OWNER:	US
LAUNCH DATE:	24.34 Oct 1978
DRY MASS (KG):	900
MAIN BODY:	Cylinder-nozzle; 2.4 m diameter by 8 m length
MAJOR APPENDAGES:	None
ATTITUDE CONTROL:	None at time of the event.
ENERGY SOURCES:	On-board propellants, range safety device

EVENT DATA

DATE:	26 Dec 1981	LOCATION:	Unknown
TIME:	Unknown	ASSESSED CAUSE:	Propulsion
ALTITUDE:	Unknown		

PRE-EVENT ELEMENTS

EPOCH:	81360.19972720	MEAN ANOMALY:	311.8261
RIGHT ASCENSION:	277.7553	MEAN MOTION:	13.85390161
INCLINATION:	99.3003	MEAN MOTION DOT/2:	.000000425
ECCENTRICITY:	.0010821	MEAN MOTION DOT DOT/6:	.0
ARG. OF PERIGEE:	48.3801	BSTAR:	.00004426123

DEBRIS CLOUD DATA

MAXIMUM ΔP:	Unknown
MAXIMUM ΔI:	0.6 deg*

*Based on uncataloged debris data

COMMENTS

Nimbus 7 R/B is designated Cameo in US Space Command Satellite Catalog in reference to scientific piggy-back payload attached to the Delta second stage. This satellite experienced an anomalous event prior to and after the event cited above (See Section 3). Most fragments decayed very rapidly, preventing an accurate assessment of the event and its resulting debris cloud. No new objects were cataloged as a result of this event. The event apparently occurred prior to 0700 GMT.

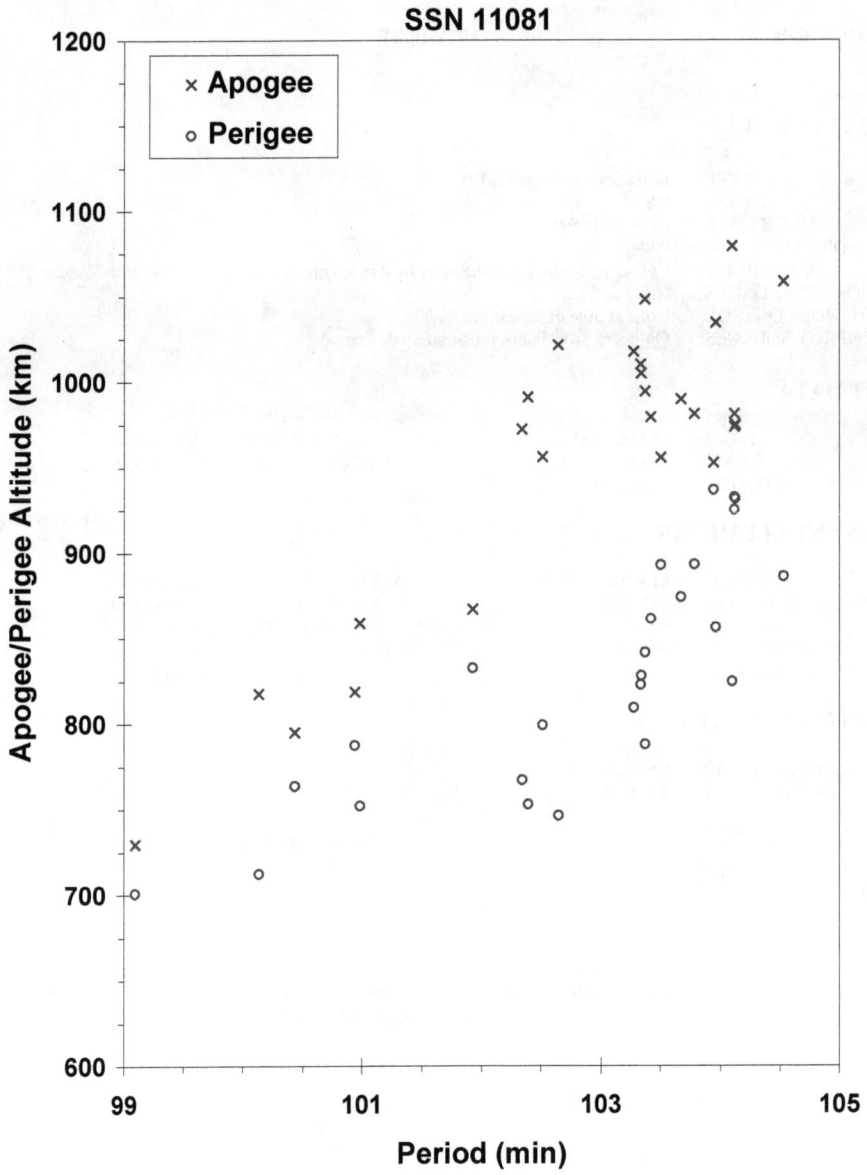

The Nimbus 7 R/B debris cloud remnant of 27 fragments a few days after the event as reconstructed from the US SSN database. Most fragments have already experienced considerable drag effects.

SATELLITE DATA

TYPE:	Tsyklon Third Stage
OWNER:	CIS
LAUNCH DATE:	26.29 Oct 1978
DRY MASS (KG):	1360
MAIN BODY:	Cone-cylinder; 2.1 m diameter by 3.3 m length
MAJOR APPENDAGES:	None
ATTITUDE CONTROL:	None at time of the event.
ENERGY SOURCES:	Unknown

EVENT DATA

DATE:	9 May 1988	LOCATION:	29S, 126E (dsc)
TIME:	1218 GMT	ASSESSED CAUSE:	Propulsion
ALTITUDE:	1705 km		

PRE-EVENT ELEMENTS

EPOCH:	88121.02005933	MEAN ANOMALY:	279.0818
RIGHT ASCENSION:	359.3059	MEAN MOTION:	11.97080974
INCLINATION:	82.5543	MEAN MOTION DOT/2:	.000000208
ECCENTRICITY:	.0011463	MEAN MOTION DOT DOT/6:	.0
ARG. OF PERIGEE:	81.1553	BSTAR:	.0

DEBRIS CLOUD DATA

MAXIMUM ΔP:	7.8 min
MAXIMUM ΔI:	0.9 deg

COMMENTS

This flight, which successfully carried three separate payloads, was the fifth orbital mission of the Tsyklon third stage. Propellants used were N_2O_4 and UDMH. Nearly 10 years elapsed from launch to breakup. A second Tsyklon third stage breakup after a similar length of time; see 1987-068B.

REFERENCE DOCUMENT

History of Soviet/Russian Satellite Fragmentations-A Joint U.S.-Russian Investigation, N. L. Johnson et al, Kaman Sciences Corporation, October 1995.

SSN 11087

Cosmos 1045 R/B debris cloud of 25 fragments as determined 1 week after the event.

SATELLITE DATA

TYPE:	Payload
OWNER:	US
LAUNCH DATE:	24.35 Feb 1979
DRY MASS (KG):	850
MAIN BODY:	Cylinder; 2.1 m diameter by 1.3 m length
MAJOR APPENDAGES:	1 solar panel
ATTITUDE CONTROL:	Spin-stabilized
ENERGY SOURCES:	None

EVENT DATA

DATE:	13 Sep 1985	LOCATION:	35N, 234E (asc)
TIME:	2043 GMT	ASSESSED CAUSE:	Deliberate
ALTITUDE:	525 km		

PRE-EVENT ELEMENTS

EPOCH:	85256.72413718	MEAN ANOMALY:	260.9644
RIGHT ASCENSION:	182.5017	MEAN MOTION:	15.11755304
INCLINATION:	97.6346	MEAN MOTION DOT/2:	.00000616
ECCENTRICITY:	.0022038	MEAN MOTION DOT DOT/6:	.0
ARG. OF PERIGEE:	99.4081	BSTAR:	.000037918

DEBRIS CLOUD DATA

MAXIMUM ΔP:	12.7 min
MAXIMUM ΔI:	1.4 deg

COMMENTS

P-78 was impacted by a sub-orbital object at high velocity as part of a planned test.

REFERENCE DOCUMENT

Postmortem of a Hypervelocity Impact: Summary, R. L. Kling, Technical Report CS86-LKD-001, Teledyne Brown Engineering, Colorado Springs, September 1986.

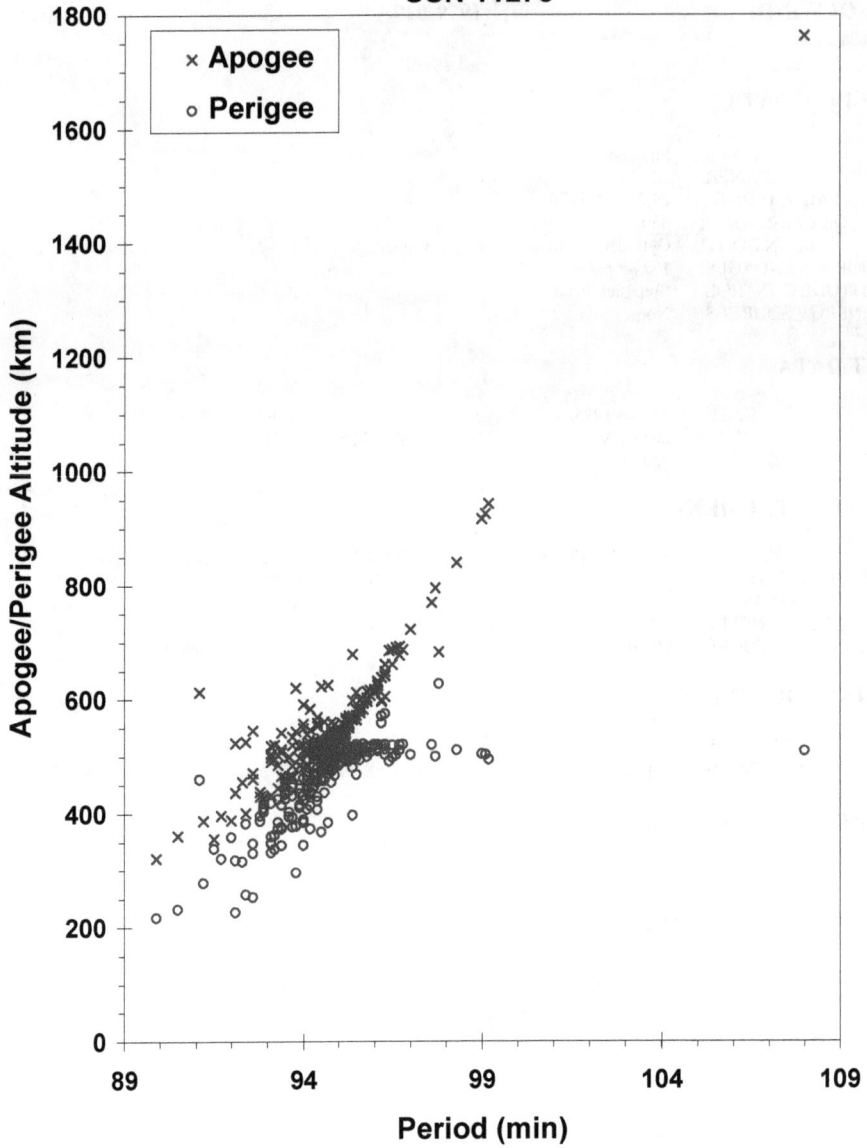

SSN 11278

P-78 debris cloud of 267 fragments seen 11 hours after the event by the US SSN PARCS radar.

COSMOS 1094   1979-033A   11333

SATELLITE DATA

TYPE:	Payload
OWNER:	CIS
LAUNCH DATE:	18.50 Apr 1979
DRY MASS (KG):	3000
MAIN BODY:	Cylinder; 1.3 m diameter by 17 m length
MAJOR APPENDAGES:	Solar panels
ATTITUDE CONTROL:	Active, 3-axis
ENERGY SOURCES:	On-board propellants, explosive charge (?)

EVENT DATA

DATE:	17 Sep 1979	LOCATION:	53S, 336E (dsc)
TIME:	1039 GMT	ASSESSED CAUSE:	Unknown
ALTITUDE:	385 km		

PRE-EVENT ELEMENTS

EPOCH:	79260.33615661	MEAN ANOMALY:	61.9566
RIGHT ASCENSION:	271.8638	MEAN MOTION:	15.58096051
INCLINATION:	65.0398	MEAN MOTION DOT/2:	.00102640
ECCENTRICITY:	.0016936	MEAN MOTION DOT DOT/6:	.0
ARG. OF PERIGEE:	297.9871	BSTAR:	.0013492

DEBRIS CLOUD DATA

MAXIMUM ΔP:	7.1 min*
MAXIMUM ΔI:	0.3 deg*

*Based on uncataloged debris data

COMMENTS

Cosmos 1094 was the fourth spacecraft of the Cosmos 699-type to experience a fragmentation. Spacecraft had been in a regime of natural decay for 4 months prior to the event. All new debris decayed before being officially cataloged.

REFERENCE DOCUMENTS

"Artificial Satellite Break-Ups (Part 1): Soviet Ocean Surveillance Satellites", N. L. Johnson, Journal of the British Interplanetary Society, February 1983, pp. 51-58.

History of Soviet/Russian Satellite Fragmentations-A Joint U.S-Russian Investigation, N. L. Johnson et al, Kaman Sciences Corporation, October 1995.

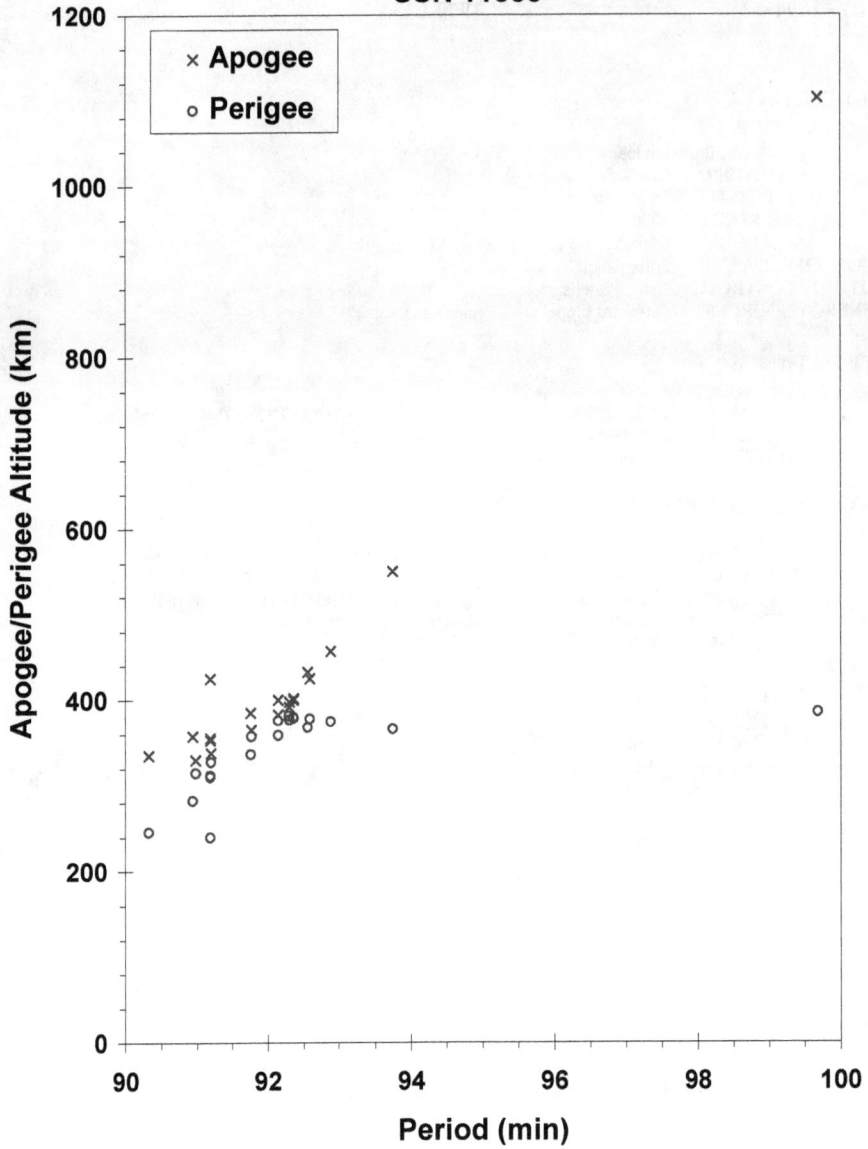

Cosmos 1094 debris cloud of 20 fragments within 1 week of the event as reconstructed from the US SSN database.

SATELLITE DATA

TYPE:	Payload
OWNER:	CIS
LAUNCH DATE:	27.76 Jun 1979
DRY MASS (KG):	1250
MAIN BODY:	Cylinder; 1.7 m diameter by 2 m length
MAJOR APPENDAGES:	Solar panels
ATTITUDE CONTROL:	Active, 3-axis
ENERGY SOURCES:	On-board propellants, explosive charge

EVENT DATA

DATE:	Mid-Feb 1980	LOCATION:	Unknown
TIME:	Unknown	ASSESSED CAUSE:	Deliberate
ALTITUDE:	Unknown		

PRE-EVENT ELEMENTS

EPOCH:	80048.26161234	MEAN ANOMALY:	5.0375
RIGHT ASCENSION:	104.4713	MEAN MOTION:	2.00453352
INCLINATION:	63.3495	MEAN MOTION DOT/2:	.0
ECCENTRICITY:	.7238911	MEAN MOTION DOT DOT/6:	.0
ARG. OF PERIGEE:	318.4445	BSTAR:	.0

DEBRIS CLOUD DATA

MAXIMUM ΔP:	3.5 min*
MAXIMUM ΔI:	0.2 deg*

*Based on uncataloged debris data

COMMENTS

Cosmos 1109 was another spacecraft of the Cosmos 862-type to experience a fragmentation. Cosmos 1109 maneuvered into an operational orbit about 19 July. The payload was "lost" after 17 February 1980 and three pieces of debris were soon found that could be traced back to that period.

REFERENCE DOCUMENT

History of Soviet/Russian Satellite Fragmentations-A Joint U.S.-Russian Investigation, N. L. Johnson et al, Kaman Sciences Corporation, October 1995.

SSN 11417

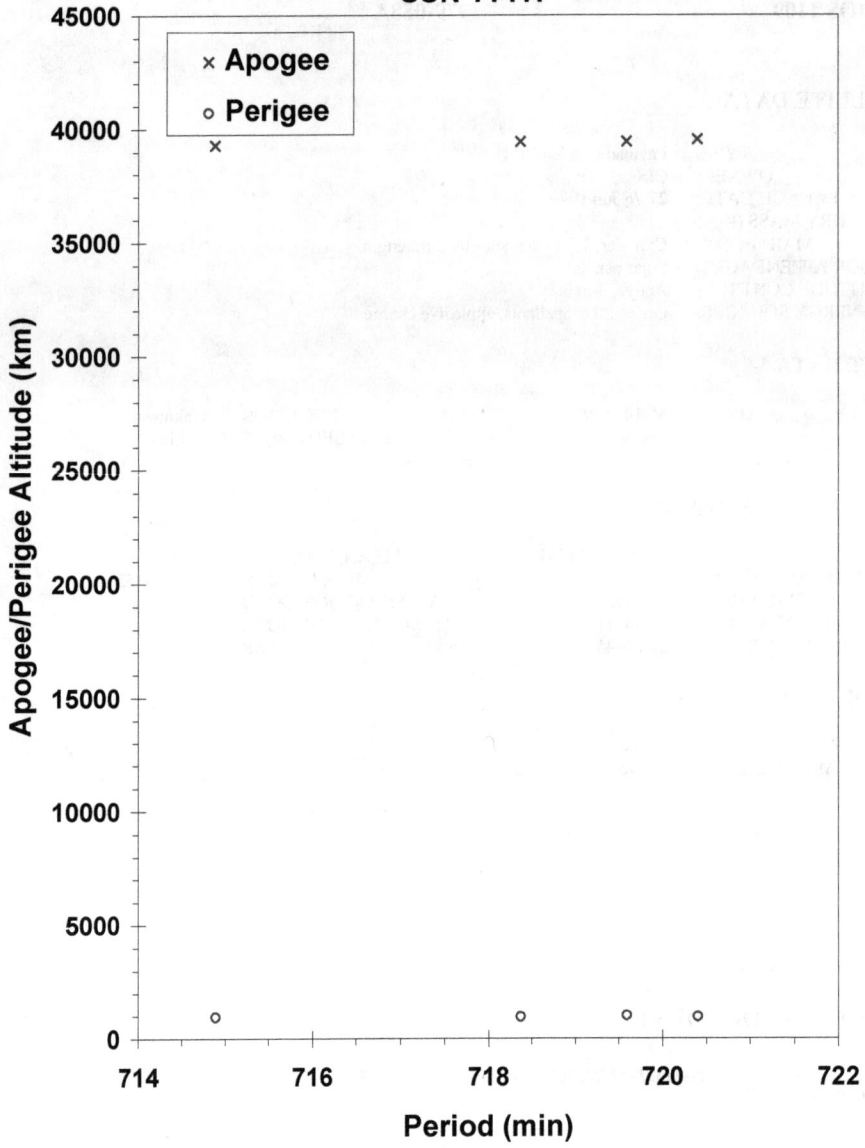

Cosmos 1109 and three fragments in February 1980 as reconstructed
from the US SSN database.

COSMOS 1124   1979-077A   11509

SATELLITE DATA

TYPE:	Payload
OWNER:	CIS
LAUNCH DATE:	28.01 Aug 1979
DRY MASS (KG):	1250
MAIN BODY:	Cylinder; 1.7 m diameter by 2 m length
MAJOR APPENDAGES:	Solar panels
ATTITUDE CONTROL:	Active, 3-axis
ENERGY SOURCES:	On-board propellants, explosive charge

EVENT DATA

DATE:	9 Sep 1979	LOCATION:	52N, 304E (asc)
TIME:	0230 GMT	ASSESSED CAUSE:	Deliberate
ALTITUDE:	8375 km		

PRE-EVENT ELEMENTS

EPOCH:	79249.09448656	MEAN ANOMALY:	3.7678
RIGHT ASCENSION:	288.1742	MEAN MOTION:	2.00548359
INCLINATION:	63.0212	MEAN MOTION DOT/2:	.0
ECCENTRICITY:	.7383335	MEAN MOTION DOT DOT/6:	.0
ARG. OF PERIGEE:	318.3799	BSTAR:	.0

DEBRIS CLOUD DATA

MAXIMUM ΔP:	4.0 min*
MAXIMUM ΔI:	0.1 deg*

*Based on uncataloged debris data

COMMENTS

Cosmos 1124 was another spacecraft of the Cosmos 862-type to experience a fragmentation. After insertion into a Molniya-type transfer orbit on 28 August, Cosmos 1124's ascending node was allowed to drift until 3 September when a maneuver placed the spacecraft into an operational, semi-synchronous orbit. The fragmentation occurred 6 days later. The spacecraft never maneuvered again and soon drifted off station.

REFERENCE DOCUMENT

History of Soviet/Russian Satellite Fragmentations-A Joint U.S.-Russian Investigation, N. L. Johnson et al, Kaman Sciences Corporation, October 1995.

149

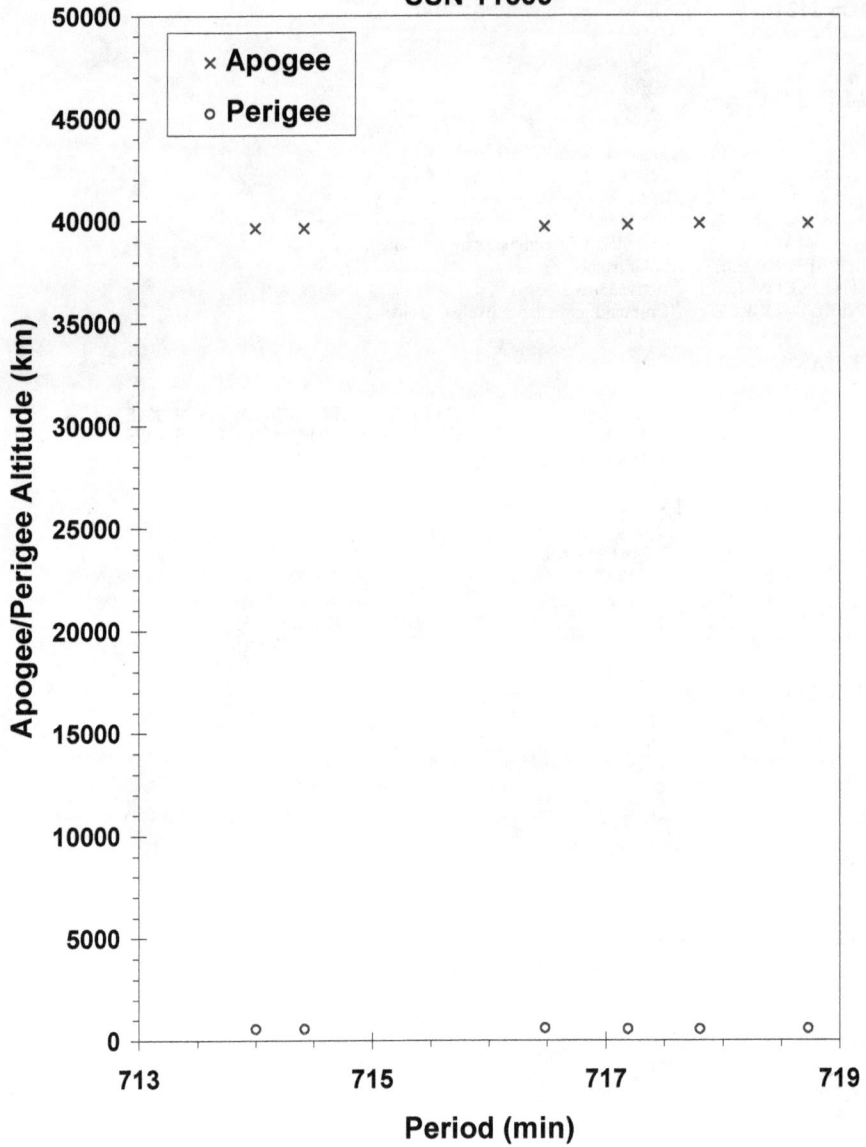

Cosmos 1124 debris cloud of 6 fragments about 1 week after the event as reconstructed from the US SSN database.

SATELLITE DATA

TYPE:	Ariane 1 Third Stage
OWNER:	ESA
LAUNCH DATE:	24.72 Dec 1979
DRY MASS (KG):	1400
MAIN BODY:	Cylinder; 2.6 m diameter by 10.3 m length
MAJOR APPENDAGES:	None
ATTITUDE CONTROL:	None at time of the event.
ENERGY SOURCES:	On-board propellants, range safety device

EVENT DATA

DATE:	Apr 1980	LOCATION:	Unknown
TIME:	Unknown	ASSESSED CAUSE:	Propulsion
ALTITUDE:	Unknown		

PRE-EVENT ELEMENTS

EPOCH:	80088.55565320	MEAN ANOMALY:	17.6019
RIGHT ASCENSION:	101.5521	MEAN MOTION:	2.48253031
INCLINATION:	17.9092	MEAN MOTION DOT/2:	.001764977
ECCENTRICITY:	.7152375	MEAN MOTION DOT DOT/6:	.0
ARG. OF PERIGEE:	264.7858	BSTAR:	.001078542

DEBRIS CLOUD DATA

MAXIMUM ΔP:	Unknown
MAXIMUM ΔI:	Unknown

COMMENTS

This mission was the inaugural flight of the Ariane 1 launch vehicle. Payload and R/B were apparently cross-tagged until mid-January 1980. Detection and tracking of debris has always been extremely difficult in part due to low inclination and highly elliptical orbit. Debris data were first developed in the second half of April, and calculations suggest the fragmentation occurred during the first week of April. The magnitude of the event and the total number of pieces created are unknown. Many debris had high decay rates.

REFERENCE DOCUMENT

A Preliminary Analysis of the Fragmentation of the Spot 1 Ariane Third Stage, N. L. Johnson, Technical Report CS87-LKD-003, Teledyne Brown Engineering, Colorado Springs, March 1987.

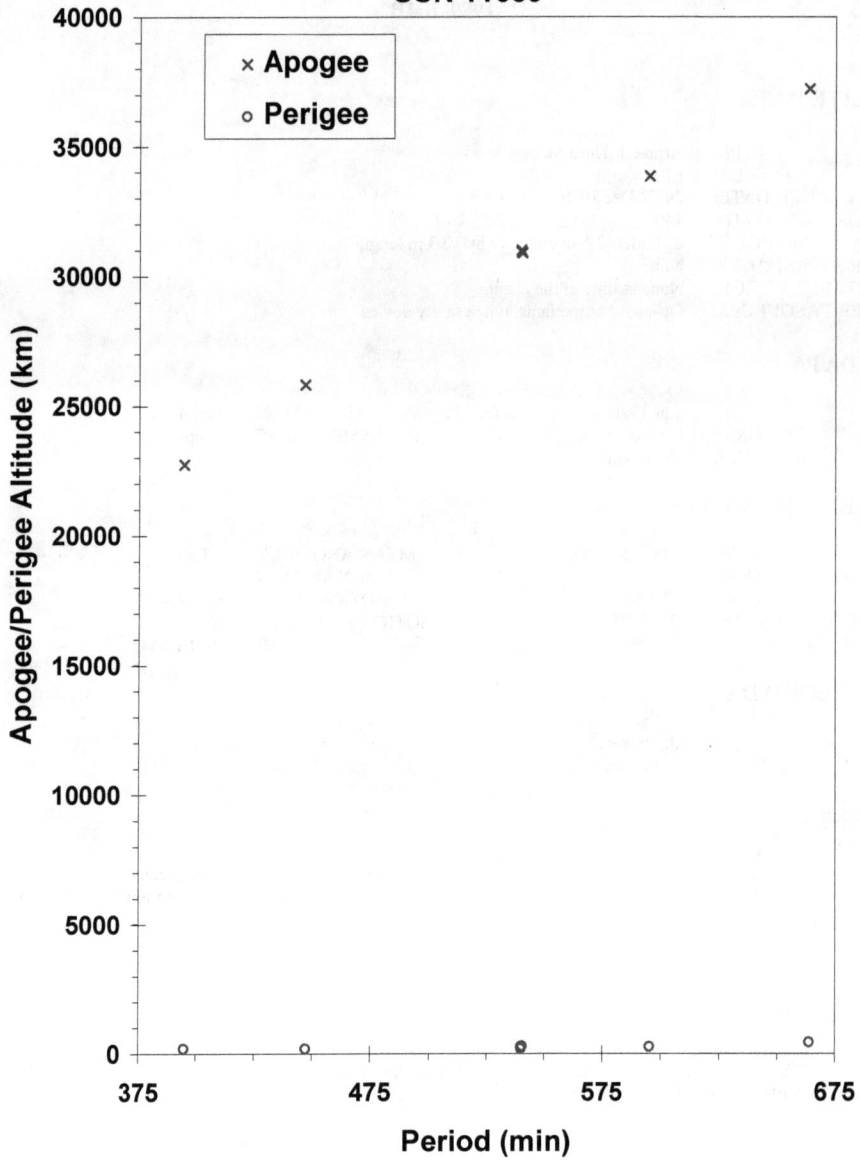

SSN 11659

CAT R/B debris cloud of 7 fragments about 8 weeks after the event as reconstructed from the US SSN database.

COSMOS 1167   1980-021A   11729

SATELLITE DATA

TYPE:	Payload
OWNER:	CIS
LAUNCH DATE:	14.44 Mar 1980
DRY MASS (KG):	3000
MAIN BODY:	Cylinder; 1.3 m diameter by 17 m length
MAJOR APPENDAGES:	Solar panels
ATTITUDE CONTROL:	Active, 3-axis
ENERGY SOURCES:	On-board propellants, explosive charge (?)

EVENT DATA

DATE:	15 Jul 1981	LOCATION:	10N, 106E (asc)
TIME:	0921 GMT	ASSESSED CAUSE:	Unknown
ALTITUDE:	430 km		

PRE-EVENT ELEMENTS

EPOCH:	81196.19449955	MEAN ANOMALY:	110.8351
RIGHT ASCENSION:	174.9184	MEAN MOTION:	15.54665775
INCLINATION:	65.0101	MEAN MOTION DOT/2:	.00025375
ECCENTRICITY:	.0068471	MEAN MOTION DOT DOT/6:	.0
ARG. OF PERIGEE:	248.6139	BSTAR:	.00034595

DEBRIS CLOUD DATA

MAXIMUM ΔP:	1.0 min*
MAXIMUM ΔI:	0.5 deg*

*Based on uncataloged debris data

COMMENTS

Cosmos 1167 was the fifth spacecraft of the Cosmos 699-type to experience a fragmentation. The spacecraft had been in a regime of natural decay for 3 months prior to the event. Most debris reentered before being officially cataloged.

REFERENCE DOCUMENTS

The Fragmentations of USSR Satellites 11729 and 12504 (U), J. R. Gabbard and P. M. Landry, Technical Memorandum 82-S-03, DCS/Plans, Hdqtrs NORAD/ADCOM, Colorado Springs, August 1982 (Secret).

"Artificial Satellite Break-Ups (Part 1): Soviet Ocean Surveillance Satellites", N. L. Johnson, Journal of the British Interplanetary Society, February 1983, pp. 51-58.

History of Soviet/Russian Satellite Fragmentations-A Joint U.S.-Russian Investigation, N. L Johnson et al, Kaman Sciences Corporation, October 1995.

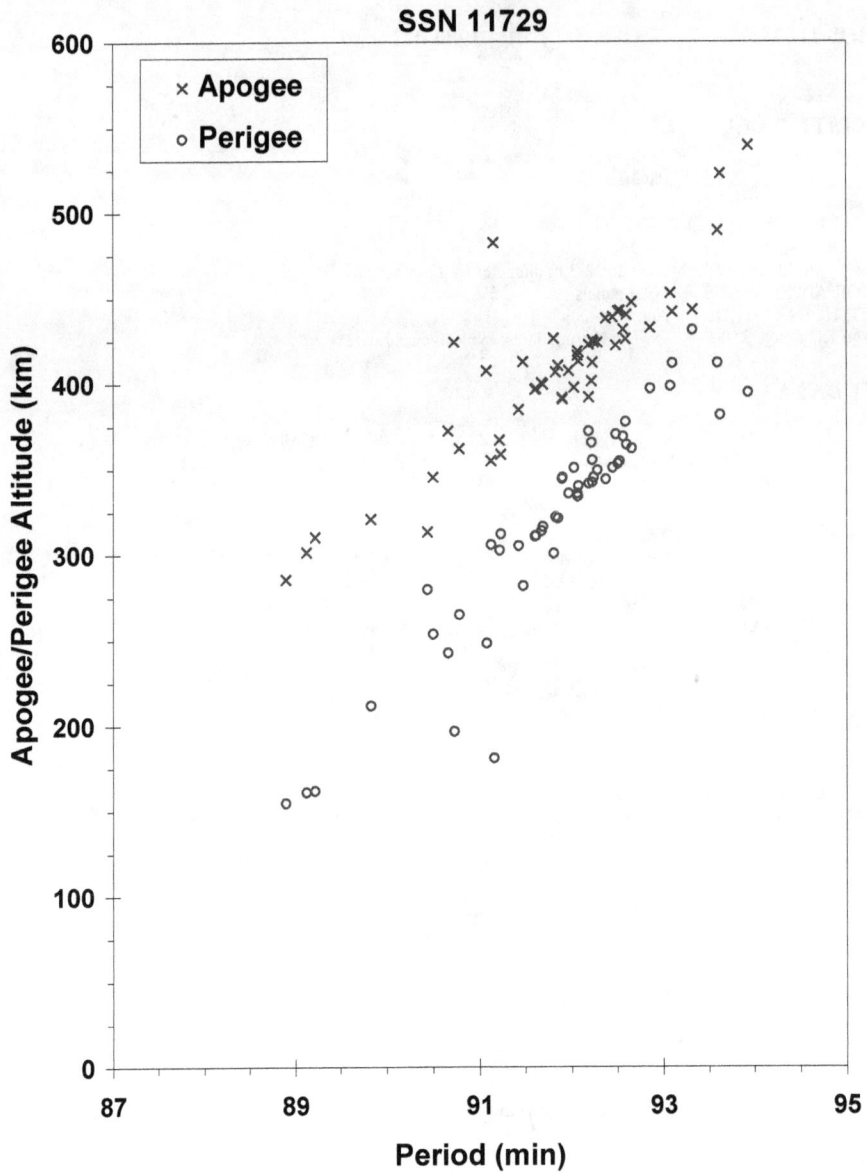

SSN 11729

Cosmos 1167 debris cloud remnant of 53 fragments about 2 weeks after the event as reconstructed from the US SSN database.

SATELLITE DATA

TYPE:	Payload
OWNER:	CIS
LAUNCH DATE:	18.04 Apr 1980
DRY MASS (KG):	1400
MAIN BODY:	Cylinder; 2 m diameter by 4 m length
MAJOR APPENDAGES:	None
ATTITUDE CONTROL:	Active, 3-axis
ENERGY SOURCES:	On-board propellants, explosive charge

EVENT DATA

DATE:	18 Apr 1980	LOCATION:	47N, 322E (asc)
TIME:	0726 GMT	ASSESSED CAUSE:	Deliberate
ALTITUDE:	1625 km		

POST-EVENT ELEMENTS

EPOCH:	80109.51771250	MEAN ANOMALY:	102.2095
RIGHT ASCENSION:	250.9679	MEAN MOTION:	13.64414319
INCLINATION:	66.1153	MEAN MOTION DOT/2:	.0
ECCENTRICITY:	.0865337	MEAN MOTION DOT DOT/6:	.0
ARG. OF PERIGEE:	248.5294	BSTAR:	.0

DEBRIS CLOUD DATA

MAXIMUM ΔP:	5.4 min
MAXIMUM ΔI:	0.6 deg

COMMENTS

Cosmos 1174 was launched on a two-revolution rendezvous with Cosmos 1171. After a close approach, Cosmos 1174 performed a final maneuver shortly before its warhead was intentionally fired. Elements above are first data available after the final maneuver but also following the fragmentation. Cosmos 1174 was part of test series begun with Cosmos 249.

REFERENCE DOCUMENTS

"Artificial Satellite Break-Ups (Part 2): Soviet Anti-Satellite Program", N.L. Johnson, Journal of the British Interplanetary Society, August 1983, pp. 357-362.

History of Soviet/Russian Satellite Fragmentations-A Joint U.S.-Russian Investigation, N. L. Johnson et al, Kaman Sciences Corporation, October 1995.

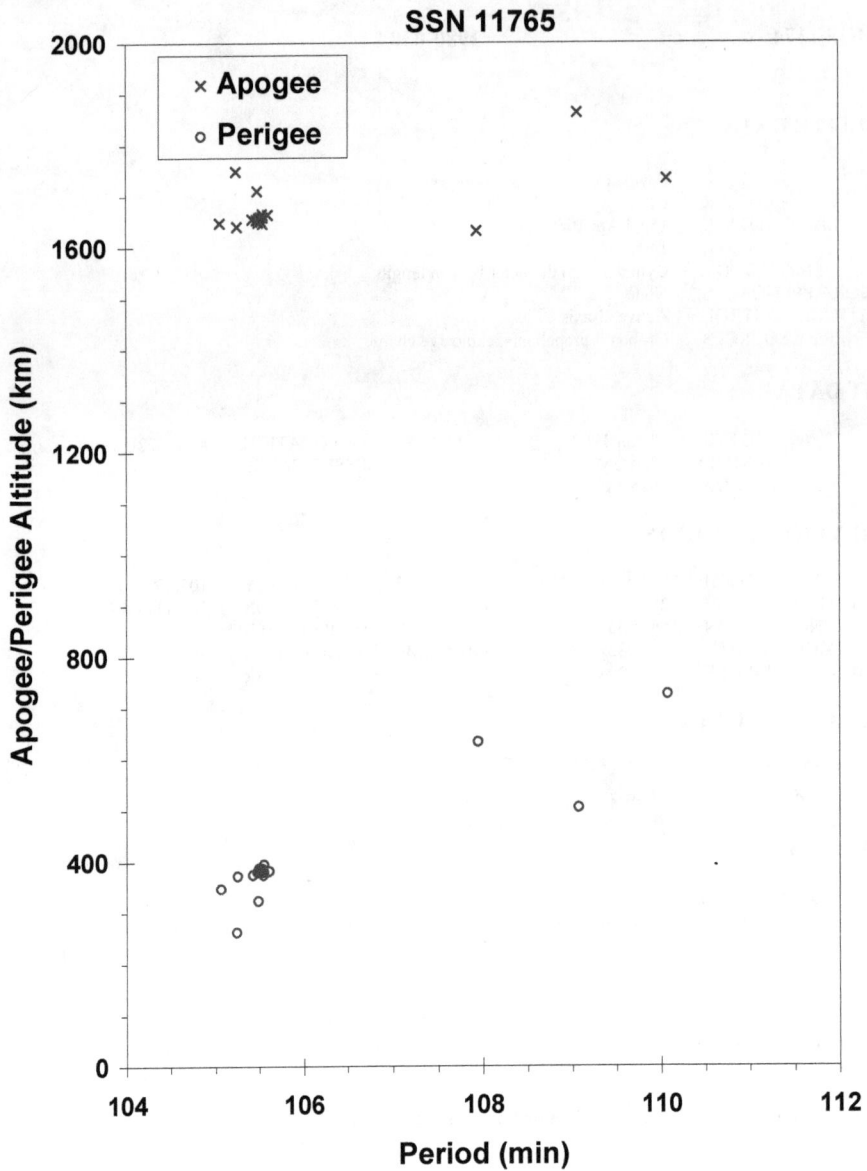

Cosmos 1174 debris cloud of 18 identified fragments about 10 days after the event as reconstructed from the US SSN database.

SATELLITE DATA

TYPE:	Payload
OWNER:	CIS
LAUNCH DATE:	2.04 Jul 1980
DRY MASS (KG):	1250
MAIN BODY:	Cylinder; 1.7 m diameter by 2 m length
MAJOR APPENDAGES:	Solar panels
ATTITUDE CONTROL:	Active, 3-axis
ENERGY SOURCES:	On-board propellants, explosive charge

EVENT DATA

DATE:	14 May 1981	LOCATION:	Unknown
TIME:	Unknown	ASSESSED CAUSE:	Deliberate
ALTITUDE:	Unknown		

PRE-EVENT ELEMENTS

EPOCH:	81133.07322634	MEAN ANOMALY:	5.1166
RIGHT ASCENSION:	198.5704	MEAN MOTION:	2.00555560
INCLINATION:	62.6448	MEAN MOTION DOT/2:	.00001257
ECCENTRICITY:	.7180863	MEAN MOTION DOT DOT/6:	.0
ARG. OF PERIGEE:	319.4330	BSTAR:	.0

DEBRIS CLOUD DATA

MAXIMUM	ΔP:	6.0 min*
MAXIMUM	ΔI:	0.1 deg*

*Based on uncataloged debris data

COMMENTS

Cosmos 1191 was another spacecraft of the Cosmos 862-type to experience a fragmentation. The first debris elements were developed for 25 May.

REFERENCE DOCUMENT

History of Soviet/Russian Satellite Fragmentations-A Joint U.S.-Russian Investigation, N. L. Johnson et al, Kaman Sciences Corporation, October 1995.

SSN 11871

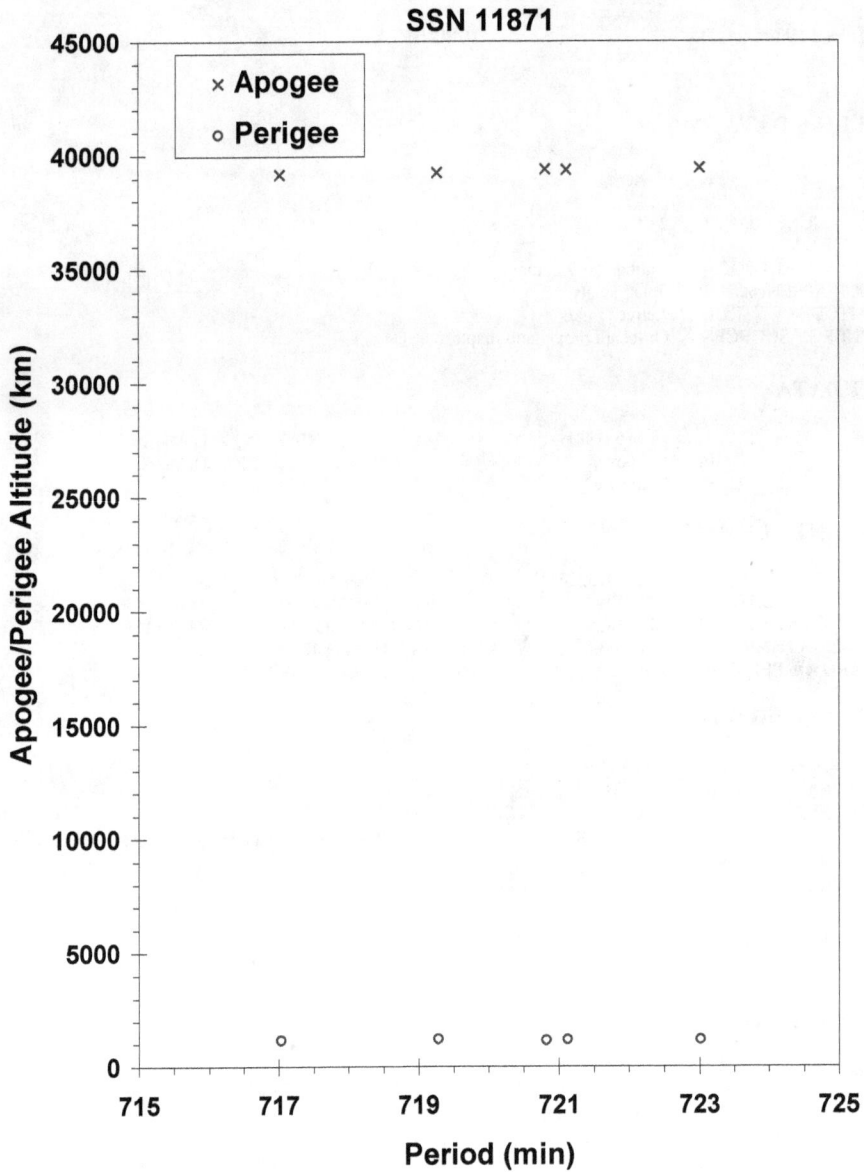

Cosmos 1191 debris cloud of 5 identified fragments 1 month after the event as reconstructed from the US SSN database.

COSMOS 1217   **1980-085A**   **12032**

SATELLITE DATA

TYPE:	Payload
OWNER:	CIS
LAUNCH DATE:	24.46 Oct 1980
DRY MASS (KG):	1250
MAIN BODY:	Cylinder; 1.7 m diameter by 2 m length
MAJOR APPENDAGES:	Solar panels
ATTITUDE CONTROL:	Active, 3-axis
ENERGY SOURCES:	On-board propellants, explosive charge

EVENT DATA

DATE:	12 Feb 1983	LOCATION:	Unknown
TIME:	Unknown	ASSESSED CAUSE:	Deliberate
ALTITUDE:	Unknown		

PRE-EVENT ELEMENTS

EPOCH:	83042.34036514	MEAN ANOMALY:	6.0502
RIGHT ASCENSION:	36.1600	MEAN MOTION:	2.00587025
INCLINATION:	65.2478	MEAN MOTION DOT/2:	0.00001154
ECCENTRICITY:	0.7021051	MEAN MOTION DOT DOT/6:	0
ARG. OF PERIGEE:	314.5975	BSTAR:	0

DEBRIS CLOUD DATA

MAXIMUM ΔP:	Unknown
MAXIMUM ΔI:	Unknown

COMMENTS

Cosmos 1217 was another member of the Cosmos 862-type to experience a fragmentation.

REFERENCE DOCUMENT

History of Soviet/Russian Satellite Fragmentation-A Joint U.S.-Russian Investigation, N. L. Johnson et al, Kaman Sciences Corporation, October 1995.

Insufficient data to construct a Gabbard diagram.

SATELLITE DATA

TYPE:	Payload
OWNER:	CIS
LAUNCH DATE:	4.63 Nov 1980
DRY MASS (KG):	3000
MAIN BODY:	Cylinder; 1.3 m diameter by 17 m length
MAJOR APPENDAGES:	Solar panels
ATTITUDE CONTROL:	Active, 3-axis
ENERGY SOURCES:	On-board propellants, explosive charge (?)

EVENT DATA (1)

DATE:	20 Jun 1982	LOCATION:	10S, 332E (dsc)
TIME:	1818 GMT	ASSESSED CAUSE:	Unknown
ALTITUDE:	875 km		

PRE-EVENT ELEMENTS (1)

EPOCH:	82171.72558670	MEAN ANOMALY:	0.2166
RIGHT ASCENSION:	330.3811	MEAN MOTION:	14.49658466
INCLINATION:	65.0033	MEAN MOTION DOT/2:	.00000066
ECCENTRICITY:	.0219432	MEAN MOTION DOT DOT/6:	.0
ARG. OF PERIGEE:	357.8883	BSTAR:	.000025640

EVENT DATA (2)

DATE:	25 Aug 1982	LOCATION:	65S, 238E (dsc)
TIME:	1231 GMT	ASSESSED CAUSE:	Unknown
ALTITUDE:	665 km		

PRE-EVENT ELEMENTS (2)

EPOCH:	82230.91714195	MEAN ANOMALY:	22.7965
RIGHT ASCENSION:	159.4489	MEAN MOTION:	14.49745561
INCLINATION:	65.0025	MEAN MOTION DOT/2:	.0
ECCENTRICITY:	.0225583	MEAN MOTION DOT DOT/6:	.0
ARG. OF PERIGEE:	336.3217	BSTAR:	.0

DEBRIS CLOUD DATA

MAXIMUM ΔP:	3.4 min*
MAXIMUM ΔI:	1.8 deg*

*Based on uncataloged debris data

COMMENTS

Cosmos 1220 was the seventh spacecraft of the Cosmos 699-type to experience a fragmentation. The spacecraft had been in a natural decay regime for more than 14 months at the time of the first event. A total of 47 fragments had been officially cataloged by the time of the second event that occurred 2 months later. See similar dual events happening in the summer of 1982 with Cosmos 1306 and Cosmos 1260.

REFERENCE DOCUMENTS

Analysis of PARCS Recorded Data on the Breakup of Satellite 12054, J.W. Rider, Technical Report MSB83-ADC-0162, Teledyne Brown Engineering, Huntsville, January 1983.

Analysis of Cosmos 1220 and Cosmos 1306 Fragments (U), D. Fennessy, Report AH-23, FTD/OLAI, Cheyenne Mountain, Colorado, 12 January 1983 (Secret).

"Artificial Satellite Break-Ups (Part 1): Soviet Ocean Surveillance Satellites", N. L. Johnson, Journal of the British Interplanetary Society, February 1983, pp. 51-58.

History of Soviet/Russian Satellite Fragmentations-A Joint U.S.-Russian Investigation, N. L. Johnson et al, Kaman Sciences Corporation, October 1995.

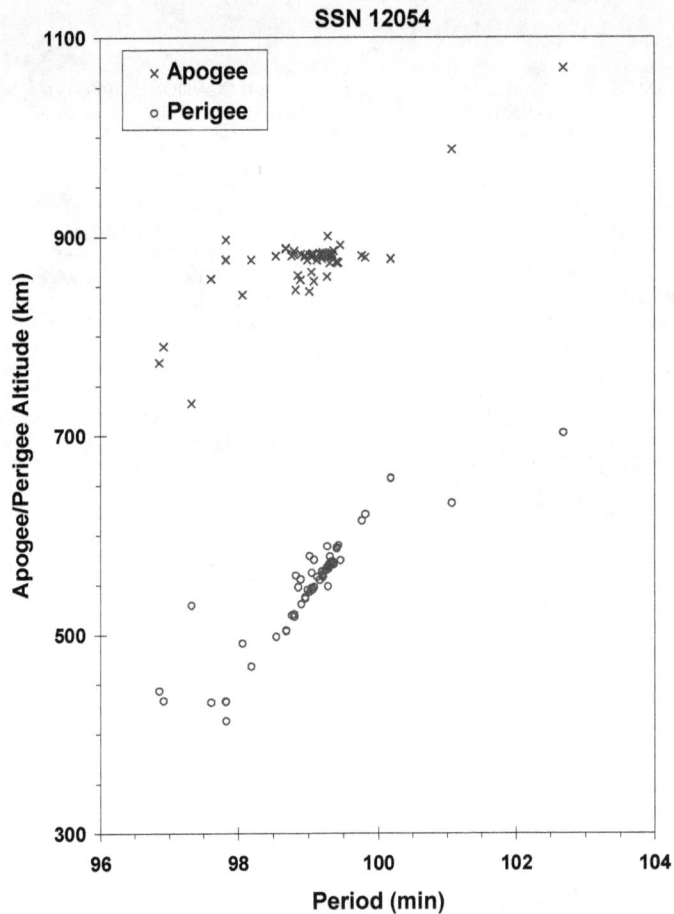

Cosmos 1220 debris cloud of 72 fragments about 1 week after the first event as reconstructed from the US SSN database.

COSMOS 1247   **1981-016A**   **12303**

SATELLITE DATA

TYPE:	Payload
OWNER:	CIS
LAUNCH DATE:	19.41 Feb 1981
DRY MASS (KG):	1250
MAIN BODY:	Cylinder; 1.7 m diameter by 2 m length
MAJOR APPENDAGES:	Solar panels
ATTITUDE CONTROL:	Active, 3-axis
ENERGY SOURCES:	On-board propellants, explosive charge

EVENT DATA

DATE:	20 Oct 1981	LOCATION:	Unknown
TIME:	Unknown	ASSESSED CAUSE:	Deliberate
ALTITUDE:	Unknown		

PRE-EVENT ELEMENTS

EPOCH:	81293.17083627	MEAN ANOMALY:	5.0298
RIGHT ASCENSION:	214.2278	MEAN MOTION:	2.00570861
INCLINATION:	62.9685	MEAN MOTION DOT/2:	.0
ECCENTRICITY:	.7233048	MEAN MOTION DOT DOT/6:	.0
ARG. OF PERIGEE:	318.2473	BSTAR:	.0

DEBRIS CLOUD DATA

MAXIMUM ΔP:	2.7 min*
MAXIMUM ΔI:	0.4 deg*

*See comments below

COMMENTS

Cosmos 1247 was another spacecraft of the Cosmos 862-type to experience a fragmentation. Cosmos 1247 appears to have completed the first burn of a 2-phase maneuver sequence on the event date, followed by debris generation. The ΔP and ΔI values above are based on the post-maneuver, 711-minute orbit of 12303 rather than the pre-maneuver, 718-minute orbit cited above.

REFERENCE DOCUMENT

History of Soviet/Russian Satellite Fragmentations-A Joint U.S.-Russian Investigation, N. L Johnson et al, Kaman Sciences Corporation, October 1995.

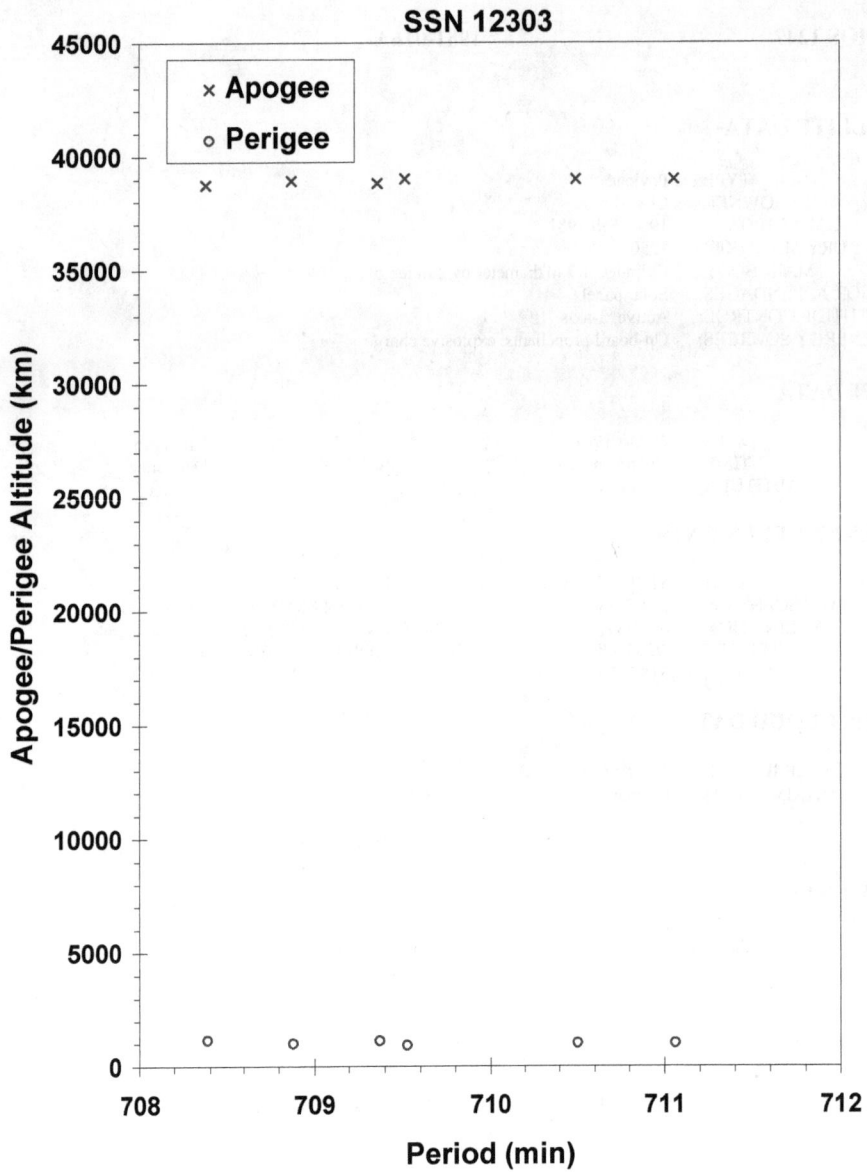

SSN 12303

Apogee/Perigee Altitude (km) vs Period (min)

Legend:
× Apogee
○ Perigee

Cosmos 1247 debris cloud of 6 fragments about 6 weeks after the event as reconstructed from the US SSN database.

COSMOS 1260   1981-028A   12364

SATELLITE DATA

TYPE:	Payload
OWNER:	CIS
LAUNCH DATE:	20.99+ Mar 1981
DRY MASS (KG):	3000
MAIN BODY:	Cylinder; 1.3 m diameter by 17 m length
MAJOR APPENDAGES:	Solar panels
ATTITUDE CONTROL:	Active, 3-axis
ENERGY SOURCES:	On-board propellants, explosive charge (?)

EVENT DATA (1)

DATE:	8 May 1982	LOCATION:	40N, 62E (asc)
TIME:	0444 GMT	ASSESSED CAUSE:	Unknown
ALTITUDE:	555 km		

PRE-EVENT ELEMENTS (1)

EPOCH:	82127.98788154	MEAN ANOMALY:	28.1726
RIGHT ASCENSION:	337.2406	MEAN MOTION:	14.88799005
INCLINATION:	65.0246	MEAN MOTION DOT/2:	.00003980
ECCENTRICITY:	.0214690	MEAN MOTION DOT DOT/6:	.0
ARG. OF PERIGEE:	330.7493	BSTAR:	.00028791

EVENT DATA (2)

DATE:	10 Aug 1982	LOCATION:	51N, 238E (dsc)
TIME:	2335 GMT	ASSESSED CAUSE:	Deliberate
ALTITUDE:	750 km		

PRE-EVENT ELEMENTS (2)

EPOCH:	82222.89259484	MEAN ANOMALY:	62.7628
RIGHT ASCENSION:	45.7388	MEAN MOTION:	14.89366232
INCLINATION:	65.0248	MEAN MOTION DOT/2:	.00004369
ECCENTRICITY:	.0219155	MEAN MOTION DOT DOT/6:	.0
ARG. OF PERIGEE:	295.0884	BSTAR:	.00030390

DEBRIS CLOUD DATA

MAXIMUM ΔP:	5.2 min
MAXIMUM ΔI:	1.0 deg

COMMENTS

Cosmos 1260 was the sixth spacecraft of the Cosmos 699-type to experience a fragmentation. The spacecraft had been in a regime of natural decay for 8 months before the first event. After the event the main remnant became satellite 13183, which then fragmented 3 months later. A total of 40 new fragments were officially cataloged prior to the second event. See also Cosmos 1220 and Cosmos 1306 for similar dual fragmentations of Cosmos 699-type spacecraft during this period.

REFERENCE DOCUMENTS

"Artificial Satellite Break-Ups (Part 1): Soviet Ocean Surveillance Satellites", N. L. Johnson, Journal of the British Interplanetary Society, February 1983, pp. 51-58.

History of Soviet/Russian Satellite Fragmentations-A Joint U.S.-Russian Investigation, N. L. Johnson et al, Kaman Sciences Corporation, October 1995.

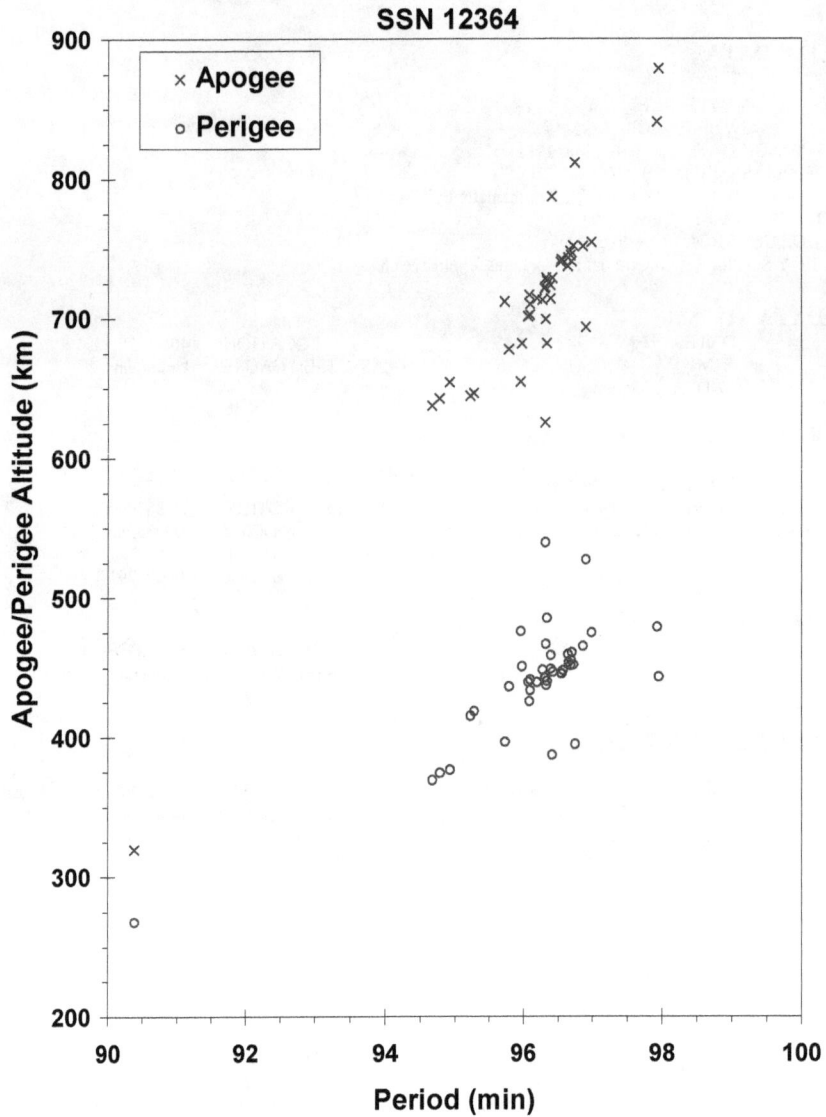

Cosmos 1260 debris cloud of 43 fragments 3 weeks after the first event from the US SSN database.

COSMOS 1261   1981-031A   12376

SATELLITE DATA

TYPE:	Payload
OWNER:	CIS
LAUNCH DATE:	31.40 Mar 1981
DRY MASS (KG):	1250
MAIN BODY:	Cylinder; 1.7 m diameter by 2 m length
MAJOR APPENDAGES:	Solar panels
ATTITUDE CONTROL:	Active, 3-axis
ENERGY SOURCES:	On-board propellants, explosive charge

EVENT DATA

DATE:	Apr-May 1981	LOCATION:	Unknown
TIME:	Unknown	ASSESSED CAUSE:	Deliberate
ALTITUDE:	Unknown		

PRE-EVENT ELEMENTS

EPOCH:	81095.90157023	MEAN ANOMALY:	4.6715
RIGHT ASCENSION:	282.6240	MEAN MOTION:	2.00494188
INCLINATION:	63.0386	MEAN MOTION DOT/2:	.0
ECCENTRICITY:	.7369210	MEAN MOTION DOT DOT/6:	.0
ARG. OF PERIGEE:	316.4347	BSTAR:	.0

DEBRIS CLOUD DATA

MAXIMUM ΔP:	2.3 min*
MAXIMUM ΔI:	0.3 deg*

*Based on uncataloged debris data

COMMENTS

Cosmos 1261 was another spacecraft of the Cosmos 862-type to experience a fragmentation. The spacecraft attempted to maneuver from its transfer orbit to an operational orbit 3 days after launch. The maneuver appears to have been unsuccessful, and the spacecraft never became groundtrack-stabilized. Some debris appeared immediately after the maneuver, while additional debris were discovered in mid-May. More than one event may have occurred. The element set above is the first available after the unsuccessful maneuver.

REFERENCE DOCUMENT

History of Soviet/Russian Satellite Fragmentations-A Joint U.S.-Russian Investigation, N. L. Johnson et al, Kaman Sciences Corporation, October 1995.

167

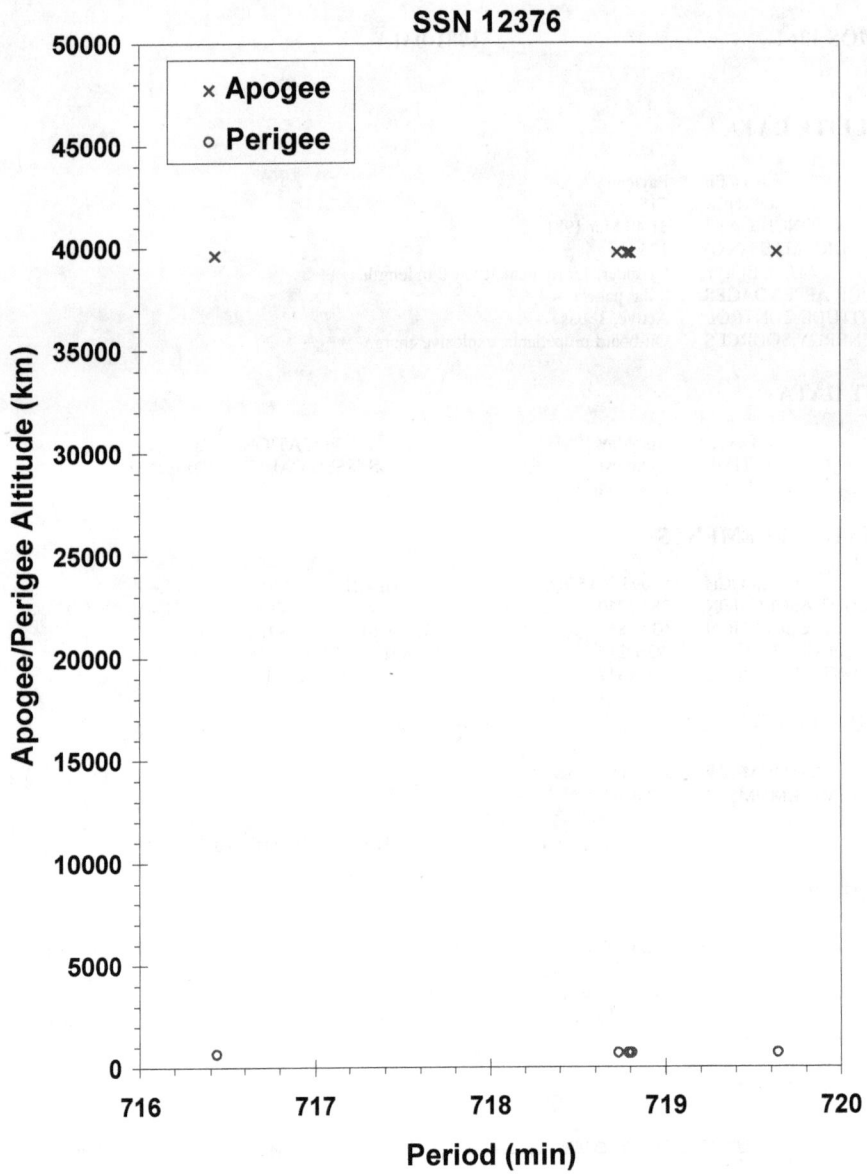

SSN 12376

Cosmos 1261 debris cloud of 6 fragments about 8 weeks after (initial) event as reconstructed from the US SSN database.

COSMOS 1275   1981-053A   12504

SATELLITE DATA

TYPE:	Payload
OWNER:	CIS
LAUNCH DATE:	4.66 Jun 1981
DRY MASS (KG):	800
MAIN BODY:	Cylinder; 2.4 m diameter by 4 m length
MAJOR APPENDAGES:	Gravity-gradient boom
ATTITUDE CONTROL:	Gravity gradient
ENERGY SOURCES:	Unknown

EVENT DATA

DATE:	24 Jul 1981	LOCATION:	68N, 197E (asc)
TIME:	2351 GMT	ASSESSED CAUSE:	Battery
ALTITUDE:	980 km		

PRE-EVENT ELEMENTS

EPOCH:	81205.39693092	MEAN ANOMALY:	221.3567
RIGHT ASCENSION:	119.8245	MEAN MOTION:	13.73455672
INCLINATION:	82.9633	MEAN MOTION DOT/2:	.000000580
ECCENTRICITY:	.0036415	MEAN MOTION DOT DOT/6:	.0
ARG. OF PERIGEE:	139.0334	BSTAR:	.00004538900

DEBRIS CLOUD DATA

MAXIMUM ΔP:	4.9 min
MAXIMUM ΔI:	0.4 deg

COMMENTS

Cosmos 1275 is the only member of its class to fragment. The satellite was only 50 days old at the time of the event. During the February 1992 Space Debris Conference in Moscow, Russian analysts discussed independent studies about the probable cause of the breakup. Later, the official Russian assessment asserted that a battery malfunction was the likely culprit.

REFERENCE DOCUMENTS

The Fragmentations of USSR Satellites 11729 and 12504 (U), J.R. Gabbard and P.M. Landry, Technical Memorandum 82-S-03, DCS/Plans, Hdqtrs NORAD/ADCOM, Colorado Springs, August 1982 (Secret).

Determining the Cause of a Satellite Breakup: A Case Study of the Kosmos 1275 Breakup, D.S. McKnight, IAA-87-573, 38th Congress of the International Astronautical Federation, Brighton, England, October 1987.

History of Soviet/Russian Satellite Fragmentations-A Joint U.S.-Russian Investigation, N. L. Johnson et al, Kaman Sciences Corporation, October 1995.

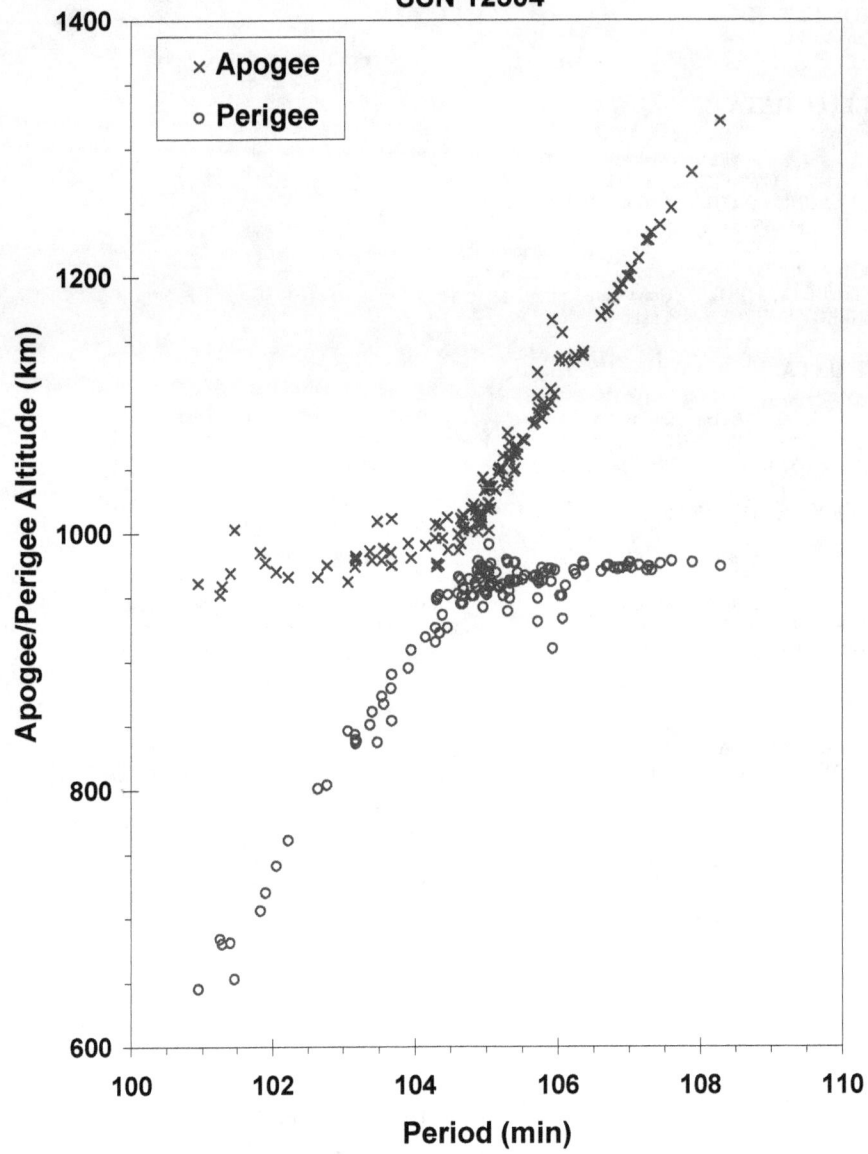

SSN 12504

Cosmos 1275 debris cloud of 136 identified fragments 1 week after the event as reconstructed from the US SSN database.

COSMOS 1278   1981-058A   12547

SATELLITE DATA

TYPE:	Payload
OWNER:	CIS
LAUNCH DATE:	19.81 Jun 1981
DRY MASS (KG):	1250
MAIN BODY:	Cylinder; 1.7 m diameter by 2 m length
MAJOR APPENDAGES:	Solar panels
ATTITUDE CONTROL:	Active, 3-axis
ENERGY SOURCES:	On-board propellants, explosive charge

EVENT DATA

DATE:	Early Dec 1986	LOCATION:	Unknown
TIME:	Unknown	ASSESSED CAUSE:	Deliberate
ALTITUDE:	Unknown		

PRE-EVENT ELEMENTS

EPOCH:	86334.22199701	MEAN ANOMALY:	12.7886
RIGHT ASCENSION:	288.0814	MEAN MOTION:	2.00618298
INCLINATION:	67.1073	MEAN MOTION DOT/2:	.0
ECCENTRICITY:	.6594262	MEAN MOTION DOT DOT/6:	.0
ARG. OF PERIGEE:	291.9890	BSTAR:	.0

DEBRIS CLOUD DATA

MAXIMUM ΔP:	0.1 min
MAXIMUM ΔI:	0.0 deg

COMMENTS

Cosmos 1278 was another spacecraft of the Cosmos 862-type to experience a fragmentation. Spacecraft had apparently been inactive since early 1984. Additional fragments may exist, but surveillance for small objects in this orbit is difficult.

REFERENCE DOCUMENT

History of Soviet/Russian Satellite Fragmentations-A Joint U.S.-Russian Investigation, N. L. Johnson et al, Kaman Sciences Corporation, October 1995.

SSN 12547

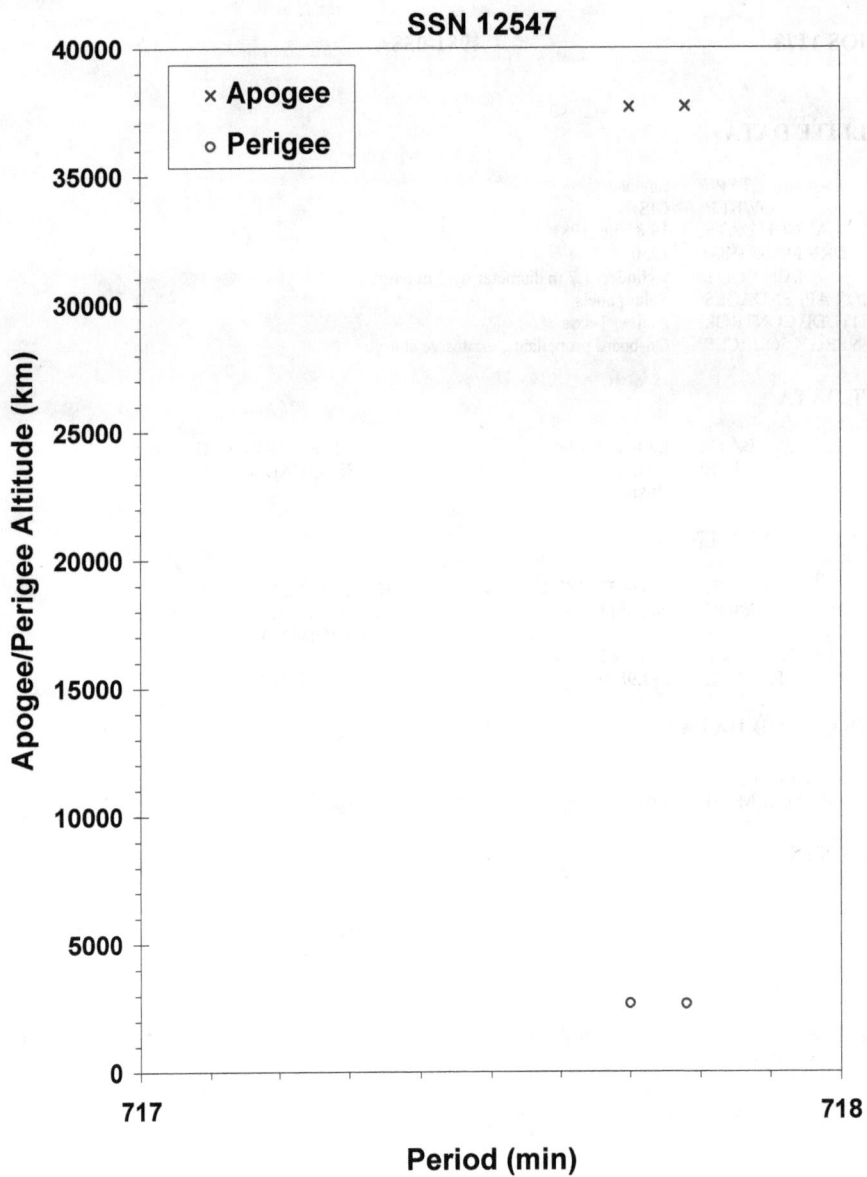

Cosmos 1278 and additional fragment in mid-December 1986. Elements from the US SSN as published by the NASA Goddard Space Flight Center.

COSMOS 1285   1981-071A   12627

SATELLITE DATA

TYPE:	Payload
OWNER:	CIS
LAUNCH DATE:	4.01 Aug 1981
DRY MASS (KG):	1250
MAIN BODY:	Cylinder; 1.7 m diameter by 2 m length
MAJOR APPENDAGES:	Solar panels
ATTITUDE CONTROL:	Active, 3-axis
ENERGY SOURCES:	On-board propellants, explosive charge

EVENT DATA

DATE:	21 Nov 1981	LOCATION:	Unknown
TIME:	Unknown	ASSESSED CAUSE:	Deliberate
ALTITUDE:	Unknown		

PRE-EVENT ELEMENTS

EPOCH:	81324.16708257	MEAN ANOMALY:	4.8196
RIGHT ASCENSION:	249.5852	MEAN MOTION:	1.98014597
INCLINATION:	63.1086	MEAN MOTION DOT/2:	.00000781
ECCENTRICITY:	.7350717	MEAN MOTION DOT DOT/6:	.0
ARG. OF PERIGEE:	317.0022	BSTAR:	.0

DEBRIS CLOUD DATA

MAXIMUM ΔP:	8.6 min*
MAXIMUM ΔI:	0.2 deg*

*Based on uncataloged debris data

COMMENTS

Cosmos 1285 was another spacecraft of the Cosmos 862-type to experience a fragmentation. Spacecraft was placed in a temporary transfer orbit on the day of launch by its launch vehicle but never maneuvered to an operational orbit, suggesting an early fatal spacecraft malfunction. Event occurred 3.5 months after the launch.

REFERENCE DOCUMENT

History of Soviet/Russian Satellite Fragmentations-A Joint U.S.-Russian Investigation, N. L. Johnson et al, Kaman Sciences Corporation, October 1995.

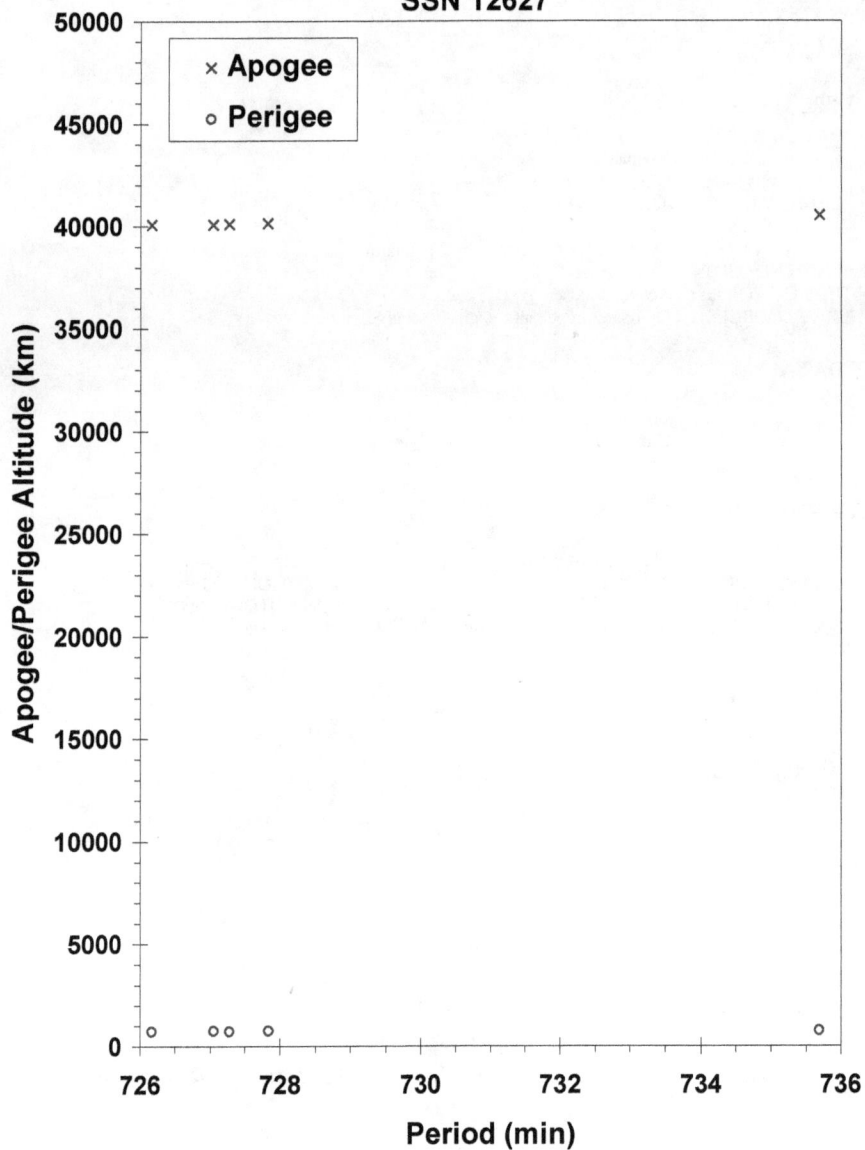

SSN 12627

Cosmos 1285 debris cloud of 5 fragments less than 1 week after the event as reconstructed from the US SSN database.

SATELLITE DATA

TYPE:	Payload
OWNER:	CIS
LAUNCH DATE:	4.35 Aug 1981
DRY MASS (KG):	3000
MAIN BODY:	Cylinder; 1.3 m diameter by 17 m length
MAJOR APPENDAGES:	Solar panels
ATTITUDE CONTROL:	Active, 3-axis
ENERGY SOURCES:	On-board propellants, explosive charge (?)

EVENT DATA

DATE:	29 Sep 1982	LOCATION:	51N, 80E (asc)
TIME:	0520 GMT	ASSESSED CAUSE:	Unknown
ALTITUDE:	325 km		

PRE-EVENT ELEMENTS

EPOCH:	82272.21193719	MEAN ANOMALY:	92.4681
RIGHT ASCENSION:	132.9736	MEAN MOTION:	15.86141247
INCLINATION:	65.0071	MEAN MOTION DOT/2:	.00400345
ECCENTRICITY:	.0017215	MEAN MOTION DOT DOT/6:	.0
ARG. OF PERIGEE:	267.4145	BSTAR:	.0015199

DEBRIS CLOUD DATA

MAXIMUM ΔP:	0.9 min*
MAXIMUM ΔI:	0.2 deg*

*Based on uncataloged debris data

COMMENTS

Cosmos 1286 was the ninth spacecraft of the Cosmos 699-type to experience a fragmentation. The spacecraft had been in a regime of natural decay for more than 6 months at the time of the event. The low altitude and high drag conditions made determination of the precise breakup time uncertain. The breakup or a precursor event may have occurred earlier on 29 September 1982. Most fragments decayed before being officially cataloged.

REFERENCE DOCUMENTS

"Artificial Satellite Break-Ups (Part 1): Soviet Ocean Surveillance Satellites", N. L. Johnson, Journal of the British Interplanetary Society, February 1983, pp. 51-58.

History of Soviet/Russian Satellite Fragmentation-A Joint U.S.-Russian Investigation, N. L. Johnson et al, Kaman Sciences Corporation, October 1995.

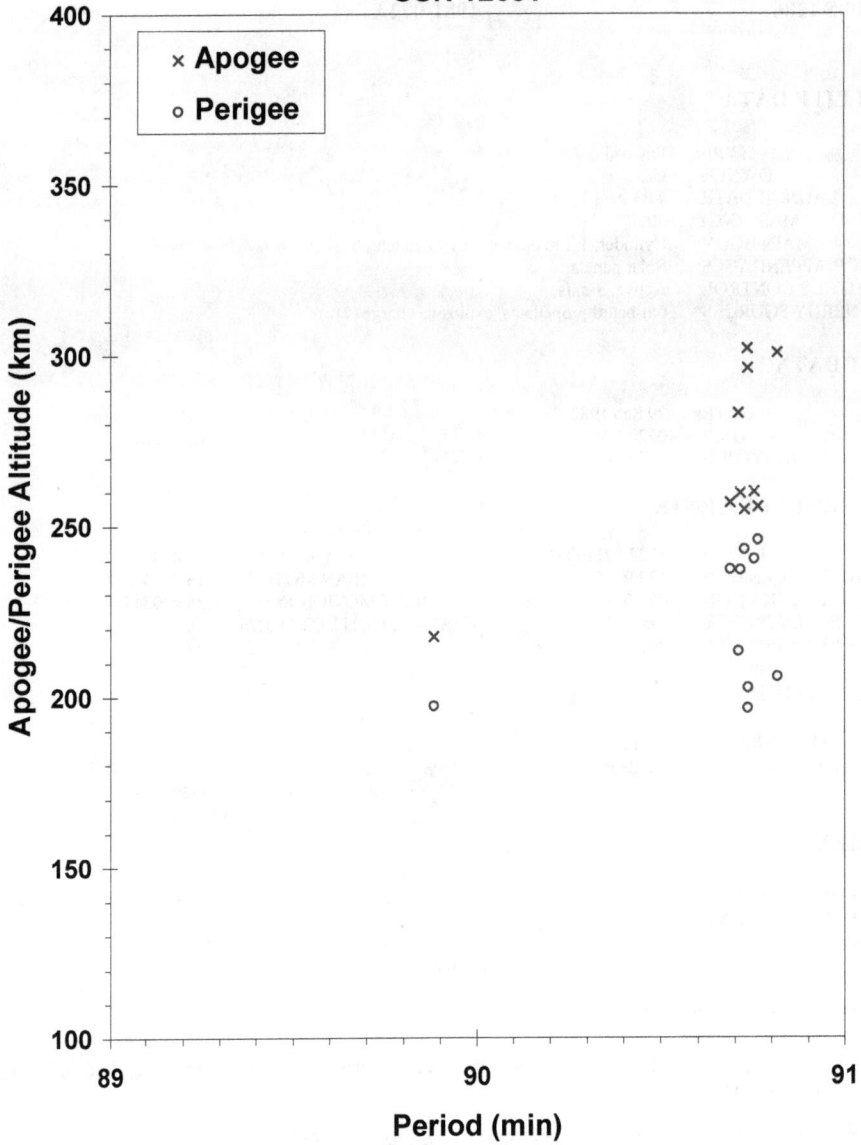

SSN 12631

Cosmos 1286 debris cloud of 10 fragments 1 day after the event as reconstructed from the US SSN database.

COSMOS 1305 R/B   1981-088F   12827

SATELLITE DATA

TYPE:	Molniya Final Stage
OWNER:	CIS
LAUNCH DATE:	11.36 Sep 1981
DRY MASS (KG):	1100
MAIN BODY:	Cylinder; 2.7 m diameter by 3 m length
MAJOR APPENDAGES:	None
ATTITUDE CONTROL:	Active, 3-axis
ENERGY SOURCES:	On-board propellants

EVENT DATA

DATE:	11 Sep 1981	LOCATION:	Unknown
TIME:	Unknown	ASSESSED CAUSE:	Propulsion
ALTITUDE:	Unknown		

POST-EVENT ELEMENTS

EPOCH:	81258.60717998	MEAN ANOMALY:	26.9249
RIGHT ASCENSION:	68.6245	MEAN MOTION:	5.48678032
INCLINATION:	62.8166	MEAN MOTION DOT/2:	.0
ECCENTRICITY:	.4855644	MEAN MOTION DOT DOT/6:	.0
ARG. OF PERIGEE:	286.6972	BSTAR:	.0

DEBRIS CLOUD DATA

MAXIMUM ΔP:	Unknown
MAXIMUM ΔI:	Unknown

COMMENTS

Cosmos 1305 R/B malfunctioned about 1 hour after launch during a maneuver from a LEO parking orbit to a Molniya-type orbit. The maneuver was initiated at approximately 0937 GMT near 58S, 245E (asc) at an altitude of 600 km. Apogee was raised to less than 14,000 km. Debris tracking after the event was limited, preventing an accurate assessment of magnitude of the event. First debris officially cataloged in June 1983. Debris generation is assumed to have occurred during or immediately after the unsuccessful maneuver. The element set above is for the rocket body after burn termination.

REFERENCE DOCUMENT

History of Soviet/Russian Satellite Fragmentations-A Joint U.S.-Russian Investigation, N. L. Johnson et al, Kaman Sciences Corporation, October 1995.

SSN 12827

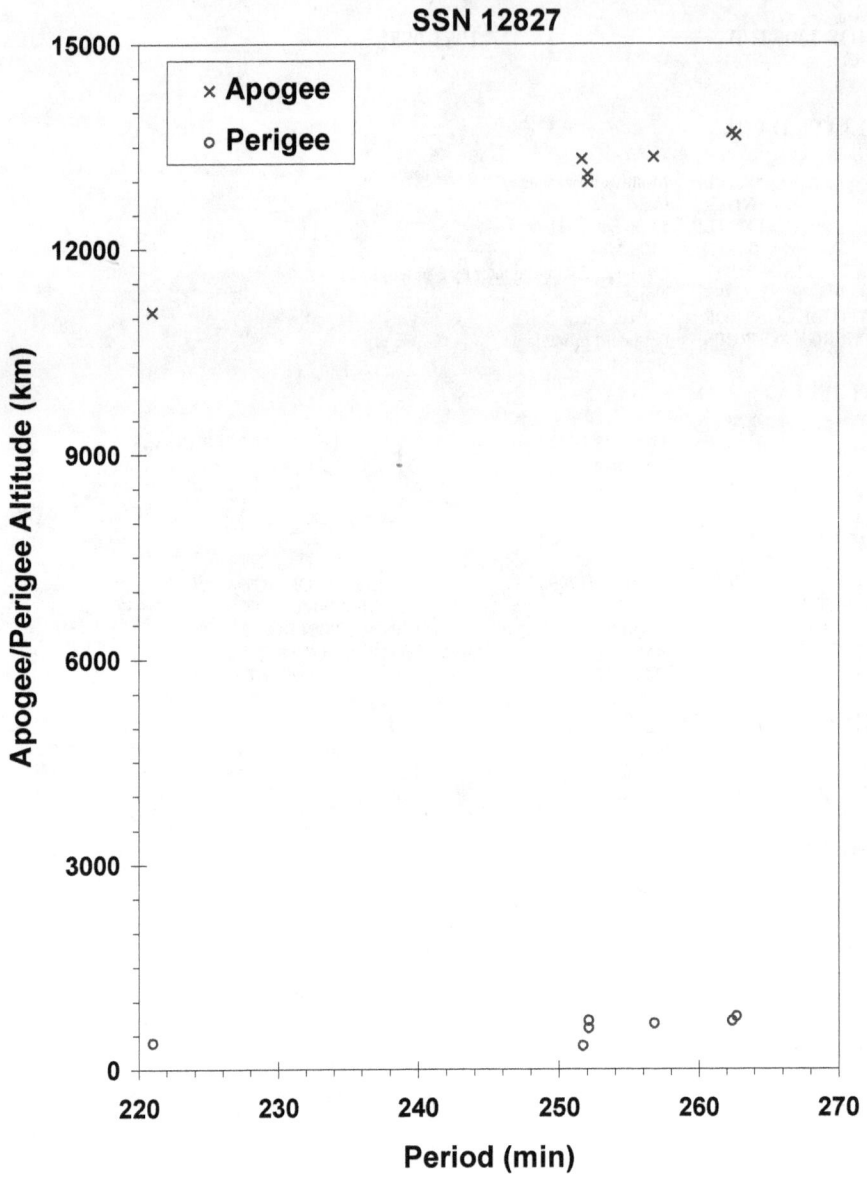

Cosmos 1305 R/B debris cloud of 7 fragments about 2 years after the event as reconstructed from the US SSN database.

COSMOS 1306   1981-089A   12828

SATELLITE DATA

TYPE:	Payload
OWNER:	CIS
LAUNCH DATE:	14.85 Sep 1981
DRY MASS (KG):	3000
MAIN BODY:	Cylinder; 1.3 m diameter by 17 m length
MAJOR APPENDAGES:	Solar panels
ATTITUDE CONTROL:	Active, 3-axis
ENERGY SOURCES:	On-board propellants, explosive charge (?)

EVENT DATA (1)

DATE:	12 Jul 1982	LOCATION:	65S, 40E (asc)
TIME:	2325 GMT	ASSESSED CAUSE:	Unknown
ALTITUDE:	380 km		

PRE-EVENT ELEMENTS (1)

EPOCH:	82193.22052182	MEAN ANOMALY:	72.7640
RIGHT ASCENSION:	43.8843	MEAN MOTION:	15.58171668
INCLINATION:	64.9399	MEAN MOTION DOT/2:	.00042116
ECCENTRICITY:	.0019953	MEAN MOTION DOT DOT/6:	.0
ARG. OF PERIGEE:	287.2390	BSTAR:	.00055055

EVENT DATA (2)

DATE:	18 Sep 1982	LOCATION:	32N, 293E (asc)
TIME:	1702 GMT	ASSESSED CAUSE:	Unknown
ALTITUDE:	370 km		

PRE-EVENT ELEMENTS (2)

EPOCH:	82260.17037940	MEAN ANOMALY:	44.8033
RIGHT ASCENSION:	173.7764	MEAN MOTION:	15.65882738
INCLINATION:	64.9408	MEAN MOTION DOT/2:	.00076164
ECCENTRICITY:	.0002181	MEAN MOTION DOT DOT/6:	.0
ARG. OF PERIGEE:	315.2578	BSTAR:	.00073994

DEBRIS CLOUD DATA

MAXIMUM ΔP:	2.1 min*
MAXIMUM ΔI:	0.2 deg

*Based on uncataloged debris data

COMMENTS

Cosmos 1306 was the eighth spacecraft of the Cosmos 699-type to experience a fragmentation. The first event occurred 5 months after the spacecraft had entered a regime of natural decay. After the event the main remnant was tagged as satellite 13369, while a piece of debris tagged as 12828 decayed on 16 July 1982. Only 5 new fragments were officially cataloged prior to the second event when satellite 13369 experienced a fragmentation. Three long-lived fragments cataloged with 1981-89 (13393, 13404, and 14837) were actually part of the breakup of 1980-89, another Cosmos 699-type satellite. Most Cosmos 1306 debris reentered quickly and elements were developed for only a few fragments.

REFERENCE DOCUMENTS

<u>Analysis of Cosmos 1220 and Cosmos 1306 Fragments (U)</u>, D. Fennessy, Report AH-23, FTD/OLAI, Cheyenne Mountain, Colorado, 12 January 1983 (Secret)

"Artificial Satellite Break-Ups (Part 1): Soviet Ocean Surveillance Satellites", N. L. Johnson, <u>Journal of the British Interplanetary Society</u>, February 1983, pp. 51-58.

<u>History of Soviet/Russian Satellite Fragmentations-A Joint U.S.-Russian Investigation</u>, N. L. Johnson et al, Kaman Sciences Corporation, October 1995.

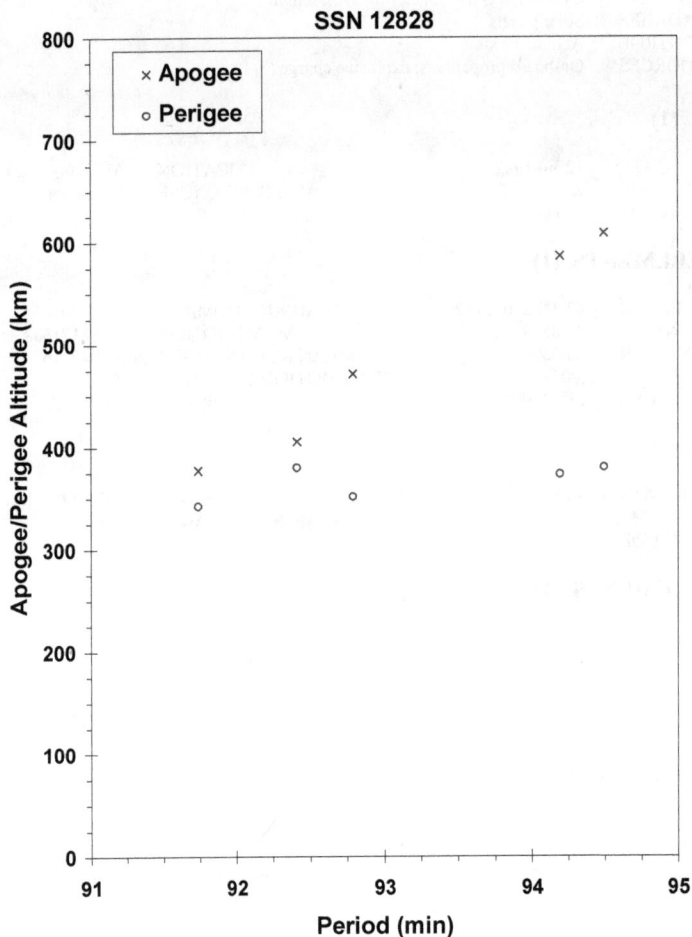

Cosmos 1306 debris cloud of 5 identified fragments 1 day after the event as reconstructed from the US SSN database.

COSMOS 1317   1981-108A   12933

SATELLITE DATA

TYPE:	Payload
OWNER:	CIS
LAUNCH DATE:	31.95 Oct 1981
DRY MASS (KG):	1250
MAIN BODY:	Cylinder; 1.7 m diameter by 2 m length
MAJOR APPENDAGES:	Solar panels
ATTITUDE CONTROL:	Active, 3-axis
ENERGY SOURCES:	On-board propellants, explosive

EVENT DATA

DATE:	25-28 Jan 1984	LOCATION:	Unknown
TIME:	Unknown	ASSESSED CAUSE:	Deliberate
ALTITUDE:	Unknown		

PRE-EVENT ELEMENTS

EPOCH:	84024.46309667	MEAN ANOMALY:	4.4900
RIGHT ASCENSION:	219.5352	MEAN MOTION:	2.00535027
INCLINATION:	62.8286	MEAN MOTION DOT/2:	.0
ECCENTRICITY:	.7103977	MEAN MOTION DOT DOT/6:	.0
ARG. OF PERIGEE:	324.1891	BSTAR:	.0

DEBRIS CLOUD DATA

MAXIMUM ΔP:	1.8 min
MAXIMUM ΔI:	0.3 deg

COMMENTS

Cosmos 1317 was another spacecraft of the Cosmos 862-type to experience a fragmentation

REFERENCE DOCUMENT

History of Soviet/Russian Satellite Fragmentations-A Joint U.S.-Russian Investigation, N. L. Johnson et al, Kaman Sciences Corporation, October 1995.

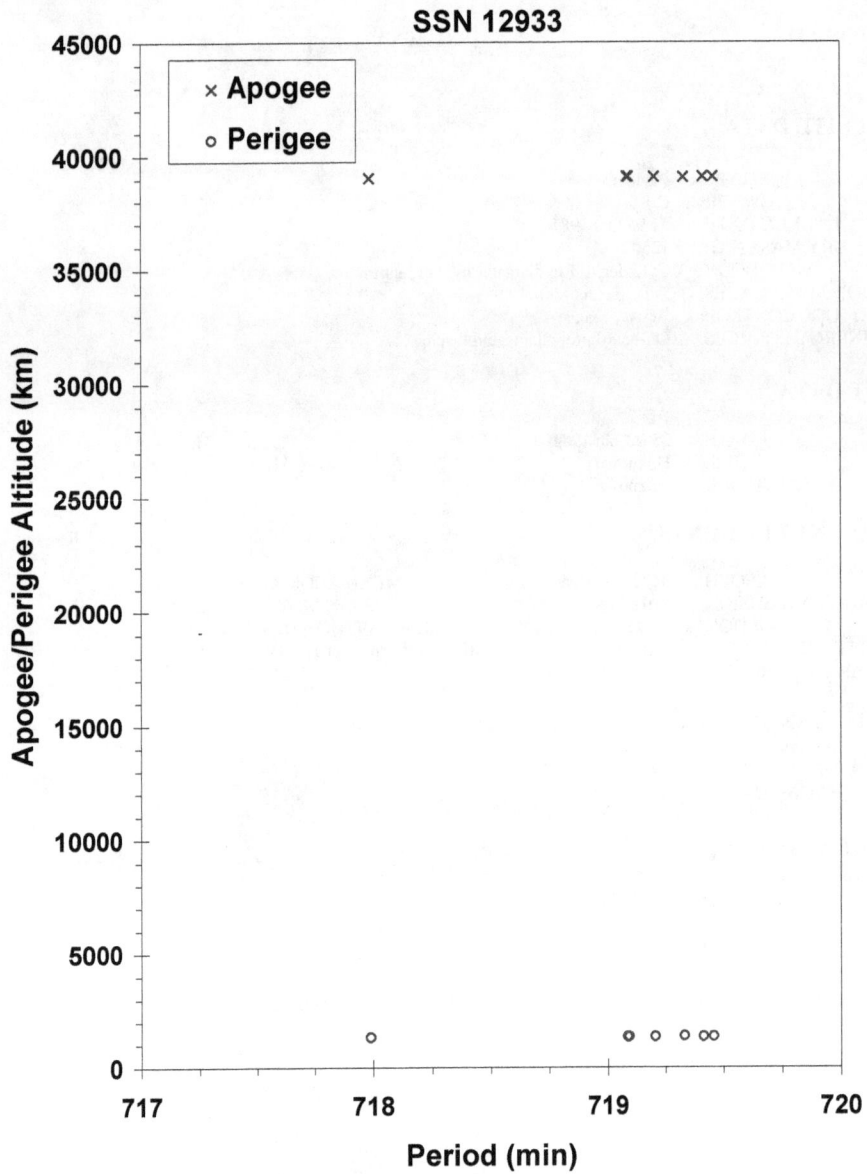

Cosmos 1317 debris cloud of 7 fragments about 2 weeks after the event as reconstructed from the US SSN database.

SATELLITE DATA

TYPE:	Payload
OWNER:	CIS
LAUNCH DATE:	29.41 Apr 1982
DRY MASS (KG):	3000
MAIN BODY:	Cylinder; 1.3 m diameter by 17 m length
MAJOR APPENDAGES:	Solar panels
ATTITUDE CONTROL:	Active, 3-axis
ENERGY SOURCES:	On-board propellants, explosive charge (?)

EVENT DATA (1)

DATE:	8 Aug 1983	LOCATION:	32S, 310E (asc)
TIME:	2331 GMT	ASSESSED CAUSE:	Unknown
ALTITUDE:	365 km		

PRE-EVENT ELEMENTS (1)

EPOCH:	83220.21851552	MEAN ANOMALY:	66.8795
RIGHT ASCENSION:	279.4096	MEAN MOTION:	15.63233551
INCLINATION:	65.0504	MEAN MOTION DOT/2:	.00048258
ECCENTRICITY:	.0024043	MEAN MOTION DOT DOT/6:	.0
ARG. OF PERIGEE:	292.8515	BSTAR:	.00051620

EVENT DATA (2)

DATE:	1 Feb 1984	LOCATION:	4S, 200E (asc)
TIME:	0322 GMT	ASSESSED CAUSE:	Unknown
ALTITUDE:	320 km		

PRE-EVENT ELEMENTS (2)

EPOCH:	84031.38369465	MEAN ANOMALY:	81.7159
RIGHT ASCENSION:	25.3553	MEAN MOTION:	15.84652631
INCLINATION:	65.0404	MEAN MOTION DOT/2:	.00119378
ECCENTRICITY:	.0017572	MEAN MOTION DOT DOT/6:	.0
ARG. OF PERIGEE:	278.1110	BSTAR:	.00050318

EVENT DATA (3)

DATE:	20 Feb 1984	LOCATION:	Unknown
TIME:	Before 0340 GMT	ASSESSED CAUSE:	Unknown
ALTITUDE:	Unknown		

PRE-EVENT ELEMENTS (3)

EPOCH:	84050.69015256	MEAN ANOMALY:	105.8772
RIGHT ASCENSION:	316.3115	MEAN MOTION:	15.97914042
INCLINATION:	65.0338	MEAN MOTION DOT/2:	.00430956
ECCENTRICITY:	.0014134	MEAN MOTION DOT DOT/6:	.000083799
ARG. OF PERIGEE:	254.0517	BSTAR:	.00093344

DEBRIS CLOUD DATA

MAXIMUM ΔP:	36.8 min*
MAXIMUM ΔI:	2.3 deg*

*Based on uncataloged debris data (Event 1)

COMMENTS

Cosmos 1355 was the tenth spacecraft of the Cosmos 699-type to experience a fragmentation. The spacecraft had been in a regime of natural decay for 6 months prior to the first event. Twenty-one fragments were cataloged following the first event, and the main body became satellite 14275. This object spawned at least seven more fragments on 1 February. The parent was then retagged to the original 13150 satellite number. The third event resulted in the development of 13 new fragment element sets, but none were cataloged and the low altitude prevented an estimate of a precise breakup location.

REFERENCE DOCUMENTS

"Artificial Satellite Break-Ups (Part 1): Soviet Ocean Surveillance Satellites", N. L. Johnson, Journal of the British Interplanetary Society, February 1983, pp. 51-58.

Analysis of the Fragmentation of Kosmos 1355, N. L. Johnson, Technical Report CS84-SPACECMD-28, Teledyne Brown Engineering, Colorado Springs, January 1985.

History of Soviet/Russian Satellite Fragmentations-A Joint U.S.-Russian Investigation, N. L. Johnson et al, Kaman Sciences Corporation, October 1995.

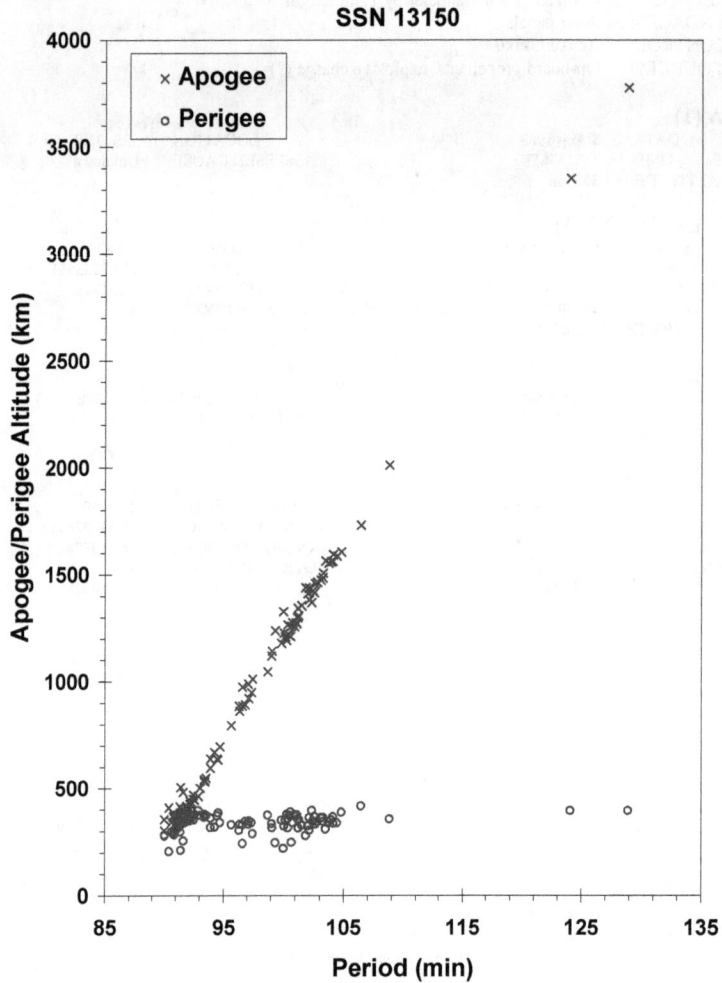

SSN 13150

Cosmos 1355 debris cloud of 150 fragments about 7 hours after the first event in August 1983 as seen by the US SSN PARCS radar. Figure from the cited reference.

184

SATELLITE DATA

TYPE:	Payload
OWNER:	CIS
LAUNCH DATE:	6.72 Jun 1982
DRY MASS (KG):	650
MAIN BODY:	Polyhedron; 1.4 m by 1.4 m
MAJOR APPENDAGES:	Solar panels, gravity-gradient boom (?)
ATTITUDE CONTROL:	Gravity gradient (?)
ENERGY SOURCES:	Battery

EVENT DATA

DATE:	21 Oct 1985	LOCATION:	66N, 351E (asc)
TIME:	0353 GMT	ASSESSED CAUSE:	Battery
ALTITUDE:	995 km		

PRE-EVENT ELEMENTS

EPOCH:	85293.85195210	MEAN ANOMALY:	333.5602
RIGHT ASCENSION:	350.2805	MEAN MOTION:	13.71079597
INCLINATION:	65.8390	MEAN MOTION DOT/2:	.00000158
ECCENTRICITY:	.0005355	MEAN MOTION DOT DOT/6:	.0
ARG. OF PERIGEE:	26.5667	BSTAR:	.00023894

DEBRIS CLOUD DATA

MAXIMUM ΔP:	2.3 min*
MAXIMUM ΔI:	0.1 deg*

*Based on uncataloged debris data

COMMENTS

Cosmos 1375 was the third spacecraft of the Cosmos 839-type to experience a fragmentation. Although these satellites are used in conjunction with the Cosmos 249-type spacecraft that are deliberately fragmented, the cause of Cosmos 839-type events appears to be unrelated. In the case of Cosmos 1375, 40 months elapsed since its test with a Cosmos 249-type spacecraft.

REFERENCE DOCUMENTS

"Artificial Satellite Break-Ups (Part 2): Soviet Anti-Satellite Program", N.L. Johnson, Journal of the British Interplanetary Society, August 1983, pp. 357-362.

Analysis of the Kosmos 1375 Fragmentation, J. M. Koskella and R. L. Kling, Technical Report CS86-USASDC-0006, Teledyne Brown Engineering, Colorado Springs, March 1986.

History of Soviet/Russian Satellite Fragmentations-A Joint U.S.-Russian Investigation, N. L. Johnson et al, Kaman Sciences Corporation, October 1995.

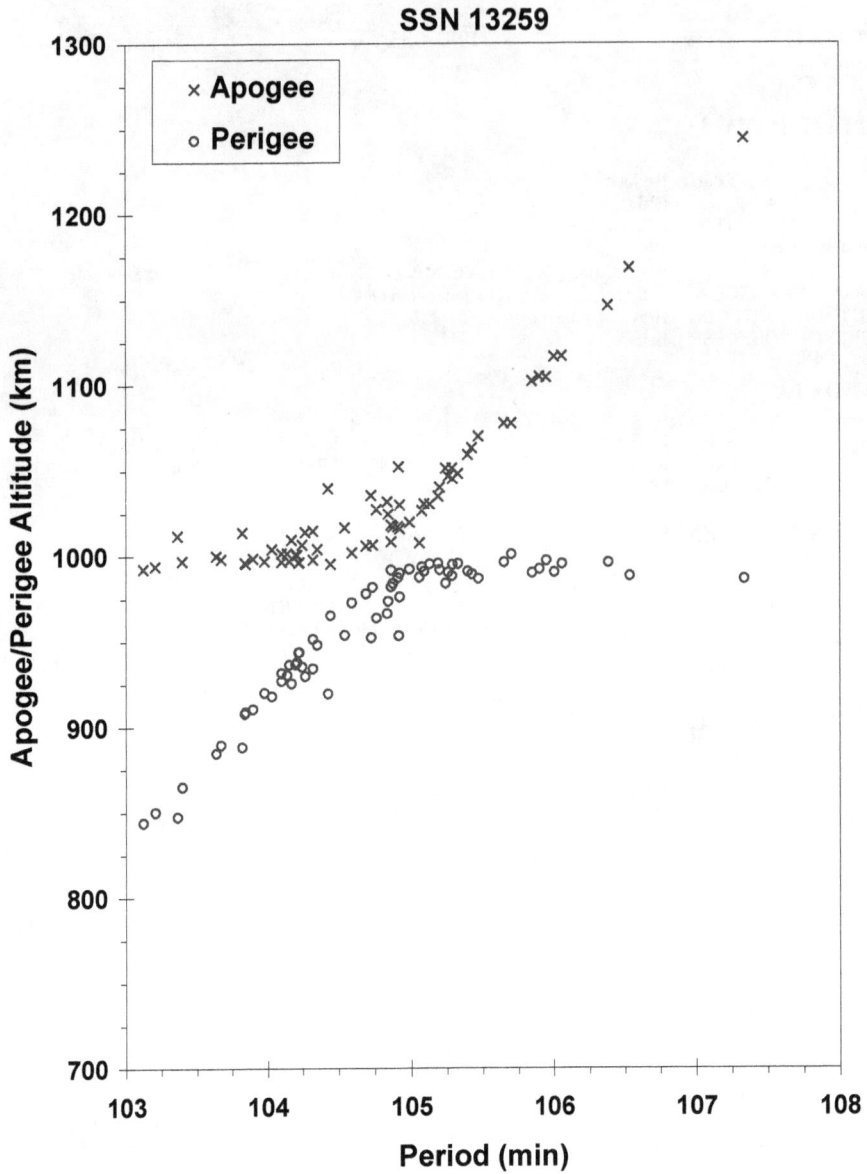

SSN 13259

Cosmos 1375 debris cloud of 68 fragments seen a few hours after the event by the US SSN PARCS radar.

COSMOS 1405   1982-088A   13508

SATELLITE DATA

TYPE:	Payload
OWNER:	CIS
LAUNCH DATE:	4.74 Sep 1982
DRY MASS (KG):	3000
MAIN BODY:	Cylinder; 1.3 m diameter by 17 m length
MAJOR APPENDAGES:	Solar panels
ATTITUDE CONTROL:	Active, 3-axis
ENERGY SOURCES:	On-board propellants, explosive charge (?)

EVENT DATA

DATE:	20 Dec 1983	LOCATION:	25S, 45E (dsc)
TIME:	1215 GMT	ASSESSED CAUSE:	Unknown
ALTITUDE:	330 km		

PRE-EVENT ELEMENTS

EPOCH:	83354.22079767	MEAN ANOMALY:	42.0375
RIGHT ASCENSION:	126.1259	MEAN MOTION:	15.81899265
INCLINATION:	65.0055	MEAN MOTION DOT/2:	.00186341
ECCENTRICITY:	.0020774	MEAN MOTION DOT DOT/6:	.0
ARG. OF PERIGEE:	318.0927	BSTAR:	.00088277

DEBRIS CLOUD DATA

MAXIMUM ΔP:	7.3 min*
MAXIMUM ΔI:	2.0 deg*

*Based on uncataloged debris data

COMMENTS

Cosmos 1405 was the eleventh spacecraft of the Cosmos 699-type to experience a fragmentation. Spacecraft had been in natural decay for 12 months prior to the event. Most debris reentered before being officially cataloged.

REFERENCE DOCUMENTS

"Artificial Satellite Break-Ups (Part 1): Soviet Ocean Surveillance Satellites", N. L. Johnson, Journal of the British Interplanetary Society, February 1983, pp. 51-58.

Separation of Objects from Cosmos 1405, F.T. Lipp, NAVSPASUR Technical Note 1-84, Naval Space Surveillance System, Dahlgren, 2 April 1984.

Analysis of the Fragmentation of Kosmos 1405, N.L. Johnson, Technical Report CS84-SPACECMD-10, Teledyne Brown Engineering, Colorado Springs, September 1984.

History of Soviet/Russian Satellite Fragmentations-A Joint U.S.-Russian Investigation, N. L. Johnson et al, Kaman Sciences Corporation, October 1995.

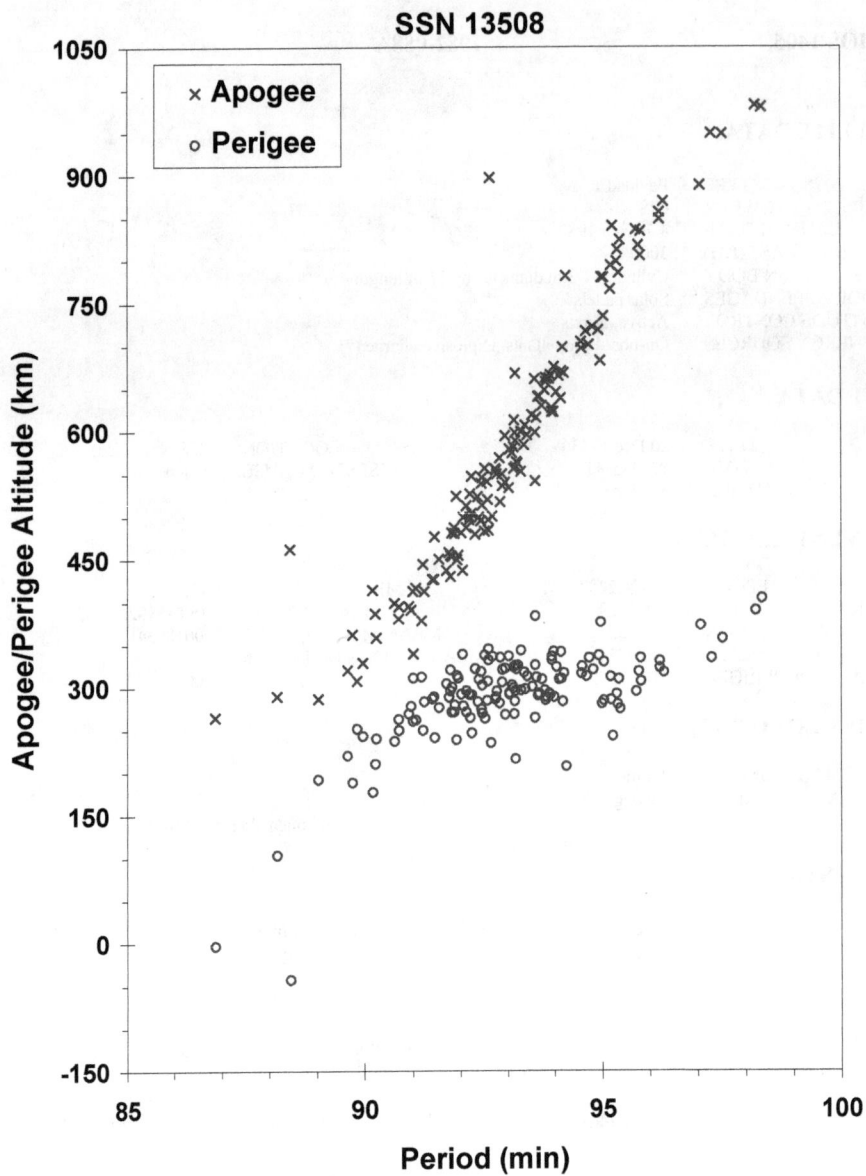

SSN 13508

Cosmos 1405 debris cloud of 143 fragments 1 hour after the event as seen by the US SSN PARCS radar.

COSMOS 1423 R/B   1982-115E   13696

SATELLITE DATA

TYPE:	Molniya Final Stage
OWNER:	CIS
LAUNCH DATE:	8.58 Dec 1982
DRY MASS (KG):	1100
MAIN BODY:	Cylinder; 2.7 m diameter by 3 m length
MAJOR APPENDAGES:	None
ATTITUDE CONTROL:	Active, 3-axis
ENERGY SOURCES:	On-board propellants

EVENT DATA

DATE:	8 Dec 1982	LOCATION:	62S, 302E (asc)
TIME:	1448 GMT	ASSESSED CAUSE:	Propulsion
ALTITUDE:	400 km		

PRE-EVENT ELEMENTS

EPOCH:	82342.56790507	MEAN ANOMALY:	305.2204
RIGHT ASCENSION:	316.3789	MEAN MOTION:	15.79849844
INCLINATION:	62.9496	MEAN MOTION DOT/2:	.0
ECCENTRICITY:	.0143321	MEAN MOTION DOT DOT/6:	.0
ARG. OF PERIGEE:	56.2493	BSTAR:	.0

DEBRIS CLOUD DATA

MAXIMUM ΔP:	4.9 min
MAXIMUM ΔI:	0.2 deg

COMMENTS

Fragmentation occurred at the time the Molniya final stage was fired to move the payload from a parking orbit to a Molniya-type transfer orbit. Pre-event elements are taken from satellite 13686 for first revolution parking orbit. A second fragmentation may have occurred on 9 December 1982.

REFERENCE DOCUMENT

History of Soviet/Russian Satellite Fragmentations-A Joint U.S.-Russian Investigation, N. L. Johnson et al, Kaman Sciences Corporation, October 1995.

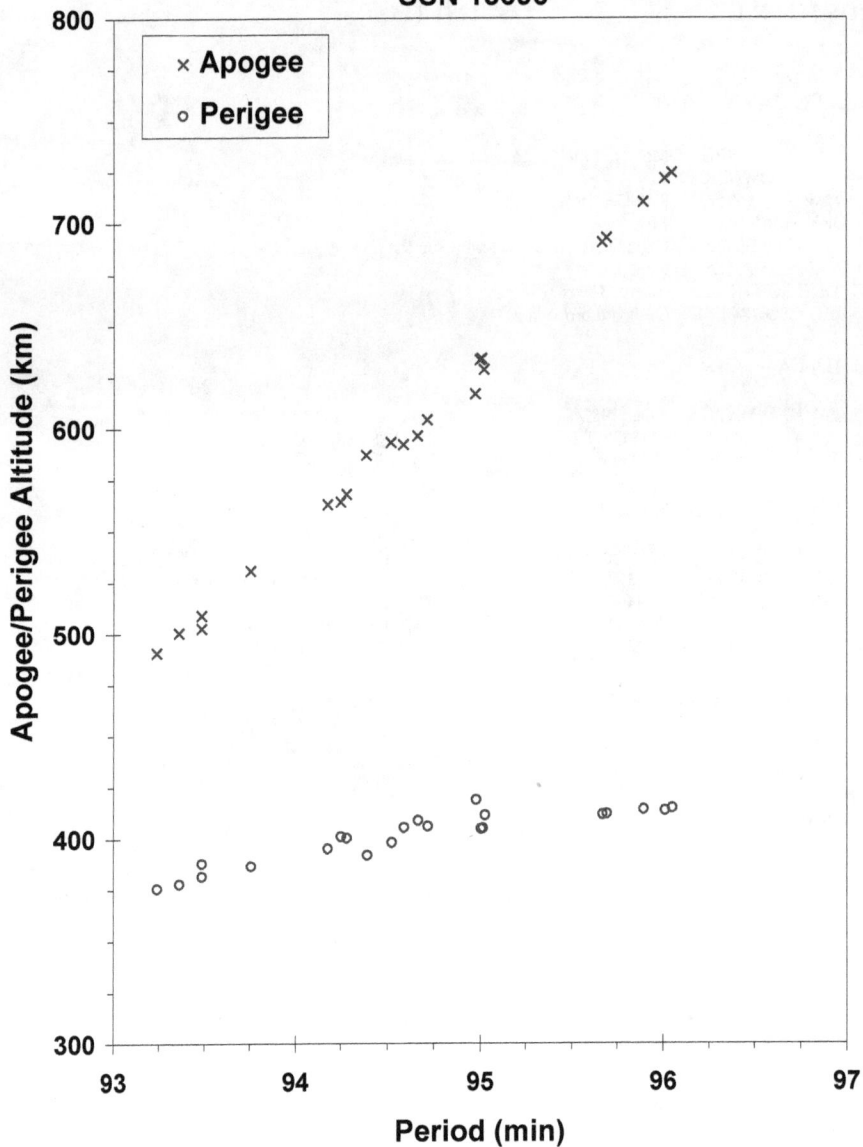

SSN 13696

Cosmos 1423 R/B debris cloud of 22 fragments soon after the event(s) as reconstructed from the US SSN database.

ASTRON ULLAGE MOTOR   1983-020B   13902

SATELLITE DATA

TYPE:	Mission Related Debris
OWNER:	CIS
LAUNCH DATE:	23.53 Mar 1983
DRY MASS (KG):	55
MAIN BODY:	Ellipsoid; 0.6 m diameter by 1.0 m length
MAJOR APPENDAGES:	None
ATTITUDE CONTROL:	None
ENERGY SOURCES:	On-board propellants

EVENT DATA

DATE:	3 Sep 1984	LOCATION:	12S, 352E (dsc)
TIME:	2023 GMT	ASSESSED CAUSE:	Propulsion
ALTITUDE:	400 km		

PRE-EVENT ELEMENTS

EPOCH:	84247.05150886	MEAN ANOMALY:	106.3279
RIGHT ASCENSION:	94.4099	MEAN MOTION:	14.50264973
INCLINATION:	51.5306	MEAN MOTION DOT/2:	.00079313
ECCENTRICITY:	.0710960	MEAN MOTION DOT DOT/6:	.0000075234
ARG. OF PERIGEE:	246.1573	BSTAR:	.00035531

DEBRIS CLOUD DATA

MAXIMUM ΔP:	2.4 min*
MAXIMUM ΔI:	0.3 deg*

*Based on uncataloged debris data

COMMENTS

Parent satellite was one of two small engine units that are routinely released after the first burn of the Proton fourth stage. The nature of these objects was identified by Dr. Boris V. Cherniatiev, Deputy Constructor for the Energiya NPO, in October 1992. The cause of this fragmentation is assumed to be related to the residual hypergolic propellants on board and failure of the membrane separating the fuel and oxidizer. Element sets on 16 fragments were developed. None were officially cataloged. This was the first in a series of fragmentations of this object type.

REFERENCE DOCUMENTS

The Fragmentation of Proton Debris, D. J. Nauer, TBE Technical Report CS93-LKD-004, Teledyne Brown Engineering, Colorado Springs, 31 December 1992.

"Identification and Resolution of an Orbital Debris Problem with the Proton Launch Vehicle", B.V. Cherniatiev et al, Proceedings of the First European Conference on Space Debris, April 1993.

History of Soviet/Russian Satellite Fragmentations-A Joint U.S.-Russian Investigation, N. L. Johnson et al, Kaman Sciences Corporation, October 1995.

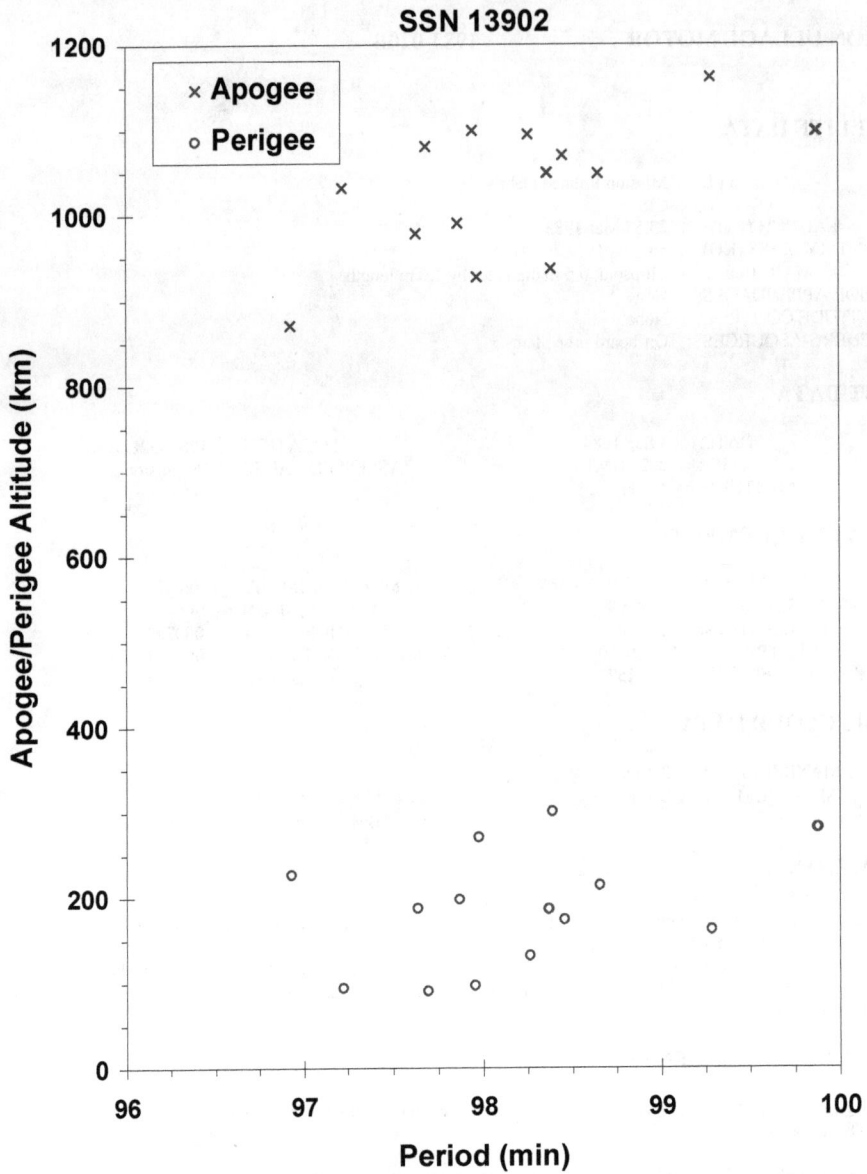

SSN 13902

Astron ullage motor debris cloud of 16 fragments as determined within a few days of the first event. Elements from the US SSN database.

SATELLITE DATA

TYPE:	Payload
OWNER:	US
LAUNCH DATE:	28.66 Mar 1983
DRY MASS (KG):	1000
MAIN BODY:	Cylinder-box; 1.9 m diameter by 7.5 m length
MAJOR APPENDAGES:	1 solar panel
ATTITUDE CONTROL:	Active, 3-axis
ENERGY SOURCES:	On-board propellants

EVENT DATA

DATE:	30 Dec 1985	LOCATION:	68S, 300E (dsc)
TIME:	1005 GMT	ASSESSED CAUSE:	Battery
ALTITUDE:	825 km		

PRE-EVENT ELEMENTS

EPOCH:	85348.40460348	MEAN ANOMALY:	83.2801
RIGHT ASCENSION:	16.9717	MEAN MOTION:	14.22481975
INCLINATION:	98.6488	MEAN MOTION DOT/2:	.00000037
ECCENTRICITY:	.0015724	MEAN MOTION DOT DOT/6:	.0
ARG. OF PERIGEE:	276.6589	BSTAR:	.000025130

DEBRIS CLOUD DATA

MAXIMUM ΔP:	4.7 min
MAXIMUM ΔI:	0.1 deg

COMMENTS

A malfunction on NOAA 8 caused a battery to overcharge, resulting in a minor explosion of the battery. The spacecraft was operational at the time of the event. Six new fragments were detected and cataloged. All decayed by February 1989, leaving the parent still in orbit.

REFERENCE DOCUMENT

"NOAA Turns Off Satellite Following Malfunction", Aviation Week and Space Technology, 13 January 1986, p. 21.

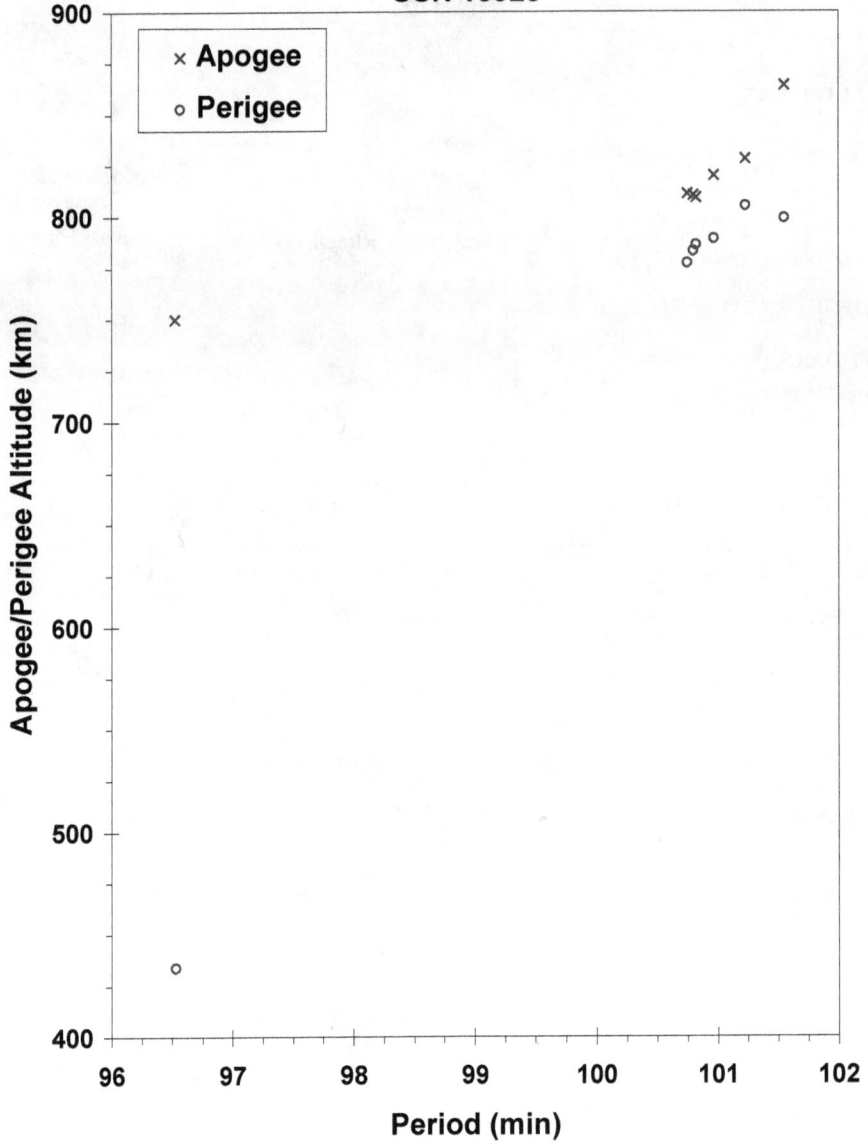

SSN 13923

NOAA 8 debris cloud of 6 fragments plus the parent satellite 1 day after the event as reconstructed from the Naval Space Surveillance System database.

COSMOS 1456   1983-038A   14034

SATELLITE DATA

TYPE:	Payload
OWNER:	CIS
LAUNCH DATE:	25.81 Apr 1983
DRY MASS (KG):	1250
MAIN BODY:	Cylinder; 1.7 m diameter by 2 m length
MAJOR APPENDAGES:	Solar panels
ATTITUDE CONTROL:	Active, 3-axis
ENERGY SOURCES:	On-board propellants, explosive charge

EVENT DATA

DATE:	13 Aug 1983	LOCATION:	Unknown
TIME:	Unknown	ASSESSED CAUSE:	Deliberate
ALTITUDE:	Unknown		

PRE-EVENT ELEMENTS

EPOCH:	83225.00107283	MEAN ANOMALY:	4.5332
RIGHT ASCENSION:	79.8630	MEAN MOTION:	2.00589678
INCLINATION:	63.3076	MEAN MOTION DOT/2:	.0
ECCENTRICITY:	.7324437	MEAN MOTION DOT DOT/6:	.0
ARG. OF PERIGEE:	320.0041	BSTAR:	.0068163

DEBRIS CLOUD DATA

MAXIMUM ΔP:	4.8 min*
MAXIMUM ΔI:	0.4 deg*

*Based on uncataloged debris data

COMMENTS

Cosmos 1456 was another spacecraft of the Cosmos 862-type to experience a fragmentation. The spacecraft may have been active at the time of the event, having last made a station-keeping maneuver on 22 June 1983. The next station-keeping maneuver should have occurred in the second half of August or early September 1983. The spacecraft began drifting off station immediately after the event and never recovered.

REFERENCE DOCUMENT

History of Soviet/Russian Satellite Fragmentations-A Joint U.S.-Russian Investigation, N. L. Johnson et al, Kaman Sciences Corporation, October 1995.

195

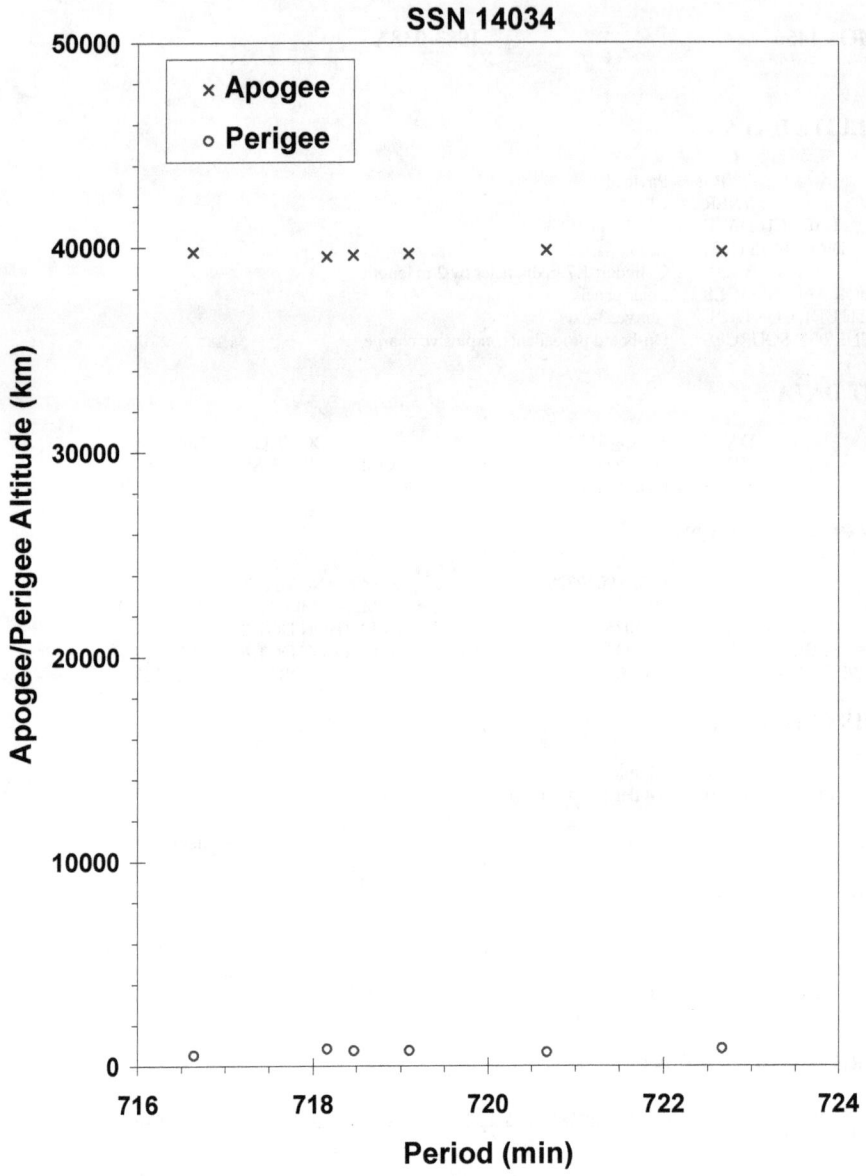

SSN 14034

Cosmos 1456 debris cloud of 6 fragments less than 3 weeks after the event as reconstructed from the US SSN database.

COSMOS 1461   1983-044A   14064

SATELLITE DATA

TYPE:	Payload
OWNER:	CIS
LAUNCH DATE:	7.44 May 1983
DRY MASS (KG):	3000
MAIN BODY:	Cylinder; 1.3 m diameter by 17 m length
MAJOR APPENDAGES:	Solar panels
ATTITUDE CONTROL:	Active, 3-axis
ENERGY SOURCES:	On-board propellants, explosive charge (?)

EVENT DATA (1)

DATE:	11 Mar 1985	LOCATION:	4S, 196E (asc)
TIME:	0940 GMT	ASSESSED CAUSE:	Unknown
ALTITUDE:	750 km		

PRE-EVENT ELEMENTS (1)

EPOCH:	85068.60956125	MEAN ANOMALY:	101.2285
RIGHT ASCENSION:	157.6403	MEAN MOTION:	14.49322542
INCLINATION:	65.0244	MEAN MOTION DOT/2:	.00000357
ECCENTRICITY:	.0224980	MEAN MOTION DOT DOT/6:	.0
ARG. OF PERIGEE:	256.3703	BSTAR:	.000080310

EVENT DATA (2)

DATE:	13 May 1985	LOCATION:	10N, 82E (asc)
TIME:	0133 GMT	ASSESSED CAUSE:	Unknown
ALTITUDE:	845 km		

PRE-EVENT ELEMENTS (2)

EPOCH:	85125.54047130	MEAN ANOMALY:	121.1528
RIGHT ASCENSION:	353.4544	MEAN MOTION:	14.49239036
INCLINATION:	65.0248	MEAN MOTION DOT/2:	.0
ECCENTRICITY:	.0222492	MEAN MOTION DOT DOT/6:	.0
ARG. OF PERIGEE:	236.8082	BSTAR:	.0

DEBRIS CLOUD DATA

MAXIMUM ΔP:	5.9 min*
MAXIMUM ΔI:	1.0 deg*

*Based on uncataloged debris data

COMMENTS

Cosmos 1461 was the twelfth spacecraft of the Cosmos 699-type to experience a fragmentation. Cosmos 1461 entered a natural decay regime more than 13 months prior to first event. After the first event as many as 20 fragments were detected but only six new objects were cataloged. The second event occurred 2 months later and produced considerably more debris. These events followed the pattern set by Cosmos 1220 and Cosmos 1260.

REFERENCE DOCUMENTS

"Artificial Satellite Break-Ups (Part 1): Soviet Ocean Surveillance Satellites", N. L. Johnson, Journal of the British Interplanetary Society, February 1983, pp. 51-58.

Analysis of the Fragmentation of Kosmos 1461, G.T. DeVere and N.L. Johnson, Technical Report CS85-BMDSC-0056, Teledyne Brown Engineering, Colorado Springs, September 1985.

History of Soviet/Russian Satellite Fragmentations-A Joint U.S.-Russian Investigation, N. L. Johnson et al, Kaman Sciences Corporation, October 1995.

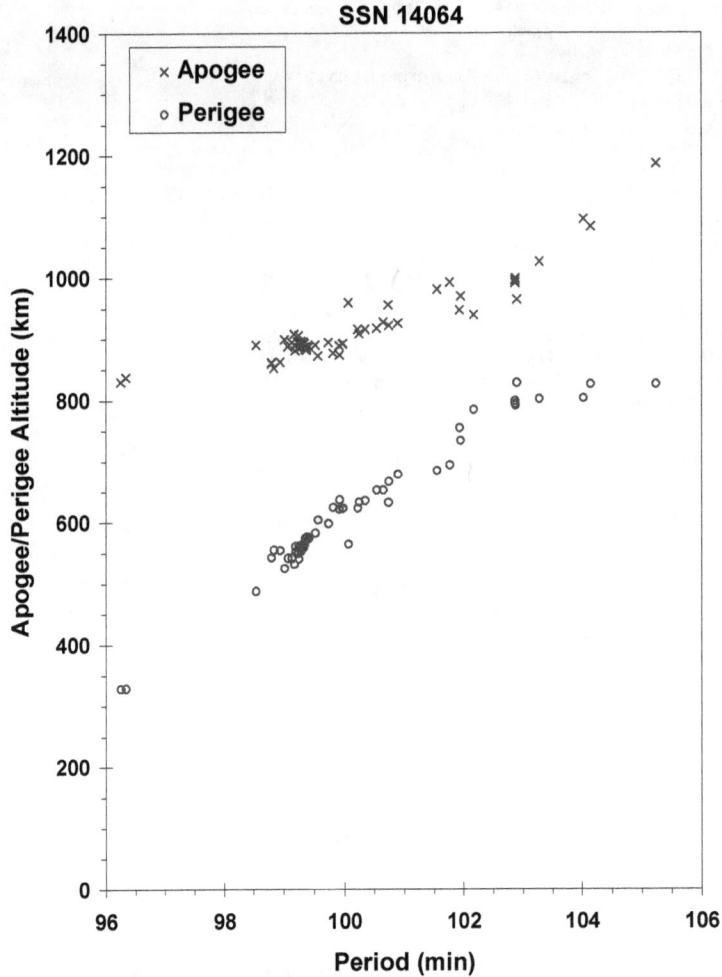

SSN 14064

Cosmos 1461 debris cloud remnant of 65 fragments 4 days after the second event as reconstructed from the US SSN database.

COSMOS 1481   1983-070A   14182

SATELLITE DATA

TYPE:	Payload
OWNER:	CIS
LAUNCH DATE:	8.80 Jul 1983
DRY MASS (KG):	1250
MAIN BODY:	Cylinder; 1.7 m diameter by 2 m length
MAJOR APPENDAGES:	Solar panels
ATTITUDE CONTROL:	Active, 3-axis
ENERGY SOURCES:	On-board propellants, explosive charge

EVENT DATA

DATE:	9 Jul 1983	LOCATION:	Unknown
TIME:	Unknown	ASSESSED CAUSE:	Deliberate
ALTITUDE:	Unknown		

PRE-EVENT ELEMENTS

EPOCH:	83189.85702098	MEAN ANOMALY:	4.6462
RIGHT ASCENSION:	166.3194	MEAN MOTION:	2.03523282
INCLINATION:	62.9394	MEAN MOTION DOT/2:	.00000702
ECCENTRICITY:	.7337681	MEAN MOTION DOT DOT/6:	.0
ARG. OF PERIGEE:	317.9301	BSTAR:	.0

DEBRIS CLOUD DATA

MAXIMUM ΔP:	8.7 min*
MAXIMUM ΔI:	0.8 deg*

*Based on uncataloged debris data

COMMENTS

Cosmos 1481 was the twelfth spacecraft of the Cosmos 862-type to experience a fragmentation. The event apparently occurred within a day of launch. An expected orbital maneuver by Cosmos 1481 to move from its transfer orbit to an operational orbit about 3 days after launch was never performed.

REFERENCE DOCUMENT

History of Soviet/Russian Satellite Fragmentations-A Joint U.S.-Russian Investigation, N. L. Johnson et al, Kaman Sciences Corporation, October 1995.

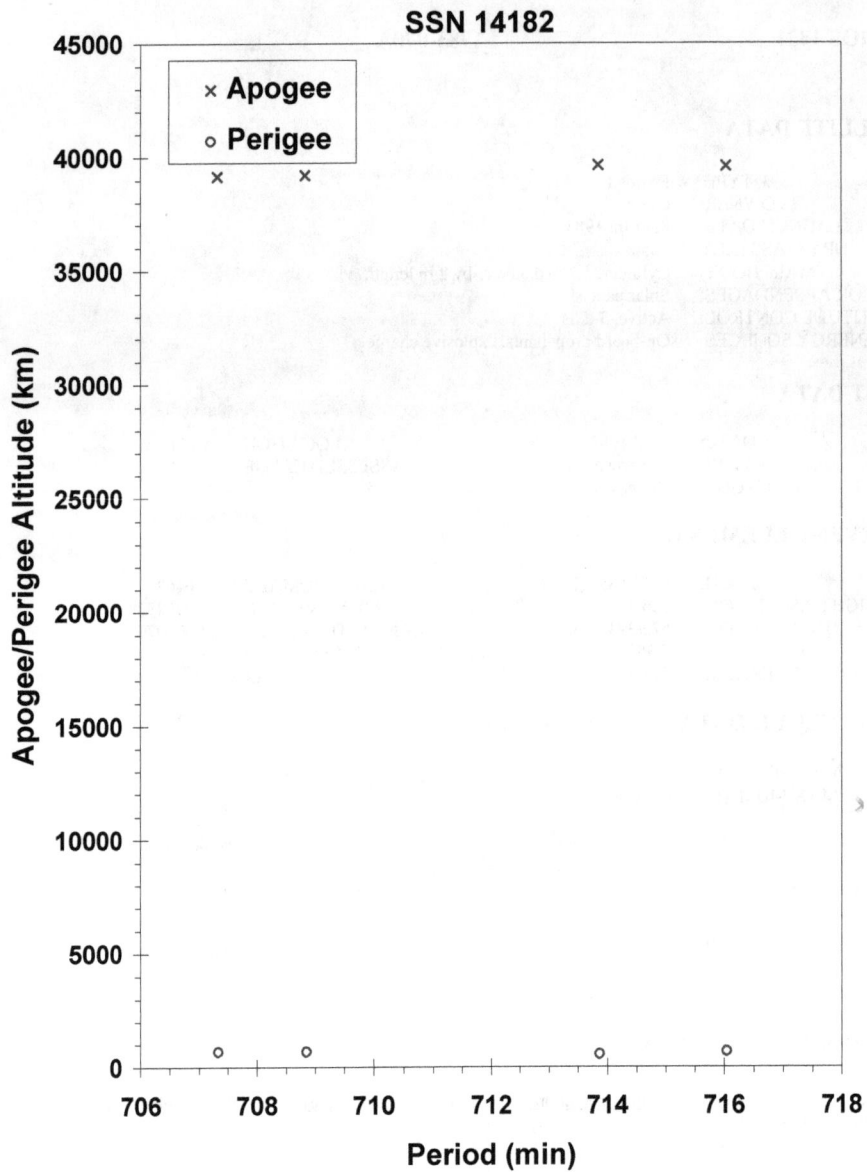

Cosmos 1481 debris cloud of 4 objects 1 month after the event as reconstructed from the
US SSN database.

SATELLITE DATA

TYPE:	Payload
OWNER:	CIS
LAUNCH DATE:	24.25 Jul 1983
DRY MASS (KG):	1800
MAIN BODY:	Cylinder; 1.5 m diameter by 5.0 m length
MAJOR APPENDAGES:	Solar panels, antenna
ATTITUDE CONTROL:	Gravity gradient; momentum wheels
ENERGY SOURCES:	Electrical system (?); pressurized vessels

EVENT DATA

DATE:	18 Oct 1993	LOCATION:	7S, 111E (asc)
TIME:	1204 GMT	ASSESSED CAUSE:	Unknown
ALTITUDE:	605 km		

PRE-EVENT ELEMENTS

EPOCH:	93289.76777232	MEAN ANOMALY:	40.8047
RIGHT ASCENSION:	316.3082	MEAN MOTION:	14.98254133
INCLINATION:	97.5219	MEAN MOTION DOT/2:	.00001299
ECCENTRICITY:	.0033451	MEAN MOTION DOT DOT/6:	.0
ARG. OF PERIGEE:	319.0655	BSTAR:	.00011294

DEBRIS CLOUD DATA

MAXIMUM ΔP:	14.0 min *
MAXIMUM ΔI:	2.5 deg *

* Based on uncataloged debris data

COMMENTS

Cosmos 1484 was the third of four Resurs-0 prototypes flown in sun-synchronous orbits, and the only one to fragment. This is the first sun-synchronous Russian satellite to ever fragment. The NAVSPOC generated 79 analyst satellites on this event.

REFERENCE DOCUMENTS

The Fragmentation of Cosmos 1484, D. J. Nauer, Technical Report CS94-LKD-003, Teledyne Brown Engineering, Colorado Springs, 17 November 1993.

The Soviet Year in Space, 1990, N. L. Johnson, Teledyne Brown Engineering, 1991.

History of Soviet/Russian Satellite Fragmentations-A Joint U.S.-Russian Investigation, N. L. Johnson et al, Kaman Sciences Corporation, October 1995.

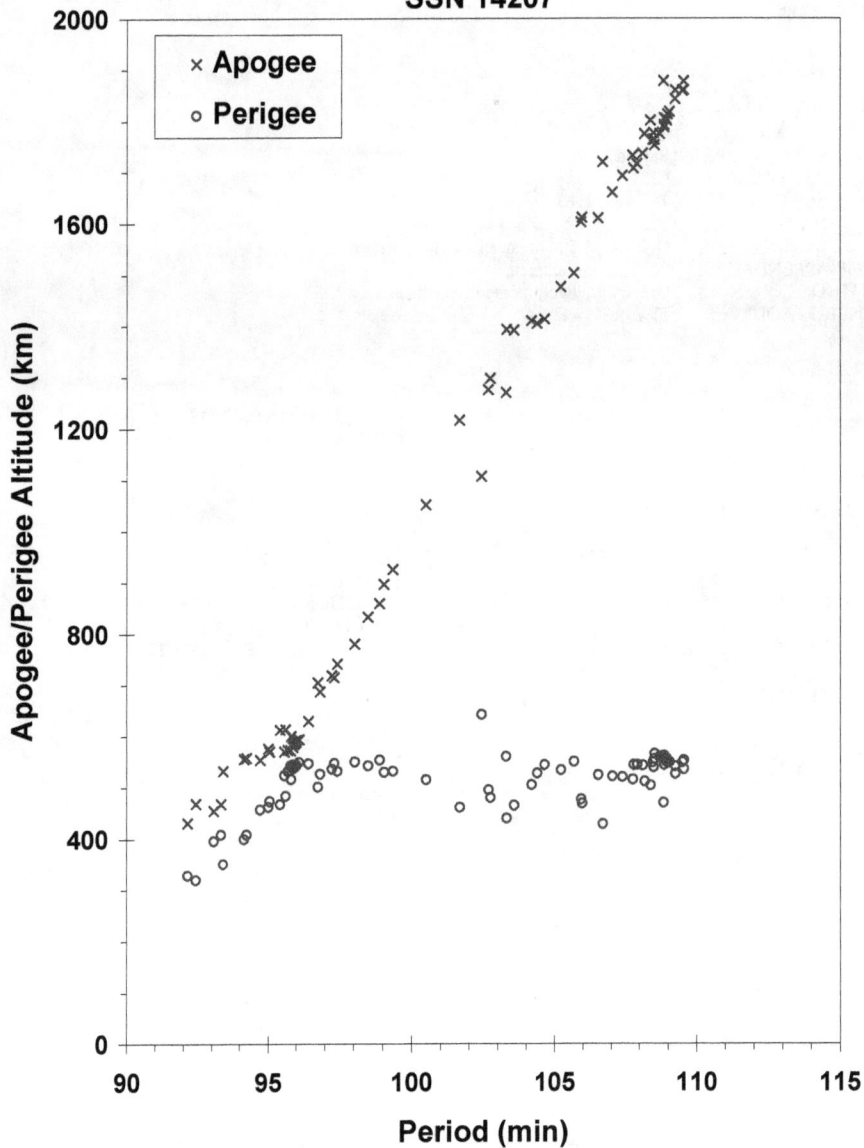

SSN 14207

Cosmos 1484 debris cloud of 79 fragments assembled by the NAVSPOC.

COSMOS 1519-1521 ULLAGE MOTOR   1983-127H   14608

SATELLITE DATA

TYPE:	Mission Related Debris
OWNER:	CIS
LAUNCH DATE:	29.04 Dec 1983
DRY MASS (KG):	55
MAIN BODY:	Ellipsoid; 0.6 m diameter by 1 m length
MAJOR APPENDAGES:	None
ATTITUDE CONTROL:	None
ENERGY SOURCES:	On-board propellants

EVENT DATA

DATE:	4 Feb 1991	LOCATION:	28N, 106E (dsc)
TIME:	0312 GMT	ASSESSED CAUSE:	Propulsion
ALTITUDE:	18550 km		

PRE-EVENT ELEMENTS

EPOCH:	91032.22560633	MEAN ANOMALY:	10.4843
RIGHT ASCENSION:	133.4557	MEAN MOTION:	4.30882556
INCLINATION:	51.9464	MEAN MOTION DOT/2:	.00004140
ECCENTRICITY:	.5787304	MEAN MOTION DOT DOT/6:	.0
ARG. OF PERIGEE:	315.5487	BSTAR:	.0018354

DEBRIS CLOUD DATA

MAXIMUM ΔP:	Unknown
MAXIMUM ΔI:	Unknown

COMMENTS

Parent satellite was one of two small engine units that are routinely released after the first burn of the Proton fourth stage. The nature of these objects was identified by Dr. Boris V. Chernlatiev, Deputy Constructor for the Energiya NPO, in October 1992. The cause of this fragmentation is assumed to be related to the residual hypergolic propellants on board and failure of the membrane separating the fuel and oxidizer. NAVSPASUR observed at least 12 fragments on the day of the event and approximately three dozen on 7 February. An element set was initially developed on only one new fragment. This was the third in a series of fragmentations of this object type.

REFERENCE DOCUMENTS

The Fragmentation of Proton Debris, D. J. Nauer, TBE Technical Report CS93-LKD-004, Teledyne Brown Engineering, Colorado Springs, 31 December 1992.

History of Soviet/Russian Satellite Fragmentations-A Joint U.S.-Russian Investigation, N. L. Johnson et al, Kaman Sciences Corporation, October 1995.

"Identification and Resolution of an Orbital Debris Problem with the Proton Launch Vehicle", B. V. Cherniatiev et al, Proceedings of the First European Conference on Space Debris, April 1993.

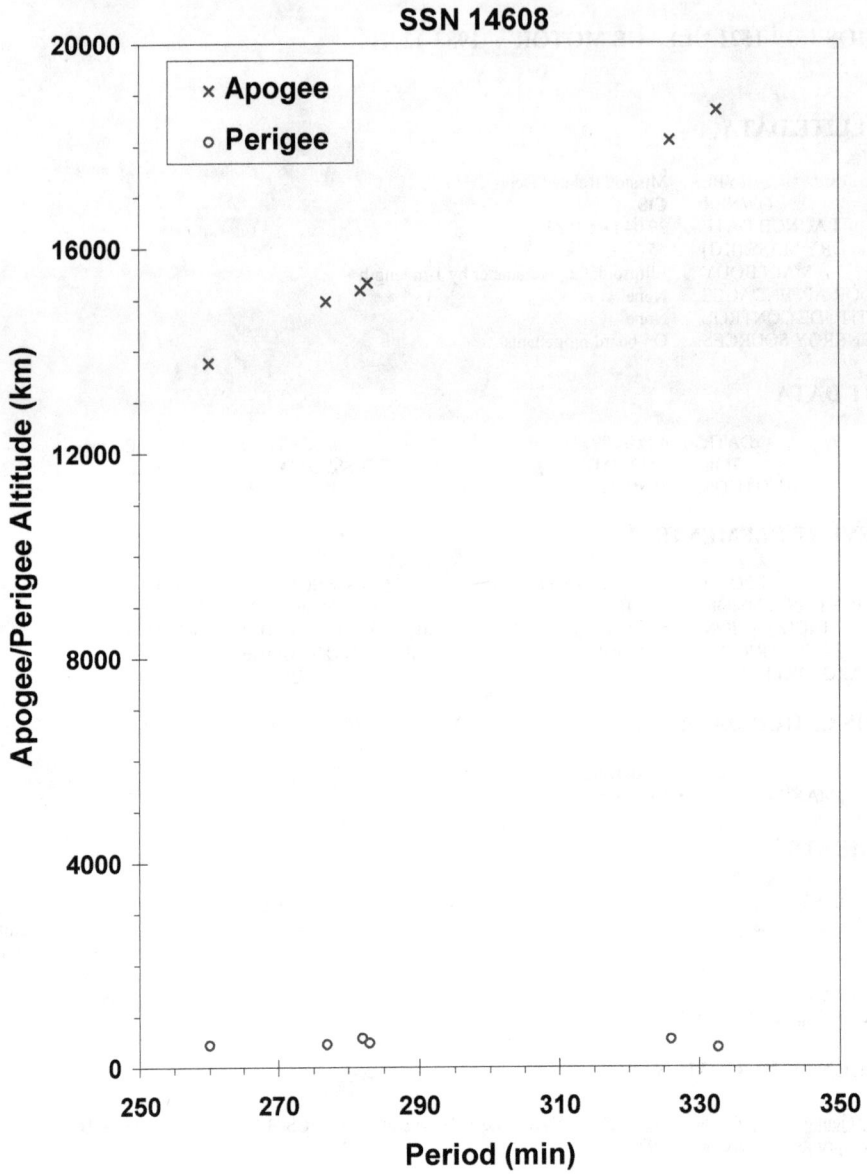

SSN 14608

Gabbard diagram of the five cataloged and single analyst satellite generated on the Cosmos 1519-21 debris cloud.

SATELLITE DATA

TYPE:	PAM-D Upper Stage (STAR 48 motor)
OWNER:	US
LAUNCH DATE:	3.54 Feb 1984
DRY MASS (KG):	2200
MAIN BODY:	Sphere-nozzle; 1.2 m by 2.1 m
MAJOR APPENDAGES:	None
ATTITUDE CONTROL:	Spin-stabilized
ENERGY SOURCES:	On-board propellants

EVENT DATA

DATE:	6 Feb 1984	LOCATION:	0N, 120E (asc)
TIME:	1600 GMT	ASSESSED CAUSE:	Propulsion
ALTITUDE:	280 km		

PRE-EVENT ELEMENTS

EPOCH:	84037.35377144	MEAN ANOMALY:	82.4657
RIGHT ASCENSION:	138.8370	MEAN MOTION:	15.97451864
INCLINATION:	28.4669	MEAN MOTION DOT/2:	.00197501
ECCENTRICITY:	.0006481	MEAN MOTION DOT DOT/6:	.0
ARG. OF PERIGEE:	277.3659	BSTAR:	.00040999

DEBRIS CLOUD DATA

MAXIMUM ΔP:	9.4 min*
MAXIMUM ΔI:	0.3 deg*

*Based on uncataloged debris data

COMMENTS

Palapa B2 and its PAM-D upper stage were deployed from the Space Shuttle Challenger at 1513 GMT, 6 February 1984. Ignition of the upper stage occurred on schedule at 1600 GMT, but the nozzle fragmented within 10 seconds. Without the nozzle the burn could not be sustained and a natural shutdown quickly followed. The PAM-D then separated from Palapa B2. The above elements are for the Shuttle prior to deployment. The Shuttle made a small posigrade evasive maneuver after deployment and before ignition of the PAM-D. See also Westar 6 R/B fragmentation.

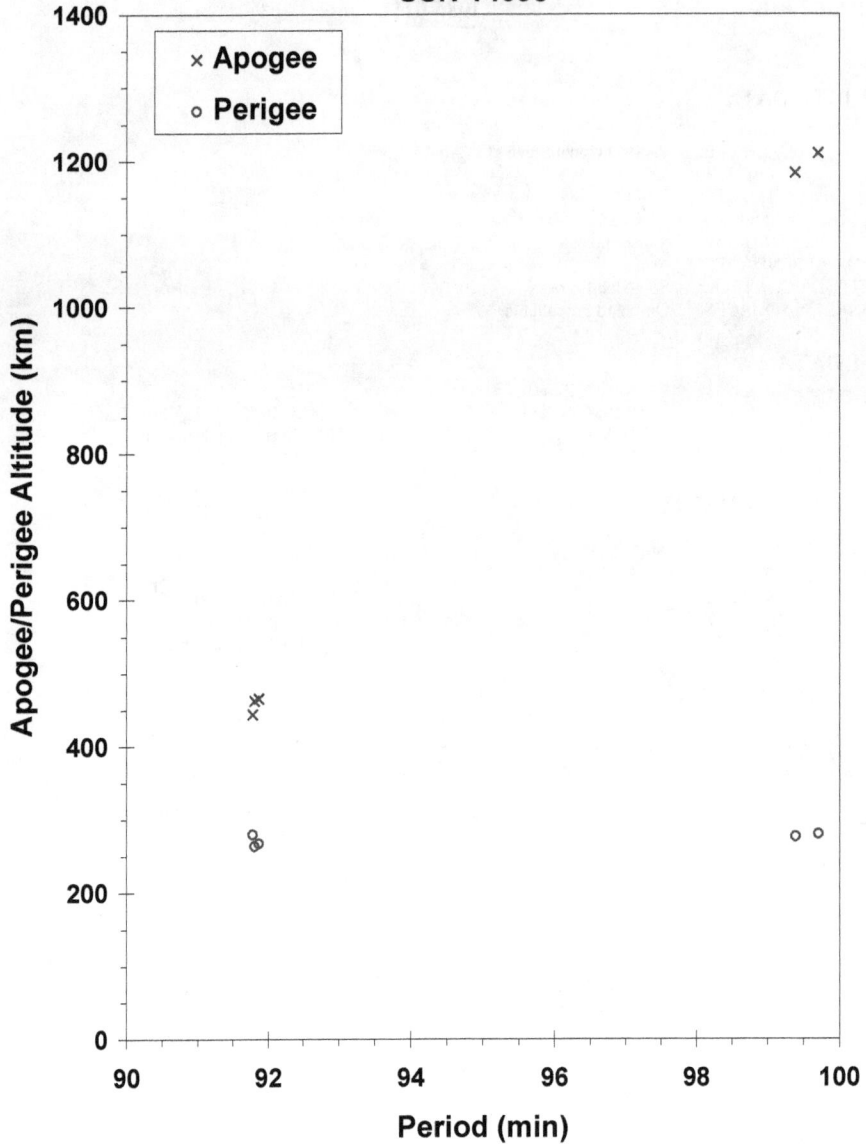

SSN 14693

Palapa B2 R/B debris cloud of 5 fragments about 3 days after the event as reconstructed from the US SSN database. The Palapa B2 R/B is the object with the second highest orbital period.

SATELLITE DATA

TYPE:	PAM-D Upper Stage (STAR 48 motor)
OWNER:	US
LAUNCH DATE:	3.54 Feb 1984
DRY MASS (KG):	2200
MAIN BODY:	Sphere-nozzle; 1.2 m by 2.1 m
MAJOR APPENDAGES:	None
ATTITUDE CONTROL:	Spin-stabilized
ENERGY SOURCES:	On-board propellants

EVENT DATA

DATE:	3 Feb 1984	LOCATION:	0N, 56E (asc)
TIME:	2145 GMT	ASSESSED CAUSE:	Propulsion
ALTITUDE:	305 km		

PRE-EVENT ELEMENTS

EPOCH:	84034.84362284	MEAN ANOMALY:	48.7355
RIGHT ASCENSION:	157.5848	MEAN MOTION:	15.88299499
INCLINATION:	28.4660	MEAN MOTION DOT/2:	.00000250
ECCENTRICITY:	.0006644	MEAN MOTION DOT DOT/6:	.0
ARG. OF PERIGEE:	311.2683	BSTAR:	.0

DEBRIS CLOUD DATA

MAXIMUM ΔP:	9.7 min
MAXIMUM ΔI:	0.8 deg

COMMENTS

Westar 6 and its PAM-D upper stage were deployed from the Space Shuttle Challenger at 2100 GMT, 3 February 1984. Ignition of the upper stage occurred on schedule at 2145 GMT but the nozzle fragmented within 10 seconds. Without the nozzle the burn could not be sustained and a natural shutdown quickly followed. The PAM-D then separated from Westar 6. See also Palapa B2 R/B fragmentation.

REFERENCE DOCUMENT

Westar Failure, Technical Memorandum from N.L. Johnson, Teledyne Brown Engineering, to Preston Landry, NORAD/ADCOM/XPYS, Colorado Springs, 7 February 1984.

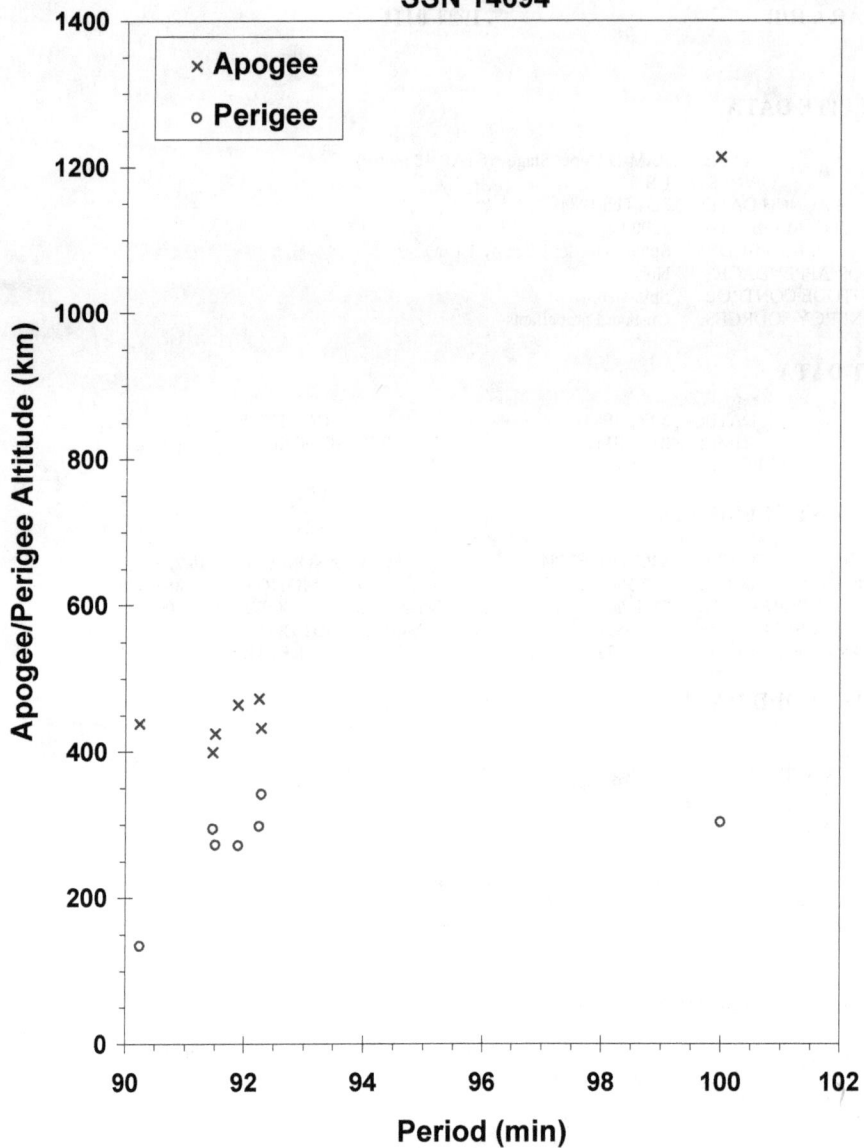

SSN 14694

Westar 6 R/B debris cloud of 7 fragments less than 2 days after the event as reconstructed from the US SSN database. The Westar 6 R/B is the object in the high, 100-min orbit.

COSMOS 1588   1984-083A   15167

SATELLITE DATA

TYPE:	Payload
OWNER:	CIS
LAUNCH DATE:	7.95 Aug 1984
DRY MASS (KG):	3000
MAIN BODY:	Cylinder; 1.3 m diameter by 17 m length
MAJOR APPENDAGES:	Solar panels
ATTITUDE CONTROL:	Active, 3-axis
ENERGY SOURCES:	On-board propellants, explosive charge (?)

EVENT DATA

DATE:	23 Feb 1986	LOCATION:	29N, 187E (asc)
TIME:	1850 GMT	ASSESSED CAUSE:	Unknown
ALTITUDE:	430 km		

PRE-EVENT ELEMENTS

EPOCH:	86048.57631415	MEAN ANOMALY:	72.5463
RIGHT ASCENSION:	268.3025	MEAN MOTION:	15.47795866
INCLINATION:	65.0271	MEAN MOTION DOT/2:	.00005888
ECCENTRICITY:	.0022403	MEAN MOTION DOT DOT/6:	.0
ARG. OF PERIGEE:	287.3230	BSTAR:	.00011680

DEBRIS CLOUD DATA

MAXIMUM ΔP:	2.0 min
MAXIMUM ΔI:	0.4 deg

COMMENTS

Cosmos 1588 was the thirteenth spacecraft of the Cosmos 699-type to experience a fragmentation. Spacecraft had been in natural decay for 7 months prior to the event.

REFERENCE DOCUMENTS

"Artificial Satellite Break-Ups (Part 1): Soviet Ocean Surveillance Satellites", N. L. Johnson, Journal of the British Interplanetary Society, February 1983, pp. 51-58.

History of Soviet/Russian Satellite Fragmentations-A Joint U.S.-Russian Investigation, N. L. Johnson et al, Kaman Sciences Corporation, October 1995.

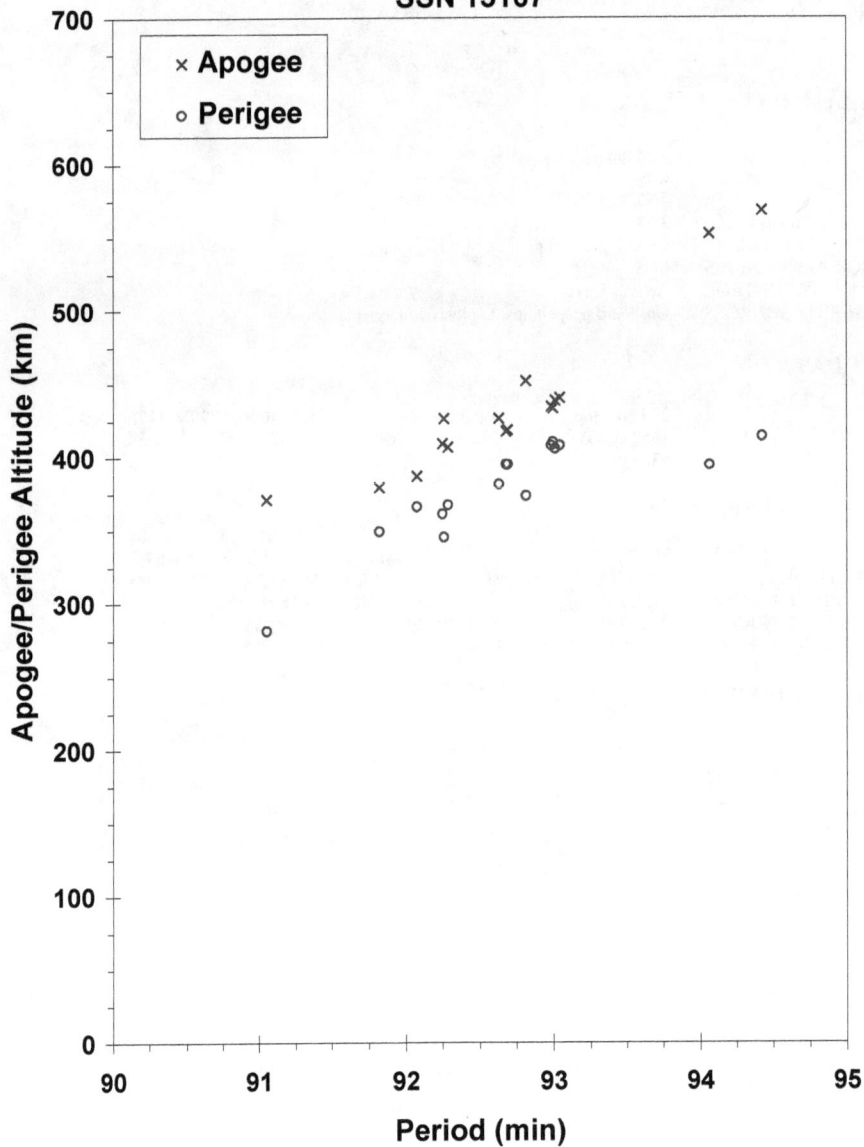

SSN 15167

Cosmos 1588 cataloged debris cloud of 16 fragments 3 weeks after the event as reconstructed from the US SSN database.

COSMOS 1603 ULLAGE MOTOR   1984-106F   15338

SATELLITE DATA

TYPE:	Mission Related Debris
OWNER:	CIS
LAUNCH DATE:	28.58 Sep 1984
DRY MASS (KG):	55
MAIN BODY:	Ellipsoid; 0.6 m diameter by 1 m length
MAJOR APPENDAGES:	None
ATTITUDE CONTROL:	None
ENERGY SOURCES:	On-board propellants

EVENT DATA

DATE:	5 Sep 1992	LOCATION:	46.1S, 351.8E
TIME:	1451 GMT	ASSESSED CAUSE:	Propulsion
ALTITUDE:	835 km		

PRE-EVENT ELEMENTS

EPOCH:	92249.36121283	MEAN ANOMALY:	6.5555
RIGHT ASCENSION:	353.4197	MEAN MOTION:	14.15474339
INCLINATION:	66.5712	MEAN MOTION DOT/2:	-.00009086
ECCENTRICITY:	.0007128	MEAN MOTION DOT DOT/6:	.0
ARG. OF PERIGEE:	353.5641	BSTAR:	-.004641

DEBRIS CLOUD DATA

MAXIMUM ΔP:	2.9 min
MAXIMUM ΔI:	0.5 deg

COMMENTS

Parent satellite was one of two small engine units that are routinely released after the first burn of the Proton fourth stage. The nature of these objects was identified by Dr. Boris V. Chernlatiev, Deputy Constructor for the Energiya NPO, in October 1992. The cause of this fragmentation is assumed to be related to the residual hypergolic propellants on board and failure of the membrane separating the fuel and oxidizer. NAVSPASUR has observed 62 objects associated with this breakup. This was the sixth in a series of fragmentations of this object type.

REFERENCE DOCUMENTS

Soviet Space Programs 1980-1985, Science and Technology Series, Volume 66, Nicholas L. Johnson, American Astronautical Society, Univelt, Inc., 1987.

The Fragmentation of Proton Debris, D. J. Nauer, TBE Technical Report CS93-LKD-004, Teledyne Brown Engineering, Colorado Springs, 31 December 1992.

History of Soviet/Russian Satellite Fragmentations-A Joint U.S.-Russian Investigation, N. L. Johnson et al, Kaman Sciences Corporation, October 1995.

"Identification and Resolution of an Orbital Debris Problem with the Proton Launch Vehicle", B. V. Cherniatiev et al, Proceedings of the First European Conference on Space Debris, April 1993.

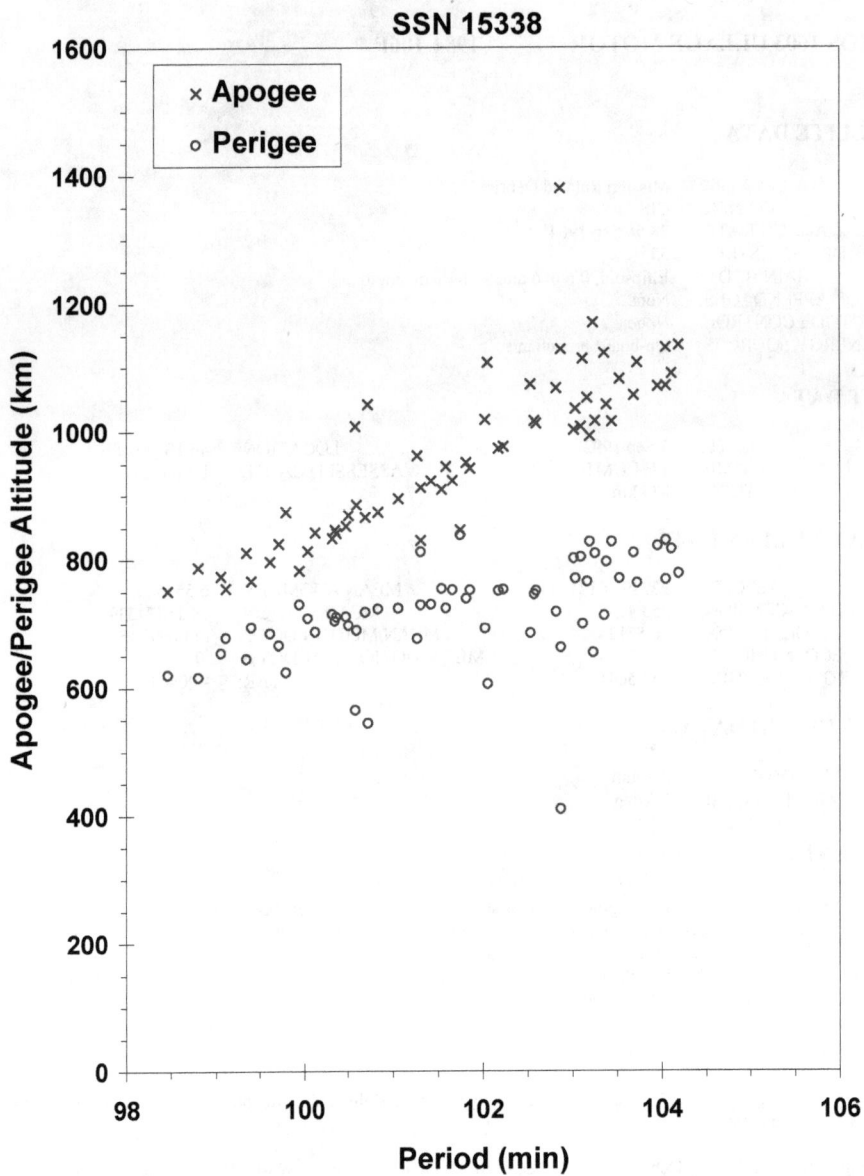

SSN 15338

Naval Space Surveillance System tracked 62 objects within the Cosmos 1603 debris cloud, with 22 appearing in the Satellite Catalog.

SPACENET 2/MARECS B2 R/B   **1984-114C**   **15388**

SATELLITE DATA

TYPE:	Ariane 3 Final Stage
OWNER:	France
LAUNCH DATE:	10.05 Nov 1984
DRY MASS (KG):	~1100
MAIN BODY:	Unknown
MAJOR APPENDAGES:	None
ATTITUDE CONTROL:	None
ENERGY SOURCES:	On-board propellants

EVENT DATA

DATE:	20 Nov 1984	LOCATION:	Unknown
TIME:	1425Z	ASSESSED CAUSE:	Propulsion
ALTITUDE:	Unknown		

PRE-EVENT ELEMENTS

EPOCH:	84325.41617	MEAN ANOMALY:	146.5463
RIGHT ASCENSION:	236.1289	MEAN MOTION:	2.26087292
INCLINATION:	7.0293	MEAN MOTION DOT/2:	.00001128
ECCENTRICITY:	.7265710	MEAN MOTION DOT DOT/6:	.0
ARG. OF PERIGEE:	187.8823	BSTAR:	.0010954

DEBRIS CLOUD DATA

MAXIMUM ΔP:	Unknown
MAXIMUM ΔI:	Unknown

COMMENTS

This Ariane R/B fragmentation occurred to 10 days after launch but not detected until 2003. This is the first Ariane Rocket Body fragmentation that is attributed to France. Previous Ariane Stages were attributed to ESA.

Insufficient data to construct a Gabbard diagram.

SATELLITE DATA

TYPE:	Payload
OWNER:	CIS
LAUNCH DATE:	18.90 Apr 1985
DRY MASS (KG):	3000
MAIN BODY:	Cylinder; 1.3 m diameter by 17 m length
MAJOR APPENDAGES:	Solar panels
ATTITUDE CONTROL:	Active, 3-axis
ENERGY SOURCES:	On-board propellants, explosive charge (?)

EVENT DATA

DATE:	20 Nov 1987	LOCATION:	65N, 300E (dsc)
TIME:	0131 GMT	ASSESSED CAUSE:	Unknown
ALTITUDE:	410 km		

PRE-EVENT ELEMENTS

EPOCH:	87323.98216942	MEAN ANOMALY:	105.3951
RIGHT ASCENSION:	286.0367	MEAN MOTION:	15.56048984
INCLINATION:	65.0306	MEAN MOTION DOT/2:	.00039428
ECCENTRICITY:	.0018658	MEAN MOTION DOT DOT/6:	.0
ARG. OF PERIGEE:	254.4728	BSTAR:	.00055895

DEBRIS CLOUD DATA

MAXIMUM ΔP:	5.5 min*
MAXIMUM ΔI:	0.2 deg*

*Based on cataloged and uncataloged debris data

COMMENTS

Cosmos 1646 was the sixteenth spacecraft of the Cosmos 699-type to experience a fragmentation. Spacecraft had been in natural decay for nearly 20 months prior to the event. Many debris reentered before being officially cataloged.

REFERENCE DOCUMENTS

"Artificial Satellite Break-Ups (Part 1): Soviet Ocean Surveillance Satellites", N. L. Johnson, Journal of the British Interplanetary Society, February 1983, pp. 51-58.

History of Soviet/Russian Satellite Fragmentations-A Joint U.S.-Russian Investigation, N. L. Johnson et al, Kaman Sciences Corporation, October 1995.

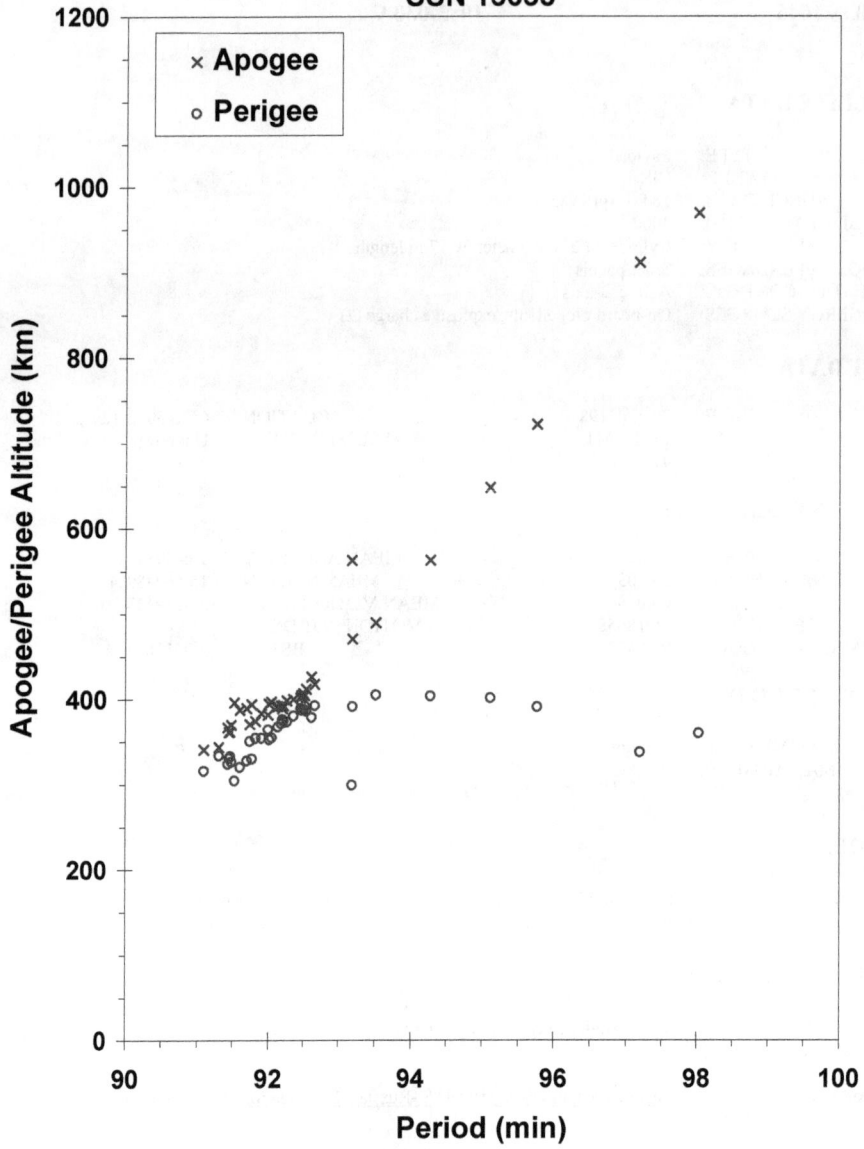

SSN 15653

Cosmos 1646 debris cloud remnant of 38 fragments about 10 days after the event as reconstructed from the US SSN database.

COSMOS 1650-1652 ULLAGE MOTOR   1985-037G

SATELLITE DATA

TYPE:	Mission Related Debris
OWNER:	CIS
LAUNCH DATE:	17 May 1985
DRY MASS (KG):	55
MAIN BODY:	Ellipsoid; 0.6 m diameter by 1 m length
MAJOR APPENDAGES:	None
ATTITUDE CONTROL:	None
ENERGY SOURCES:	On-board propellants

EVENT DATA

DATE:	29 Nov 1998	LOCATION:	38.3N, 172.6E
TIME:	0925 GMT	ASSESSED CAUSE:	Propulsion
ALTITUDE:	16420 km		

POST-EVENT ELEMENTS

EPOCH:	98332.38151447	MEAN ANOMALY:	98.9018
RIGHT ASCENSION:	344.4719	MEAN MOTION:	4.35077855212150
INCLINATION:	52.0277	MEAN MOTION DOT/2:	.00009109
ECCENTRICITY:	.5772516	MEAN MOTION DOT/6:	00000-0
ARG. OF PERIGEE:	209.7130	BSTAR:	.030939

DEBRIS CLOUD DATA

MAXIMUM ΔP:	91.18 min
MAXIMUM ΔI:	.76 deg

COMMENTS

This is the 18th event of the Proton Block DM SOZ Ullage Motor class identified to date; it is the seventh associated with a GLONASS mission. This mission was conducted before the engineering defect was identified and passivation measures implemented. In this orbit, debris may be long-lived but hard to track. A total of 60 debris objects were detected.

REFERENCE DOCUMENT

"1998 Ends with Eighth Satellite Breakup", The Orbital Debris Quarterly News, NASA JSC, January 1999. Available online at http://www.orbitaldebris.jsc nasa.gov/newsletter/pdfs/ODQNv4i1.pdf.

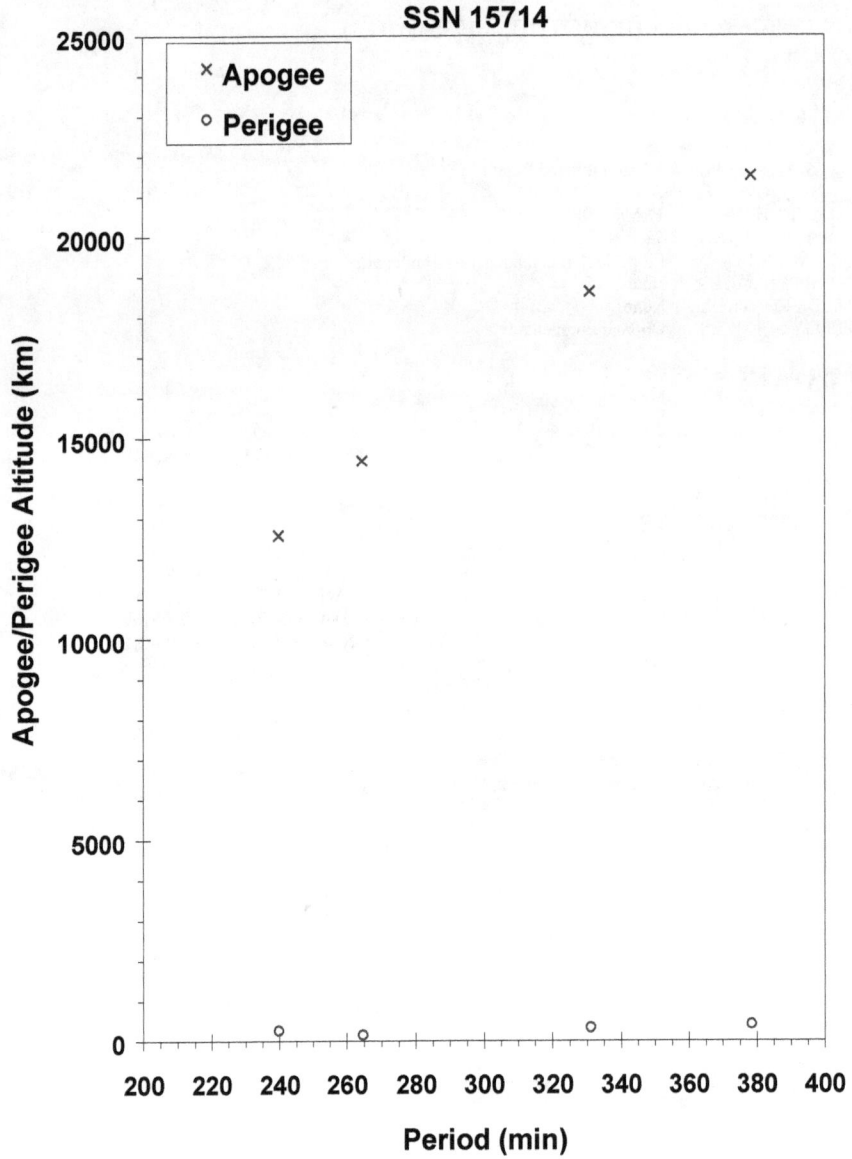

Cosmos 1650-1652 ullage motor debris cloud of 4 fragments within 1 day of the event as reconstructed from the US SSN database.

COSMOS 1654   1985-039A   15734

SATELLITE DATA

TYPE:	Payload
OWNER:	CIS
LAUNCH DATE:	23.53 May 1985
DRY MASS (KG):	5700
MAIN BODY:	Cone-cylinder; 2.7 m diameter by 6.3 m length
MAJOR APPENDAGES:	Solar panels
ATTITUDE CONTROL:	Active, 3-axis
ENERGY SOURCES:	On-board propellants, explosive charge

EVENT DATA

DATE:	21 Jun 1985	LOCATION:	8N, 292E (asc)
TIME:	1047 GMT	ASSESSED CAUSE:	Deliberate
ALTITUDE:	200 km		

PRE-EVENT ELEMENTS

EPOCH:	85172.01363851	MEAN ANOMALY:	313.0734
RIGHT ASCENSION:	1.2391	MEAN MOTION:	16.11890623
INCLINATION:	64.8566	MEAN MOTION DOT/2:	.00311214
ECCENTRICITY:	.0086971	MEAN MOTION DOT DOT/6:	.000034493
ARG. OF PERIGEE:	47.8764	BSTAR:	.00015520

DEBRIS CLOUD DATA

MAXIMUM ΔP:	22.1 min*
MAXIMUM ΔI:	1.5 deg*

*Based on uncataloged debris data

COMMENTS

Spacecraft was destroyed after a malfunction prevented controlled reentry and landing in the Soviet Union. Most debris reentered before being officially cataloged.

REFERENCE DOCUMENTS

Analysis of the Fragmentation of Kosmos 1654, G.T. DeVere, Technical Report CS86-BMDSC-0003, Teledyne Brown Engineering, Colorado Springs, October 1985.

History of Soviet/Russian Satellite Fragmentations-A Joint U.S.-Russian Investigation, N. L. Johnson et al, Kaman Sciences Corporation, October 1995.

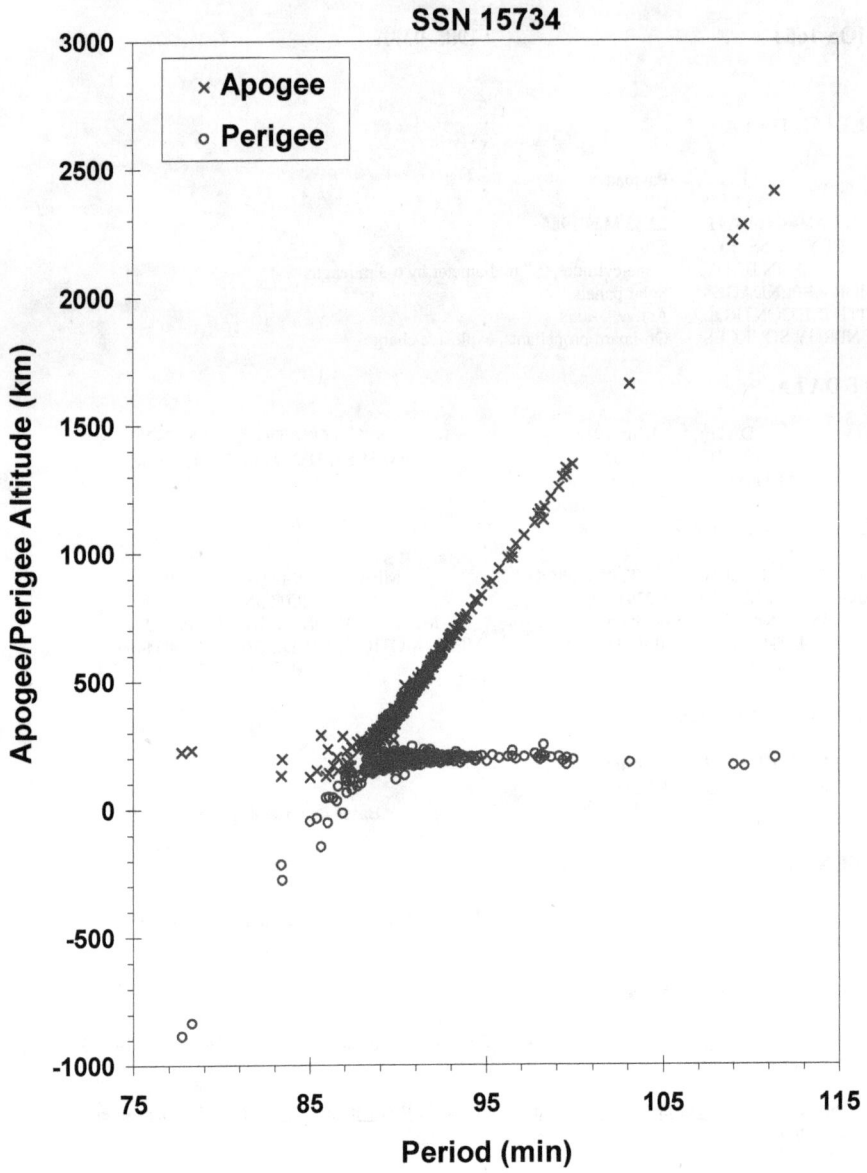

Cosmos 1654 debris cloud remnant of 543 fragments seen 9 hours after the event by the US SSN PARCS radar.

COSMOS 1656 ULLAGE MOTOR   **1985-042E**   **15773**

SATELLITE DATA

TYPE:	Mission Related Debris
OWNER:	CIS
LAUNCH DATE:	30.62 May 1985
DRY MASS (KG):	55
MAIN BODY:	Ellipsoid; 0.6 m diameter by 1 m length
MAJOR APPENDAGES:	None
ATTITUDE CONTROL:	None
ENERGY SOURCES:	On-board propellants

EVENT DATA

DATE:	5 Jan 1988	LOCATION:	66N, 151E (asc)
TIME:	0147 GMT	ASSESSED CAUSE:	Propulsion
ALTITUDE:	860 km		

PRE-EVENT ELEMENTS

EPOCH:	88002.58690356	MEAN ANOMALY:	91.9605
RIGHT ASCENSION:	205.7335	MEAN MOTION:	14.17143400
INCLINATION:	66.5867	MEAN MOTION DOT/2:	.00000144
ECCENTRICITY:	.0034143	MEAN MOTION DOT DOT/6:	.0
ARG. OF PERIGEE:	267.7562	BSTAR:	.000088961

DEBRIS CLOUD DATA

MAXIMUM	ΔP:	3.3 min
MAXIMUM	ΔI:	0.0 deg

COMMENTS

Parent satellite was one of two small engine units that are routinely released after the first burn of the Proton fourth stage. The nature of these objects was identified by Dr. Boris V. Chernlatiev, Deputy Constructor for the Energiya NPO, in October 1992. The cause of this fragmentation is assumed to be related to the residual hypergolic propellants on board and failure of the membrane separating the fuel and oxidizer. NAVSPASUR observed two additional, uncataloged fragments associated with this event. This was the second in a series of fragmentations of this object type.

REFERENCE DOCUMENTS

The Fragmentation of Proton Debris, D. J. Nauer, TBE Technical Report CS93-LKD-004, Teledyne Brown Engineering, Colorado Springs, 31 December 1992.

History of Soviet/Russian Satellite Fragmentations-A Joint U.S.-Russian Investigation, N. L. Johnson et al, Kaman Sciences Corporation, October 1995.

"Identification and Resolution of an Orbital Debris Problem with the Proton Launch Vehicle", B. V. Cherniatiev et al, Proceedings of the First European Conference on Space Debris, April 1993.

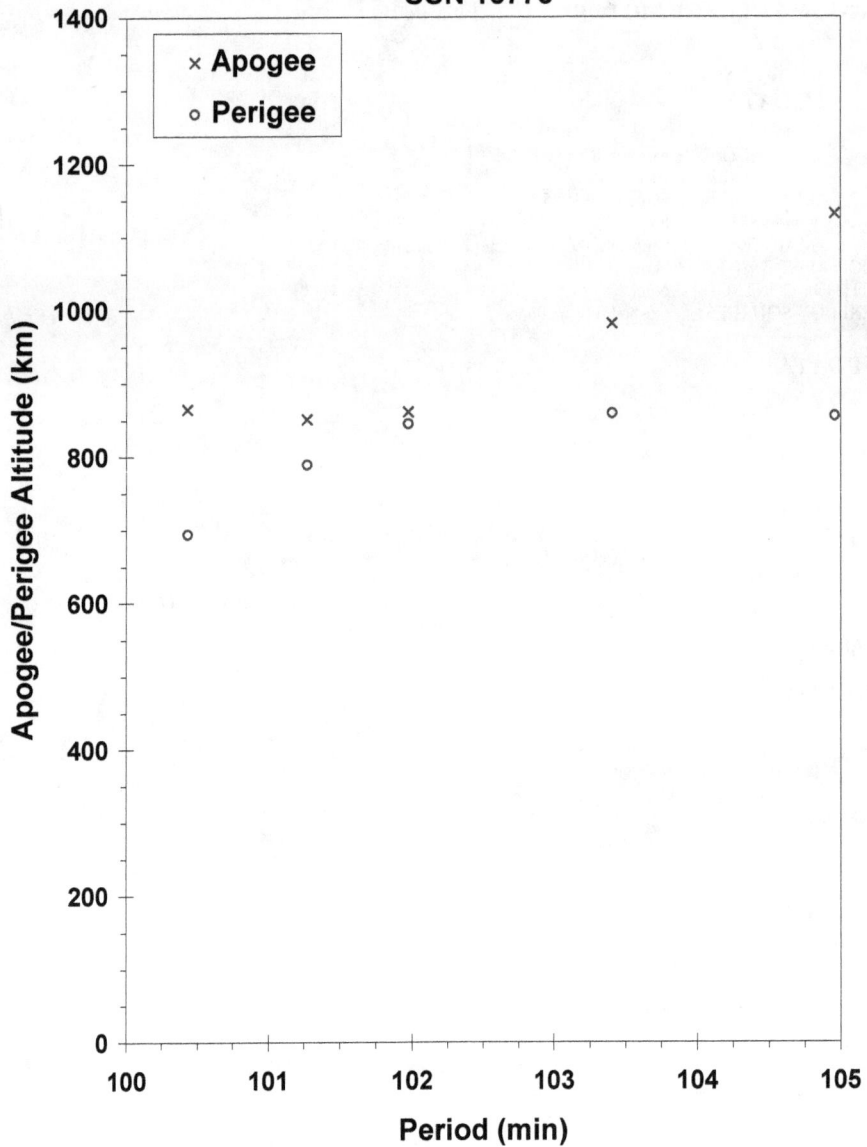

SSN 15773

Fragments from Cosmos 1656 debris as determined 2 weeks after the event. Elements from the US SSN database as published by NASA Goddard Space Flight Center.

SATELLITE DATA

TYPE:	Payload
OWNER:	CIS
LAUNCH DATE:	19.07 Sep 1985
DRY MASS (KG):	3000
MAIN BODY:	Cylinder; 1.3 m diameter by 17 m length
MAJOR APPENDAGES:	Solar panels
ATTITUDE CONTROL:	Active, 3-axis
ENERGY SOURCES:	On-board propellants, explosive charge (?)

EVENT DATA

DATE:	18 Dec 1986	LOCATION:	22S, 292 E (asc)
TIME:	2017 GMT	ASSESSED CAUSE:	Unknown
ALTITUDE:	415 km		

PRE-EVENT ELEMENTS

EPOCH:	86351. 87879723	MEAN ANOMALY:	315.5258
RIGHT ASCENSION:	337.4852	MEAN MOTION:	15.45249396
INCLINATION:	65.0089	MEAN MOTION DOT/2:	.00011076
ECCENTRICITY:	.0068048	MEAN MOTION DOT DOT/6:	.0
ARG. OF PERIGEE:	45.1423	BSTAR:	.00021714

DEBRIS CLOUD DATA

MAXIMUM ΔP:	2.3 min*
MAXIMUM ΔI:	0.7 deg*

*Based on uncataloged debris data

COMMENTS

Cosmos 1682 was the fourteenth spacecraft of the Cosmos 699-type to experience a fragmentation. Spacecraft had been in natural decay for 2 months prior to the event. Many debris reentered before being officially cataloged.

REFERENCE DOCUMENTS

"Artificial Satellite Break-Ups (Part 1): Soviet Ocean Surveillance Satellites", N. L. Johnson, Journal of the British Interplanetary Society, February 1983, p. 51-58.

History of Soviet/Russian Satellite Fragmentations-A Joint U.S.-Russian Investigation, N. L. Johnson et al, Kaman Sciences Corporation, October 1995.

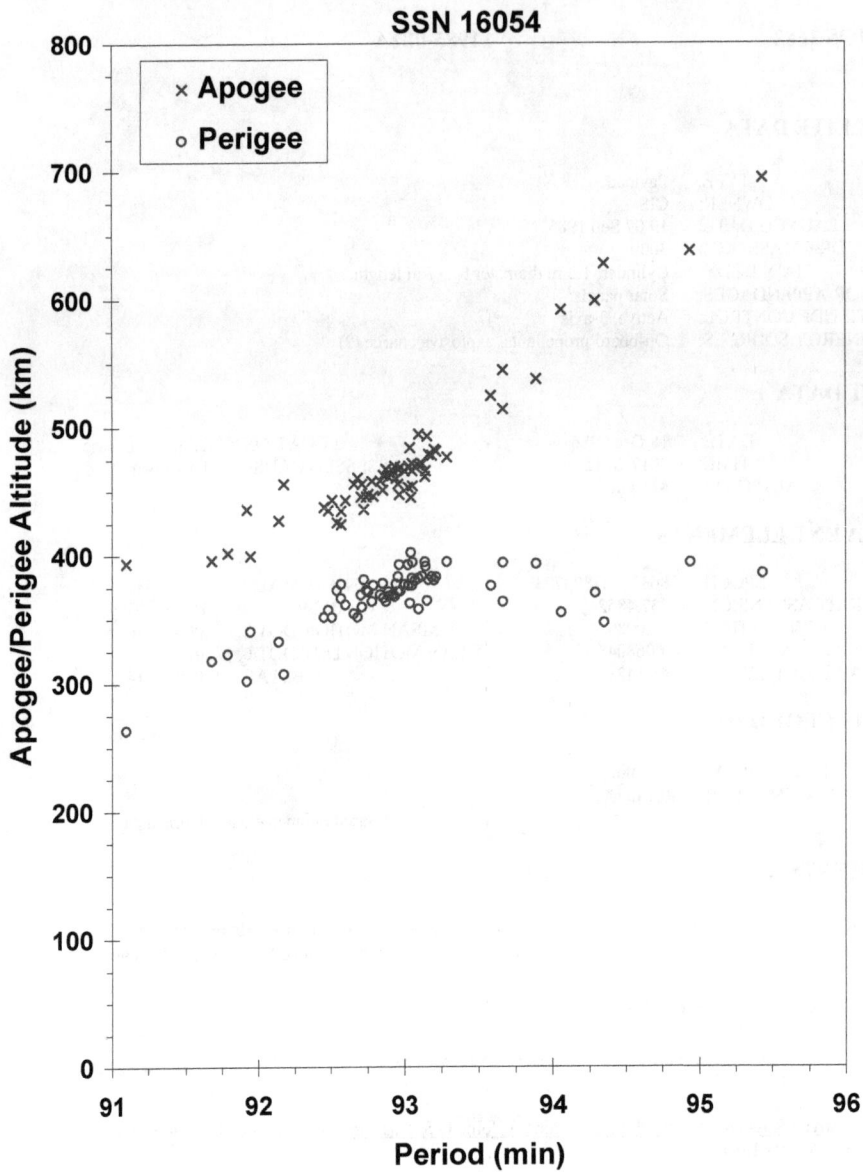

Cosmos 1682 debris cloud remnant of 66 fragments about 1 week after the event as reconstructed from the US SSN database.

COSMOS 1691   1985-094B   16139

SATELLITE DATA

TYPE:	Payload
OWNER:	CIS
LAUNCH DATE:	9.90 Oct 1985
DRY MASS (KG):	220
MAIN BODY:	Cylinder; 1.0 m diameter by 1.5 m length
MAJOR APPENDAGES:	Gravity gradient boom
ATTITUDE CONTROL:	Gravity gradient
ENERGY SOURCES:	Battery

EVENT DATA

DATE:	22 Nov 1985	LOCATION:	31N, 326E (dsc)
TIME:	0840 GMT	ASSESSED CAUSE:	Battery
ALTITUDE:	1415 km		

PRE-EVENT ELEMENTS

EPOCH:	85320.62059878	MEAN ANOMALY:	91.0897
RIGHT ASCENSION:	345.1807	MEAN MOTION:	12.62038878
INCLINATION:	82.6124	MEAN MOTION DOT/2:	.00000022
ECCENTRICITY:	.0002812	MEAN MOTION DOT DOT/6:	.0
ARG. OF PERIGEE:	268.9870	BSTAR:	.000099999

DEBRIS CLOUD DATA

MAXIMUM ΔP:	1.0 min
MAXIMUM ΔI:	0.1 deg

COMMENTS

Cosmos 1691 was one of six independent payloads on this launch, which was only the second in this program. Cosmos 1691 was the last payload deployed and may be referred to as Cosmos 1695 in the former Soviet Union. One fragment was administratively decayed in February 1989. No other payloads in this program have fragmented. This event is assessed to be the second known NiH_2 battery failure as indicated by Dr. K. M. Suitnshev during the early 1992 Space Debris Conference in Moscow. See also reference below.

REFERENCE DOCUMENT

History of Soviet/Russian Satellite Fragmentations-A Joint U.S.-Russian Investigation, N. L. Johnson et al, Kaman Sciences Corporation, October 1995.

SSN 16139

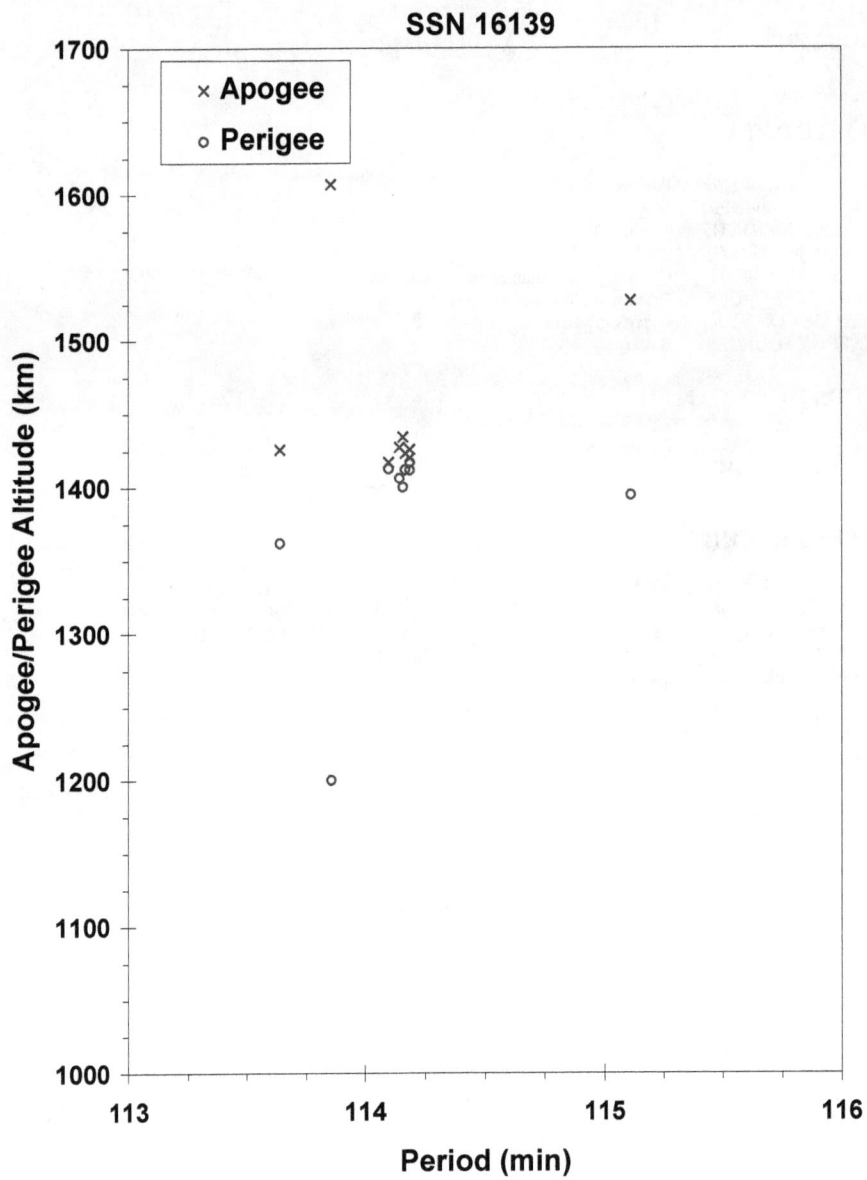

Cosmos 1691 debris cloud of 9 fragments 2 days after the event as reconstructed from Naval Space Surveillance System database.

COSMOS 1703 R/B   1985-108B   16263

SATELLITE DATA

TYPE:	Rocket Body
OWNER:	CIS
LAUNCH DATE:	22.93 Nov 1985
DRY MASS (KG):	1360
MAIN BODY:	Cone-cylinder; 2.1 m diameter by 3.3 m length
MAJOR APPENDAGES:	None
ATTITUDE CONTROL:	None
ENERGY SOURCES:	Unknown

EVENT DATA

DATE:	4 May 2006	LOCATION:	67N, 17E (dsc)
TIME:	1604 GMT	ASSESSED CAUSE:	Propulsion
ALTITUDE:	635 km		

PRE-EVENT ELEMENTS

EPOCH:	06123.63561455	MEAN ANOMALY:	329.9835
RIGHT ASCENSION:	319.0019	MEAN MOTION:	14.8137473
INCLINATION:	82.5005	MEAN MOTION DOT/2:	0.00000107
ECCENTRICITY:	.0021969	MEAN MOTION DOT DOT/6:	0.0
ARG. OF PERIGEE:	30.2640	BSTAR:	0.000010657

DEBRIS CLOUD DATA

MAXIMUM ΔP:	1.6 min
MAXIMUM ΔI:	0.2 deg

COMMENTS

This is the 5th event of the Tsyklon third stage (SL-14) identified to date.

REFERENCE DOCUMENT

"First Satellite Breakups of 2006", The Orbital Debris Quarterly News, NASA JSC, July 2006.
Available online at http://www.orbitaldebris.jsc.nasa.gov/newsletter/pdfs/ODQNv10i3.pdf.

SSN 16263

Tsyklon third stage debris cloud of 44 fragments six days after the event as reconstructed from the US SSN database.

COSMOS 1714 R/B   1985-121F   16439

SATELLITE DATA

TYPE:	Zenit Second Stage
OWNER:	USSR
LAUNCH DATE:	28.40 Dec 1985
DRY MASS (KG):	9000
MAIN BODY:	Cylinder; 3.9 m diameter by 12 m length
MAJOR APPENDAGES:	None
ATTITUDE CONTROL:	Active, 3-axis
ENERGY SOURCES:	On-board propellants

EVENT DATA

DATE:	28 Dec 1985	LOCATION:	Unknown
TIME:	Unknown	ASSESSED CAUSE:	Propulsion
ALTITUDE:	Unknown		

PRE-EVENT ELEMENTS

EPOCH:	85363.19328410	MEAN ANOMALY:	84.6199
RIGHT ASCENSION:	281.3886	MEAN MOTION:	14.77971051
INCLINATION:	71.0178	MEAN MOTION DOT/2:	0.00065991
ECCENTRICITY:	0.0306365	MEAN MOTION DOT DOT/6:	0
ARG. OF PERIGEE:	271.9949	BSTAR:	0.0041108

DEBRIS CLOUD DATA

MAXIMUM ΔP:	Unknown
MAXIMUM ΔI:	Unknown

COMMENTS

The Zenit second stage low thrust engine used to perform final orbit insertion exploded. Four pieces of debris cataloged with this mission are probably not associated with the breakup.

REFERENCE DOCUMENT

History of Soviet/Russian Satellite Fragmentations-A Joint U.S.-Russian Investigation, N. L. Johnson et al, Kaman Sciences Corporation, October 1995.

Insufficient data to construct a Gabbard diagram.

COSMOS 1710-1712 ULLAGE MOTOR   1985-118L   16446

SATELLITE DATA

TYPE:	Mission Related Debris
OWNER:	CIS
LAUNCH DATE:	24.91 Dec 1985
DRY MASS (KG):	55
MAIN BODY:	Ellipsoid; 0.6 m diameter by 1 m length
MAJOR APPENDAGES:	None
ATTITUDE CONTROL:	None
ENERGY SOURCES:	On-board propellants

EVENT DATA

DATE:	29 Dec 1991	LOCATION:	25.3N, 331.9E
TIME:	0903 GMT	ASSESSED CAUSE:	Propulsion
ALTITUDE:	4730 km		

PRE-EVENT ELEMENTS

EPOCH:	91333.40579226	MEAN ANOMALY:	46.8976
RIGHT ASCENSION:	48.0333	MEAN MOTION:	4.23089679
INCLINATION:	65.2547	MEAN MOTION DOT/2:	.00000167
ECCENTRICITY:	.5645362	MEAN MOTION DOT DOT/6:	.0
ARG. OF PERIGEE:	245.7447	BSTAR:	.0012603

DEBRIS CLOUD DATA

MAXIMUM ΔP:	5.7 min *
MAXIMUM ΔI:	0.8 deg *

                                        * based upon uncataloged debris data

COMMENTS

Parent satellite was one of two small engine units that are routinely released after the first burn of the Proton fourth stage. The nature of these objects was identified by Dr. Boris V. Chernlatiev, Deputy Constructor for the Energiya NPO, in October 1992. The cause of this fragmentation is assumed to be related to the residual hypergolic propellants on board and failure of the membrane separating the fuel and oxidizer. There were 26 objects associated with this event on 30 December per a telecon with NAVSPASUR (Edna Jenkins). Only 2 analyst satellites were generated and insufficient data was available for a Gabbard diagram. This was the fourth in a series of fragmentations of this object type.

REFERENCE DOCUMENTS

The Fragmentation of Proton Debris, D. J. Nauer, TBE Technical Report CS93-LKD-004, Teledyne Brown Engineering, Colorado Springs, 31 December 1992.

History of Soviet/Russian Satellite Fragmentations-A Joint U.S.-Russian Investigation, N. L. Johnson et al, Kaman Sciences Corporation, October 1995.

"Identification and Resolution of an Orbital Debris Problem with the Proton Launch Vehicle", B. V. Cherniatiev et al, Proceedings of the First European Conference on Space Debris, April 1993.

Insufficient data to construct a Gabbard diagram.

SPOT 1 R/B   1986-019C   16615

SATELLITE DATA

TYPE:	Ariane 1 Third Stage
OWNER:	France
LAUNCH DATE:	22.07 Feb 1986
DRY MASS (KG):	1400
MAIN BODY:	Cylinder; 2.6 m diameter by 10.3 m length
MAJOR APPENDAGES:	None
ATTITUDE CONTROL:	None at time of the event.
ENERGY SOURCES:	On-board propellants, range safety package

EVENT DATA

DATE:	13 Nov 1986	LOCATION:	7N, 42E (asc)
TIME:	1940 GMT	ASSESSED CAUSE:	Propulsion
ALTITUDE:	805 km		

PRE-EVENT ELEMENTS

EPOCH:	86305.08337689	MEAN ANOMALY:	300.1947
RIGHT ASCENSION:	18.0087	MEAN MOTION:	14.22163662
INCLINATION:	98.6973	MEAN MOTION DOT/2:	.00000203
ECCENTRICITY:	.0021203	MEAN MOTION DOT DOT/6:	.0
ARG. OF PERIGEE:	60.1312	BSTAR:	.000099999

DEBRIS CLOUD DATA

MAXIMUM ΔP:	6.2 min
MAXIMUM ΔI:	1.2 deg

COMMENTS

Event occurred approximately 9 months after the rocket body had successfully deployed the SPOT 1 and Viking payloads. First use of Ariane launch vehicle for low Earth orbit. May be related to other Ariane fragmentations.

REFERENCE DOCUMENTS

A Preliminary Analysis of the Fragmentation of the Spot 1 Ariane Third Stage, N. L. Johnson, Technical Report CS87-LKD-003, Teledyne Brown Engineering, Colorado Springs, March 1987.

Orbital Debris from Upper Stage Breakup, J.P. Loftus, Jr., ed., Vol. 121, Progress in Astronautics and Aeronautics, AIAA, 1989.

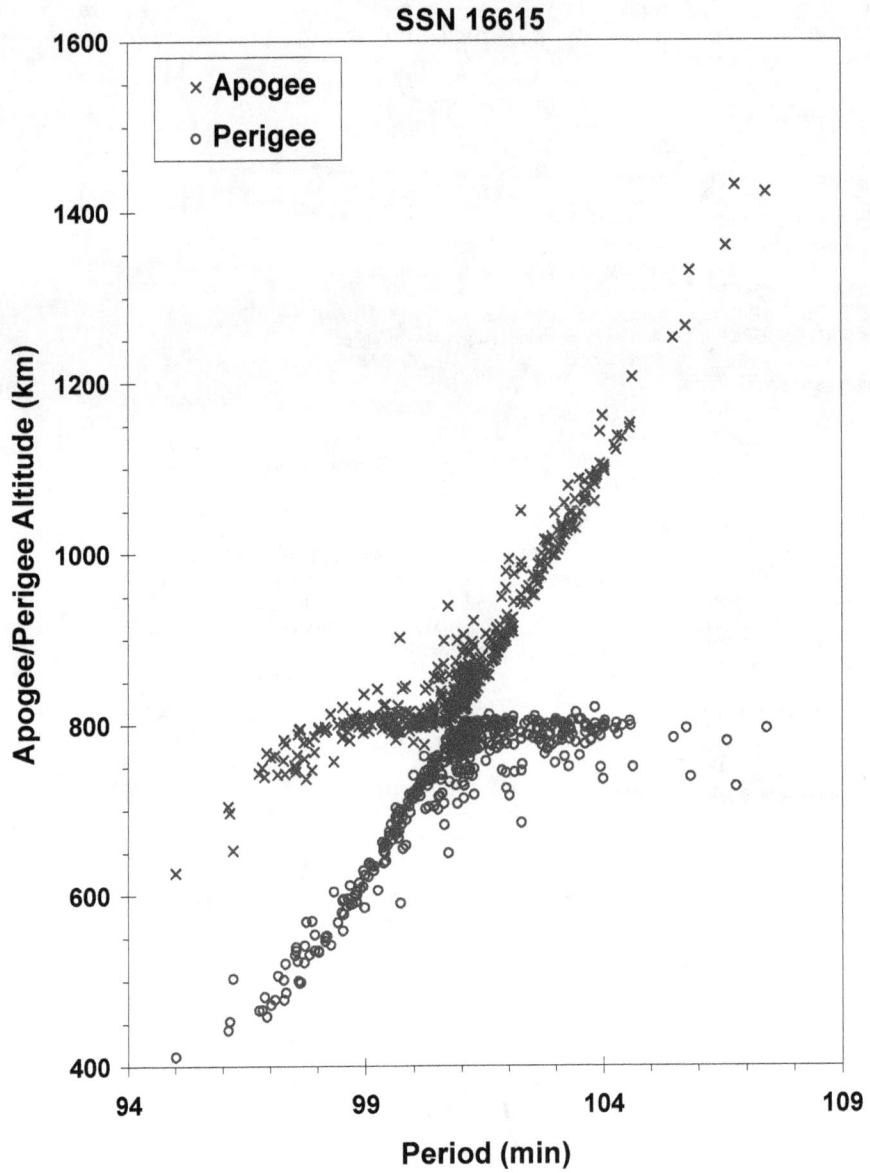

SSN 16615

Spot 1 R/B debris cloud of 463 fragments 3 months after the event as reconstructed from the US SSN database.

COSMOS 1769   1986-059A   16895

SATELLITE DATA

TYPE:	Payload
OWNER:	CIS
LAUNCH DATE:	4.21 Aug 1986
DRY MASS (KG):	3000
MAIN BODY:	Cylinder; 1.3 m diameter by 17 m length
MAJOR APPENDAGES:	Solar panels
ATTITUDE CONTROL:	Active, 3-axis
ENERGY SOURCES:	On-board propellants, explosive charge (?)

EVENT DATA

DATE:	21 Sep 1987	LOCATION:	60S, 174E (dsc)
TIME:	1205 GMT	ASSESSED CAUSE:	Unknown
ALTITUDE:	320 km		

PRE-EVENT ELEMENTS

EPOCH:	87263.81808697	MEAN ANOMALY:	70.4851
RIGHT ASCENSION:	122.5376	MEAN MOTION:	15.63167584
INCLINATION:	65.0147	MEAN MOTION DOT/2:	.00078200
ECCENTRICITY:	.0099296	MEAN MOTION DOT DOT/6:	.0
ARG. OF PERIGEE:	288.4915	BSTAR:	.00065556

DEBRIS CLOUD DATA

MAXIMUM ΔP:	1.9 min*
MAXIMUM ΔI:	0.0 deg*

*Based on uncataloged debris data

COMMENTS

Cosmos 1769 was the fifteenth spacecraft of the Cosmos 699-type to experience a fragmentation. Spacecraft was regularly maneuvered until 17 Sep 1987 when the vehicle began to decay naturally. Most debris reentered before being officially cataloged.

REFERENCE DOCUMENTS

"Artificial Satellite Break-Ups (Part 1): Soviet Ocean Surveillance Satellites", N. L. Johnson, Journal of the British Interplanetary Society, February 1983, pp. 51-58.

History of Soviet/Russian Satellite Fragmentations-A Joint U.S.-Russian Investigation, N. L. Johnson et al, Kaman Sciences Corporation, October 1995.

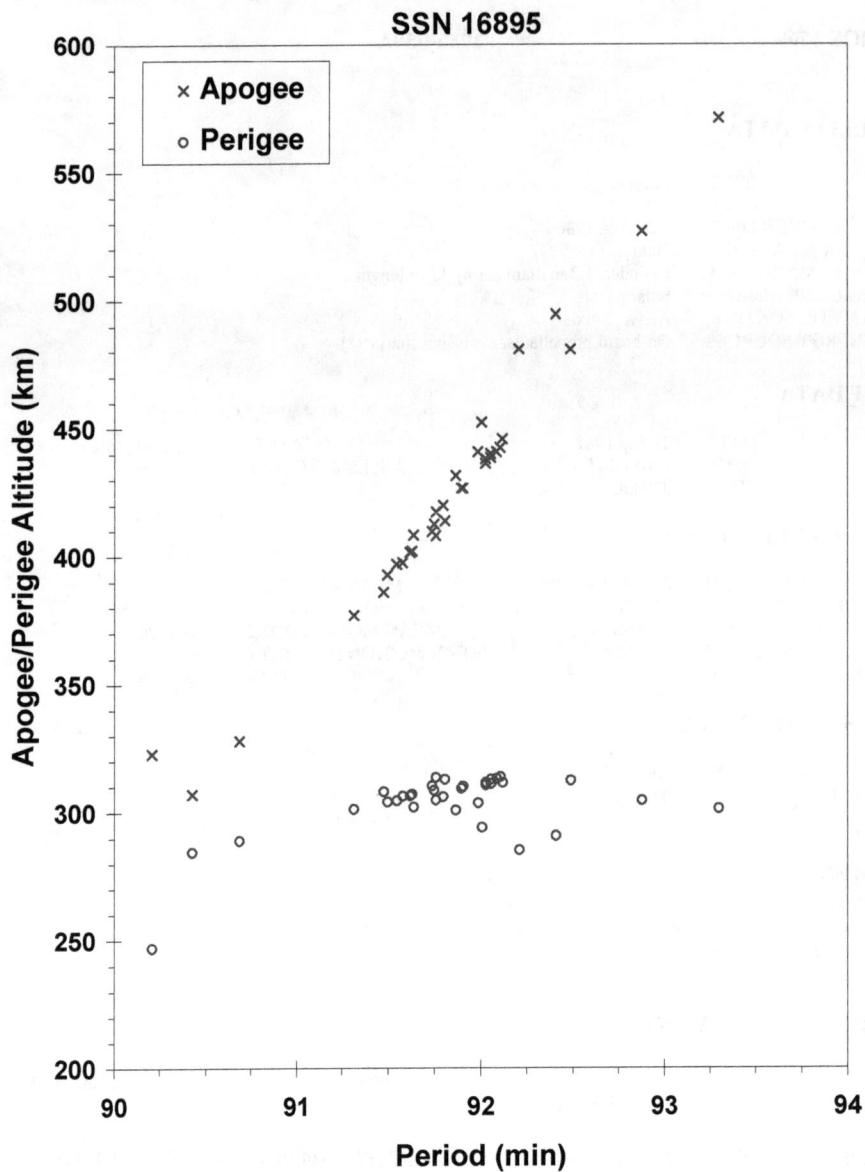

SSN 16895

Cosmos 1769 debris cloud remnant of 34 fragments 3 days after the event as reconstructed from Naval Space Surveillance System database.

SATELLITE DATA

TYPE:	Payload
OWNER:	US
LAUNCH DATE:	5.63 Sep 1986
DRY MASS (KG):	930
MAIN BODY:	Cylinder-cone; 1.2 m diameter by 4.6 m length
MAJOR APPENDAGES:	None
ATTITUDE CONTROL:	Active, 3-axis
ENERGY SOURCES:	On-board propellants, explosive charge (?)

EVENT DATA

DATE:	5 Sep 1986	LOCATION:	15N, 166E (asc)
TIME:	1752 GMT	ASSESSED CAUSE:	Deliberate
ALTITUDE:	220 km		

POST-EVENT ELEMENTS

EPOCH:	86250.63774662	MEAN ANOMALY:	335.3264
RIGHT ASCENSION:	28.1524	MEAN MOTION:	15.28976390
INCLINATION:	39.0665	MEAN MOTION DOT/2:	.01159823
ECCENTRICITY:	.0390567	MEAN MOTION DOT DOT/6:	.0000050922
ARG. OF PERIGEE:	26.7075	BSTAR:	.0028192

DEBRIS CLOUD DATA

MAXIMUM ΔP: 424.1 min*

MAXIMUM ΔI: 4.4 deg*

*Based on uncataloged debris data

COMMENTS

USA 19 deliberately collided with USA 19 R/B at high relative velocity. Both satellites were thrusting at the time of impact. Element set above is post-event and is best estimate of orbit at time of the event. Most debris reentered before being officially cataloged.

REFERENCE DOCUMENTS

The Collision of Satellites 16937 and 16938: A Preliminary Report, N. L. Johnson, Technical Report CS87-LKD-002, Teledyne Brown Engineering, Colorado Springs, 3 December 1986.

The Collision of Satellites 16937 and 16938: Debris Characterization, R. L. Kling, Technical Report CS87-LKD-005, Teledyne Brown Engineering, Colorado Springs, 15 May 1987.

Hazard Analysis of the Breakup of Satellites 16937 and 16938, Technical Report JSC 22471(U), NASA Lyndon B. Johnson Space Center, Houston, 27 February 1987.

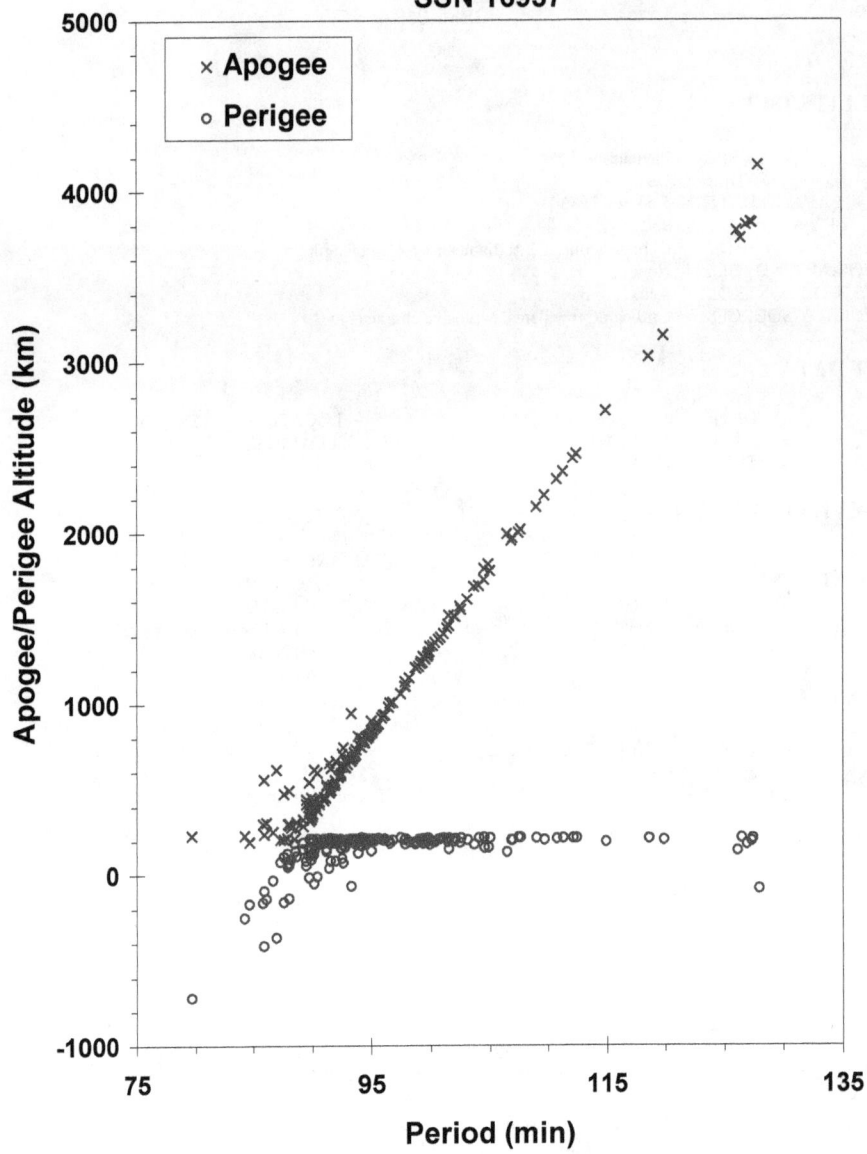

SSN 16937

USA 19 debris cloud remnant of 187 fragments 1 day after the event as seen by the US SSN radar FPS-85 at Eglin AFB, Florida.

SATELLITE DATA

TYPE:	Delta Second Stage (3920) with auxiliary payload
OWNER:	US
LAUNCH DATE:	5.63 Sep 1986
DRY MASS (KG):	1455
MAIN BODY:	Cylinder-nozzle; 1.4 m diameter by 4.8 m length
MAJOR APPENDAGES:	Mini-skirt; 2.4 m by 0.3 m
ATTITUDE CONTROL:	Active
ENERGY SOURCES:	On-board propellants

EVENT DATA

DATE:	5 Sep 1986	LOCATION:	15N, 166E (asc)
TIME:	1752 GMT	ASSESSED CAUSE:	Deliberate
ALTITUDE:	220 km		

POST-EVENT ELEMENTS

EPOCH:	86249.96053354	MEAN ANOMALY:	307.9381
RIGHT ASCENSION:	10.4654	MEAN MOTION:	15.50608380
INCLINATION:	22.7830	MEAN MOTION DOT/2:	.00138611
ECCENTRICITY:	.0288474	MEAN MOTION DOT DOT/6:	.0
ARG. OF PERIGEE:	54.7772	BSTAR:	.00033298

DEBRIS CLOUD DATA

MAXIMUM ΔP:	53.6 min*
MAXIMUM ΔI:	2.5 deg*

*Based on uncataloged debris data

COMMENTS

USA 19 R/B was deliberately struck by USA 19 at high relative velocity. Both satellites were thrusting at the time of impact. Element set above is post-event and is best estimate of orbit at time of the event. Most debris reentered before being officially cataloged.

REFERENCE DOCUMENTS

The Collision of Satellites 16937 and 16938: A Preliminary Report, N. L. Johnson, Technical Report CS87-LKD-002, Teledyne Brown Engineering, Colorado Springs, 3 December 1986.

The Collision of Satellites 16937 and 16938: Debris Characterization, R. L. Kling, Technical Report CS87-LKD-005, Teledyne Brown Engineering, Colorado Springs, 15 May 1987.

Hazard Analysis of the Breakup of Satellites 16937 and 16938, Technical Report JSC 22471(U), NASA Lyndon B. Johnson Space Center, Houston, 27 February 1987.

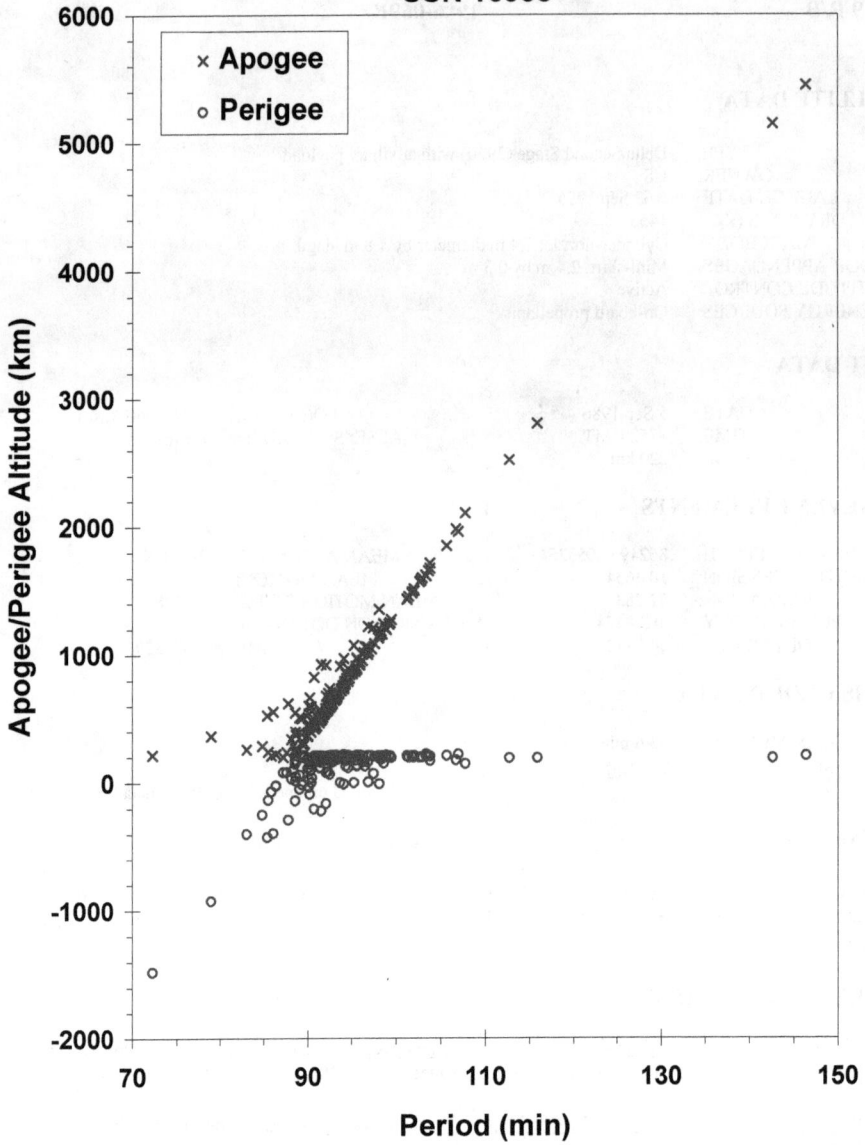

SSN 16938

USA 19 R/B debris cloud of 190 fragments 1 day after the event as seen
by the US SSN FPS-85 radar at Eglin AFB, Florida.

SATELLITE DATA

TYPE:	Payload
OWNER:	CIS
LAUNCH DATE:	15.47 Jan 1987
DRY MASS (KG):	6300
MAIN BODY:	Sphere-cylinder; 2.4 m diameter by 6.5 m length
MAJOR APPENDAGES:	None
ATTITUDE CONTROL:	Active, 3-axis
ENERGY SOURCES:	On-board propellants, explosive charge

EVENT DATA

DATE:	29 Jan 1987	LOCATION:	73N, 122E (asc)
TIME:	0555 GMT	ASSESSED CAUSE:	Deliberate
ALTITUDE:	390 km		

PRE-EVENT ELEMENTS

EPOCH:	87028.91020168	MEAN ANOMALY:	178.1696
RIGHT ASCENSION:	256.7724	MEAN MOTION:	15.60427146
INCLINATION:	72.8163	MEAN MOTION DOT/2:	.00008569
ECCENTRICITY:	.0043147	MEAN MOTION DOT DOT/6:	.0
ARG. OF PERIGEE:	182.0100	BSTAR:	.000099999

DEBRIS CLOUD DATA

MAXIMUM ΔP:	9.1 min*
MAXIMUM ΔI:	0.1 deg*

*Based on PARCS observations

COMMENTS

Spacecraft apparently destroyed after a malfunction prevented controlled reentry and landing in the Soviet Union. A total of 846 separate fragments were observed during one pass over a U.S. Space Surveillance Network radar (PARCS) 2 days after the event.

REFERENCE DOCUMENTS

The Fragmentation of Kosmos 1813, R. L. Kling and J. S. Dowdy, Technical Report CS87-LKD-004, Teledyne Brown Engineering, Colorado Springs, 8 May 1987.

History of Soviet/Russian Satellite Fragmentations-A Joint U.S.-Russian Investigation, N. L. Johnson et al, Kaman Sciences Corporation, October 1995.

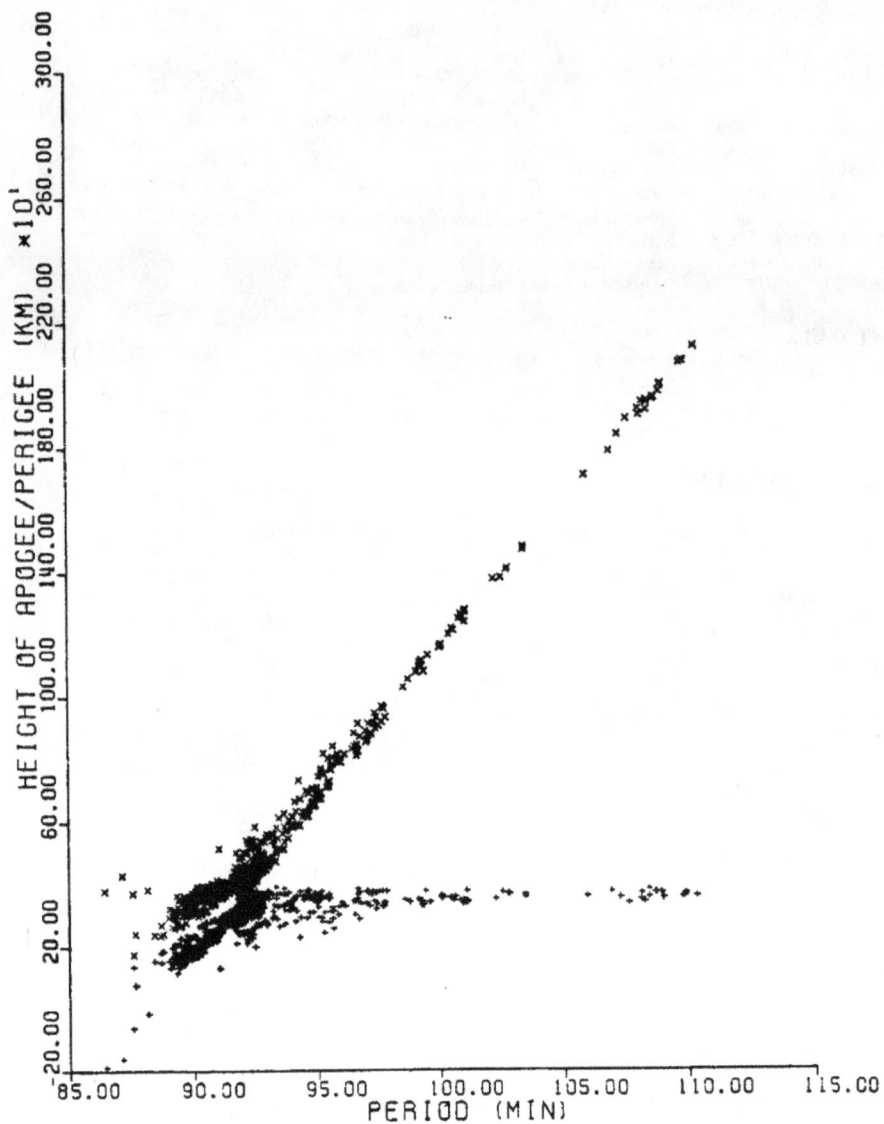

Cosmos 1813 debris cloud as reconstructed from PARCS radar observations taken about 10 hours after the breakup. A total of 846 fragments were identified with Cosmos 1813. This diagram is taken from the cited reference document.

COSMOS 1823   **1987-020A**   **17535**

SATELLITE DATA

TYPE:	Payload
OWNER:	CIS
LAUNCH DATE:	20.20 Feb 1987
DRY MASS (KG):	1500
MAIN BODY:	Cylinder; 2.4 m diameter by 4 m length
MAJOR APPENDAGES:	Gravity-gradient boom; 10 small solar panels
ATTITUDE CONTROL:	Gravity gradient
ENERGY SOURCES:	Battery

EVENT DATA

DATE:	17 Dec 1987	LOCATION:	15S, 18E (dsc)
TIME:	1739 GMT	ASSESSED CAUSE:	Battery
ALTITUDE:	1485 km		

PRE-EVENT ELEMENTS

EPOCH:	87351.61079422	MEAN ANOMALY:	147.6712
RIGHT ASCENSION:	184.5746	MEAN MOTION:	12.40947361
INCLINATION:	73.6064	MEAN MOTION DOT/2:	.0
ECCENTRICITY:	.0028819	MEAN MOTION DOT DOT/6:	.0
ARG. OF PERIGEE:	212.2988	BSTAR:	.0

DEBRIS CLOUD DATA

MAXIMUM ΔP:	4.9 min
MAXIMUM ΔI:	1.4 deg

COMMENTS

Cosmos 1823 has been acknowledged by the Soviet Union as a geodetic spacecraft, the eighth in a series that debuted in 1981. The spacecraft is known to have been operating 3 months before the event. USSR acknowledged mission termination as of 19 December 1987. Unusually strong radial velocity components are evident in cloud analyses over a period of many months. This event has been confirmed to be the third known failure of the NiH_2 battery as reported by Dr. K. M. Suitashev at the February, 1992 Space Debris Conference held in Moscow.

REFERENCE DOCUMENT

History of Soviet/Russian Satellite Fragmentations-A Joint U.S.-Russian Investigation, N. L. Johnson et al, Kaman Sciences Corporation, October 1995.

243

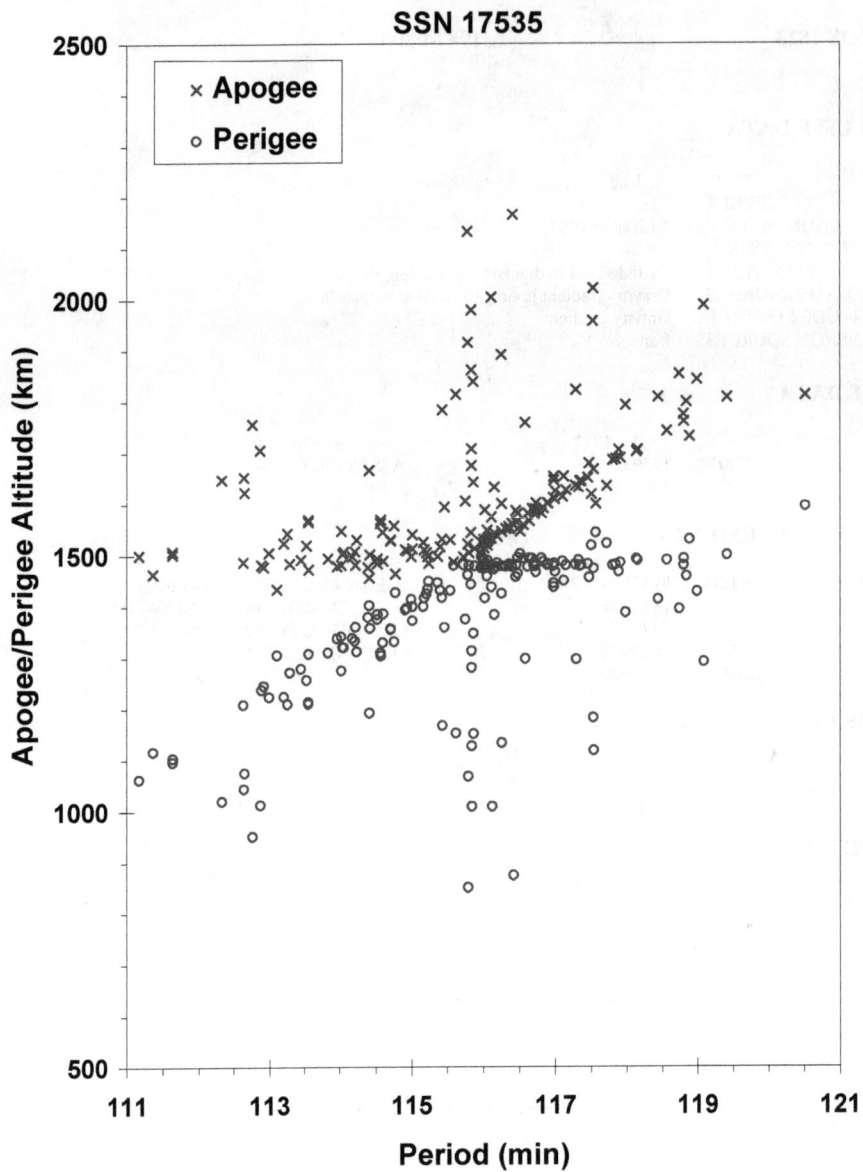

SSN 17535

Cosmos 1823 debris cloud of 165 fragments 2 weeks after the event as reconstructed from Naval Space Surveillance System database.

SATELLITE DATA

TYPE:	Payload
OWNER:	CIS
LAUNCH DATE:	9.67 Jul 1987
DRY MASS (KG):	5700
MAIN BODY:	Cone-cylinder; 2.7 m diameter by 6.3 m length
MAJOR APPENDAGES:	Solar panels
ATTITUDE CONTROL:	Active, 3-axis
ENERGY SOURCES:	On-board propellants, explosive charge

EVENT DATA

DATE:	26 Jul 1987	LOCATION:	57S, 239E (asc)
TIME:	1539 GMT	ASSESSED CAUSE:	Deliberate
ALTITUDE:	245 km		

PRE-EVENT ELEMENTS

EPOCH:	87207.60199851	MEAN ANOMALY:	300.9577
RIGHT ASCENSION:	98.7735	MEAN MOTION:	16.25421506
INCLINATION:	67.1494	MEAN MOTION DOT/2:	.01099941
ECCENTRICITY:	.0073576	MEAN MOTION DOT DOT/6:	.000028662
ARG. OF PERIGEE:	61.7654	BSTAR:	.00016423

DEBRIS CLOUD DATA

MAXIMUM ΔP:	17.3 min
MAXIMUM ΔI:	0.5 deg

COMMENTS

Spacecraft was destroyed after a malfunction prevented controlled reentry and landing in the Soviet Union. Hundreds of fragments were detected but most reentered before being officially cataloged.

REFERENCE DOCUMENT

History of Soviet/Russian Satellite Fragmentations-A Joint U.S.-Russian Investigation, N. L. Johnson et al, Kaman Sciences Corporation, October 1995.

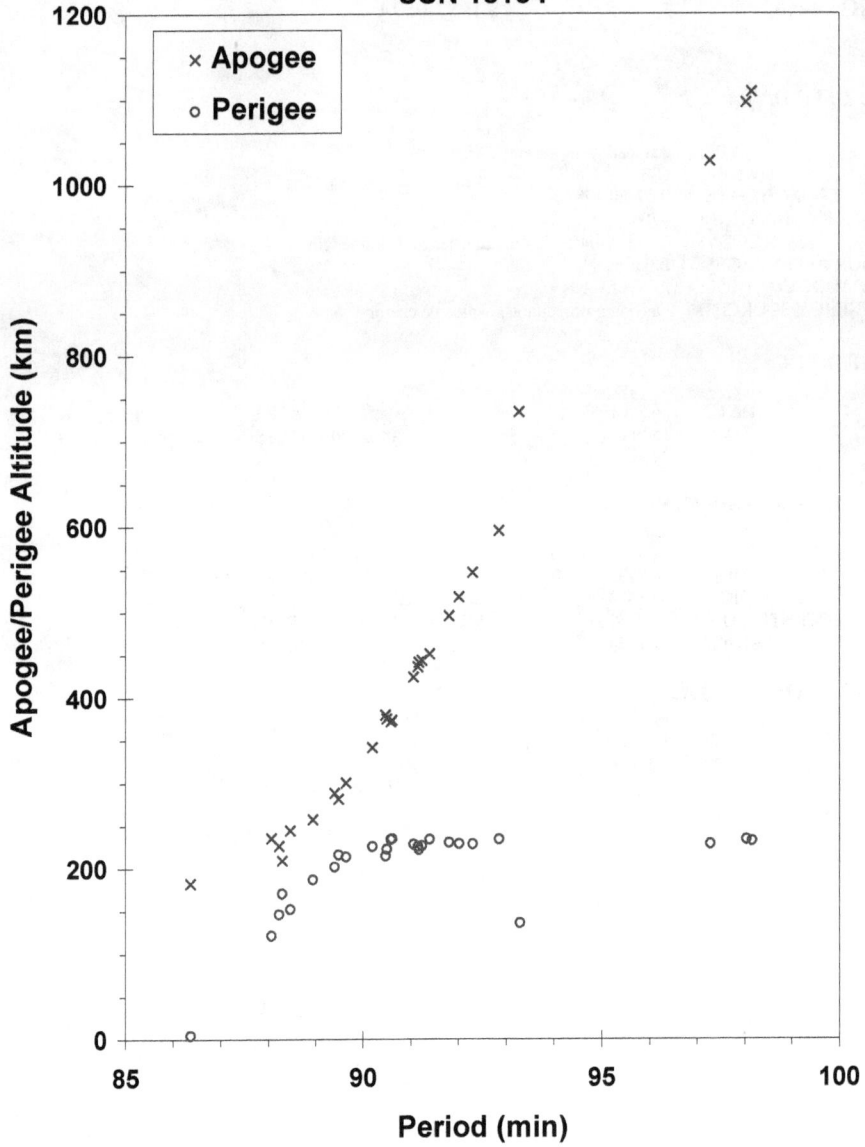

SSN 18184

Cosmos 1866 debris cloud of 27 fragments one to 2 days after the event as reconstructed from the US SSN database. Two fragments with orbital periods greater than 103 minutes were cataloged in mid-August 1987.

COSMOS 1869   1987-062A   18214

SATELLITE DATA

TYPE:	Payload
OWNER:	CIS
LAUNCH DATE:	16.18 Jul 1987
DRY MASS (KG):	1900
MAIN BODY:	Cylinder; 0.8-1.4 m diameter by 3 m length
MAJOR APPENDAGES:	Solar arrays, radar and other payload systems
ATTITUDE CONTROL:	Gravity gradient
ENERGY SOURCES:	Battery, pressurized vessels

EVENT DATA

DATE:	27 Nov 1997	LOCATION:	Unknown
TIME:	0006-0040 GMT?	ASSESSED CAUSE:	Unknown
ALTITUDE:	~630 km		

PRE-EVENT ELEMENTS

EPOCH:	97329.88487815	MEAN ANOMALY:	245.1014
RIGHT ASCENSION:	97.7878	MEAN MOTION:	14.83337853
INCLINATION:	82.5131	MEAN MOTION DOT/2:	0.00000439
ECCENTRICITY:	0.0021357	MEAN MOTION DOT DOT/6:	0
ARG. OF PERIGEE:	115.2417	BSTAR:	0.000050420

DEBRIS CLOUD DATA

MAXIMUM ΔP:	Unknown
MAXIMUM ΔI:	Unknown

COMMENTS

Cosmos 1869 suffered a failure of its radar antenna to deploy immediately after launch. The spacecraft carried other optical sensors, but the vehicle appears to have become non-operational by 1988. At least 20 debris were detected. Virtually all the debris associated with the breakup event exhibited very large area-to-mass ratios, resulting in exceptionally rapid orbital decay. By 1 December 1997 only one debris object was still being tracked by the US Space Surveillance Network.

REFERENCE DOCUMENT

"Recent Satellite Fragmentation Investigations", N. Johnson, The Orbital Debris Quarterly News, NASA JSC, January 1998, p. 3. Available online at http://www.orbitaldebris.jsc.nasa.gov/newsletter/pdfs/ODQNv3i1.pdf.

Insufficient data to construct a Gabbard diagram.

METEOR 2-16 R/B   1987-068B   18313

SATELLITE DATA

TYPE:	Tsyklon Third Stage
OWNER:	CIS
LAUNCH DATE:	18.10 Aug 1987
DRY MASS (KG):	1360
MAIN BODY:	Cylinder; 2.1 m diameter by 2.4 m length
MAJOR APPENDAGES:	None
ATTITUDE CONTROL:	None at time of the event.
ENERGY SOURCES:	Unknown

EVENT DATA

DATE:	15 Feb 1998	LOCATION:	67.8 N, 125.6 E (asc.)
TIME:	2224 GMT	ASSESSED CAUSE:	Propulsion
ALTITUDE:	945 km		

PRE-EVENT ELEMENTS

EPOCH:	98044.02783074	MEAN ANOMALY:	25.0628
RIGHT ASCENSION:	230.9724	MEAN MOTION:	13.84031596
INCLINATION:	82.5526	MEAN MOTION DOT/2:	0.00000025
ECCENTRICITY:	0.0011144	MEAN MOTION DOT DOT/6:	0
ARG. OF PERIGEE:	334.9992	BSTAR:	0.0000096468

DEBRIS CLOUD DATA

MAXIMUM ΔP:	8.2 min
MAXIMUM ΔI:	0.6 deg

COMMENTS

This is the second time a Ukrainian Tsyklon third stage has experienced a significant breakup. The previous incident in 1988 involved the Cosmos 1045 rocket body at a higher altitude. In both cases, the vehicle was approximately 10 years old. The debris from the current breakup were ejected with a wide range of velocities, from about 15 m/s to more than 250 m/s. Some debris were thrown to altitudes below 500 km, and some exhibited high area-to-mass ratios. Naval Space Command ran COMBO to determine if a tracked object was in vicinity of Meteor 2-16 R/B at the time of the event, and the results were negative.

REFERENCE DOCUMENT

"Three Upper Stage Breakups in One Week Top February Debris Activity", The Orbital Debris Quarterly News, NASA JSC, April 1998, p. 1. Available online at
http://www.orbitaldebris.jsc.nasa.gov/newsletter/pdfs/ODQNv3i2.pdf.

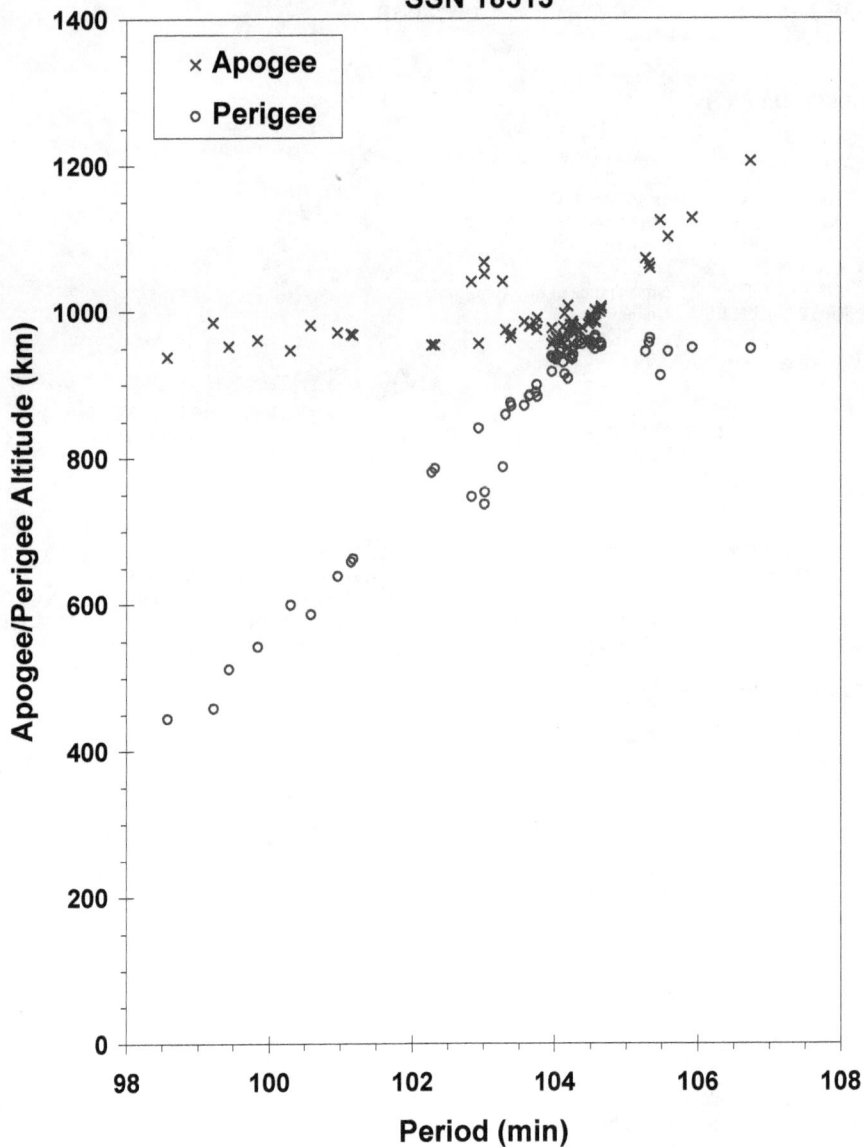

Meteor 2-16 R/B debris cloud of 67 fragments 1 week after the event
as reconstructed from the US SSN database.

AUSSAT K3/ECS 4 R/B   **1987-078C**   **18352**

SATELLITE DATA

TYPE:	Ariane 3 Third Stage
OWNER:	France
LAUNCH DATE:	16.03 Sep 1987
DRY MASS (KG):	1200
MAIN BODY:	Cylinder; 2.6 m diameter by 9.9 m length
MAJOR APPENDAGES:	None
ATTITUDE CONTROL:	None
ENERGY SOURCES:	On-board propellants

EVENT DATA

DATE:	16-19 Sep 1987	LOCATION:	Unknown
TIME:	Unknown	ASSESSED CAUSE:	Propulsion
ALTITUDE:	Unknown		

POST-EVENT ELEMENTS

EPOCH:	87264.18031994	MEAN ANOMALY:	170.9704
RIGHT ASCENSION:	176.7680	MEAN MOTION:	2.22860839
INCLINATION:	6.8720	MEAN MOTION DOT/2:	.00014489
ECCENTRICITY:	.7324768	MEAN MOTION DOT DOT/6:	.0
ARG. OF PERIGEE:	182.0665	BSTAR:	.0038829

DEBRIS CLOUD DATA

MAXIMUM ΔP:	29.1 min*
MAXIMUM ΔI:	0.9 deg*

*Based on uncataloged debris data

COMMENTS

Above elements are initial published values for the rocket body but are after the event.

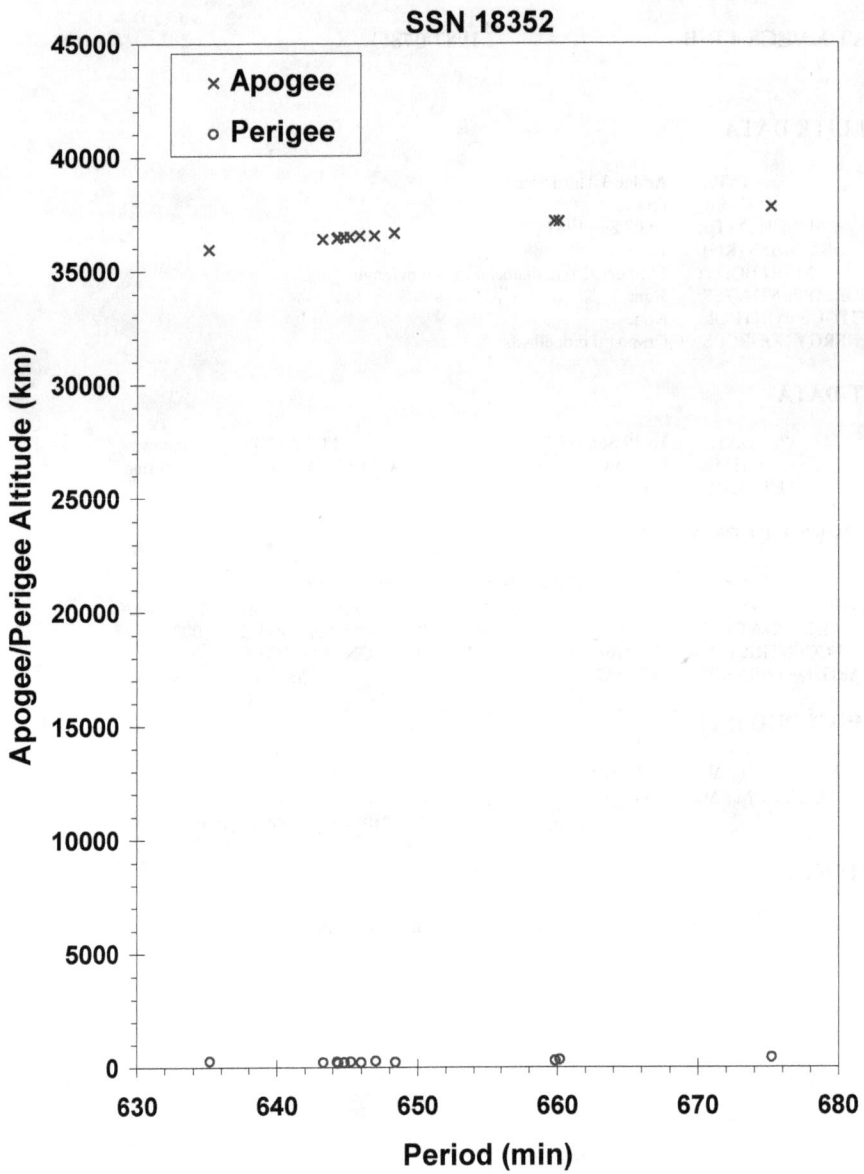

SSN 18352

AUSSAT K3/ECS 4 R/B debris cloud of 12 fragments about 4 days after launch as reconstructed from the US SSN database.

COSMOS 1883-1885 ULLAGE MOTOR   1987-079G

SATELLITE DATA

TYPE:	Mission Related Debris
OWNER:	CIS
LAUNCH DATE:	16.12 Sep 1987
DRY MASS (KG):	55
MAIN BODY:	Ellipsoid; 0.6 m diameter by 1 m length
MAJOR APPENDAGES:	None
ATTITUDE CONTROL:	None
ENERGY SOURCES:	On-board propellants

EVENT DATA

DATE:	~ 01 December 1996	LOCATION:	Unknown
TIME:	Unknown	ASSESSED CAUSE:	Propulsion
ALTITUDE:	Unknown		

PRE-EVENT ELEMENTS

EPOCH:	96335.26612005	MEAN ANOMALY:	175.6198
RIGHT ASCENSION:	300.4954	MEAN MOTION:	4.24439384
INCLINATION:	64.9068	MEAN MOTION DOT/2:	0.00015773
ECCENTRICITY:	0.5826382	MEAN MOTION DOT DOT/6:	0
ARG. OF PERIGEE:	181.3565	BSTAR:	0

DEBRIS CLOUD DATA

MAXIMUM ΔP:	234.1 min
MAXIMUM ΔI:	2.6 deg

COMMENTS

This is the 14th event of this class identified to date.

REFERENCE DOCUMENTS

"Identification and Resolution of an Orbital Debris Problem with the Proton Launch Vehicle", Cherniatiev, Chernyavskiy, Johnson, and McKnight, First European Conference on Space Debris, 5-7 April 1993.

"The Fragmentation of Proton Debris", Nauer, Teledyne Brown Engineering Technical Report CS93LKD-004, 31 December 1992.

History of Soviet/Russian Satellite Fragmentations-A Joint U.S.-Russian Investigation, N. L. Johnson et al, Kaman Sciences Corporation, October 1995.

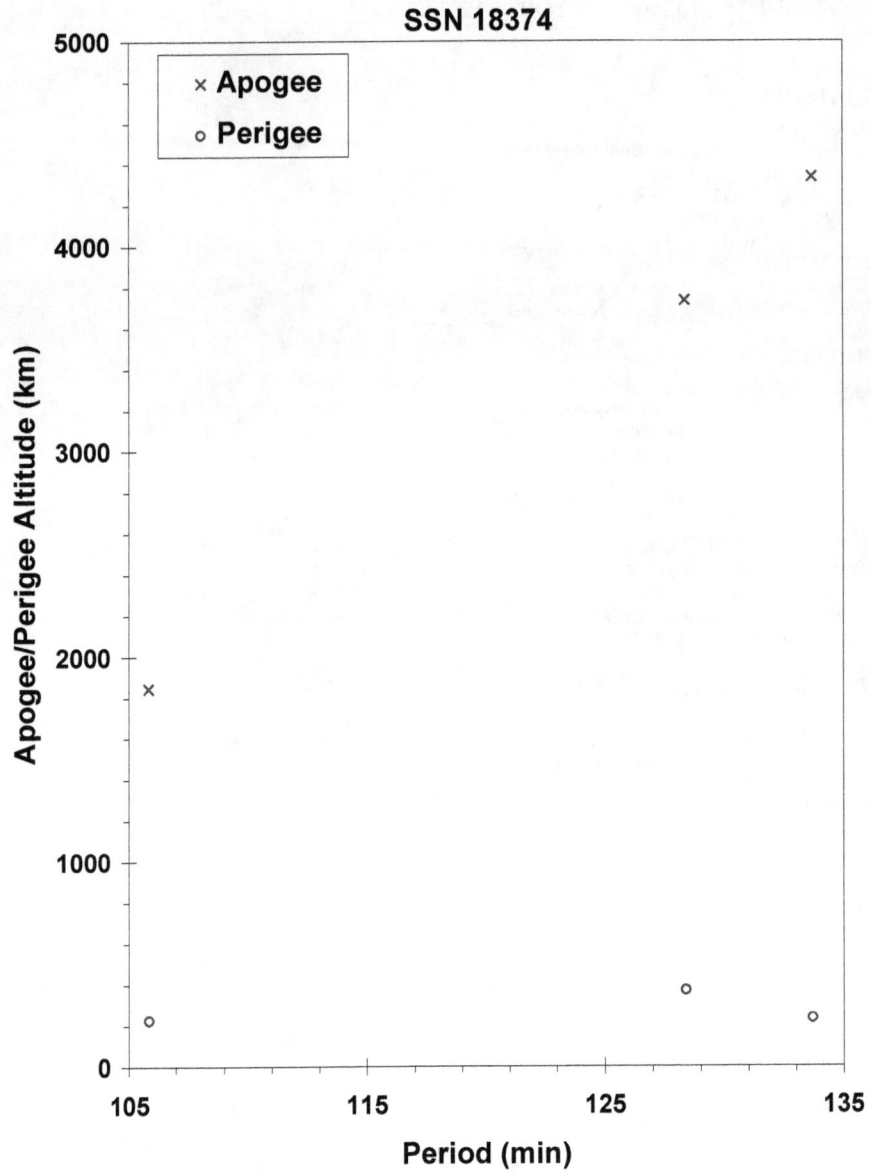

Cosmos 1883-1885 auxiliary motor debris cloud of 3 fragments 11 to 14 months after the event as reconstructed from the US SSN database.

COSMOS 1883-1885 ULLAGE MOTOR   1987-079H

18375

SATELLITE DATA

TYPE:	Mission Related Debris
OWNER:	CIS
LAUNCH DATE:	16.12 Sep 1987
DRY MASS (KG):	~55
MAIN BODY:	Ellipsoid; 0.6 m diameter by 1.0 m length
MAJOR APPENDAGES:	None
ATTITUDE CONTROL:	None
ENERGY SOURCES:	On-board propellants

EVENT DATA

DATE:	23 Apr 2003	LOCATION:	Unknown
TIME:	~1800Z	ASSESSED CAUSE:	Propulsion
ALTITUDE:	Unknown		

PRE-EVENT ELEMENTS

EPOCH:	03113.46108488	MEAN ANOMALY:	332.8061		
RIGHT ASCENSION:	156.9474	MEAN MOTION:	4.27871903		
INCLINATION:	65.2438	MEAN MOTION DOT/2:	.00000068	ECCENTRICITY:	
.5548829	MEAN MOTION DOT DOT/6:	.0			
ARG. OF PERIGEE:	85.3049	BSTAR:	.00025672		

DEBRIS CLOUD DATA

MAXIMUM ΔP:	26.0 min*
MAXIMUM ΔI:	1.19 deg*

* Based on uncataloged debris data

COMMENTS

This event marks the 27th known breakup of a Proton Block DM SOZ ullage motor since 1984. This ullage motor was launched before implementation of breakup preventive measures. 31 debris objects were cataloged from this breakup.

REFERENCE DOCUMENT

"Satellite Fragmentations in 2003", The Orbital Debris Quarterly News, NASA JSC, January 2004. Available online at http://www.orbitaldebris.jsc nasa.gov/newsletter/pdfs/ODQNv8i1.pdf.

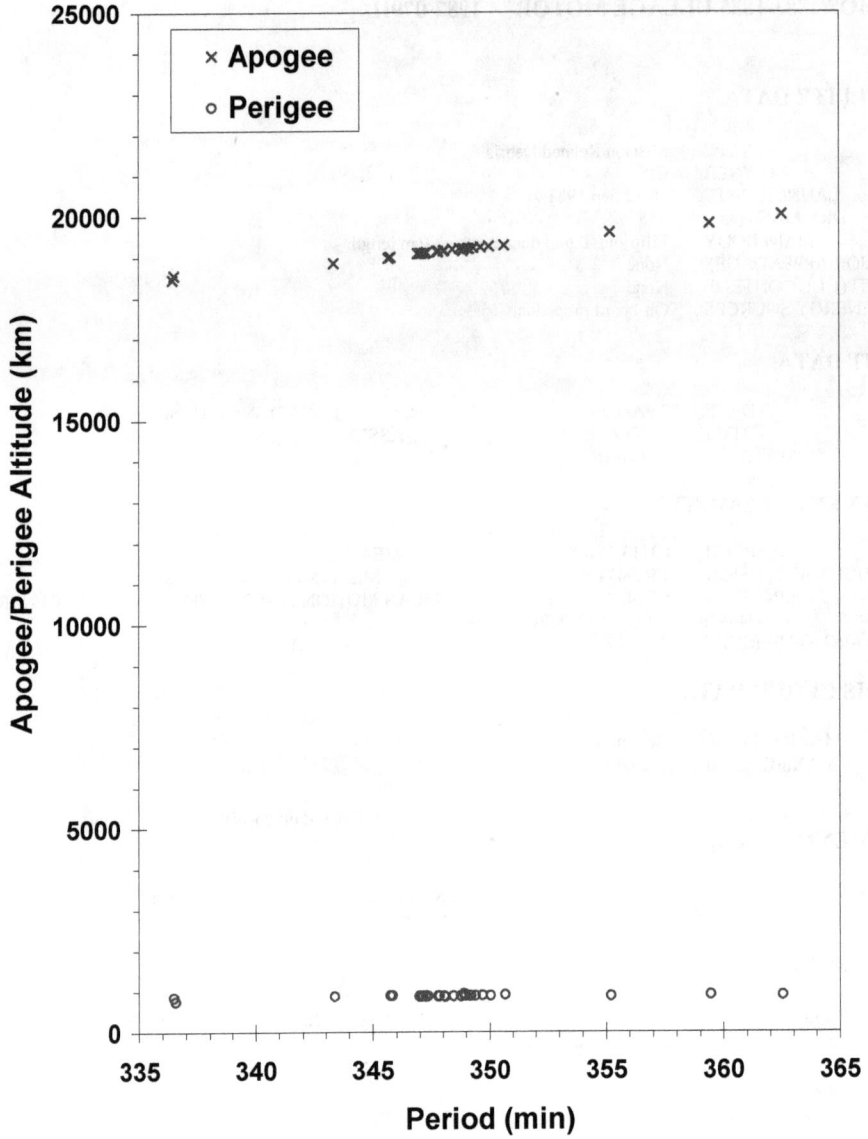

Cosmos 1883-85 auxiliary motor debris cloud of 31 fragments 2 days after the event as reconstructed from the US SSN database.

SATELLITE DATA

TYPE:	Payload
OWNER:	CIS
LAUNCH DATE:	26.48 Dec 1987
DRY MASS (KG):	6300
MAIN BODY:	Sphere-Cylinder; 2.4 m diameter by 6.5 m length
MAJOR APPENDAGES:	Solar panels
ATTITUDE CONTROL:	Active, 3-axis
ENERGY SOURCES:	On-board propellants, explosive charge

EVENT DATA

DATE:	31 Jan 1988	LOCATION:	11S, 138E (dsc)
TIME:	1109 GMT	ASSESSED CAUSE:	Deliberate
ALTITUDE:	250 km		

PRE-EVENT ELEMENTS

EPOCH:	88030.87152193	MEAN ANOMALY:	208.0352
RIGHT ASCENSION:	254.6565	MEAN MOTION:	16.07089398
INCLINATION:	82.5872	MEAN MOTION DOT/2:	.00174892
ECCENTRICITY:	.0015551	MEAN MOTION DOT DOT/6:	.000012805
ARG. OF PERIGEE:	152.1926	BSTAR:	.00022253

DEBRIS CLOUD DATA

MAXIMUM ΔP:	5.7 min*
MAXIMUM ΔI:	1.7 deg*

*Based on cataloged and uncataloged debris data

COMMENTS

Spacecraft destroyed after a malfunction prevented controlled reentry and landing in the Soviet Union. Elements for 83 objects remaining in orbit about 10 days after the event were developed. Other debris reentered before being officially cataloged.

REFERENCE DOCUMENTS

The Soviet Year in Space 1988, N. L. Johnson, Teledyne Brown Engineering, 1989, p. 27.

History of Soviet/Russian Satellite Fragmentations-A Joint U.S.-Russian Investigation, N. L. Johnson et al, Kaman Sciences Corporation, October 1995.

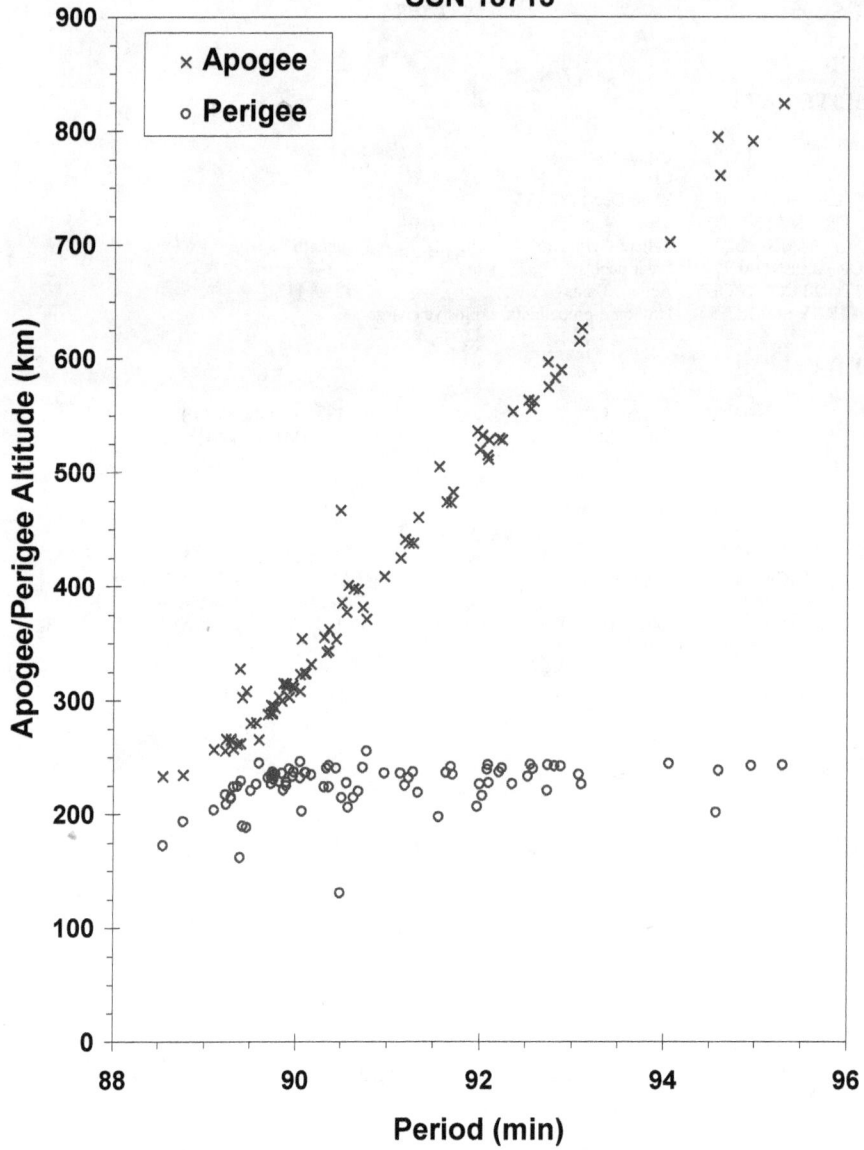

SSN 18713

Legend:
- × Apogee
- ○ Perigee

Y-axis: Apogee/Perigee Altitude (km)
X-axis: Period (min)

Cosmos 1906 debris cloud remnant of 83 objects 10 days after the event as reconstructed from Naval Space Surveillance System database.

EKRAN 17 ULLAGE MOTOR   **1987-109E**   **18719**

SATELLITE DATA

TYPE:	Mission Related Debris
OWNER:	CIS
LAUNCH DATE:	27.48 Dec 1987
DRY MASS (KG):	55
MAIN BODY:	Ellipsoid; 0.6 m diameter by 1.0 m length
MAJOR APPENDAGES:	None
ATTITUDE CONTROL:	None at time of the event.
ENERGY SOURCES:	On-board propellants

EVENT DATA

DATE:	22 May 1997	LOCATION:	Unknown
TIME:	Unknown	ASSESSED CAUSE:	Propulsion
ALTITUDE:	Unknown		

PRE-EVENT ELEMENTS

EPOCH:	97141.34020043	MEAN ANOMALY:	1.8603
RIGHT ASCENSION:	253.0389	MEAN MOTION:	3.58845480
INCLINATION:	46.6273	MEAN MOTION DOT/2:	-0.00000117
ECCENTRICITY:	0.6287941	MEAN MOTION DOT DOT/6:	0
ARG. OF PERIGEE:	349.7051	BSTAR:	0

DEBRIS CLOUD DATA

MAXIMUM ΔP:	Unknown
MAXIMUM ΔI:	Unknown

COMMENTS

This is the 15th event of this class identified to date. At least 72 debris were detected.

REFERENCE DOCUMENTS

"Three Satellite Breakups During May-June," The Orbital Debris Quarterly News, NASA JSC, July 1997, p. 2. Available online at http://www.orbitaldebris.jsc nasa.gov/newsletter/pdfs/ODQNv2i3.pdf.

"Identification and Resolution of an Orbital Debris Problem with Proton Launch Vehicle", Cherniatiev, Chernyavskiy, Johnson, and McKnight, First European Conference on Space Debris, 5-7 April 1993.

"The Fragmentation of Proton Debris", Nauer, Teledyne Brown Engineering Technical Report CS93-LKD-004, 31 Dec 1992.

History of Soviet/Russian Satellite Fragmentations-A Joint U.S.-Russian Investigation, N. L. Johnson et al, Kaman Sciences Corporation, October 1995.

Insufficient data to construct a Gabbard diagram.

COSMOS 1916   1988-007A   18823

SATELLITE DATA

TYPE:	Payload
OWNER:	CIS
LAUNCH DATE:	3.15 Feb 1988
DRY MASS (KG):	5700
MAIN BODY:	Cone-cylinder; 2.7 m diameter by 6.3 m length
MAJOR APPENDAGES:	Solar panels
ATTITUDE CONTROL:	Active, 3-axis
ENERGY SOURCES:	On-board propellants, explosive charge

EVENT DATA

DATE:	27 Feb 1988	LOCATION:	62N, 98E (asc)
TIME:	0444 GMT	ASSESSED CAUSE:	Deliberate
ALTITUDE:	155 km		

PRE-EVENT ELEMENTS

EPOCH:	88058.12322153	MEAN ANOMALY:	309.0154
RIGHT ASCENSION:	264.6529	MEAN MOTION:	16.30989909
INCLINATION:	64.8359	MEAN MOTION DOT/2:	.03233928
ECCENTRICITY:	.0060041	MEAN MOTION DOT DOT/6:	.00003669
ARG. OF PERIGEE:	51.6410	BSTAR:	.00025587

DEBRIS CLOUD DATA

MAXIMUM ΔP:	4.2 min*
MAXIMUM ΔI:	1.1 deg*

*Based on uncataloged debris data

COMMENTS

Spacecraft destroyed after a malfunction prevented controlled reentry and landing in the Soviet Union. Early elements on only 6 objects available. All debris reentered before being officially cataloged.

REFERENCE DOCUMENTS

The Soviet Year in Space 1988, N. L. Johnson, Teledyne Brown Engineering, 1989, p. 31.

History of Soviet/Russian Satellite Fragmentations-A Joint U.S.-Russian Investigation, N. L. Johnson et al, Kaman Sciences Corporation, October 1995.

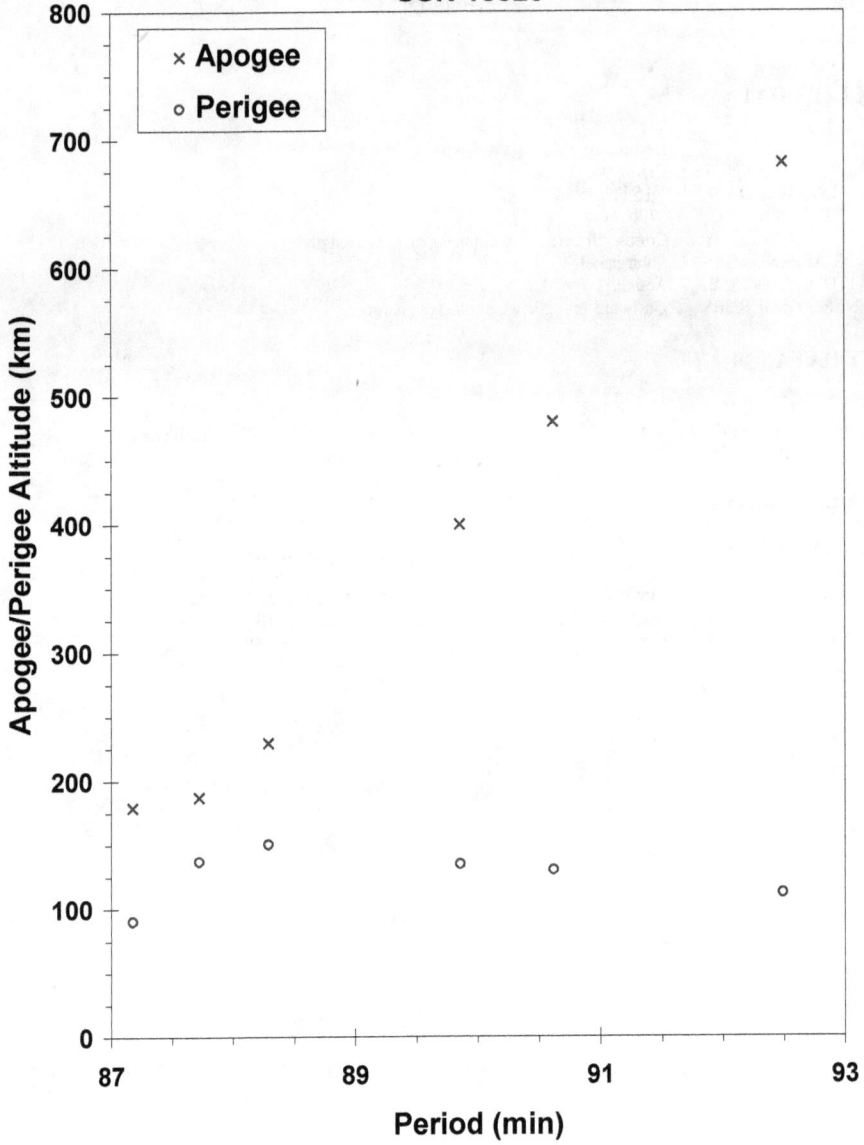

Cosmos 1916 debris cloud remnant of 6 objects within 1 day of the event as reconstructed from the US SSN database.

COSMOS 1934   1988-023A   18985

SATELLITE DATA

TYPE:	Payload
OWNER:	CIS
LAUNCH DATE:	22.59 Mar 1988
DRY MASS (KG):	800
MAIN BODY:	Cylinder; 2.035 m diameter
MAJOR APPENDAGES:	Several short booms
ATTITUDE CONTROL:	Gravity-gradient (passive)
ENERGY SOURCES:	Batteries

EVENT DATA

DATE:	23 Dec 2001	LOCATION:	Unknown
TIME:	Unknown	ASSESSED CAUSE:	Accidental Collision
ALTITUDE:	Unknown		

PRE-EVENT ELEMENTS

EPOCH:	91356.93360267	MEAN ANOMALY:	291.3330
RIGHT ASCENSION:	126.2142	MEAN MOTION:	13.75709229
INCLINATION:	82.9564	MEAN MOTION DOT/2:	0.00000135
ECCENTRICITY:	.0041502	MEAN MOTION DOT DOT/6:	0.0
ARG. OF PERIGEE:	69.2265	BSTAR:	0.00012752

DEBRIS CLOUD DATA

MAXIMUM ΔP:	Unknown
MAXIMUM ΔI:	Unknown

COMMENTS

The collision occurred with a piece of launch debris from Cosmos 926. The debris piece (Sat. No. 13475) was still incorrectly identified in the US SSC as of 1 December 2007. There were several very close conjunctions during the day in question, the exact time of the collision is unknown. Two pieces of debris were cataloged from the event long before the event was recognized as a collision.

REFERENCE DOCUMENT

"Accidental Collisions of Cataloged Satellites Identified", The Orbital Debris Quarterly News, NASA JSC, April 2005. Available online at http://www.orbitaldebris.jsc nasa.gov/newsletter/pdfs/ODQNv9i2.pdf.

Insufficient data to construct a Gabbard diagram.

INTELSAT 513 R/B   **1988-040B**   **19122**

SATELLITE DATA

TYPE:	Ariane 2 Third Stage
OWNER:	France
LAUNCH DATE:	17.99 May 1988
DRY MASS (KG):	~1480
MAIN BODY:	Cylinder; 2.6 m diameter by 11.7 m length
MAJOR APPENDAGES:	None
ATTITUDE CONTROL:	None
ENERGY SOURCES:	On-board propellants

EVENT DATA

DATE:	9 Jul 2002	LOCATION:	4.3 N, 5.7 E
TIME:	1930Z	ASSESSED CAUSE:	Propulsion
ALTITUDE:	21,500		

PRE-EVENT ELEMENTS

EPOCH:	02190.22071506	MEAN ANOMALY:	172.0370
RIGHT ASCENSION:	187.4675	MEAN MOTION:	2.28211164
INCLINATION:	7.0311	MEAN MOTION DOT/2:	.00000024
ECCENTRICITY:	.7162572	MEAN MOTION DOT DOT/6:	.0
ARG. OF PERIGEE:	181.6723	BSTAR:	.0

DEBRIS CLOUD DATA

MAXIMUM ΔP:	6.60 min*
MAXIMUM ΔI:	0.33 deg*

* Based on uncataloged debris data

COMMENTS

This is the second breakup of an Ariane 2 third stage officially recognized and the 11th overall breakup of an Ariane upper stage. This stage was launched prior to the implementation of passivation measures. The age of the stage at the time of the breakup was 14 years. Six pieces of debris were initially seen by the SSN, while four were cataloged.

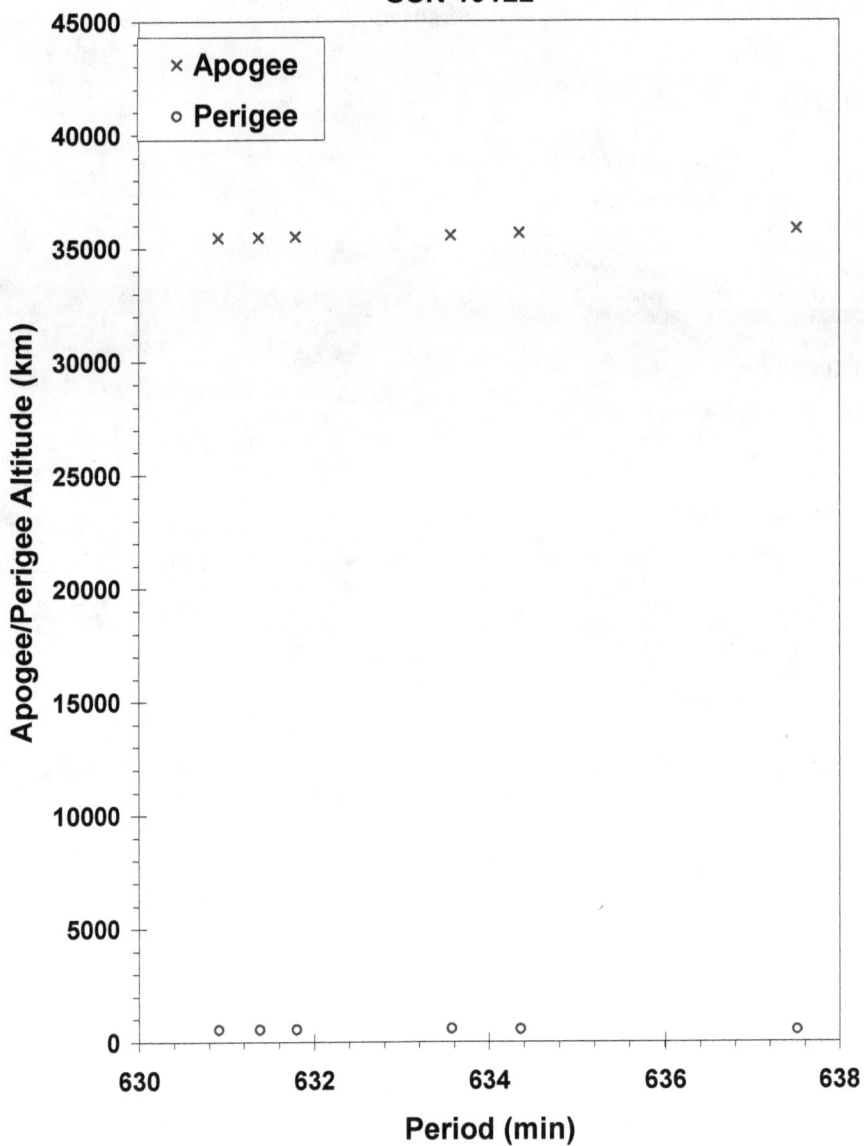

SSN 19122

Intelsat 513 R/B debris cloud of 6 fragments 2 weeks after the breakup as reconstructed from the US SSN database.

COSMOS 1970-1972 ULLAGE MOTOR   **1988-085F**   **19535**

SATELLITE DATA

TYPE:	Mission Related Debris
OWNER:	CIS
LAUNCH DATE:	16.08 Sep 1988
DRY MASS (KG):	~55
MAIN BODY:	Ellipsoid; 0.6 m diameter by 1.0 m length
MAJOR APPENDAGES:	None
ATTITUDE CONTROL:	None
ENERGY SOURCES:	On-board propellants

EVENT DATA

DATE:	04 Aug 2003	LOCATION:	Unknown
TIME:	~0725Z	ASSESSED CAUSE:	Propulsion
ALTITUDE:	Unknown		

PRE-EVENT ELEMENTS

EPOCH:	03214.47923598	MEAN ANOMALY:	334.9286		
RIGHT ASCENSION:	239.4643	MEAN MOTION:	4.29128214		
INCLINATION:	65.3341	MEAN MOTION DOT/2:	.00007107	ECCENTRICITY:	
.5561230	MEAN MOTION DOT DOT/6:	.0			
ARG. OF PERIGEE:	85.1870	BSTAR:	.071402		

DEBRIS CLOUD DATA

MAXIMUM ΔP:	18.8 min*
MAXIMUM ΔI:	2.79 deg*

* Based on uncataloged debris data

COMMENTS

This event marks the 28th known breakup of a Proton Block DM SOZ ullage motor since 1984. This ullage motor was launched before implementation of breakup preventive measures. Approximately 175 objects were initially seen by the SSN 1 week after the event. 76 debris objects were cataloged.

REFERENCE DOCUMENT

"Satellite Fragmentations in 2003", The Orbital Debris Quarterly News, NASA JSC, January 2004. Available online at http://www.orbitaldebris.jsc nasa.gov/newsletter/pdfs/ODQNv8i1.pdf.

267

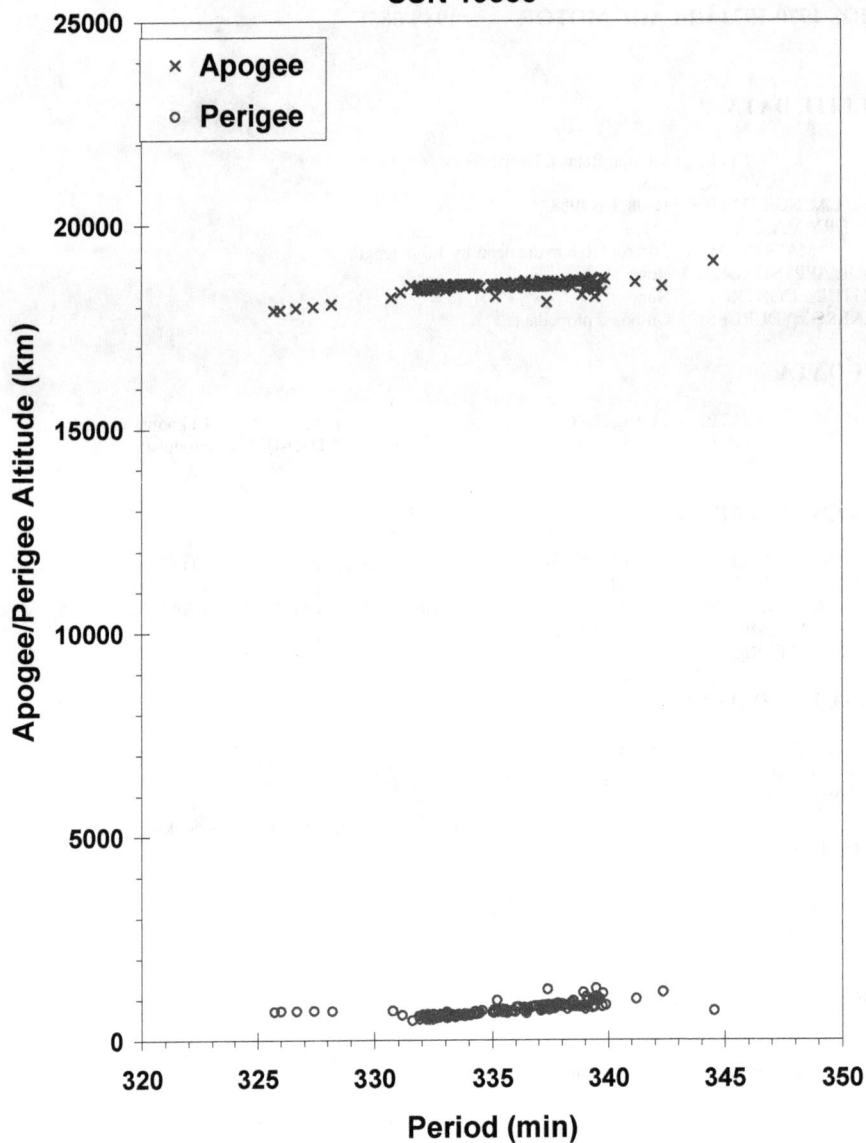

SSN 19535

Cosmos 1970-72 auxiliary motor debris cloud of 175 fragments 1 week after the event as reconstructed from the US SSN database.

COSMOS 1970-1972 ULLAGE MOTOR   1988-085G

SATELLITE DATA

TYPE:	Mission Related Debris
OWNER:	CIS
LAUNCH DATE:	16 Sep 1988
DRY MASS (KG):	55
MAIN BODY:	Ellipsoid; 0.6 m diameter by 1.0 m length
MAJOR APPENDAGES:	None
ATTITUDE CONTROL:	None
ENERGY SOURCES:	On-board propellants

EVENT DATA

DATE:	9 Mar 1999	LOCATION:	Unknown
TIME:	Unknown	ASSESSED CAUSE:	Propulsion
ALTITUDE:	Unknown		

POST-EVENT ELEMENTS

EPOCH:	99067.36656961	MEAN ANOMALY:	189.8576
RIGHT ASCENSION:	108.7309	MEAN MOTION:	4.28860956162171
INCLINATION:	64.6425	MEAN MOTION DOT/2:	.00000813
ECCENTRICITY:	.5827119	MEAN MOTION DOT DOT/6:	00000-0
ARG. OF PERIGEE:	176.8483	BSTAR:	.0022335

DEBRIS CLOUD DATA

MAXIMUM ΔP:	Unknown
MAXIMUM ΔI:	Unknown

COMMENTS

This is the 19th event of the Proton Block DM SOZ Ullage Motor class identified to date; it is the eighth associated with a GLONASS mission. This mission was conducted before the engineering defect was identified and passivation measures implemented. In this orbit, debris may be long-lived but hard to track. 17 debris objects were detected.

REFERENCE DOCUMENT

"Abandoned Proton Ullage Motors Continue to Create Debris", The Orbital Debris Quarterly News, NASA JSC, April 1999. Available online at http://www.orbitaldebris.jsc nasa.gov/newsletter/pdfs/ODQNv4i2.pdf.

Insufficient data to construct a Gabbard diagram.

SKYNET 4B/ASTRA 1A R/B   **1988-109C**   **19689**

SATELLITE DATA

TYPE:	Ariane 4 H-10 Third Stage
OWNER:	France
LAUNCH DATE:	11.02 Dec 1988
DRY MASS (KG):	1760
MAIN BODY:	Cylinder; 2.6 m diameter by 11.4 m length
MAJOR APPENDAGES:	None
ATTITUDE CONTROL:	None at time of the event.
ENERGY SOURCES:	On-board propellants

EVENT DATA

DATE:	17 Feb 1998	LOCATION:	6.9 N, 157.2 E (dsc)
TIME:	1235 GMT	ASSESSED CAUSE:	Propulsion
ALTITUDE:	19630 km		

PRE-EVENT ELEMENTS

EPOCH:	98047.29326560	MEAN ANOMALY:	25.3394
RIGHT ASCENSION:	23.7998	MEAN MOTION:	2.25942020
INCLINATION:	7.3381	MEAN MOTION DOT/2:	0.00000046
ECCENTRICITY:	0.7222736	MEAN MOTION DOT DOT/6:	0
ARG. OF PERIGEE:	248.1711	BSTAR:	0.00057969

DEBRIS CLOUD DATA

MAXIMUM ΔP:	Unknown
MAXIMUM ΔI:	Unknown

COMMENTS

This mission was the second for the Ariane 4 series and occurred prior to implementation of passivation measures. Using observations from the Eglin radar, specialists at Millstone radar found four new pieces from the upper stage. Naval Space Command personnel generated the first two debris element sets and calculated the approximate breakup time noted above.

REFERENCE DOCUMENT

"Three Upper Stage Breakups in One Week Top February Debris Activity", <u>The Orbital Debris Quarterly News</u>, NASA JSC, April 1998, p. 1. Available online at http://www.orbitaldebris.jsc nasa.gov/newsletter/pdfs/ODQNv3i2.pdf.

Insufficient data to construct a Gabbard diagram.

COSMOS 1987-1989 ULLAGE MOTOR   1989-001G

SATELLITE DATA

TYPE:	Mission Related Debris
OWNER:	CIS
LAUNCH DATE:	10 Jan 1989
DRY MASS (KG):	55
MAIN BODY:	Ellipsoid; 0.6 m diameter by 1 m length
MAJOR APPENDAGES:	None
ATTITUDE CONTROL:	None
ENERGY SOURCES:	On-board propellants

EVENT DATA

DATE:	3 Aug 1998	LOCATION:	Unknown
TIME:	Unknown	ASSESSED CAUSE:	Propulsion
ALTITUDE:	Unknown		

POST-EVENT ELEMENTS

EPOCH:	98211.80543118	MEAN ANOMALY:	172.2753
RIGHT ASCENSION:	16.7694	MEAN MOTION:	4.24137167
INCLINATION:	64.9243	MEAN MOTION DOT/2:	.00000287
ECCENTRICITY:	.5776927	MEAN MOTION DOT DOT/6:	00000-0
ARG. OF PERIGEE:	182.6029	BSTAR:	.0041366

DEBRIS CLOUD DATA

MAXIMUM ΔP:	162.64 min
MAXIMUM ΔI:	3.78 deg

COMMENTS

This is the 17th event of the Proton Block DM SOZ Ullage Motor class identified to date; it is the sixth associated with a GLONASS mission. This mission was conducted before the engineering defect was identified and passivation measures implemented. In this orbit, debris may be long-lived but hard to track. More than 110 debris objects were detected.

REFERENCE DOCUMENT

"Solitary Breakup and Anomalous Events in Third Quarter are Familiar", The Orbital Debris Quarterly News, NASA JSC, October 1998. Available online at: http://www.orbitaldebris.jsc.nasa.gov/newsletter/pdfs/ODQNv3i4.pdf.

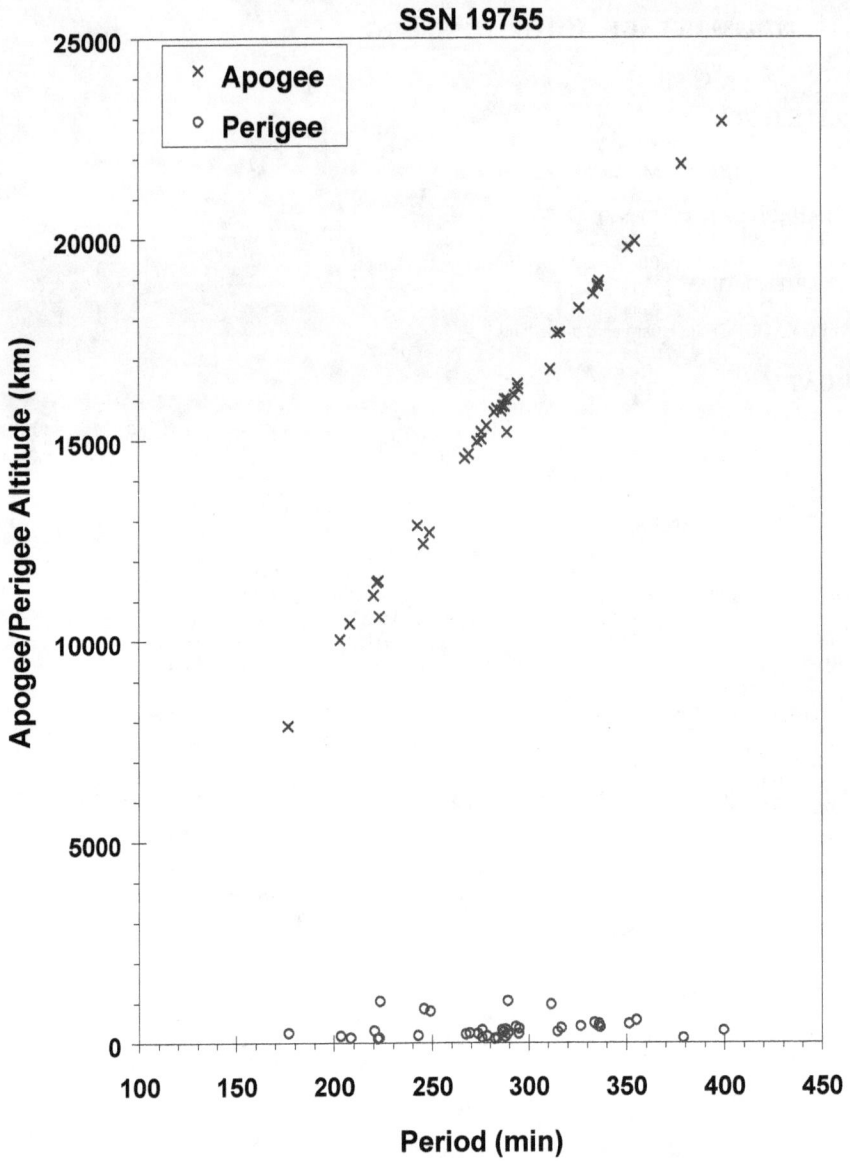

Cosmos 1987-1989 ullage motor debris cloud of 39 fragments 7 days after the event as reconstructed from the US SSN database.

COSMOS 1987-1989 ULLAGE MOTOR   1989-001H   **19856**

SATELLITE DATA

TYPE:	Mission Related Debris
OWNER:	CIS
LAUNCH DATE:	10 Jan 1989
DRY MASS (KG):	~55
MAIN BODY:	Ellipsoid; 0.6 m diameter by 1.0m length
MAJOR APPENDAGES:	None
ATTITUDE CONTROL:	None
ENERGY SOURCES:	On-board propellants

EVENT DATA

DATE:	13 Nov 2003	LOCATION:	Unknown
TIME:	Unknown	ASSESSED CAUSE:	Propulsion
ALTITUDE:	Unknown		

PRE-EVENT ELEMENTS

EPOCH:	03317.76135862	MEAN ANOMALY:	339.1502	
RIGHT ASCENSION:	52.9695	MEAN MOTION:	4.24824637	
INCLINATION:	65.4357	MEAN MOTION DOT/2:	.00000161	ECCENTRICITY:
.5599025	MEAN MOTION DOT DOT/6:		.0	
ARG. OF PERIGEE:	72.44443	BSTAR:	.0017638	

DEBRIS CLOUD DATA

MAXIMUM ΔP:	Unknown
MAXIMUM ΔI:	Unknown

COMMENTS

This event marks the 29th known breakup of a Proton Block DM SOZ ullage motor since 1984, although the event went undetected for over 3 months. This ullage motor was launched before implementation of breakup preventive measures. No debris objects were cataloged from this breakup.

Insufficient data to construct a Gabbard diagram.

GORIZONT 17 ULLAGE MOTOR   **1989-004E**   **19771**

SATELLITE DATA

TYPE:	Mission Related Debris
OWNER:	CIS
LAUNCH DATE:	26.39 Jan 1989.
DRY MASS (KG):	55
MAIN BODY:	Ellipsoid; 0.6 m diameter by 1 m diameter
MAJOR APPENDAGES:	None
ATTITUDE CONTROL:	None
ENERGY SOURCES:	On-board propellants

EVENT DATA

DATE:	17-18 Dec 1992	LOCATION:	Unknown
TIME:	Unknown	ASSESSED CAUSE:	Propulsion
ALTITUDE:	Unknown		

PRE-EVENT ELEMENTS

EPOCH:	92351.90838995	MEAN ANOMALY:	1.4295
RIGHT ASCENSION:	266.2338	MEAN MOTION:	4.60309514
INCLINATION:	46.7001	MEAN MOTION DOT/2:	.00060784
ECCENTRICITY:	.5692927	MEAN MOTION DOT DOT/6:	.0000093219
ARG. OF PERIGEE:	353.9854	BSTAR:	.0015056

DEBRIS CLOUD DATA

MAXIMUM ΔP:	Unknown
MAXIMUM ΔI:	Unknown

COMMENTS

Parent satellite was one of two small engine units that are routinely released after the first burn of the Proton fourth stage. The nature of these objects was identified by Dr. Boris V. Chernlatiev, Deputy Constructor for the Energiya NPO, in October 1992. The cause of this fragmentation is assumed to be related to the residual hypergolic propellants on board and failure of the membrane separating the fuel and oxidizer. NAVSPASUR observed between 30-40 objects that were associated with this breakup. Only 4 element sets were generated, insufficient for a Gabbard Diagram or BLAST point. This was the seventh in a series of fragmentations of this object type, and was the second located in a geosynchronous transfer orbit.

REFERENCE DOCUMENTS

The Fragmentation of Proton Debris, D. J. Nauer, TBE Technical Report CS93-LKD-004, Teledyne Brown Engineering, Colorado Springs, 31 December 1992.

Analysis of Fragmentations From December 1992 - February 1993, TBE Technical Report CS93-LKD-010, Teledyne Brown Engineering, Colorado Springs, 30 March 1993.

History of Soviet/Russian Satellite Fragmentations-A Joint U.S.-Russian Investigation, N. L. Johnson et al, Kaman Sciences Corporation, October 1995.

"Identification and Resolution of an Orbital Debris Problem with the Proton Launch Vehicle", B. V. Cherniatiev et al, Proceedings of the First European Conference on Space Debris, April 1993.

Insufficient data to construct a Gabbard diagram.

ARIANE 2 R/B   **1989-006B**   **19773**

SATELLITE DATA

TYPE:	Ariane 2 third stage with VEB
OWNER:	France
LAUNCH DATE:	27.06 Jan 1989
DRY MASS (KG):	~1480 kg
MAIN BODY:	2.6 m diameter by 11.7 m length
MAJOR APPENDAGES:	None
ATTITUDE CONTROL:	None
ENERGY SOURCES:	On-board propellants?

EVENT DATA

DATE:	~1 Jan 2001	LOCATION:	Unknown
TIME:	Unknown	ASSESSED CAUSE:	Propulsion
ALTITUDE:	Unknown		

PRE-EVENT ELEMENTS

EPOCH:	00366.06151127	MEAN ANOMALY:	45.8970
RIGHT ASCENSION:	73.3900	MEAN MOTION:	2.26500973
INCLINATION:	8.3781	MEAN MOTION DOT/2:	.00000580
ECCENTRICITY:	.7188412	MEAN MOTION DOT DOT/6:	.0000000
ARG. OF PERIGEE:	225.8250	BSTAR:	.0040973

DEBRIS CLOUD DATA

MAXIMUM ΔP:	Unknown*
MAXIMUM ΔI:	Unknown*

* Not calculated due to provisional nature
of orbital data.

COMMENTS

This is the first breakup of an Ariane 2 third stage officially recognized. One Ariane 3 third stage (same as Ariane 2) is known to have broken-up within a few days of launch in 1987. Both vehicles were launched before passivation measures were incorporated with Ariane third stages. Ariane third stage passivation was introduced in January 1990 and has been employed on all Ariane missions since October 1993. The age of the Ariane 2 third stage at the time of the breakup was nearly 12 years.

SSN 19773

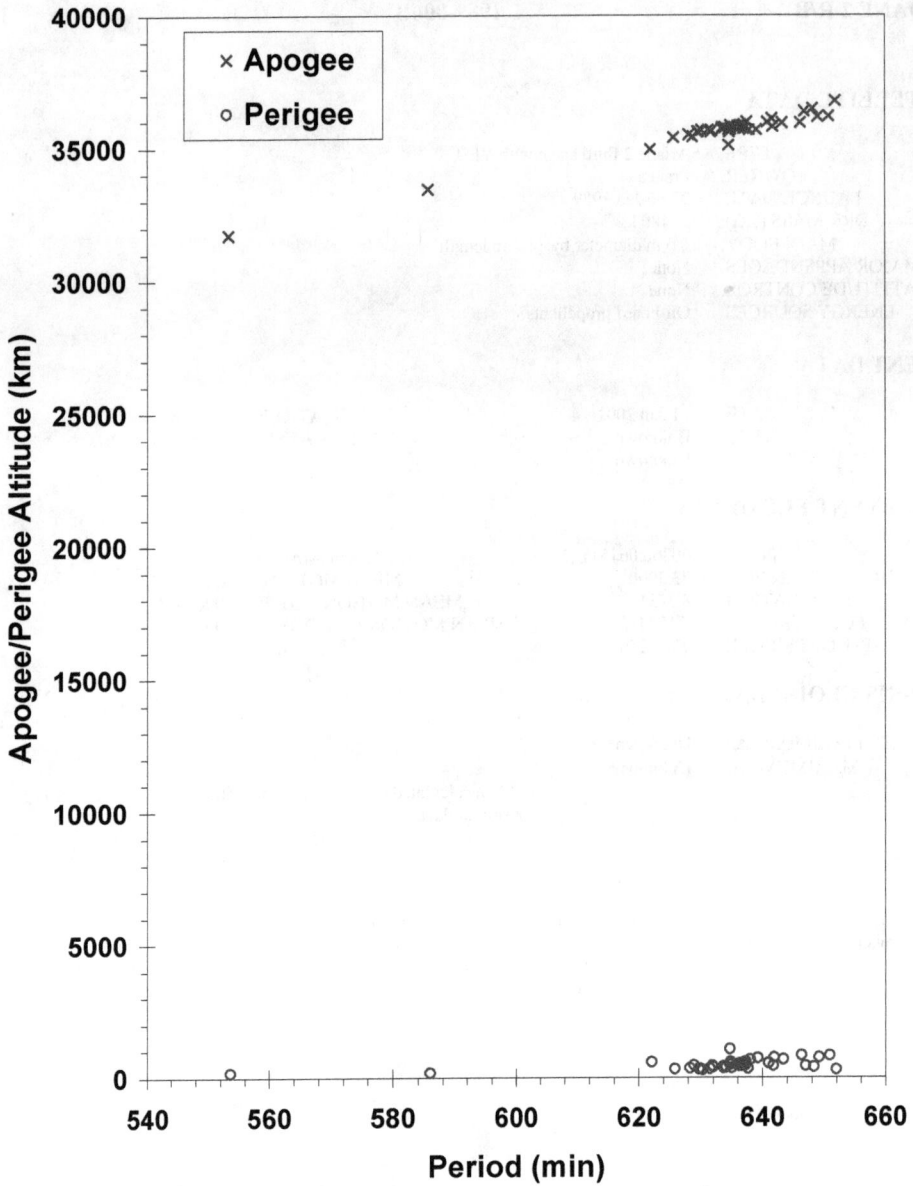

INTELSAT V F-15 R/B (Ariane 2) debris cloud as constructed using SSN 8XXXX series elements sets (10 January 2001 and before).

SATELLITE DATA

TYPE:	Mission Related Debris
OWNER:	CIS
LAUNCH DATE:	31.36 May 1989
DRY MASS (KG):	55
MAIN BODY:	Ellipsoid; 0.6 m diameter by 1 m length
MAJOR APPENDAGES:	None
ATTITUDE CONTROL:	None
ENERGY SOURCES:	On-board propellant

EVENT DATA

DATE:	10 Jun 2006	LOCATION:	65S, 100E (dsc)
TIME:	1320 GMT	ASSESSED CAUSE:	Propulsion
ALTITUDE:	17,375 km		

PRE-EVENT ELEMENTS

EPOCH:	06161.39815228	MEAN ANOMALY:	338.3349
RIGHT ASCENSION:	289.1150	MEAN MOTION:	4.32576815
INCLINATION:	65.0603	MEAN MOTION DOT/2:	-0.00002708
ECCENTRICITY:	.5578964	MEAN MOTION DOT DOT/6:	0.0
ARG. OF PERIGEE:	74.2422	BSTAR:	-0.018697

DEBRIS CLOUD DATA

MAXIMUM ΔP:	12.9 min
MAXIMUM ΔI:	2.1 deg

COMMENTS

This event marks the 34th known breakup of a Proton Block DM SOZ ullage motor since 1984. There were two more fragmentation events for this object during July 2006; the second event was on July 3, the third event was July 27. These events resulted in the most debris cataloged of any SOZ ullage motor in history with over 100 pieces cataloged. The majority of debris (>75) were created during the second event on 3 July 2006.

REFERENCE DOCUMENTS

History of Soviet/Russian Satellite Fragmentations-A Joint U.S.-Russian Investigation, N. L. Johnson et al, Kaman Sciences Corporation, October 1995.

"First Satellite Breakups of 2006", The Orbital Debris Quarterly News, NASA JSC, July 2006. Available online at http://www.orbitaldebris.jsc nasa.gov/newsletter/pdfs/ODQNv10i3.pdf.

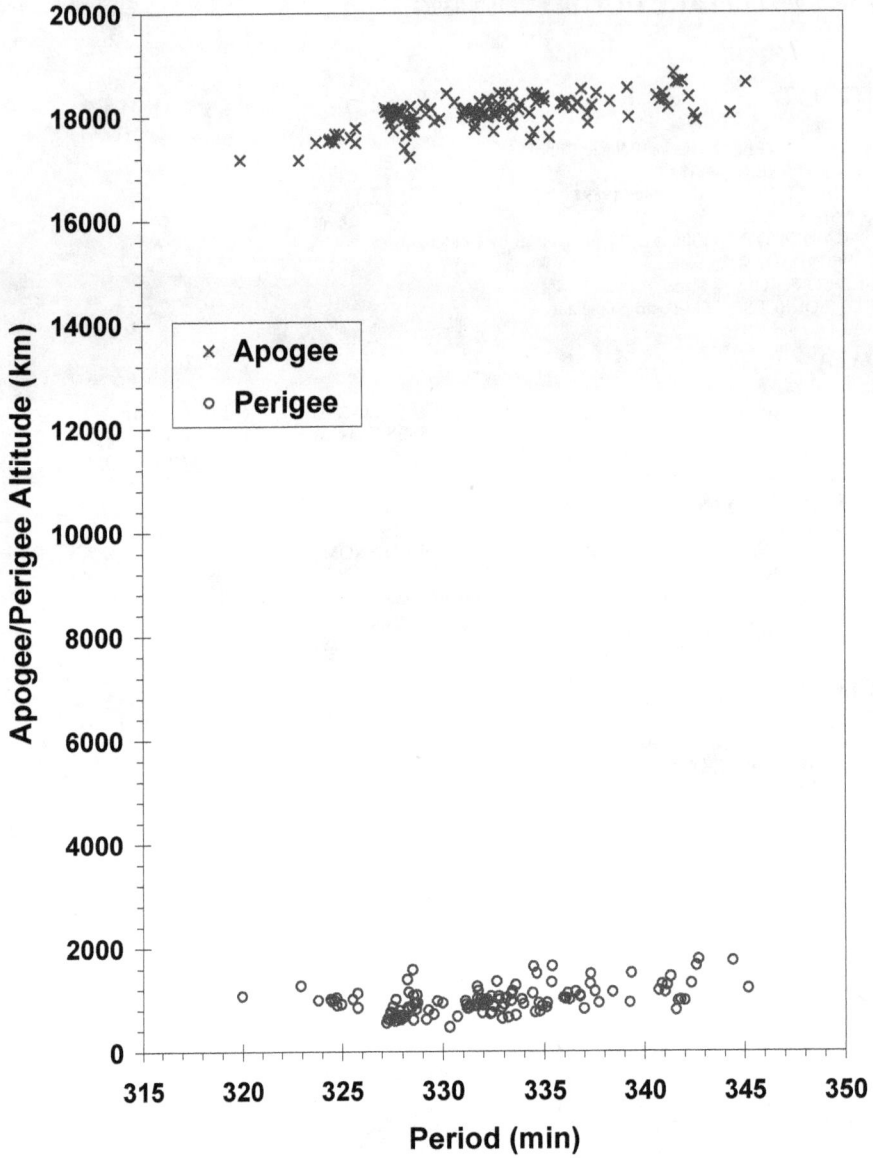

SSN 20081

SOZ Motor debris cloud around September 2006 with 131 objects, reflecting debris from all three fragmentation events as reconstructed from the US SSN database.

SATELLITE DATA

TYPE:	Mission Related Debris
OWNER:	CIS
LAUNCH DATE:	5.95 Jul 1989
DRY MASS (KG):	55
MAIN BODY:	Ellipsoid; 0.6 m diameter by 1 m length
MAJOR APPENDAGES:	None
ATTITUDE CONTROL:	None
ENERGY SOURCES:	On-board propellants

EVENT DATA

DATE:	12 Jan 1993	LOCATION:	Unknown
TIME:	Unknown	ASSESSED CAUSE:	Propulsion
ALTITUDE:	Unknown		

PRE-EVENT ELEMENTS

EPOCH:	93004.96424013	MEAN ANOMALY:	353.7659
RIGHT ASCENSION:	215.2912	MEAN MOTION:	2.68234049
INCLINATION:	46.7556	MEAN MOTION DOT/2:	.00007021
ECCENTRICITY:	.6967354	MEAN MOTION DOT DOT/6:	.0
ARG. OF PERIGEE:	45.1358	BSTAR:	.0017532

DEBRIS CLOUD DATA

MAXIMUM ΔP:	Unknown
MAXIMUM ΔI:	Unknown

COMMENTS

Parent satellite was one of two small engine units that are routinely released after the first burn of the Proton fourth stage. The nature of these objects was identified by Dr. Boris V. Chernlatiev, Deputy Constructor for the Energiya NPO, in October 1992. The cause of this fragmentation is assumed to be related to the residual hypergolic propellants on board and failure of the membrane separating the fuel and oxidizer. NAVSPASUR observed 18 objects that were associated with this breakup. Only 5 element sets were generated, and were of insufficient quality for a credible Gabbard Diagram or BLAST point. This was the eighth in a series of fragmentations of this object type, and was the third located in a geosynchronous transfer orbit.

REFERENCE DOCUMENTS

The Fragmentation of Proton Debris, D. J. Nauer, TBE Technical Report CS93-LKD-004, Teledyne Brown Engineering, Colorado Springs, 31 December 1992.

Analysis of Fragmentations From December 1992 - February 1993, TBE Technical Report CS93-LKD-010, Teledyne Brown Engineering, Colorado Springs, 30 March 1993.

History of Soviet/Russian Satellite Fragmentations-A Joint U.S.-Russian Investigation, N. L. Johnson et al, Kaman Sciences Corporation, October 1995.

"Identification and Resolution of an Orbital Debris Problem with the Proton Launch Vehicle", B. V. Cherniatiev et al, Proceedings of the First European Conference on Space Debris, April 1993.

Insufficient data to construct a Gabbard diagram.

SATELLITE DATA

TYPE:	Payload
OWNER:	CIS
LAUNCH DATE:	12.63 Jul 1989
DRY MASS (KG):	5700
MAIN BODY:	Cone-cylinder; 2.7 m diameter by 6.3 m length
MAJOR APPENDAGES:	Solar panels
ATTITUDE CONTROL:	Active, 3-axis
ENERGY SOURCES:	On-board propellants, explosive charge

EVENT DATA

DATE:	28 Jul 1989	LOCATION:	35-65N, 95-140E (asc)
TIME:	0410-0420 GMT	ASSESSED CAUSE:	Deliberate
ALTITUDE:	150 km		

PRE-EVENT ELEMENTS

EPOCH:	89208.98384568	MEAN ANOMALY:	302.7810
RIGHT ASCENSION:	89.7470	MEAN MOTION:	16.33519268
INCLINATION:	67.1441	MEAN MOTION DOT/2:	.03079561
ECCENTRICITY:	.0048139	MEAN MOTION DOT DOT/6:	.000029506
ARG. OF PERIGEE:	57.9032	BSTAR:	.00023479

DEBRIS CLOUD DATA

MAXIMUM ΔP:	7.1 min*
MAXIMUM ΔI:	1.3 deg*

*Based on uncataloged debris data

COMMENTS

Spacecraft was destroyed after a malfunction prevented controlled reentry and landing in the Soviet Union. Early element sets on only 20 objects available. Rapid decay of objects made calculation of breakup time and location difficult.

REFERENCE DOCUMENTS

The Fragmentation of Kosmos 2030, N. L. Johnson, Technical Report CS89-TR-JSC-002, Teledyne Brown Engineering, Colorado Springs, Colorado, September 1989.

History of Soviet/Russian Satellite Fragmentations-A Joint U.S.-Russian Investigation, N. L. Johnson et al, Kaman Sciences Corporation, October 1995.

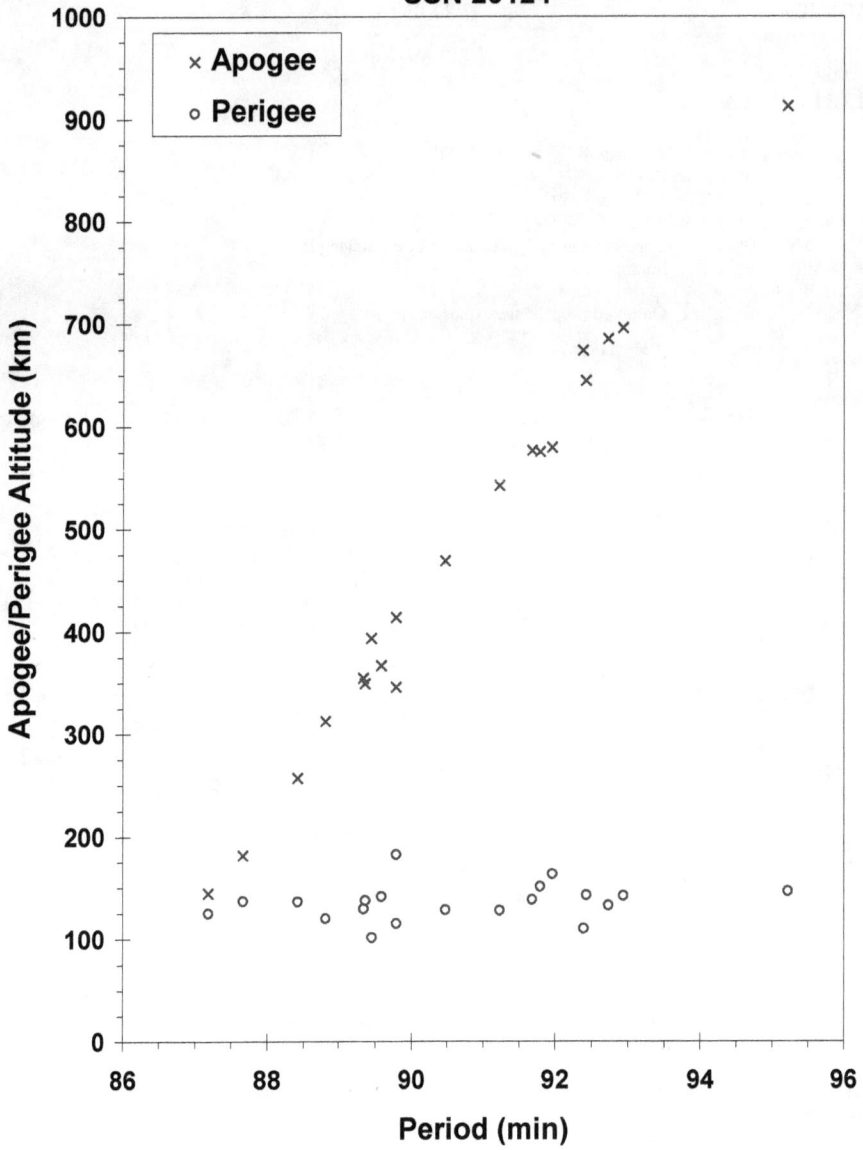

SSN 20124

Cosmos 2030 debris cloud remnant of 20 objects two to 3 days after the event as reconstructed from the US SSN database. This diagram is taken from the cited reference.

COSMOS 2031   1989-056A   20136

SATELLITE DATA

TYPE:	Payload
OWNER:	CIS
LAUNCH DATE:	18.51 Jul 1989
DRY MASS (KG):	6000
MAIN BODY:	Cylinder; 2.4 m diameter by 7 m length
MAJOR APPENDAGES:	Solar panels
ATTITUDE CONTROL:	Active, 3-axis
ENERGY SOURCES:	On-board propellants, explosive charge

EVENT DATA

DATE:	31 Aug 1989	LOCATION:	43N, 111E (dsc)
TIME:	1851 GMT	ASSESSED CAUSE:	Deliberate
ALTITUDE:	270 km		

PRE-EVENT ELEMENTS

EPOCH:	89243.76468690	MEAN ANOMALY:	305.4386
RIGHT ASCENSION:	242.9132	MEAN MOTION:	15.89273241
INCLINATION:	50.5464	MEAN MOTION DOT/2:	.00196451
ECCENTRICITY:	.0093577	MEAN MOTION DOT DOT/6:	.00002154
ARG. OF PERIGEE:	55.5300	BSTAR:	.00045172

DEBRIS CLOUD DATA

MAXIMUM ΔP:	7.4 min*
MAXIMUM ΔI:	0.9 deg*

*Based on uncataloged debris data

COMMENTS

Spacecraft was destroyed with a planned detonation. Cosmos 2031 was the first of a new series of spacecraft that employs end-of-mission detonation as standard operating procedure. Early elements on 43 objects available. Most debris reentered before being officially cataloged.

REFERENCE DOCUMENTS

The Fragmentation of Kosmos 2031, N. L. Johnson, Technical Report CS89-TR-JSC-003, Teledyne Brown Engineering, Colorado Springs, Colorado, September 1989.

History of Soviet/Russian Satellite Fragmentations-A Joint U.S.-Russian Investigation, N. L. Johnson et al, Kaman Sciences Corporation, October 1995.

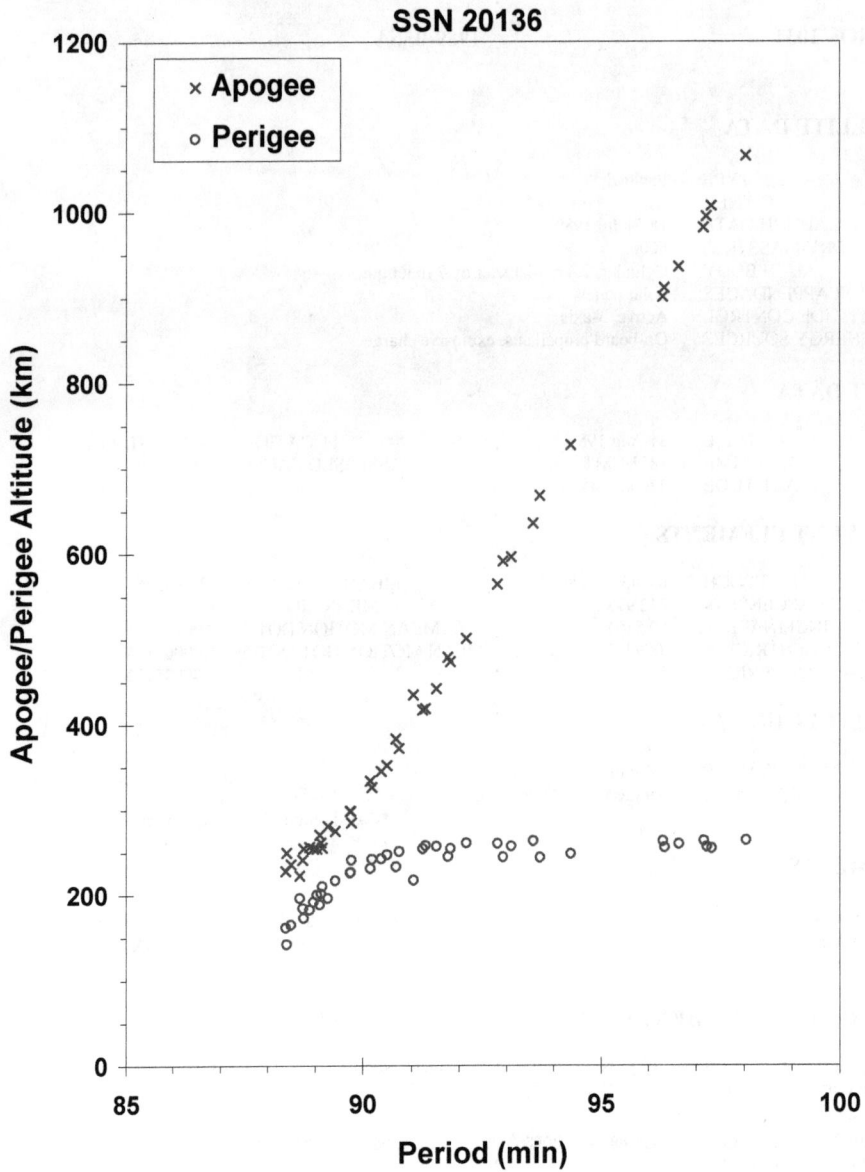

SSN 20136

Cosmos 2031 debris cloud remnant of 43 objects 3 days after the event as reconstructed from Naval Space Surveillance System database. This diagram is taken from the cited reference.

SATELLITE DATA

TYPE:	Rocket Body
OWNER:	US
LAUNCH DATE:	18.61 Nov 1989
DRY MASS (KG):	920
MAIN BODY:	Cylinder; 2.4 m diameter by 6.0 m length
MAJOR APPENDAGES:	None
ATTITUDE CONTROL:	None at time of event
ENERGY SOURCES:	None at time of event

EVENT DATA

DATE:	3 Dec 2006	LOCATION:	52S, 168E (dsc)
TIME:	0337 GMT	ASSESSED CAUSE:	Unknown
ALTITUDE:	730 km		

PRE-EVENT ELEMENTS

EPOCH:	06336.48315357	MEAN ANOMALY:	65.8381
RIGHT ASCENSION:	123.6830	MEAN MOTION:	14.46527792
INCLINATION:	97.0839	MEAN MOTION DOT/2:	0.00000076
ECCENTRICITY:	.0073269	MEAN MOTION DOT DOT/6:	0.0
ARG. OF PERIGEE:	293.5127	BSTAR:	0.000029963

DEBRIS CLOUD DATA

MAXIMUM ΔP:	2.1 min
MAXIMUM ΔI:	0.3 deg

COMMENTS

The Delta 2 rocket body had been passivated and dormant for 17 years. Observation of the object showed a high tumbling rate after the event. A collision with a smaller particle has not been ruled out. For the most part, the cataloged objects from this event were short-lived, i.e., less than 1 month.

REFERENCE DOCUMENT

"Significant Increase in Satellite Breakups During 2006", The Orbital Debris Quarterly News, NASA JSC, January 2007. Available online at http://www.orbitaldebris.jsc.nasa.gov/newsletter/pdfs/ODQNv11i1.pdf.

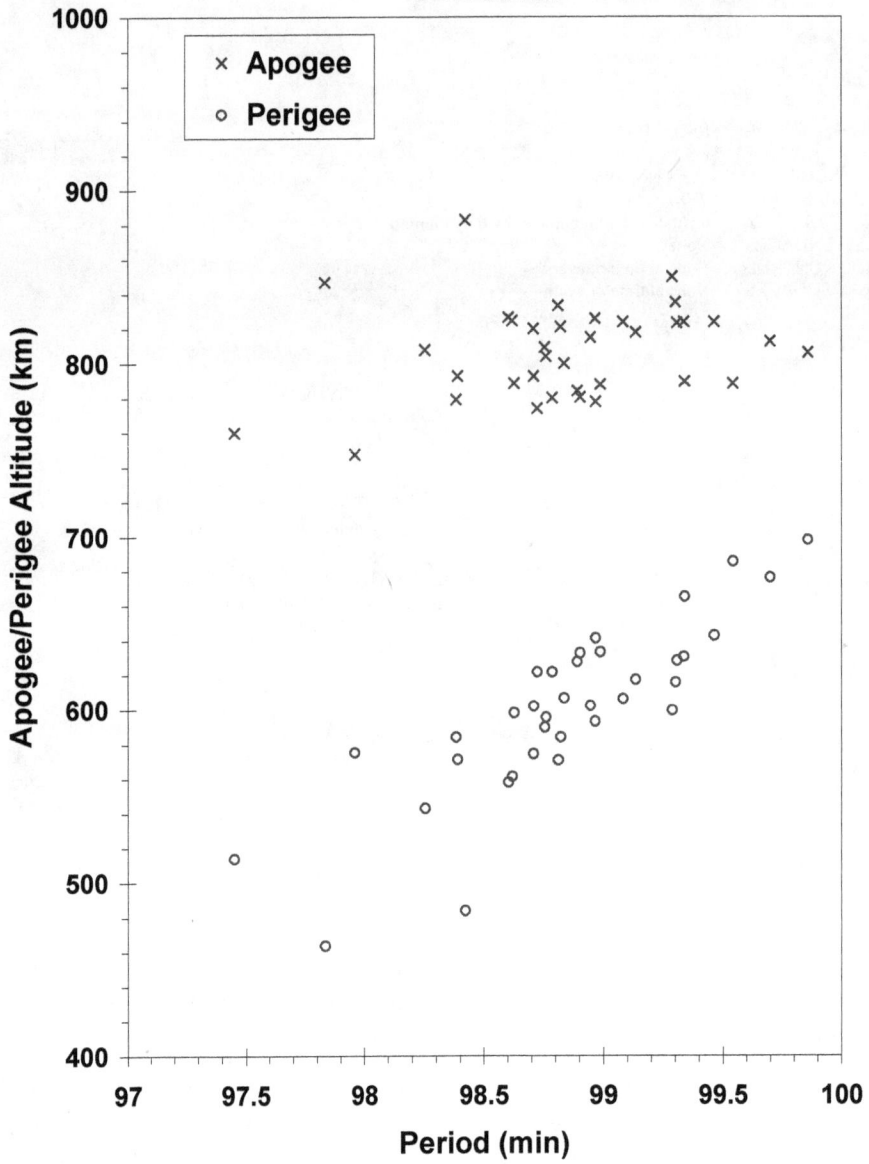

SSN 20323

COBE R/B debris cloud with 36 fragments, 6 days after the event
as reconstructed from the US SSN database.

COSMOS 2053 R/B   **1989-100B**   **20390**

SATELLITE DATA

TYPE:	Tsyklon Third Stage
OWNER:	CIS
LAUNCH DATE:	27 Dec 1989
DRY MASS (KG):	1360
MAIN BODY:	Cone-cylinder; 2.1 m diameter by 3.3 m length
MAJOR APPENDAGES:	None
ATTITUDE CONTROL:	None
ENERGY SOURCES:	Unknown

EVENT DATA

DATE:	18 Apr 1999	LOCATION:	16.9S, 234.1E
TIME:	0119 GMT	ASSESSED CAUSE:	Propulsion
ALTITUDE:	485 km		

POST-EVENT ELEMENTS

EPOCH:	99107.56102679	MEAN ANOMALY:	26.3814
RIGHT ASCENSION:	275.5509	MEAN MOTION:	15.29126555517603
INCLINATION:	73.5159	MEAN MOTION DOT/2:	.00003667
ECCENTRICITY:	.0010450	MEAN MOTION DOT DOT/6:	00000-0
ARG. OF PERIGEE:	333.6852	BSTAR:	.0013164

DEBRIS CLOUD DATA

MAXIMUM ΔP:	4.18 min
MAXIMUM ΔI:	.66 deg

COMMENTS

This is the 3rd event of the Tsyklon third stage (SL-14) identified to date, and the second within 14 months. All stages have been about 10 years old at the time of breakup. The vehicle is a Ukrainian-produced stage with unknown end-of-mission passivation. Its propellants are UDMH and N204. More than 60 debris objects were detected.

REFERENCE DOCUMENT

"Third Tsyklon Upper Stage Breaks Up", The Orbital Debris Quarterly News, NASA JSC, July 1999. Available online at http://www.orbitaldebris.jsc nasa.gov/newsletter/pdfs/ODQNv4i3.pdf

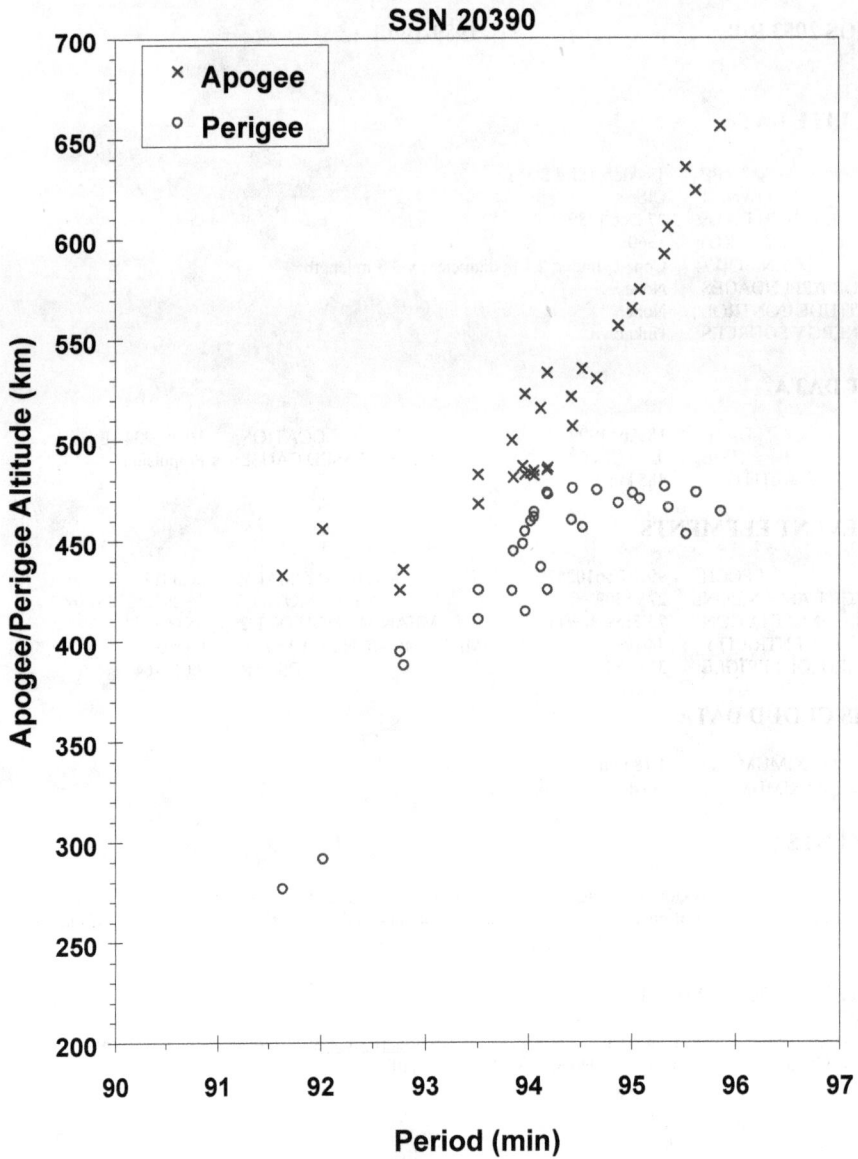

Cosmos 2053 rocket body debris cloud of 31 fragments 9 days after the event as reconstructed from the US SSN database.

COSMOS 2054 ULLAGE MOTOR   1989-101E   20399

SATELLITE DATA

TYPE:	Mission Related Debris
OWNER:	CIS
LAUNCH DATE:	27.47 Dec 1989
DRY MASS (KG):	55
MAIN BODY:	Ellipsoid; 0.6 m diameter by 1 m length
MAJOR APPENDAGES:	None
ATTITUDE CONTROL:	None
ENERGY SOURCES:	On-board propellants

EVENT DATA

DATE:	Jul 1992 (?)	LOCATION:	Unknown
TIME:	Unknown	ASSESSED CAUSE:	Propulsion
ALTITUDE:	Unknown		

PRE-EVENT ELEMENTS

EPOCH:	92182.661921495	MEAN ANOMALY:	6.2737
RIGHT ASCENSION:	305.7529	MEAN MOTION:	2.98492104
INCLINATION:	47.1115	MEAN MOTION DOT/2:	.00001757
ECCENTRICITY:	.6700939	MEAN MOTION DOT DOT/6:	.0
ARG. OF PERIGEE:	319.3202	BSTAR:	.0014976

DEBRIS CLOUD DATA

MAXIMUM ΔP:	Unknown
MAXIMUM ΔI:	Unknown

COMMENTS

Parent satellite was one of two small engine units that are routinely released after the first burn of the Proton fourth stage. The nature of these objects was identified by Dr. Boris V. Chernlatiev, Deputy Constructor for the Energiya NPO, in October 1992. The cause of this fragmentation is assumed to be related to the residual hypergolic propellants on board and failure of the membrane separating the fuel and oxidizer. NAVSPASUR observed 18 objects that were associated with this breakup. Twelve element sets were generated, but were of insufficient quality for a credible Gabbard Diagram or BLAST point. One object was cataloged on this event in early August 1992. This was the fifth in a series of fragmentations of this object type, and was the first located in a geosynchronous transfer orbit.

REFERENCE DOCUMENTS

The Fragmentation of Proton Debris, D. J. Nauer, TBE Technical Report CS93-LKD-004, Teledyne Brown Engineering, Colorado Springs, 31 December 1992.

Analysis of Fragmentations From December 1992 - February 1993, TBE Technical Report CS93-LKD-010, Teledyne Brown Engineering, Colorado Springs, 30 March 1993.

History of Soviet/Russian Satellite Fragmentations-A Joint U.S.-Russian Investigation, N. L. Johnson et al, Kaman Sciences Corporation, October 1995.

"Identification and Resolution of an Orbital Debris Problem with the Proton Launch Vehicle", B. V. Cherniatiev et al, Proceedings of the First European Conference on Space Debris, April 1993.

Insufficient data to construct a Gabbard diagram.

COSMOS 2079-2081 ULLAGE MOTOR   1990-045G   **20631**

SATELLITE DATA

TYPE:	Mission Related Debris
OWNER:	CIS
LAUNCH DATE:	19 May 1990
DRY MASS (KG):	55
MAIN BODY:	Ellipsoid; 0.6 m diameter by 1 m length
MAJOR APPENDAGES:	None
ATTITUDE CONTROL:	None
ENERGY SOURCES:	On-board propellants

EVENT DATA

DATE:	~28 Mar 1999	LOCATION:	Unknown
TIME:	Unknown	ASSESSED CAUSE:	Propulsion
ALTITUDE:	Unknown		

POST-EVENT ELEMENTS

EPOCH:	99087.88291821	MEAN ANOMALY:	123.5812
RIGHT ASCENSION:	319.9610	MEAN MOTION:	04.24414150137202
INCLINATION:	64.8090	MEAN MOTION DOT/2:	.00000311
ECCENTRICITY:	.5789417	MEAN MOTION DOT DOT/6:	00000-0
ARG. OF PERIGEE:	199.4305	BSTAR:	.0040281

DEBRIS CLOUD DATA

MAXIMUM ΔP:	Unknown
MAXIMUM ΔI:	Unknown

COMMENTS

This is the 20th event of the Proton Block DM SOZ Ullage Motor class identified to date; it is the ninth associated with a GLONASS mission. This mission was conducted before the engineering defect was identified and passivation measures implemented. In this orbit, debris may be long-lived but hard to track. 76 debris objects were detected.

REFERENCE DOCUMENT

"Abandoned Proton Ullage Motors Continue to Create Debris", The Orbital Debris Quarterly News, NASA JSC, April 1999. Available online at http://www.orbitaldebris.jsc.nasa.gov/newsletter/pdfs/ODQNv4i2.pdf.

Insufficient data to construct a Gabbard diagram.

SATELLITE DATA

TYPE:	CZ-4A Final Stage
OWNER:	PRC
LAUNCH DATE:	3.04 Sep 1990
DRY MASS (KG):	1000
MAIN BODY:	Cylinder-Nozzle; 2.9 m diameter by ~5 m length
MAJOR APPENDAGES:	none
ATTITUDE CONTROL:	none
ENERGY SOURCES:	On-board propellants

EVENT DATA

DATE:	4 Oct 1990	LOCATION:	81S, 68E (asc)
TIME:	2014 GMT	ASSESSED CAUSE:	Propulsion
ALTITUDE:	895 km		

PRE-EVENT ELEMENTS

EPOCH:	90276.6451544	MEAN ANOMALY:	162.6773
RIGHT ASCENSION:	310.6975	MEAN MOTION:	14.01192890
INCLINATION:	98.9340	MEAN MOTION DOT/2:	.000003118
ECCENTRICITY:	.0010179	MEAN MOTION DOT DOT/6:	.0
ARG. OF PERIGEE:	197.4122	BSTAR:	.0002183343

DEBRIS CLOUD DATA

MAXIMUM	ΔP:	5.8 min
MAXIMUM	ΔI:	0.1 deg

COMMENTS

This second flight of the CZ-4 final stage successfully deployed three payloads (one weather satellite and two inflated balloons) into a sun-synchronous orbit. Propellants used were N_2O_4 and UDMH. An estimated 70-75 fragments were detected soon after the event.

REFERENCE DOCUMENTS

The Fragmentation of Fengyun 1-2 R/B, N. L. Johnson, Technical Report CS90-TR-JSC-013, Teledyne Brown Engineering, Colorado Springs, Colorado, November 1990.

"Analyzing the Cause of LM-4 (A)'s Upper Stage's Disintegration and the Countermeasures", W. X. Zhang and S. Y. Liao, 5th International Conference of Pacific Basin Societies, 6-9 Jun 1993, Shanghai.

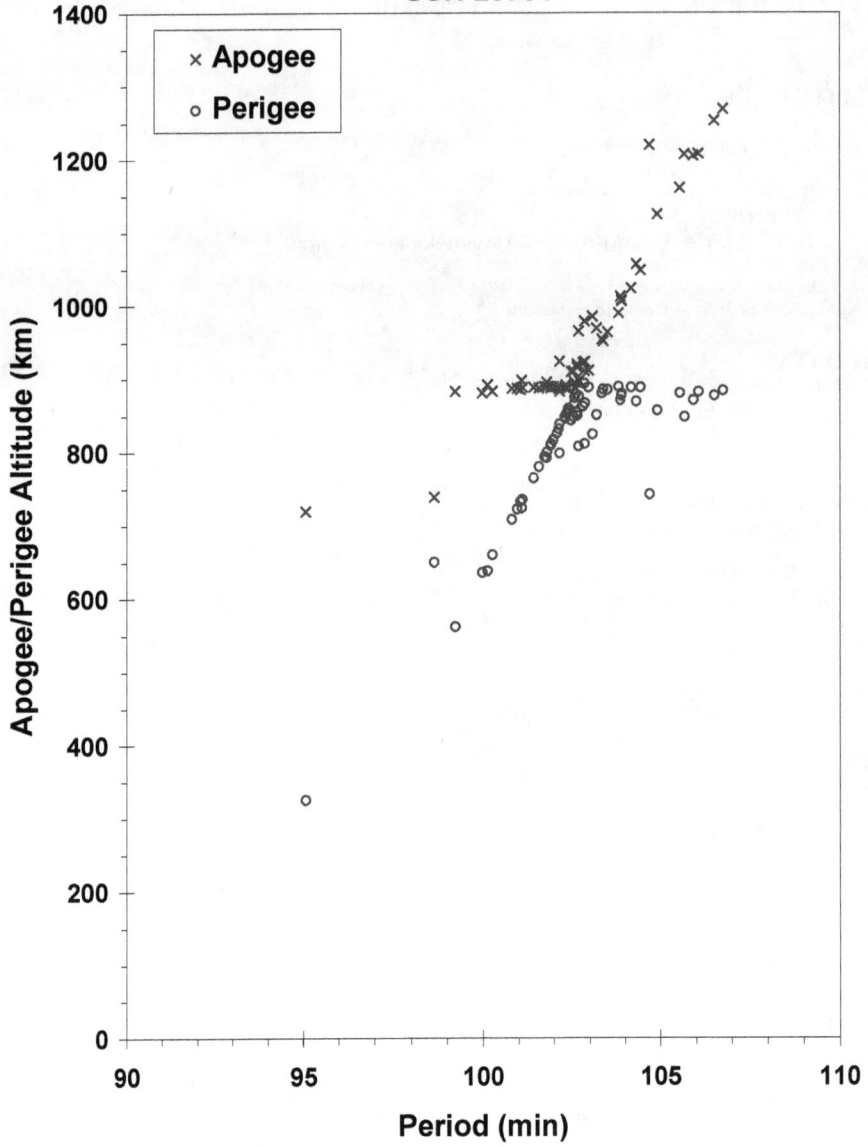

SSN 20791

Fengyun 1-2 R/B debris cloud remnant of 65 objects 5 days after the event as reconstructed from Naval Space Surveillance System database.

COSMOS 2101   1990-087A   20828

SATELLITE DATA

TYPE:	Payload
OWNER:	CIS
LAUNCH DATE:	1.46 Oct 1990
DRY MASS (KG):	6000
MAIN BODY:	Cylinder; 2.4 m diameter by 7 m length
MAJOR APPENDAGES:	Solar panels
ATTITUDE CONTROL:	Active, 3-axis
ENERGY SOURCES:	On-board propellants, explosive charge

EVENT DATA

DATE:	30 Nov 1990	LOCATION:	54N, 157E (dsc)
TIME:	1720 GMT	ASSESSED CAUSE:	Deliberate
ALTITUDE:	210 km		

PRE-EVENT ELEMENTS

EPOCH:	90334.45391019	MEAN ANOMALY:	205.3252
RIGHT ASCENSION:	347.9431	MEAN MOTION:	16.12811753
INCLINATION:	64.7547	MEAN MOTION DOT/2:	.00671617
ECCENTRICITY:	.0065418	MEAN MOTION DOT DOT/6:	.000035339
ARG. OF PERIGEE:	155.2258	BSTAR:	.00040815

DEBRIS CLOUD DATA

MAXIMUM ΔP: >7.3 min*
MAXIMUM ΔI: 0.3 deg*

*Based on uncataloged debris data

COMMENTS

Spacecraft was destroyed with a planned detonation. Second fragmentation of the Cosmos 2031 subclass. Early elements on only 7 objects available. Most debris reentered before being officially cataloged.

REFERENCE DOCUMENTS

The Fragmentation of Kosmos 2101, N. L. Johnson, Technical Report CS91-TR-JSC-002, Teledyne Brown Engineering, Colorado Springs, Colorado, January 1991.

History of Soviet/Russian Satellite Fragmentations-A Joint U.S.-Russian Investigation, N. L. Johnson et al, Kaman Sciences Corporation, October 1995.

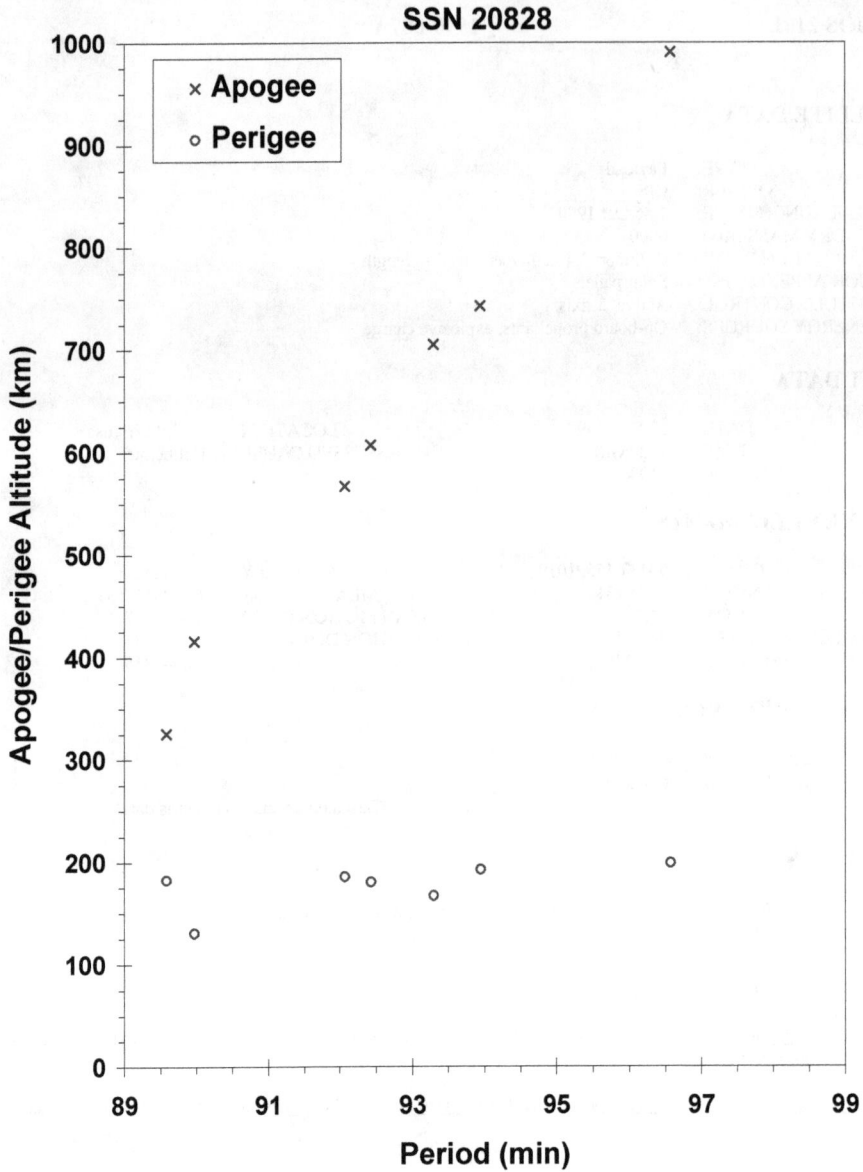

SSN 20828

Apogee/Perigee Altitude (km) vs **Period (min)**

Legend:
- × Apogee
- ○ Perigee

Cosmos 2101 debris cloud remnant of 7 objects 3 days after the event was reconstructed from Naval Space Surveillance System database. This diagram is taken from the cited reference.

GORIZONT 22 ULLAGE MOTOR   **1990-102E**   **20957**

SATELLITE DATA

TYPE:	Mission Related Debris
OWNER:	CIS
LAUNCH DATE:	23 Nov 1990
DRY MASS (KG):	55
MAIN BODY:	Ellipsoid; 0.6 m diameter by 1 m length
MAJOR APPENDAGES:	None
ATTITUDE CONTROL:	None
ENERGY SOURCES:	On-board propellants

EVENT DATA

DATE:	Approx. 14 Dec 1995	LOCATION:	Unknown
TIME:	Unknown	ASSESSED CAUSE:	Propulsion
ALTITUDE:	Unknown		

PRE-EVENT ELEMENTS

EPOCH:	95348.79476448	MEAN ANOMALY:	300.3633
RIGHT ASCENSION:	140.3319	MEAN MOTION:	5.84898259
INCLINATION:	46.4887	MEAN MOTION DOT/2:	.00111293
ECCENTRICITY:	.4967539	MEAN MOTION DOT DOT/6:	.00000006
ARG. OF PERIGEE:	117.7610	BSTAR:	.00074791

DEBRIS CLOUD DATA

MAXIMUM ΔP:	Unknown
MAXIMUM ΔI:	Unknown

COMMENTS

Parent satellite was one of two small engine units that are routinely released after the first burn of the Proton fourth stage. The nature of these objects was identified by Dr. Boris V. Chernlatiev, Deputy Constructor for the Energiya NPO, in October 1992. The cause of this fragmentation is assumed to be related to the residual hypergolic propellants on board and failure of the membrane separating the fuel and oxidizer. NAVSPASUR observed 69 objects that were associated with this breakup.

REFERENCE DOCUMENTS

The Fragmentation of Proton Debris, D. J. Nauer, TBE Technical Report CS93-LKD-004, Teledyne Brown Engineering, Colorado Springs, 31 December 1992.

Analysis of Fragmentations From December 1992 - February 1993, TBE Technical Report CS93-LKD-010, Teledyne Brown Engineering, Colorado Springs, 30 March 1993.

History of Soviet/Russian Satellite Fragmentations-A Joint U.S.-Russian Investigation, N. L. Johnson et al, Kaman Sciences Corporation, October 1995.

"Identification and Resolution of an Orbital Debris Problem with the Proton Launch Vehicle", B. V. Cherniatiev et al, Proceedings of the First European Conference on Space Debris, April 1993.

Insufficient data to construct a Gabbard diagram.

USA 68   1990-105A   20978

SATELLITE DATA

TYPE:	Payload
OWNER:	US
LAUNCH DATE:	1.66 Dec 1990
DRY MASS (KG):	855
MAIN BODY:	Cylinder; 1.1 m diameter by 3.7 m length
MAJOR APPENDAGES:	1 solar panel
ATTITUDE CONTROL:	Active, 3 axis
ENERGY SOURCES:	On-board propellants

EVENT DATA

DATE:	1 Dec 1990	LOCATION:	6N, 232E (dsc)
TIME:	1610 GMT	ASSESSED CAUSE:	Propulsion
ALTITUDE:	850 km		

POST-EVENT ELEMENTS

EPOCH:	90335.71008487	MEAN ANOMALY:	0.9090
RIGHT ASCENSION:	4.0350	MEAN MOTION:	14.29892145
INCLINATION:	98.8600	MEAN MOTION DOT/2:	-.00000049
ECCENTRICITY:	.0080986	MEAN MOTION DOT DOT/6:	.0
ARG. OF PERIGEE:	359.1948	BSTAR:	-0.000010171

DEBRIS CLOUD DATA

MAXIMUM ΔP:	>2.0 min*
MAXIMUM ΔI:	1.0 deg*

*Based on uncataloged debris data

COMMENTS

During the burn of USA 68's solid-fuel apogee kick motor (STAR-37S, TE-M-364-15), the 20 kg nozzle came apart, terminating thrust. At shutdown USA 68 was in an orbit of 610 km by 850 km. Immediately, a hydrazine orbit make-up system was activated, providing an additional 32.3 m/s DV. More than 40 pieces of non-Mission related debris were observed within a day of the event. The observed debris may include components of the USA 68 sun shield and AKM nozzle shield (total mass 2 kg). Most debris decayed very rapidly. The payload remained operational.

REFERENCE DOCUMENT

The Fragmentation of USA 68, N.L. Johnson, Technical Report CS91-TR-JSC-005, Teledyne Brown Engineering, Colorado Springs, Colorado, March 1991.

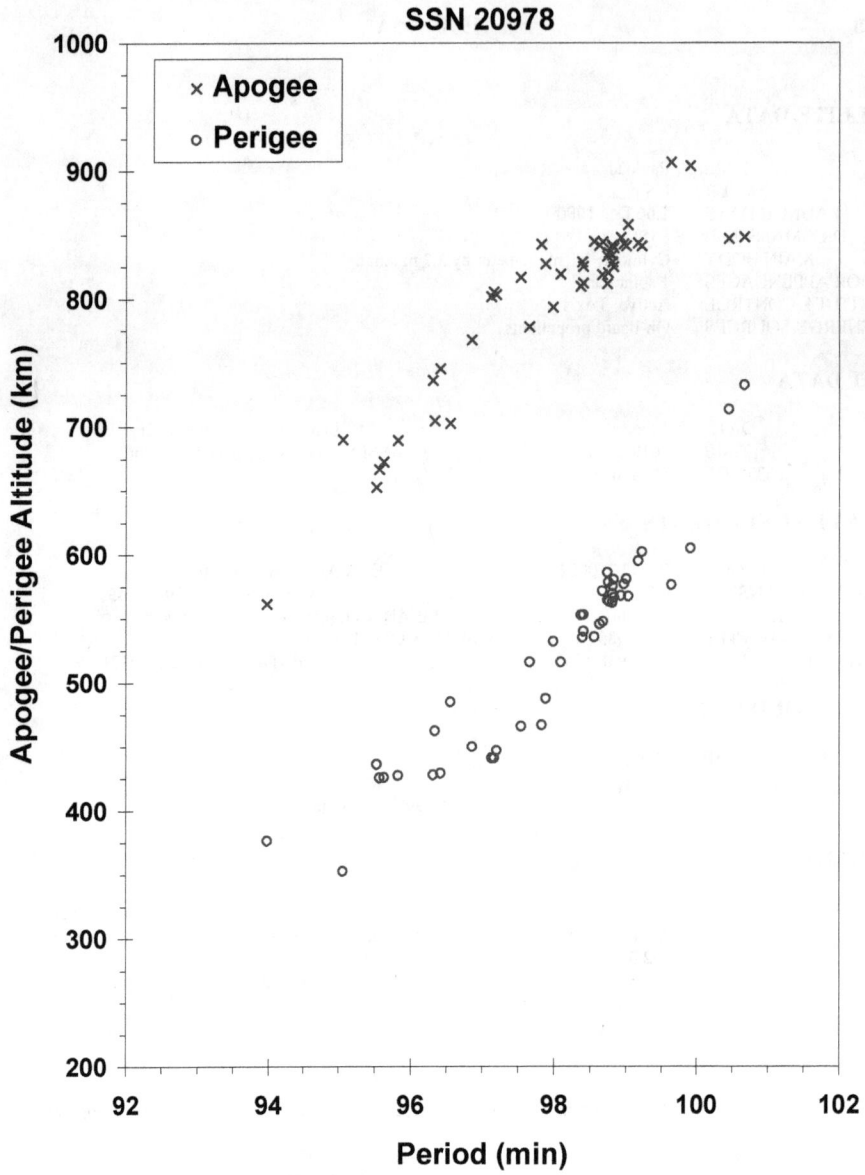

SSN 20978

USA 68 debris cloud remnant of 48 fragments 12 days after the event as reconstructed from the US SSN database.

SATELLITE DATA

TYPE:	Mission Related Debris
OWNER:	CIS
LAUNCH DATE:	8.11 Dec 1990
DRY MASS (KG):	~55
MAIN BODY:	Ellipsoid; 0.6 m diameter by 1.0 m length
MAJOR APPENDAGES:	None
ATTITUDE CONTROL:	None
ENERGY SOURCES:	On-board propellants

EVENT DATA

DATE:	21 Feb 2003	LOCATION:	34.11 S, 151.39 W
TIME:	~0300Z	ASSESSED CAUSE:	Propulsion
ALTITUDE:	~17650 km		

PRE-EVENT ELEMENTS

EPOCH:	03051.93857279	MEAN ANOMALY:	300.1330
RIGHT ASCENSION:	5.3297	MEAN MOTION:	4.24855437
INCLINATION:	65.3642	MEAN MOTION DOT/2:	-.00000082
ECCENTRICITY:	.5638383	MEAN MOTION DOT DOT/6:	.0
ARG. OF PERIGEE:	126.1785	BSTAR:	-.00014197

DEBRIS CLOUD DATA

MAXIMUM	ΔP:	Unknown
MAXIMUM	ΔI:	Unknown

COMMENTS

No debris was cataloged for this breakup. This is the 26th event of the Proton Block DM SOZ ullage motor since 1984. This ullage motor was launched prior to Russian recognition of the problem and before implementation of preventive measures.

REFERENCE DOCUMENT

"Satellite Fragmentations in 2003", The Orbital Debris Quarterly News, NASA JSC, January 2004. Available online at http://www.orbitaldebris.jsc nasa.gov/newsletter/pdfs/ODQNv8i1.pdf.

Insufficient data to construct a Gabbard diagram.

COSMOS 2109-2111 ULLAGE MOTOR   1990-110H   21013

SATELLITE DATA

TYPE:	Mission Related Debris
OWNER:	CIS
LAUNCH DATE:	8.11 Dec 1990
DRY MASS (KG):	55
MAIN BODY:	Ellipsoid; 0.6 m diameter by 1.0 m length
MAJOR APPENDAGES:	None
ATTITUDE CONTROL:	None
ENERGY SOURCES:	On-board propellants

EVENT DATA

DATE:	14 Mar 1998	LOCATION:	Unknown
TIME:	Unknown	ASSESSED CAUSE:	Propulsion
ALTITUDE:	Unknown		

PRE-EVENT ELEMENTS

EPOCH:	98072.07217599	MEAN ANOMALY:	85.4178
RIGHT ASCENSION:	306.4512	MEAN MOTION:	4.23530449
INCLINATION:	65.0803	MEAN MOTION DOT/2:	0.00000895
ECCENTRICITY:	0.5724061	MEAN MOTION DOT DOT/6:	0
ARG. OF PERIGEE:	216.7168	BSTAR:	0.0025728

DEBRIS CLOUD DATA

MAXIMUM ΔP:	Unknown
MAXIMUM ΔI:	Unknown

COMMENTS

This is the 16th event of this class identified to date; it is the fifth associated with a GLONASS mission. More than 110 debris detected, but element sets developed for only a few.

REFERENCE DOCUMENTS

"Identification and Resolution of an Orbital Debris Problem with the Proton Launch Vehicle", Cherniatiev, Chernyavskiy, Johnson, and McKnight, First European Conference on Space Debris, 5-7 April 1993.

"The Fragmentation of Proton Debris", Nauer, Teledyne Brown Engineering Technical Report CS93-LKD-004, 31 December 1992.

History of Soviet/Russian Satellite Fragmentations-A Joint U.S.-Russian Investigation, N. L. Johnson et al, Kaman Sciences Corporation, October 1995.

"Three Upper Stage Breakups in One Week Top February Debris Activity", The Orbital Debris Quarterly News, NASA JSC, April 1998, p. 1-2. Available online at http://www.orbitaldebris.jsc.nasa.gov/newsletter/pdfs/ODQNv3i2.pdf.

Insufficient data to construct a Gabbard diagram.

SATELLITE DATA

TYPE:	Ariane 4 H-10 Third Stage
OWNER:	France
LAUNCH DATE:	15.97 Jan 1991
DRY MASS (KG):	1760
MAIN BODY:	Cylinder; 2.6 m diameter by 11.4 m length
MAJOR APPENDAGES:	None
ATTITUDE CONTROL:	None at time of the event.
ENERGY SOURCES:	On-board propellants

EVENT DATA

DATE:	Late Apr-early May 1996	LOCATION:	Unknown
TIME:	Unknown	ASSESSED CAUSE:	Propulsion
ALTITUDE:	Unknown		

PRE-EVENT ELEMENTS

EPOCH:	96106.15481796	MEAN ANOMALY:	312.6005
RIGHT ASCENSION:	104.8696	MEAN MOTION:	2.66496263
INCLINATION:	6.7146	MEAN MOTION DOT/2:	0.00007071
ECCENTRICITY:	0.6989841	MEAN MOTION DOT DOT/6:	0
ARG. OF PERIGEE:	132.7372	BSTAR:	0.0012265

DEBRIS CLOUD DATA

MAXIMUM ΔP:	147.3 min
MAXIMUM ΔI:	1.3 deg

COMMENTS

The event was first recognized by Naval Space Command analysts in early May 1996. Element sets for as many as 20 debris were developed. Since deliberate passivation of Ariane GTO stages was not implemented until 1993, the vehicle was not purged of its residual propellants or pressurants.

REFERENCE DOCUMENT

"Newly Recognized 1996 Breakup", N. L. Johnson, The Orbital Debris Quarterly News, April 1997, p. 2. Available online at http://www.orbitaldebris.jsc.nasa.gov/newsletter/pdfs/ODQNv2i2.pdf.

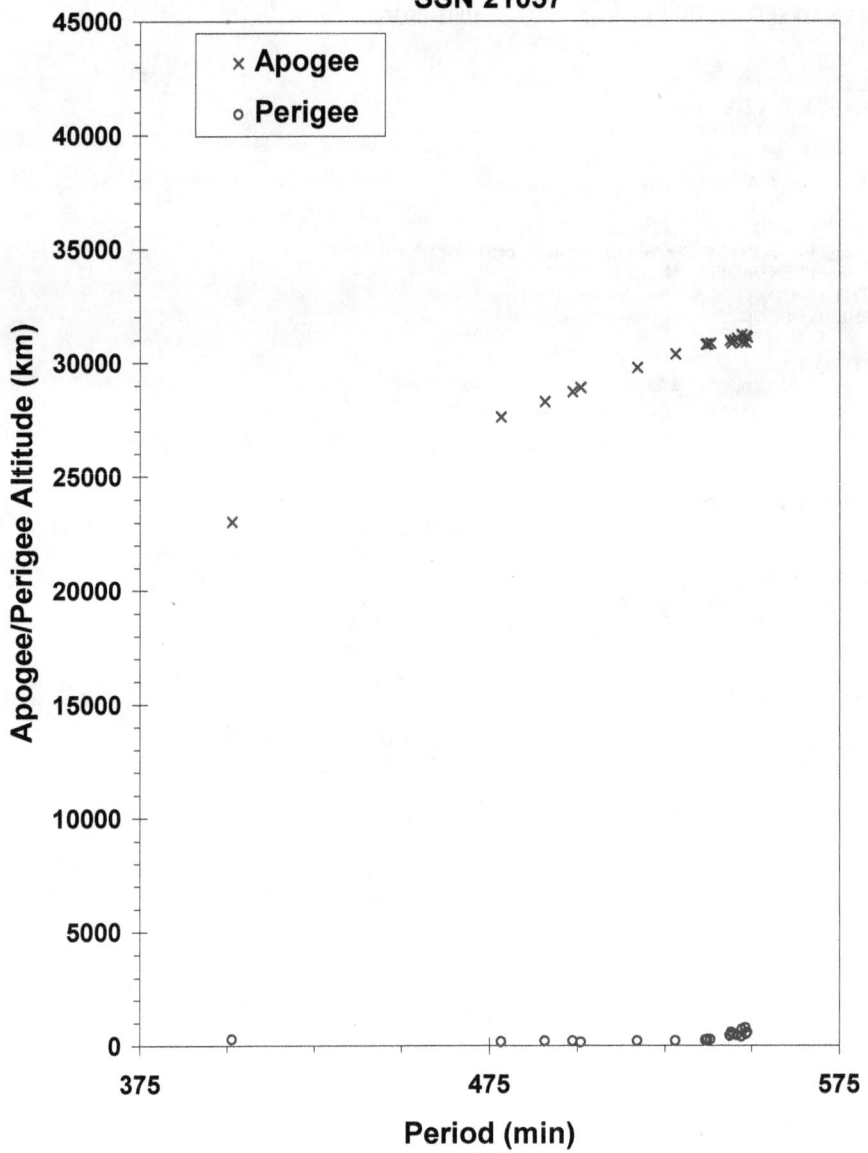

Italsat 1/Eutelsat 2 F2 R/B debris cloud of 20 fragments 1 year after
the event as reconstructed from the US SSN database.

COSMOS 2125-2132 R/B   **1991-009J**   **21108**

SATELLITE DATA

TYPE:	Cosmos Second Stage
OWNER:	CIS
LAUNCH DATE:	12.12 Feb 1991
DRY MASS (KG):	1435
MAIN BODY:	Cylinder; 2.4 m diameter by 6.6 m length
MAJOR APPENDAGES:	Payload deployment mechanism
ATTITUDE CONTROL:	None at time of the event.
ENERGY SOURCES:	Unknown

EVENT DATA

DATE:	5 Mar 1991	LOCATION:	43S, 140E (asc)
TIME:	1345 GMT	ASSESSED CAUSE:	Propulsion
ALTITUDE:	1560 km		

PRE-EVENT ELEMENTS

EPOCH:	91062.94236834	MEAN ANOMALY:	112.8991
RIGHT ASCENSION:	166.0317	MEAN MOTION:	12.19552620
INCLINATION:	74.0386	MEAN MOTION DOT/2:	.00000005
ECCENTRICITY:	.0166507	MEAN MOTION DOT DOT/6:	.0
ARG. OF PERIGEE:	245.0348	BSTAR:	.000099999

DEBRIS CLOUD DATA

MAXIMUM ΔP:	4.3 min*
MAXIMUM ΔI:	0.3 deg*

*Based on uncataloged debris data

COMMENTS

This is the second known fragmentation of the Cosmos second stage and the first in more than 25 years and 370 missions. Like the earlier event (Cosmos 61-63 R/B), this rocket body successfully completed its multiple payload delivery before breakup. NAVSPASUR determined that several minor separations occurred both prior to and after the main breakup cited above (see NAVSPASUR report referenced below).

REFERENCE DOCUMENTS

Cosmos 2125-2132 Rocket Body (U), Fragmentation and Breakup Report (U), E.L. Jenkins and R.E. Farmer, Naval Space Surveillance Center, Dahlgren, Virginia, April, 1991.

A Preliminary Analysis of the Fragmentations of the Kosmos 2125-2132 Rocket Body, N.L. Johnson, Technical Report CS91-TR-JSC-007, Teledyne Brown Engineering, Colorado Springs, Colorado, April 1991.

History of Soviet/Russian Satellite Fragmentations-A Joint U.S.-Russian Investigation, N. L. Johnson et al, Kaman Sciences Corporation, October 1995.

"The Recent Fragmentations of LEO Upper Stages", G. Chernyavskiy et al, 45th IAF Congress, 1994.

SSN 21108

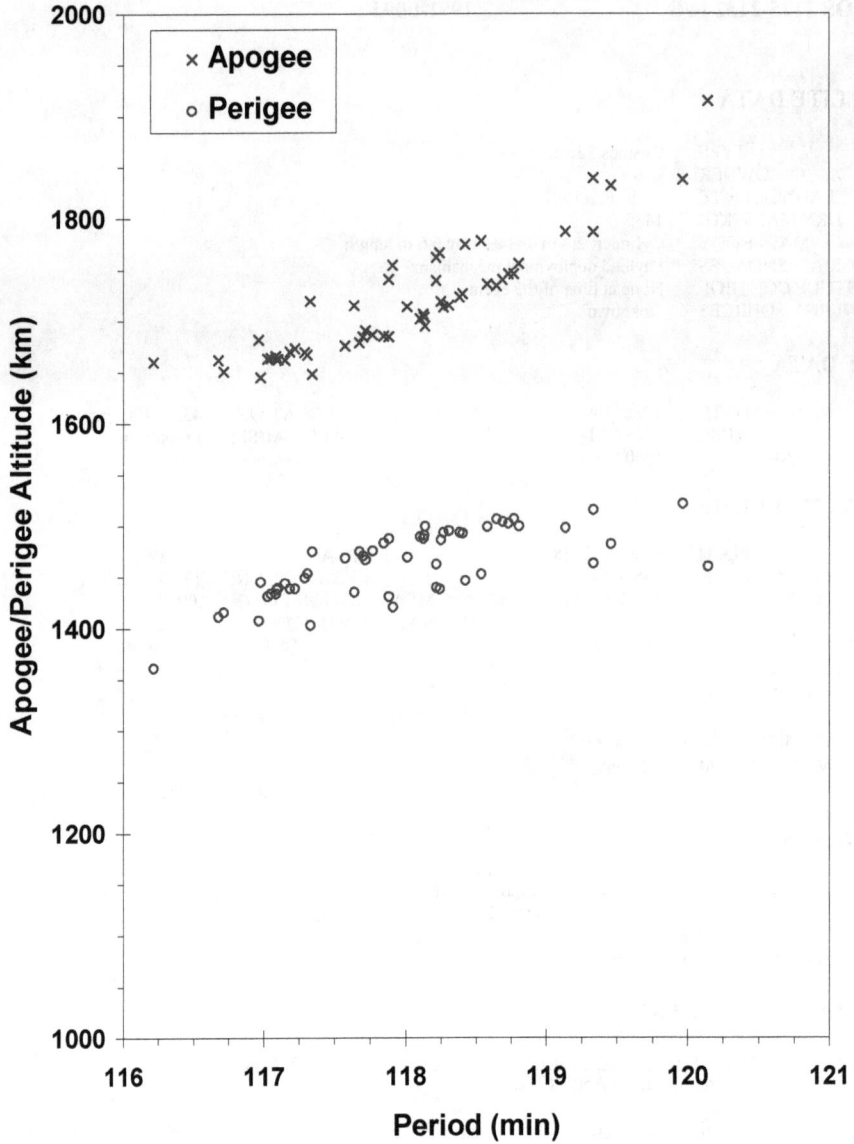

Cosmos 2125-32 R/B debris cloud of 54 objects 5 days after the major breakup event as reconstructed from Naval Space Surveillance System database. This diagram is taken from the reference cited at the top of this page.

COSMOS 2133 ULLAGE MOTOR   1991-010D   21114

SATELLITE DATA

TYPE:	Mission Related Debris
OWNER:	CIS
LAUNCH DATE:	12 Feb 1991
DRY MASS (KG):	55
MAIN BODY:	Ellipsoid; 0.6 m diameter by 1 m length
MAJOR APPENDAGES:	None
ATTITUDE CONTROL:	Unknown
ENERGY SOURCES:	On-board propellants

EVENT DATA

DATE:	7 May 1994	LOCATION:	10N, 112E
TIME:	0930 GMT	ASSESSED CAUSE:	Propulsion
ALTITUDE:	16195 km		

PRE-EVENT ELEMENTS

EPOCH:	94126.31580012	MEAN ANOMALY:	240.6661
RIGHT ASCENSION:	110.6447	MEAN MOTION:	3.78477656
INCLINATION:	46.6223	MEAN MOTION DOT/2:	.000127656
ECCENTRICITY:	0.6204369	MEAN MOTION DOT DOT/6:	00000-0
ARG. OF PERIGEE:	160.8637	BSTAR:	0.00086951

DEBRIS CLOUD DATA

MAXIMUM ΔP:	Unknown
MAXIMUM ΔI:	Unknown

COMMENTS

Parent satellite was one of two small engine units that are routinely released after the first burn of the Proton fourth stage. The nature of these objects was identified by Dr. Boris V. Chernlatiev, Deputy Constructor for the Energiya NPO, in October 1992. The cause of this fragmentation appears to be related to the residual hypergolic propellants on board and failure of the membrane separating the fuel and oxidizer. NAVSPASUR observed 38 objects that were associated with this breakup. Only 6 element sets were generated. This was the ninth in a series of fragmentations of this object type, and was the fourth located in a geosynchronous transfer orbit. Two possible fragmentation locations were calculated by the NAVSPOC. The numbers above represent the first possible calculated location.

REFERENCE DOCUMENTS

The Fragmentation of Proton Debris, D. J. Nauer, TBE Technical Report CS93-LKD-004, Teledyne Brown Engineering, Colorado Springs, 31 December 1992.

Analysis of Fragmentations From December 1992 - February 1993, TBE Technical Report CS93-LKD-010, Teledyne Brown Engineering, Colorado Springs, 30 March 1993.

The Fragmentation of Cosmos 2133 Debris, I. W. Grissom and D. J. Nauer, TBE Technical Report CS94-LKD-016, Teledyne Brown Engineering, Colorado Springs, 30 June 1994.

History of Soviet/Russian Satellite Fragmentations-A Joint U.S.-Russian Investigation, N. L. Johnson et al, Kaman Sciences Corporation, October 1995.

"Identification and Resolution of an Orbital Debris Problem with the Proton Launch Vehicle", B. V. Cherniatiev et al, Proceedings of the First European Conference on Space Debris, April 1993.

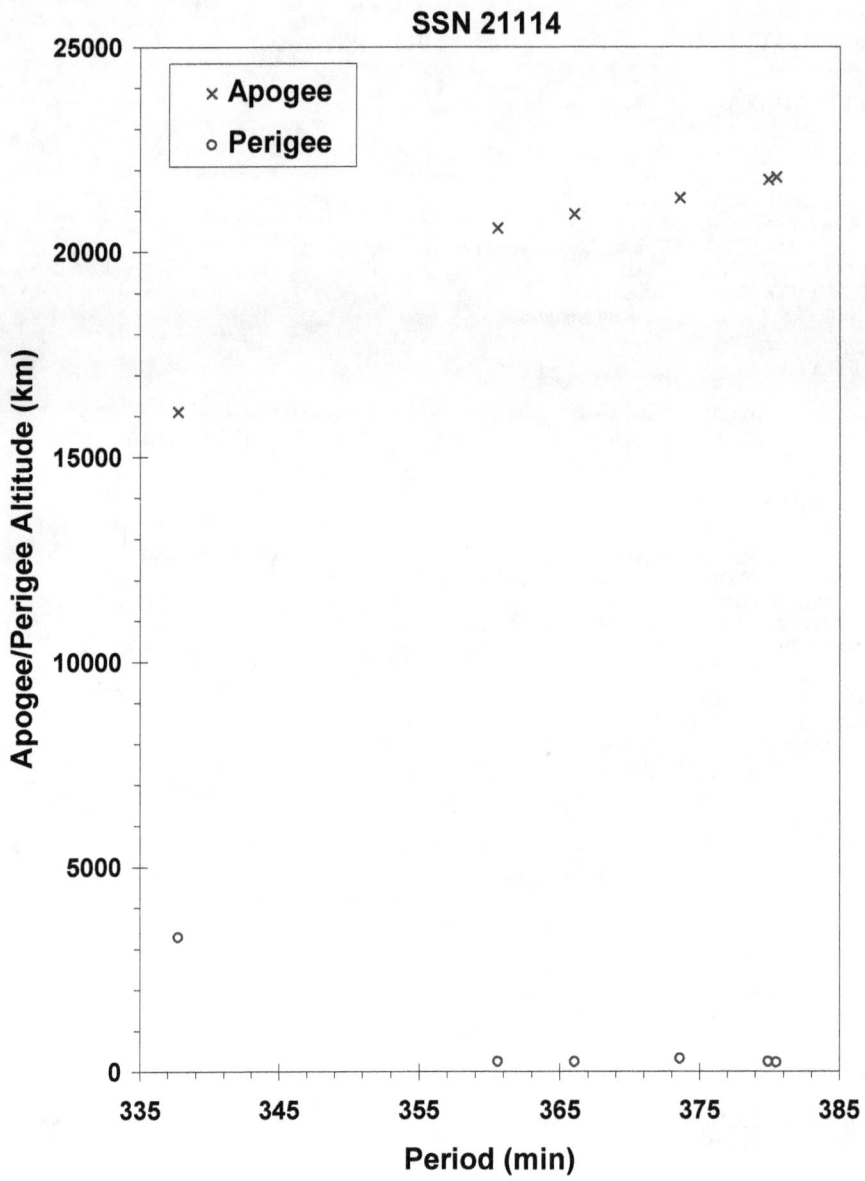

Gabbard diagram of six objects from the Cosmos 2133 debris fragmentation.

314

SATELLITE DATA

TYPE:	Ariane 4 H10 Third Stage
OWNER:	France
LAUNCH DATE:	2.98 Mar 1991
DRY MASS (KG):	1760
MAIN BODY:	Cylinder; 2.6 m diameter by 10 m length
MAJOR APPENDAGES:	None
ATTITUDE CONTROL:	None at time of the event.
ENERGY SOURCES:	On-board propellants

EVENT DATA

DATE:	27 Apr 1994	LOCATION:	0.5S, 79E (dsc)
TIME:	0144 GMT	ASSESSED CAUSE:	Propulsion
ALTITUDE:	270 km		

PRE-EVENT ELEMENTS

EPOCH:	94116.17965845	MEAN ANOMALY:	157.2349
RIGHT ASCENSION:	136.1778	MEAN MOTION:	2.86975555
INCLINATION:	6.5808	MEAN MOTION DOT/2:	.00006058
ECCENTRICITY:	.6829164	MEAN MOTION DOT DOT/6:	.0
ARG. OF PERIGEE:	185.9406	BSTAR:	.001267

DEBRIS CLOUD DATA

MAXIMUM ΔP:	148 min
MAXIMUM ΔI:	1.1 deg

COMMENTS

The fragmentation of this R/B occurred over 37 months after launch. Since deliberate passivation of Ariane GTO stages was not implemented until 1993, the vehicle was not purged of its residual propellants or pressurants. As many as 30 debris were detected.

REFERENCE DOCUMENTS

TRW Space Log 1957-1991. TRW Space and Defense Sector, Space and Technology Group. Redondo Beach, CA. 1992.

Space News, "Ariane Rocket Flies After Electrical Glitch Delay, Volume 2, Number 8, 11-17 March 1991.

Rockets of the World. Peter Alway, Ann Arbor, MI, 1993.

The Fragmentation of the Astra 1B/MOP 2 (1) Rocket Body, I. W. Grissom and D. J. Nauer, TBE Technical Report CS94-LKD-014, Teledyne Brown Engineering, Colorado Springs, 15 May 1994.

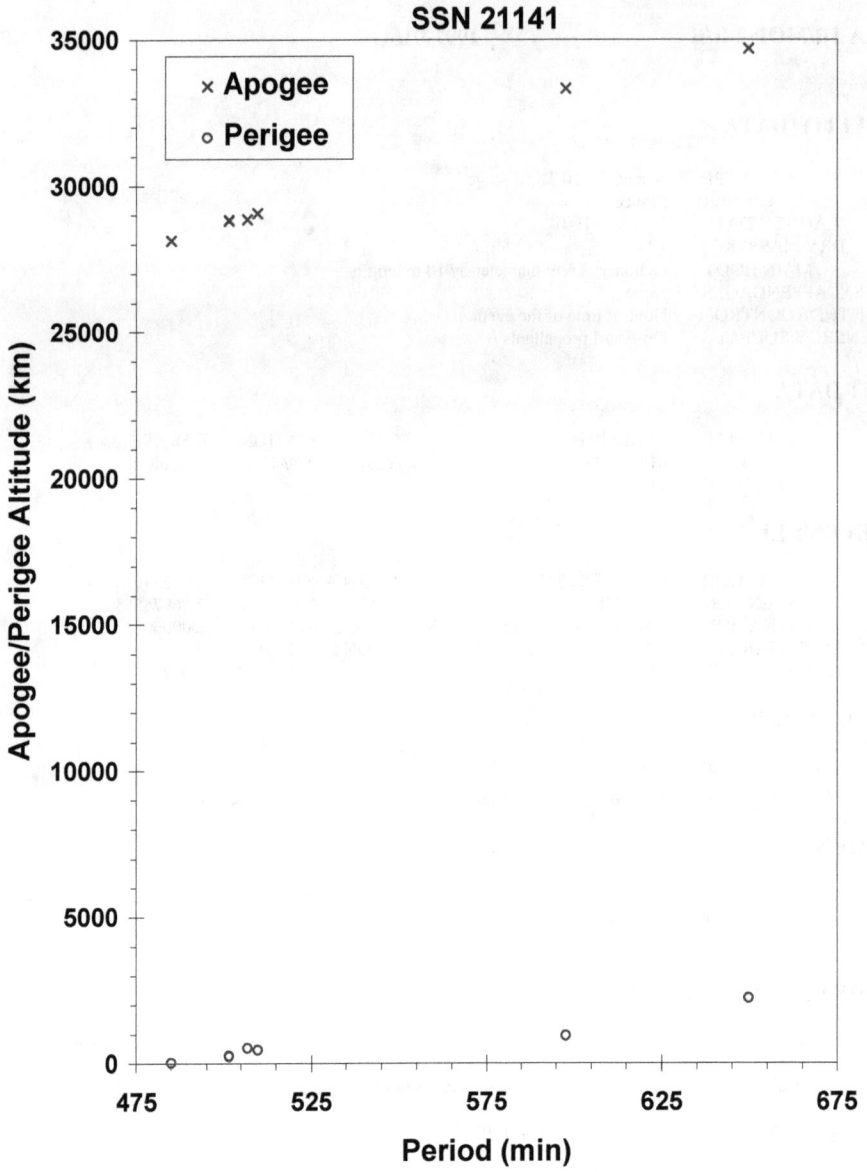

SSN 21141

Astra 1B/MOP 2 R/B debris cloud of 7 fragments as reconstructed from the US SSN database.

COSMOS 2139-2141 ULLAGE MOTOR   1991-025G   21226

SATELLITE DATA

TYPE:	Mission Related Debris
OWNER:	CIS
LAUNCH DATE:	4.45 Apr 1991
DRY MASS (KG):	~55
MAIN BODY:	Ellipsoid; 0.6 m diameter by 1.0 m length
MAJOR APPENDAGES:	None
ATTITUDE CONTROL:	None
ENERGY SOURCES:	On-board propellants

EVENT DATA

DATE:	16 Jun 2001	LOCATION:	Unknown
TIME:	~0700Z	ASSESSED CAUSE:	Propulsion
ALTITUDE:	Unknown		

PRE-EVENT ELEMENTS

EPOCH:	01165.32627059	MEAN ANOMALY:	158.6980		
RIGHT ASCENSION:	118.5521	MEAN MOTION:	4.28587592		
INCLINATION:	64.5545	MEAN MOTION DOT/2:	.00004370	ECCENTRICITY:	
	.5826262	MEAN MOTION DOT DOT/6:	.0		
ARG. OF PERIGEE:	187.0212	BSTAR:	.0011075		

DEBRIS CLOUD DATA

MAXIMUM	ΔP:	236.7 min*
MAXIMUM	ΔI:	2.13 deg*

* Based on uncataloged debris data

COMMENTS

The debris from this breakup were difficult for the US Space Surveillance Network to track. Although over 100 debris were initially tracked, none were cataloged. This is the 24th event of the Proton Block DM SOZ ullage motor since 1984. This ullage motor was launched prior to Russian recognition of the problem and before implementation of preventive measures.

REFERENCE DOCUMENT

"Two More Satellite Breakups Detected", The Orbital Debris Quarterly News, NASA JSC, July 2001. Available online at http://www.orbitaldebris.jsc nasa.gov/newsletter/pdfs/ODQNv6i3.pdf.

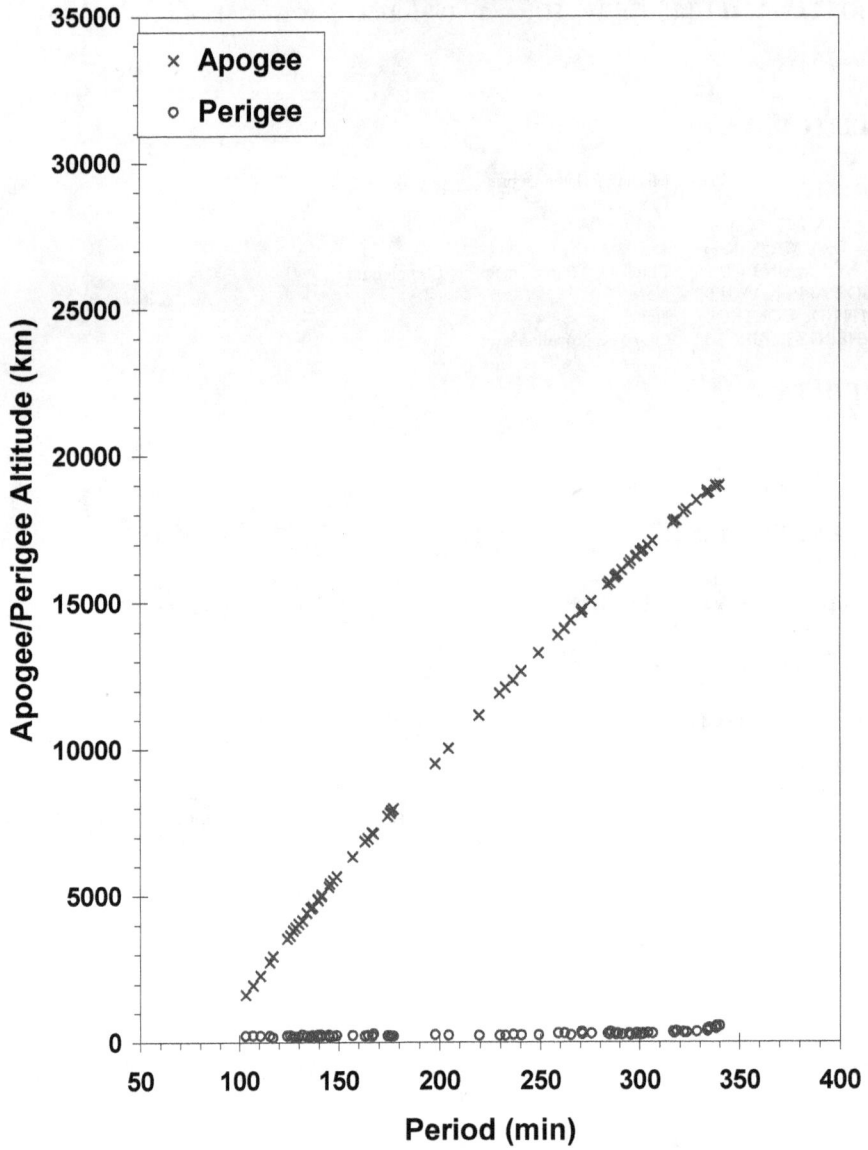

Cosmos 2139-41 auxiliary motor debris cloud of 77 fragments within 2 weeks after the event as reconstructed from the US SSN database.

COSMOS 2157-2162 R/B   1991-068G   21734

SATELLITE DATA

TYPE:	Tsyklon Third Stage
OWNER:	CIS
LAUNCH DATE:	28.30 Sep 1991
DRY MASS (KG):	1360
MAIN BODY:	Cone-cylinder; 2.1 m diameter by 2.4 m length
MAJOR APPENDAGES:	None
ATTITUDE CONTROL:	None
ENERGY SOURCES:	Unknown

EVENT DATA

DATE:	09 Oct 1999	LOCATION:	41.5N, 217.9E
TIME:	1508 GMT	ASSESSED CAUSE:	Propulsion
ALTITUDE:	1460 km		

PRE-EVENT ELEMENTS

EPOCH:	99281.98318497	MEAN ANOMALY:	220.2415
RIGHT ASCENSION:	96.5043	MEAN MOTION:	12.54216420
INCLINATION:	82.5731	MEAN MOTION DOT/2:	.00000027
ECCENTRICITY:	.0046780	MEAN MOTION DOT DOT/6:	.0
ARG. OF PERIGEE:	140.1600	BSTAR:	.00010000

DEBRIS CLOUD DATA

MAXIMUM ΔP:	2.716 min
MAXIMUM ΔI:	0.79 deg

COMMENTS

This is the 4th event of this class identified to date and the second of 1999. All stages have been about 8-10 years old at the time of breakup. The vehicle is a Ukrainian-produced stage, using UDMH and N_2O_4 as propellants. To date these stages have not been passivated at end of mission and may contain up to 300 kg of residual propellants. The issue of Tsyklon orbital stage breakups was discussed with representatives of the National Space Agency of Ukraine during 11-13 October 1999 in Darmstadt, Germany. More than 100 of these stages are currently in Earth orbit. Although the exact cause of these breakups remains unknown, all four events have occurred during periods of high solar flux, i.e., near solar maximum.

REFERENCE DOCUMENT

"Third Tsyklon Upper Stage Breaks Up", The Orbital Debris Quarterly News, NASA JSC, July 1999. Available online at http://www.orbitaldebris.jsc.nasa.gov/newsletter/pdfs/ODQNv4i3.pdf.

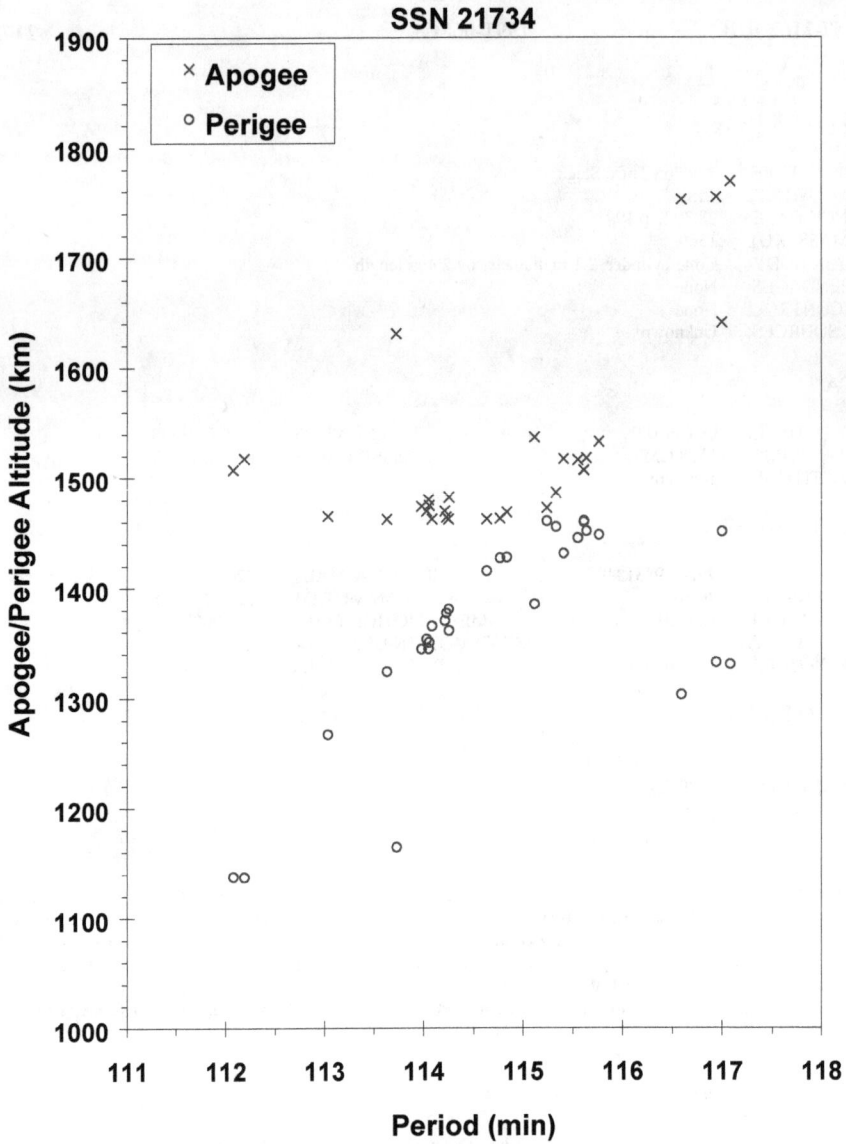

SSN 21734

Apogee/Perigee Altitude (km) vs **Period (min)**

Legend:
× Apogee
○ Perigee

Cosmos 2157-2162 rocket body debris cloud of 31 fragments within 1 day of the event as reconstructed from the US SSN database.

SATELLITE DATA

TYPE:	Payload
OWNER:	CIS
LAUNCH DATE:	9.55 October 1991
DRY MASS (KG):	6000
MAIN BODY:	Cylinder; 2.4 m diameter by 7 m length
MAJOR APPENDAGES:	Solar panels
ATTITUDE CONTROL:	Active, 3-axis
ENERGY SOURCES:	On-board propellants, explosive charge

EVENT DATA

DATE:	6 December 1991	LOCATION:	55N, 154E (dsc)
TIME:	2021 GMT	ASSESSED CAUSE:	Deliberate
ALTITUDE:	210 km		

PRE-EVENT ELEMENTS

EPOCH:	91340.51933896	MEAN ANOMALY:	213.3470
RIGHT ASCENSION:	37.7884	MEAN MOTION:	16.18797546
INCLINATION:	64.7678	MEAN MOTION DOT/2:	.00862876
ECCENTRICITY:	.0054670	MEAN MOTION DOT DOT/6:	.000035685
ARG. OF PERIGEE:	147.5032	BSTAR:	.00035926

DEBRIS CLOUD DATA

MAXIMUM ΔP: >9.8 min*

MAXIMUM ΔI: 0.2 deg*

*Based on uncataloged debris data

COMMENTS

Spacecraft was destroyed with a planned detonation. Third fragmentation of the Cosmos 2031 subclass. Early elements on only 8 objects available. All debris reentered before being officially cataloged.

REFERENCE DOCUMENTS

The Fragmentation of Kosmos 2163, Technical Report CS92-TR-JSC-002, Teledyne Brown Engineering, Colorado Springs, Colorado, January 1992.

History of Soviet/Russian Satellite Fragmentations-A Joint U.S.-Russian Investigation, N. L. Johnson et al, Kaman Sciences Corporation, October 1995.

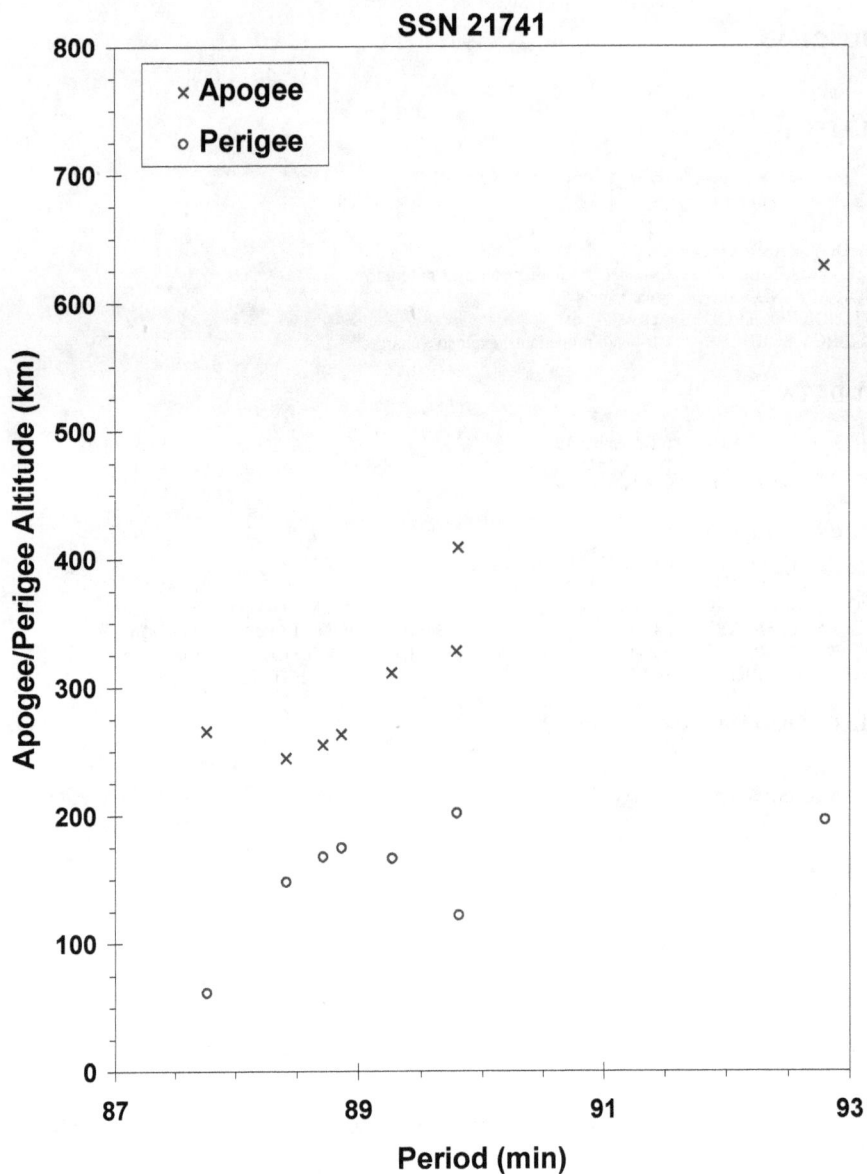

Cosmos 2163 debris cloud remnant of 8 objects 1 day after the event as reconstructed from the US SSN database. This diagram is taken from the cited reference.

INTELSAT 601 R/B   **1991-075B**   **21766**

SATELLITE DATA

TYPE:	Ariane 4 Third Stage
OWNER:	France
LAUNCH DATE:	29.96 Oct 1991
DRY MASS (KG):	~1760
MAIN BODY:	Cylinder: 2.6 m diameter by 11.4 m length
MAJOR APPENDAGES:	None
ATTITUDE CONTROL:	None
ENERGY SOURCES:	On-board propellants

EVENT DATA

DATE:	24 Dec 2001	LOCATION:	4.0 N, 344.4 E
TIME:	~2212Z	ASSESSED CAUSE:	Propulsion
ALTITUDE:	22,100 km		

PRE-EVENT ELEMENTS

EPOCH:	01358.15120659	MEAN ANOMALY:	1.1953
RIGHT ASCENSION:	264.6850	MEAN MOTION:	2.90501578
INCLINATION:	7.1968	MEAN MOTION DOT/2:	.00042976
ECCENTRICITY:	.6814056	MEAN MOTION DOT DOT/6:	.0
ARG. OF PERIGEE:	351.9651	BSTAR:	.0055981

DEBRIS CLOUD DATA

MAXIMUM ΔP:	4.45 min*
MAXIMUM ΔI:	0.08 deg*

                                        * Based on uncataloged debris data

COMMENTS

This 10-year-old Ariane 4 third stage appears to have suffered a minor fragmentation. The low inclination and high eccentricity of the orbit made debris detection and tracking difficult. Three pieces were initially detected by the SSN and ten objects cataloged a month after the event. The object was seen intact at about 2030 UTC, 24 December. Approximately 25 hours later a debris cloud of eight objects was seen by the same sensor. The perigee of the vehicle prior to breakup was sufficiently high that aerodynamic forces should not have been a factor in the event.

REFERENCE DOCUMENT

"Two Major Satellite Breakups Near End of 2001," The Orbital Debris Quarterly News, NASA JSC, January 2002. Available online at http://www.orbitaldebris.jsc.nasa.gov/newsletter/pdfs/ODQNv7i1.pdf.

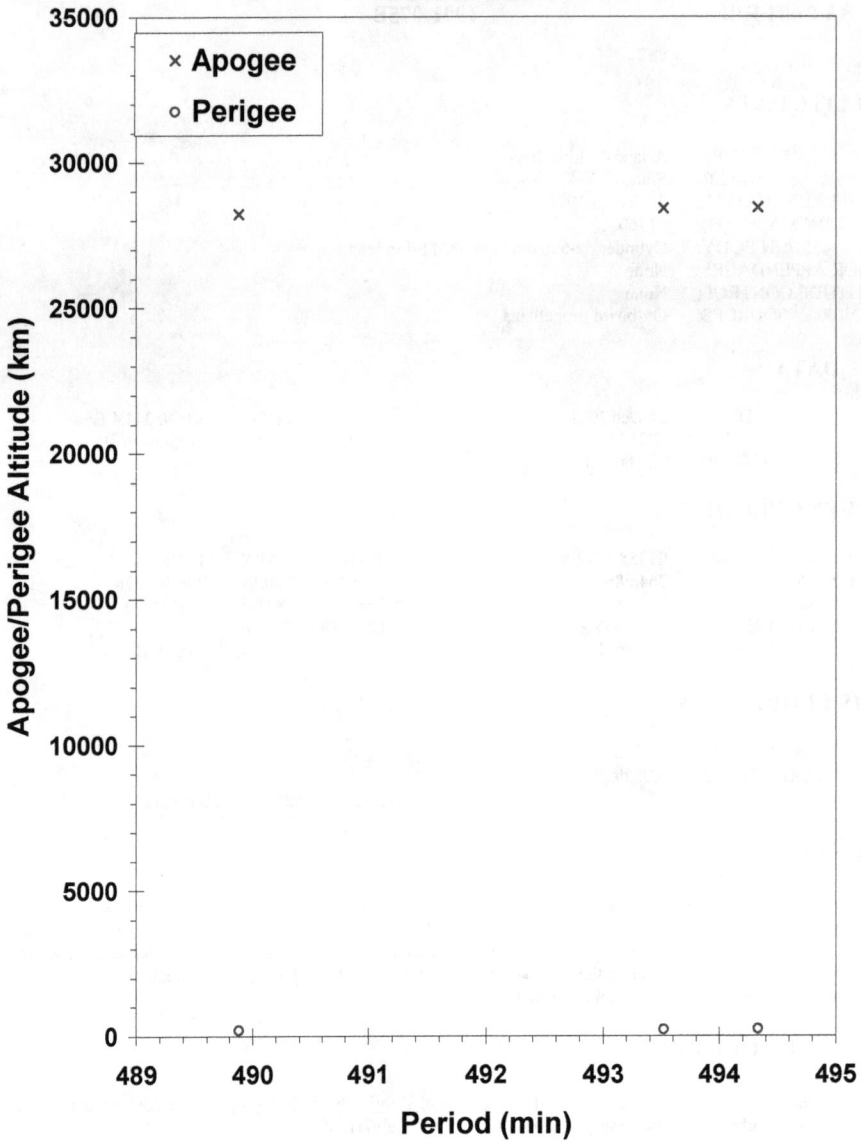

SSN 21766

Intelsat 601 R/B debris cloud of 3 fragments 3 days after the breakup as reconstructed from the US SSN database.

SATELLITE DATA

TYPE:	Payload
OWNER:	US
LAUNCH DATE:	28.56 Nov 1991
DRY MASS (KG):	~850
MAIN BODY:	Cylinder; 1.2 m diameter by 6.4 m length
MAJOR APPENDAGES:	1 solar panel
ATTITUDE CONTROL:	None at time of event
ENERGY SOURCES:	On-board propellant

EVENT DATA

DATE:	15 Apr 2004	LOCATION:	31N, 56E (asc)
TIME:	1454 GMT	ASSESSED CAUSE:	Propulsion
ALTITUDE:	835 km		

POST-EVENT ELEMENTS

EPOCH:	04106.47330773	MEAN ANOMALY:	346.2855
RIGHT ASCENSION:	129.0059	MEAN MOTION:	14.15516644
INCLINATION:	98.6744	MEAN MOTION DOT/2:	0.00000244
ECCENTRICITY:	.0012890	MEAN MOTION DOT DOT/6:	0.0
ARG. OF PERIGEE:	13.8671	BSTAR:	0.0014668

DEBRIS CLOUD DATA

MAXIMUM ΔP:	2.8 min
MAXIMUM ΔI:	0.4 deg

COMMENTS

The spacecraft was non-operational at the time of the event. The electrical power generation system had been passivated by discharging the batteries and disconnecting them from the charging circuit. Virtually no nitrogen remained on board due to a leak detected early in the mission. The only energy source assessed to be on the spacecraft at the time of the event was approximately 6 kg of hydrazine.

REFERENCE DOCUMENT

"Recent Satellite Breakups", The Orbital Debris Quarterly News, NASA JSC, October 2004.
Available online at http://www.orbitaldebris.jsc nasa.gov/newsletter/pdfs/ODQNv8i4.pdf.

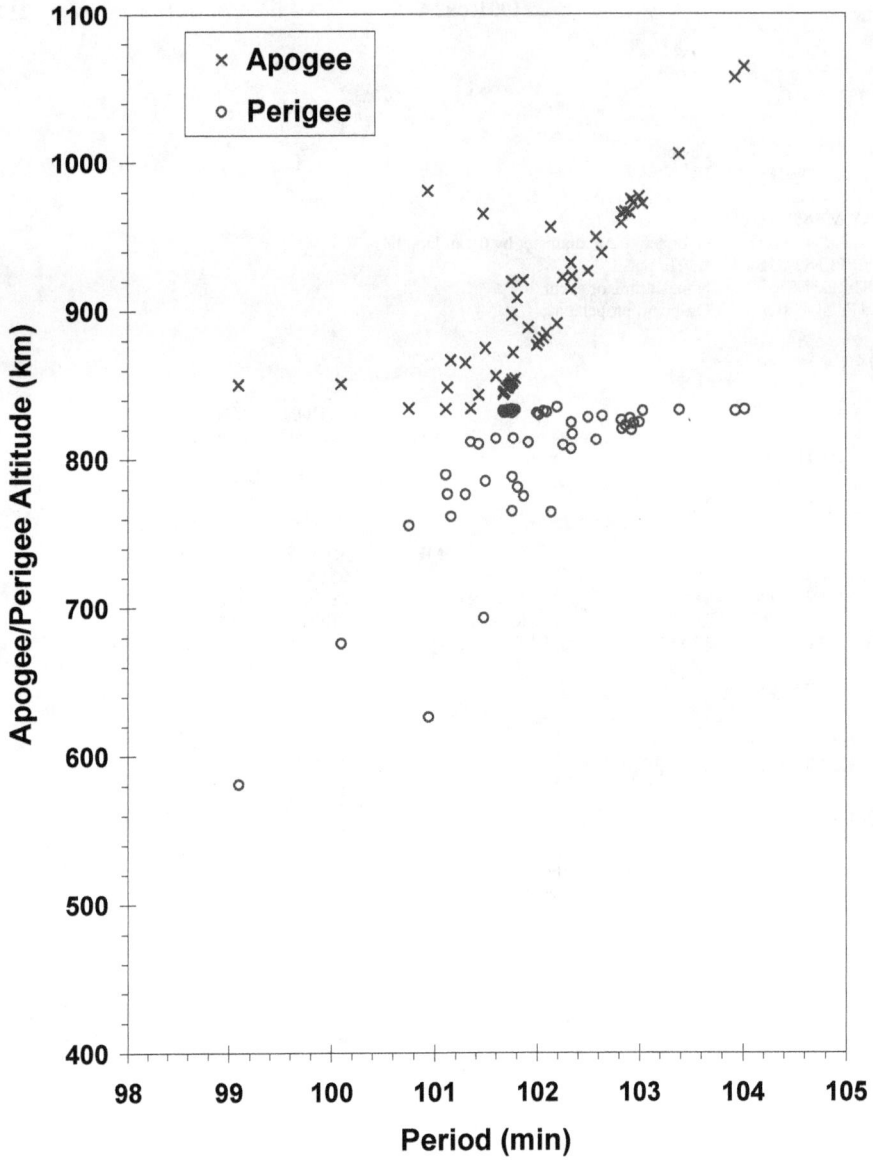

SSN 21798

USA 73 debris cloud of 56 cataloged fragments 6 weeks after the event as reconstructed from the US SSN database.

TELECOM 2B/INMARSAT 2 R/B   **1992-021C**   **21941**

SATELLITE DATA

TYPE:	Ariane 4 H10$^+$ Third Stage
OWNER:	France
LAUNCH DATE:	15.98 Apr 1992
DRY MASS (KG):	1800
MAIN BODY:	Cylinder; 2.6 m diameter by 10 m length
MAJOR APPENDAGES:	None
ATTITUDE CONTROL:	None at time of the event.
ENERGY SOURCES:	On-board propellants

EVENT DATA

DATE:	21 Apr 1993 (EST)	LOCATION:	Unknown
TIME:	Unknown	ASSESSED CAUSE:	Propulsion
ALTITUDE:	Unknown		

PRE-EVENT ELEMENTS

EPOCH:	93110.33659871	MEAN ANOMALY:	335.5551		
RIGHT ASCENSION:	224.3451	MEAN MOTION:	2.28914093		
INCLINATION:	4.03	MEAN MOTION DOT/2:	.000024	ECCENTRICITY:	
.7248434	MEAN MOTION DOT DOT/6:	.0			
ARG. OF PERIGEE:	110.6851	BSTAR:	.0020699		

DEBRIS CLOUD DATA

MAXIMUM ΔP:	Unknown
MAXIMUM ΔI:	Unknown

COMMENTS

Although analysis indicates that 92-021C fragmented around 21 April 1993, the event was not recognized until eight pieces were cataloged during the period from 30 August 1994 to 2 September 1994. Naval Space Command reported that the eight objects were discovered during the period from July 1993 to July 1994.

Insufficient data to construct a Gabbard diagram.

INSAT 2A/EUTELSAT 2F4 R/B   1992-041C   22032

SATELLITE DATA

TYPE:	Ariane 4 H10 Final Stage
OWNER:	France
LAUNCH DATE:	9.95 Jul 1992
DRY MASS (KG):	~1760
MAIN BODY:	Cylinder: 2.6 m diameter by 11.4 m length
MAJOR APPENDAGES:	None
ATTITUDE CONTROL:	None
ENERGY SOURCES:	On-board propellants

EVENT DATA

DATE:	Feb 2002	LOCATION:	Unknown
TIME:	Unknown	ASSESSED CAUSE:	Propulsion
ALTITUDE:	Unknown		

PRE-EVENT ELEMENTS

EPOCH:	02032.98792301	MEAN ANOMALY:	23.4497
RIGHT ASCENSION:	196.5922	MEAN MOTION:	3.10885568
INCLINATION:	7.0154	MEAN MOTION DOT/2:	.00036687
ECCENTRICITY:	.6663885	MEAN MOTION DOT DOT/6:	.0
ARG. OF PERIGEE:	261.6162	BSTAR:	.0

DEBRIS CLOUD DATA

MAXIMUM ΔP:	41.05 min*
MAXIMUM ΔI:	0.25 deg*

                                        * Based on uncataloged debris data

COMMENTS

This marks the sixth known fragmentation of an Ariane 4 third stage. The last three vehicles involved in such events (1988-109C, 1991-075C, and 1992-041C) had been in orbit 9-10 years at the time of their respective breakups. All flights were conducted prior to the implementation of passivation measures for Ariane GTO missions in September 1993. No Ariane launch vehicle launched since that time is known to have experienced an on-orbit fragmentation.

REFERENCE DOCUMENT

"Second Identified Satellite Breakup of 2002", The Orbital Debris Quarterly News, NASA JSC, July 2002. Available online at http://www.orbitaldebris.jsc.nasa.gov/newsletter/pdfs/ODQNv7i3.pdf.

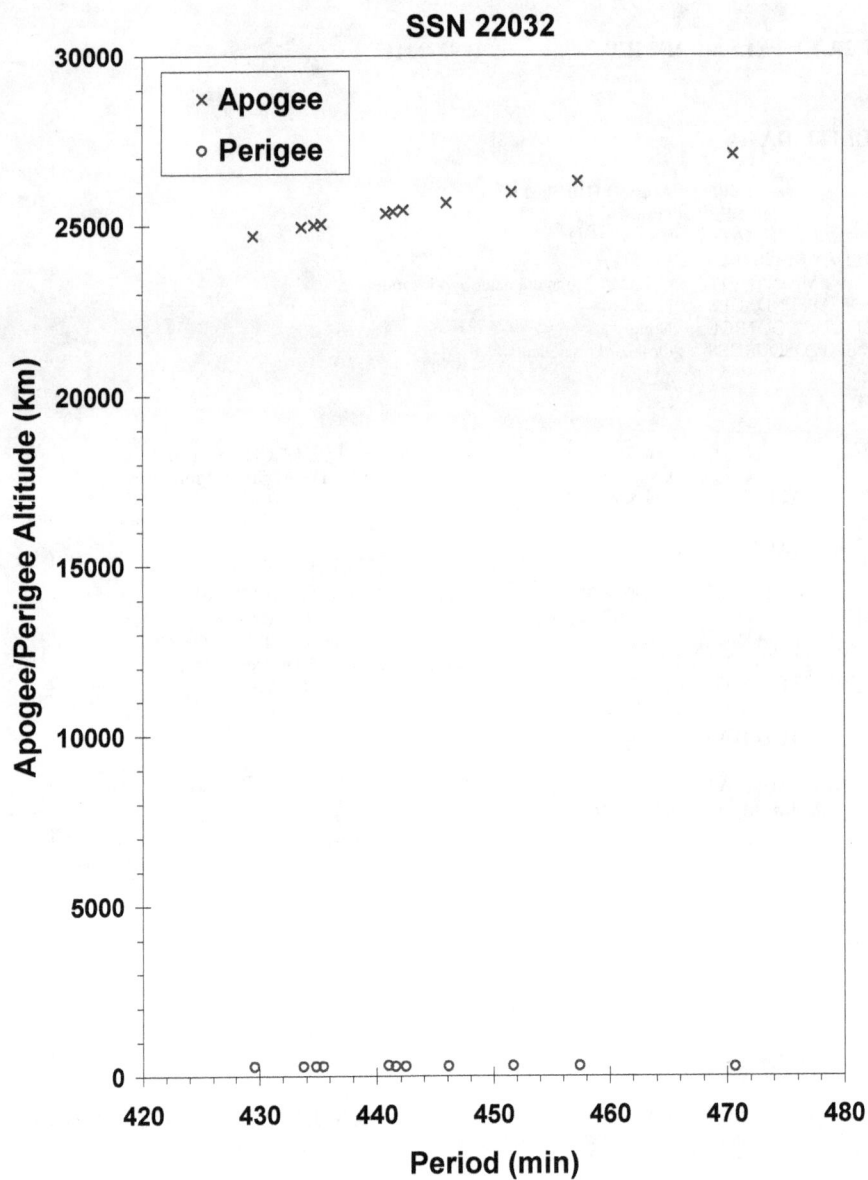

Insat 2A/Eutelsat 2F4 R/B debris cloud of 11 fragments 4 months after the breakup as reconstructed from the US SSN database.

COSMOS 2204-2206 ULLAGE MOTOR   1992-047G

SATELLITE DATA

TYPE:	Mission Related Debris
OWNER:	CIS
LAUNCH DATE:	30.08 Jul 1992
DRY MASS (KG):	55
MAIN BODY:	Ellipsoid; 0.6 m diameter by 1 m length
MAJOR APPENDAGES:	None
ATTITUDE CONTROL:	None
ENERGY SOURCES:	On-board propellant

EVENT DATA

DATE:	10 Jul 2004	LOCATION:	34S, 120W (asc)
TIME:	0240 GMT (est.)	ASSESSED CAUSE:	Propulsion
ALTITUDE:	18,525 km		

POST-EVENT ELEMENTS

EPOCH:	04190.17957430	MEAN ANOMALY:	263.0579
RIGHT ASCENSION:	223.5224	MEAN MOTION:	4.29145980
INCLINATION:	64.8832	MEAN MOTION DOT/2:	0.00005202
ECCENTRICITY:	.5757033	MEAN MOTION DOT DOT/6:	0.0
ARG. OF PERIGEE:	149.3553	BSTAR:	0.0055262

DEBRIS CLOUD DATA

MAXIMUM ΔP:	37.6 min
MAXIMUM ΔI:	0.5 deg

COMMENTS

The object was in a decaying, eccentric orbit; this event marks the 30th known breakup of a Proton Block DM SOZ ullage motor since 1984. More than 100 debris were detected from Sat. No. 22066 by the US SSN. The sister SOZ motor for this launch broke up in an unrelated event ten years earlier (22067).

REFERENCE DOCUMENT

History of Soviet/Russian Satellite Fragmentations-A Joint U.S.-Russian Investigation, N. L. Johnson et al, Kaman Sciences Corporation, October 1995.

"Recent Satellite Breakups", The Orbital Debris Quarterly News, NASA JSC, October 2004.
Available online at http://www.orbitaldebris.jsc.nasa.gov/newsletter/pdfs/ODQNv8i4.pdf.

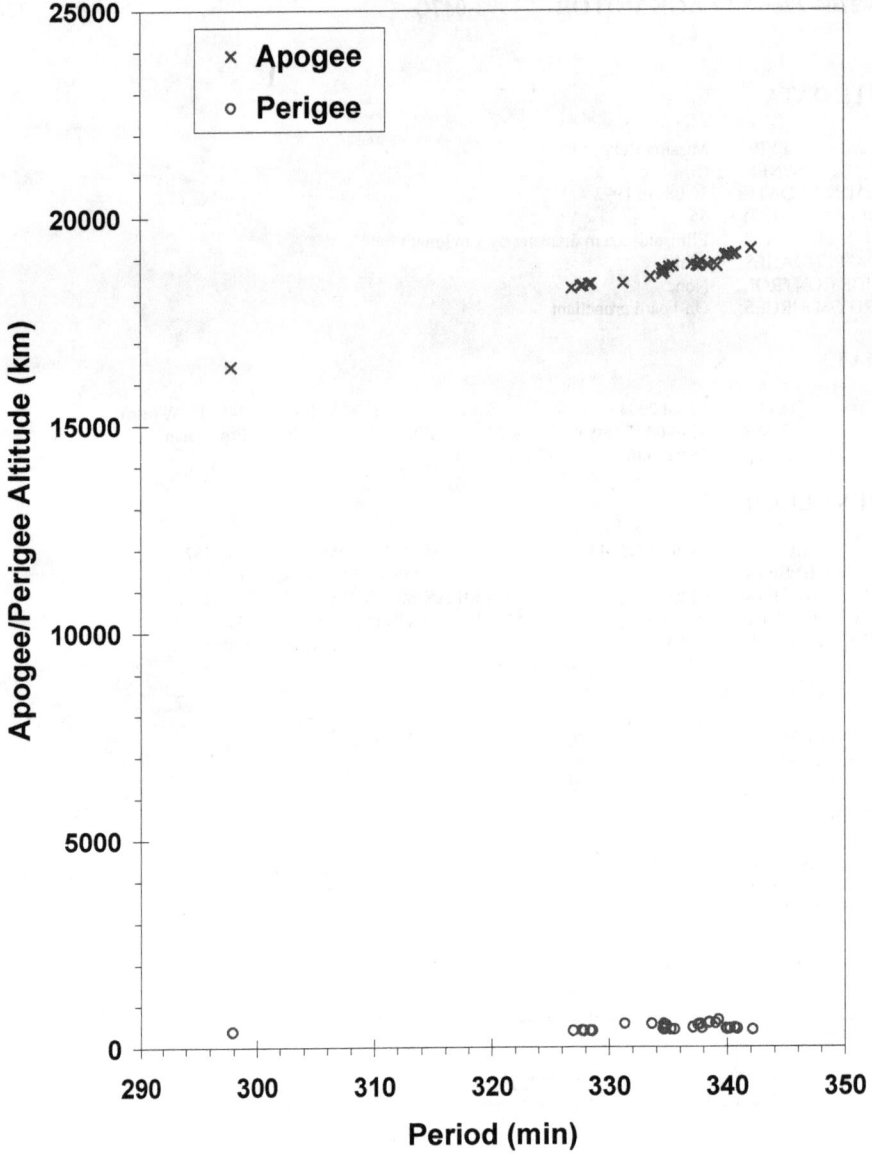

SOZ motor debris cloud of 31 fragments less than 1 week after the event as reconstructed
from the US SSN database.

COSMOS 2204-2206 ULLAGE MOTOR   1992-047H

SATELLITE DATA

TYPE:	Mission Related Debris
OWNER:	CIS
LAUNCH DATE:	30.08 Jul 1992
DRY MASS (KG):	55
MAIN BODY:	Ellipsoid; 0.6 m diameter by 1.0 m length
MAJOR APPENDAGES:	None
ATTITUDE CONTROL:	None
ENERGY SOURCES:	On-board propellants

EVENT DATA

DATE:	Prior to 0824, 8 Nov 1994	LOCATION:	Unknown
TIME:	Unknown	ASSESSED CAUSE:	Propulsion
ALTITUDE:	Unknown		

PRE-EVENT ELEMENTS

EPOCH:	94310.28602258	MEAN ANOMALY:	9.8460
RIGHT ASCENSION:	65.2049	MEAN MOTION:	4.23571466
INCLINATION:	64.8556	MEAN MOTION DOT/2:	.00001002
ECCENTRICITY:	0.5708388	MEAN MOTION DOT DOT/6:	00000-0
ARG. OF PERIGEE:	316.7786	BSTAR:	0.0033777

DEBRIS CLOUD DATA

MAXIMUM ΔP:	4.2 min
MAXIMUM ΔI:	0.9 deg

COMMENTS

Parent satellite was one of two small engine units that are routinely released after the first burn of the Proton fourth stage. The nature of these objects was identified by Dr. Boris V. Cherniatiev, Deputy Constructor for the Energiya NPO and Mr. Nicholas Johnson of Kaman Sciences, in October, 1992. The cause of this fragmentation appears to be related to the residual hypergolic propellants on board and failure of the membrane separating the fuel and oxidizer. NAVSPASUR observed 32 objects that were associated with this breakup on 8 Nov 94, 36 objects on 9 Nov and 31 objects on 10 Nov. This was the tenth in a series of fragmentations of this object type.

REFERENCE DOCUMENTS

The Fragmentation of Proton Debris, D. J. Nauer, TBE Technical Report CS93-LKD-004, Teledyne Brown Engineering, Colorado Springs, 31 December 1992.

Identification and Resolution of an Orbital Debris Problem with the Proton Launch Vehicle, B. V. Cherniatiev, et al, First European Conference on Space Debris, 5-7 April 1993.

History of Soviet/Russian Satellite Fragmentations-A Joint U.S.-Russian Investigation, N. L. Johnson et al, Kaman Sciences Corporation, October 1995.

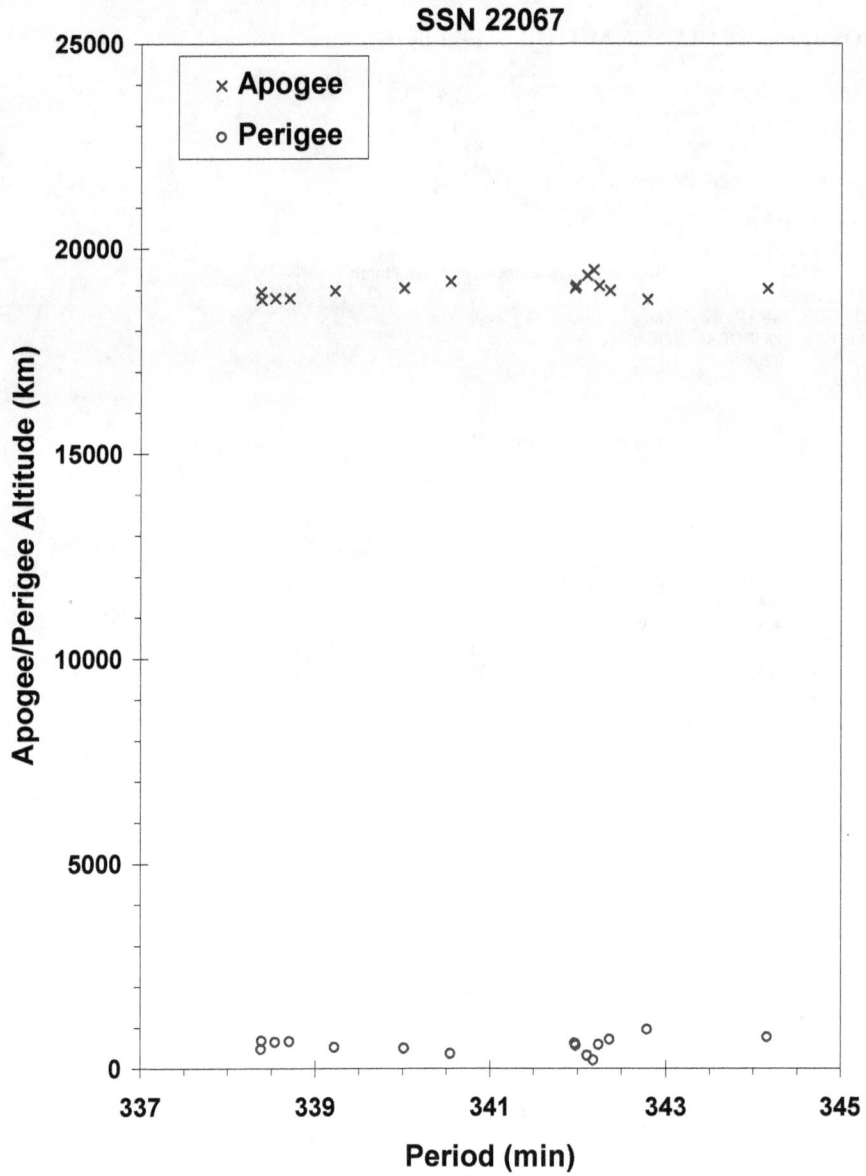

Gabbard Diagram from the Cosmos 2204-2206 debris fragmentation.

GORIZONT 27 ULLAGE MOTOR   1992-082F   22250

SATELLITE DATA

TYPE:	Mission Related Debris
OWNER:	CIS
LAUNCH DATE:	27.55 Nov 1992
DRY MASS (KG):	~55
MAIN BODY:	Ellipsoid; 0.6 m diameter by 1.0 m length
MAJOR APPENDAGES:	None
ATTITUDE CONTROL:	None
ENERGY SOURCES:	On-board propellants

EVENT DATA

DATE:	14 Jul 2001	LOCATION:	Unknown
TIME:	~1800Z	ASSESSED CAUSE:	Propulsion
ALTITUDE:	Unknown		

PRE-EVENT ELEMENTS

EPOCH:	01194.12977223	MEAN ANOMALY:	232.1640	
RIGHT ASCENSION:	101.3588	MEAN MOTION:	9.96766196	
INCLINATION:	46.4697	MEAN MOTION DOT/2:	.01023324	ECCENTRICITY:
.2850325	MEAN MOTION DOT DOT/6:	.0000014714		
ARG. OF PERIGEE:	148.6125	BSTAR:	.00050789	

DEBRIS CLOUD DATA

MAXIMUM ΔP:	Unknown
MAXIMUM ΔI:	Unknown

COMMENTS

This event marks the 25th known breakup of a Proton Block DM SOZ ullage motor since 1984. Due to the moderate eccentricity and altitude of the orbits, the debris were difficult for the US Space Surveillance Network to track. This ullage motor was launched before implementation of breakup preventive measures. No debris were cataloged from this breakup.

REFERENCE DOCUMENT

"New Satellite Breakups Detected", The Orbital Debris Quarterly News, NASA JSC, October 2001. Available online at http://www.orbitaldebris.jsc nasa.gov/newsletter/pdfs/ODQNv6i4.pdf.

Insufficient data to construct a Gabbard diagram.

SATELLITE DATA

TYPE:	Mission Related Debris
OWNER:	CIS
LAUNCH DATE:	17.53 Dec 1992
DRY MASS (KG):	55
MAIN BODY:	Ellipsoid; 0.6 m diameter by 1 m length
MAJOR APPENDAGES:	None
ATTITUDE CONTROL:	None
ENERGY SOURCES:	On-board propellant

EVENT DATA

DATE:	~22 Apr 2005	LOCATION:	Unknown
TIME:	Unknown	ASSESSED CAUSE:	Propulsion
ALTITUDE:	Unknown		

PRE-EVENT ELEMENTS

EPOCH:	05112.46798568	MEAN ANOMALY:	2.1228
RIGHT ASCENSION:	268.8209	MEAN MOTION:	3.90056983
INCLINATION:	46.7270	MEAN MOTION DOT/2:	0.00018984
ECCENTRICITY:	.6142562	MEAN MOTION DOT DOT/6:	0.00000003
ARG. OF PERIGEE:	348.3165	BSTAR:	0.0006277

DEBRIS CLOUD DATA

MAXIMUM ΔP:	Unknown
MAXIMUM ΔI:	Unknown

COMMENTS

This event marks the 32nd known breakup of a Proton Block DM SOZ ullage motor since 1984. This object had a perigee near 200 km. The event happened sometime between 1100 GMT on April 22, and 1200 GMT on April 23. Only about a dozen pieces were detected by the US SSN.

REFERENCE DOCUMENT

History of Soviet/Russian Satellite Fragmentations-A Joint U.S.-Russian Investigation, N. L. Johnson et al, Kaman Sciences Corporation, October 1995.

"Recent Satellite Breakups", The Orbital Debris Quarterly News, NASA JSC, July 2005.
Available online at http://www.orbitaldebris.jsc nasa.gov/newsletter/pdfs/ODQNv9i3.pdf.

Insufficient data to construct a Gabbard diagram.

COSMOS 2225   1992-091A   22280

SATELLITE DATA

TYPE:	Payload
OWNER:	CIS
LAUNCH DATE:	22.50 Dec 1992
DRY MASS (KG):	6000
MAIN BODY:	Cylinder; 2.4 m diameter by 7 m length
MAJOR APPENDAGES:	Solar panels
ATTITUDE CONTROL:	Active, 3-axis
ENERGY SOURCES:	On-board propellants, explosive charge

EVENT DATA

DATE:	18 Feb 1993	LOCATION:	55N, 157E (dsc)
TIME:	1856 GMT	ASSESSED CAUSE:	Deliberate
ALTITUDE:	220 km		

PRE-EVENT ELEMENTS

EPOCH:	93040.89217375	MEAN ANOMALY:	244.5776		
RIGHT ASCENSION:	125.1196	MEAN MOTION:	16.07940666		
INCLINATION:	64.8919	MEAN MOTION DOT/2:	.00301303	ECCENTRICITY:	
	.0039285	MEAN MOTION DOT DOT/6:	.000049705		
ARG. OF PERIGEE:	115.8892	BSTAR:	.00032572		

DEBRIS CLOUD DATA

MAXIMUM ΔP:	> 2.3 min
MAXIMUM ΔI:	> 0.5 deg

COMMENTS

Spacecraft was destroyed with a planned detonation. Fourth fragmentation of the Cosmos 2031 subclass. Early elements on only 10 objects (including the parent) available; 21 objects were observed by Flyingdales soon after the event.

REFERENCE DOCUMENTS

Analysis of Fragmentations from December 1992 - February 1993, Technical Report CS93-LKD-010, Teledyne Brown Engineering, Colorado Springs, Colorado, 30 March 1993.

History of Soviet/Russian Satellite Fragmentations-A Joint U.S.-Russian Investigation, N. L. Johnson et al, Kaman Sciences Corporation, October 1995.

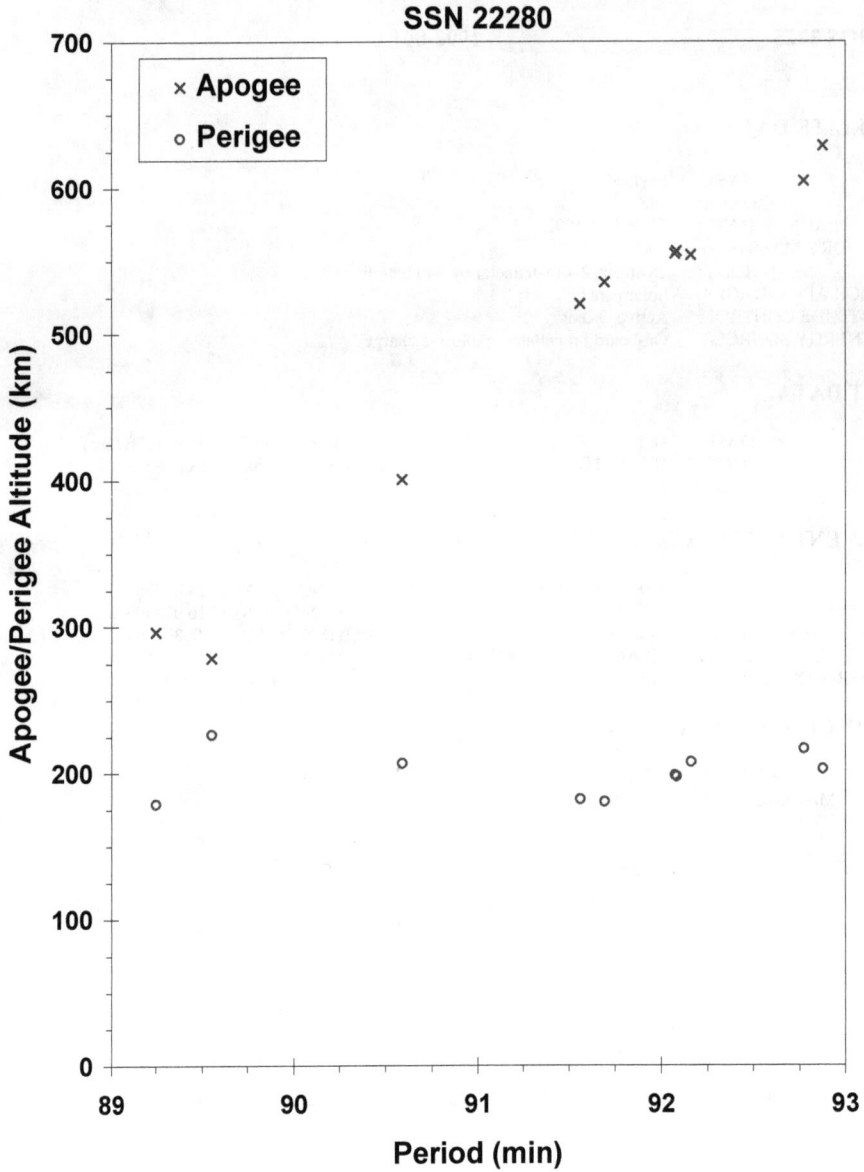

SSN 22280

Cosmos 2225 debris cloud remnant of 10 objects 4 days after the event as reconstructed
from the data provided by Naval Space Surveillance System
in a 22 February 1993 Satellite Support message.

COSMOS 2227 R/B   **1992-093B**   **22285**

SATELLITE DATA

TYPE:	Zenit Second Stage
OWNER:	CIS
LAUNCH DATE:	25.25 Dec 1992
DRY MASS (KG):	8300
MAIN BODY:	Cylinder; 3.9 m diameter by 12 m length
MAJOR APPENDAGES:	None
ATTITUDE CONTROL:	None at time of the event.
ENERGY SOURCES:	On-board propellants (~900 kg), explosive charge (?)

EVENT DATA (1)

DATE:	26 Dec 1992	LOCATION:	63 N, 60 E (asc)
TIME:	0738 GMT	ASSESSED CAUSE:	Propulsion
ALTITUDE:	830 km		

Note: NAVSPASUR could not correlate with 22285, but 22285 was closest object.

EVENT DATA (2)

DATE:	26 Dec 1992	LOCATION:	44 N, 168 E (asc)
TIME:	2249 GMT	ASSESSED CAUSE:	Unknown
ALTITUDE:	830 km		

EVENT DATA (3)

DATE:	26 Dec 1992	LOCATION:	52 N, 63 E (dsc)
TIME:	2310 GMT	ASSESSED CAUSE:	Unknown
ALTITUDE:	835 km		

EVENT DATA (4)

DATE:	30 Dec 1992	LOCATION:	22 S, 172 E (dsc)
TIME:	0903 GMT	ASSESSED CAUSE:	Propulsion
ALTITUDE:	~ 830 km		

PRE-EVENT ELEMENTS TO EVENT 1

EPOCH:	92361.30431818	MEAN ANOMALY:	289.8749		
RIGHT ASCENSION:	227.4354	MEAN MOTION:	14.1258288		
INCLINATION:	71.0274	MEAN MOTION DOT/2:	-.00061925	ECCENTRICITY:	
.0005311	MEAN MOTION DOT DOT/6:		.0		
ARG. OF PERIGEE:	71.7543	BSTAR:	-.034134		

DEBRIS CLOUD DATA

MAXIMUM ΔP:	4.3 min *
MAXIMUM ΔI:	1.4 deg *

* Based upon comparison to pre-event elements,
includes all four events, based upon cataloged elements only.

COMMENTS

Four separate events were reported by NAVSPASUR. The first observed event was accompanied by an initial 18 objects, but could not be correlated with the rocket body element set. The rocket body was the closest object to the BLAST point. The second event followed 15 hours later with 96 objects. The third event followed the second by less than 20 minutes and was based upon 51 pieces. The fourth event was accompanied by 3 objects. Element data on 164 objects has been combined into a single

Gabbard Diagram. NAVSPASUR initially generated 164 element sets on the combined debris from these 4 events. On 24 April 1995 object 22366, 1992-093BF, fragmented liberating 1 associated piece.

REFERENCE DOCUMENTS

Cosmos 2227 Rocket Body Fragmentation Event, E. L. Jenkins, et. al., NAVSPASUR, Dahlgren, VA.

Analysis of Fragmentations From December 1992 - February 1993, TBE Technical Report CS93-LKD-010, Teledyne Brown Engineering, Colorado Springs, 30 March 1993.

History of Soviet/Russian Satellite Fragmentations-A Joint U.S.-Russian Investigation, N. L. Johnson et al, Kaman Sciences Corporation, October 1995.

"The Recent Fragmentations of LEO Upper Stages", G. Chernyavskiy et al, 45th IAF Congress, October 1994.

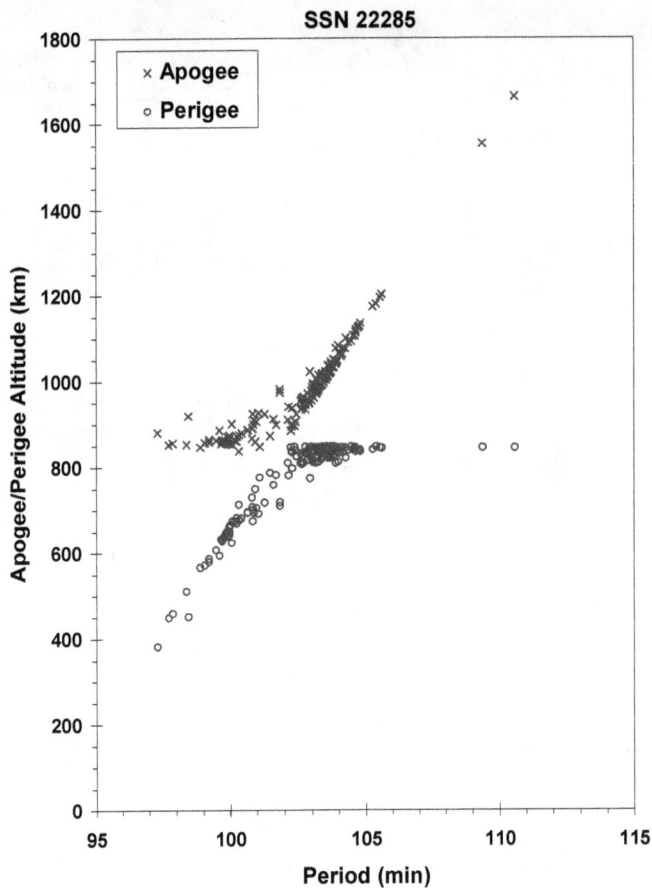

Naval Space Surveillance System generated 164 initial element sets on the four events that are plotted on the single Gabbard diagram above.

COSMOS 2237 R/B   1993-016B   22566

SATELLITE DATA

TYPE:	Zenit Second Stage
OWNER:	CIS
LAUNCH DATE:	26.10 Mar 1993
DRY MASS (KG):	8300
MAIN BODY:	Cylinder; 3.9 m diameter by 12 m length
MAJOR APPENDAGES:	None
ATTITUDE CONTROL:	None
ENERGY SOURCES:	On-board propellants (~900 kg), explosive charge (?)

EVENT DATA

DATE:	28 Mar 1993	LOCATION:	70N, 37E (dsc)
TIME:	0716 GMT	ASSESSED CAUSE:	Propulsion
ALTITUDE:	840 km		

PRE-EVENT ELEMENTS

EPOCH:	93088.27687915	MEAN ANOMALY:	84.1791
RIGHT ASCENSION:	258.8192	MEAN MOTION:	14.14093359
INCLINATION:	70.9947	MEAN MOTION DOT/2:	.00255882
ECCENTRICITY:	.0006748	MEAN MOTION DOT DOT/6:	.0
ARG. OF PERIGEE:	275.8565	BSTAR:	.12879

DEBRIS CLOUD DATA

MAXIMUM ΔP:	3.5 min
MAXIMUM ΔI:	0.1 deg

COMMENTS

At least 26 initial element sets were generated on this event by NAVSPASUR. The BLAST point was calculated from 12 objects.

REFERENCE DOCUMENTS

The Fragmentation of the Cosmos 2237 & 2243 Rocket Bodies, D. J. Nauer, TBE Technical Report CS93-LKD-016, Teledyne Brown Engineering, Colorado Springs, 15 June 1993.

History of Soviet/Russian Satellite Fragmentations-A Joint U.S.-Russian Investigation, N. L. Johnson et al, Kaman Sciences Corporation, October 1995.

"The Recent Fragmentations of LEO Upper Stages", G. Chernyavskiy et al, 45th IAF Congress, October 1994.

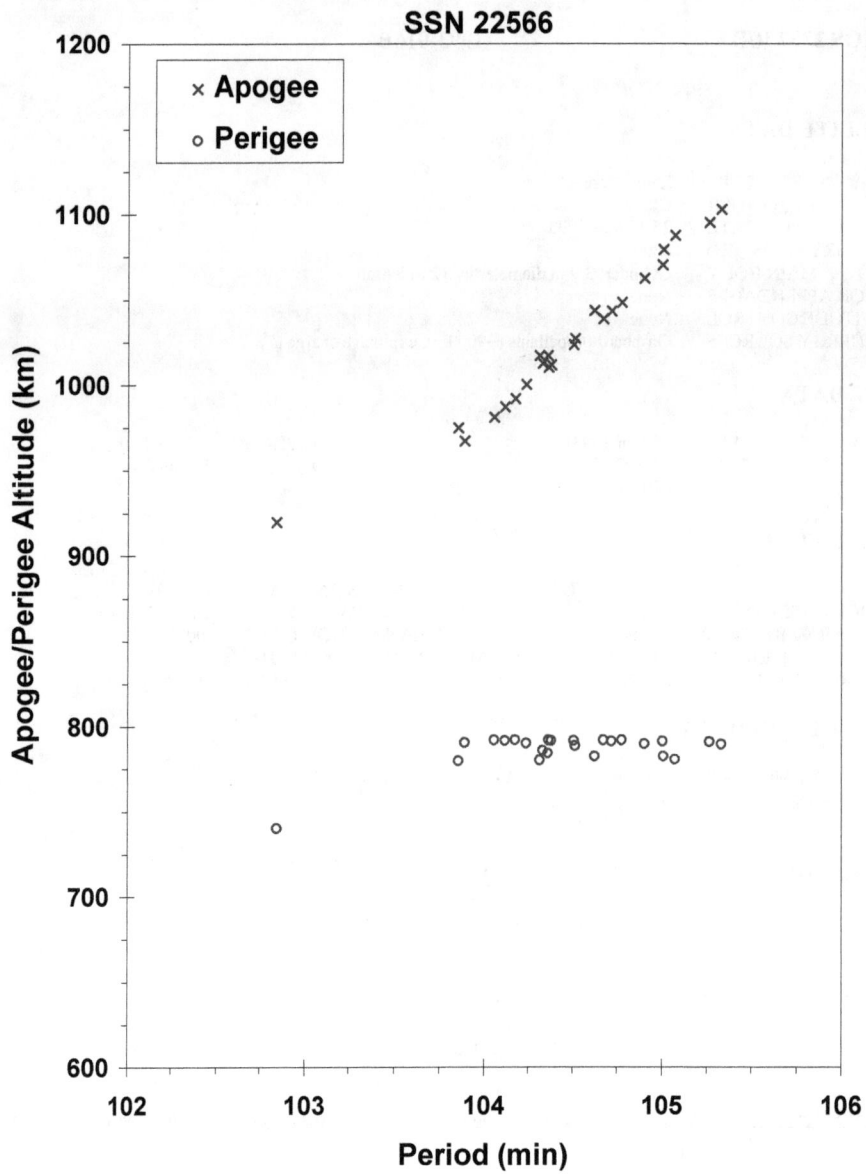

SSN 22566

Naval Space Surveillance System generated 24 initial element sets that are plotted on the Gabbard diagram above.

COSMOS 2238   **1993-018A**   **22585**

SATELLITE DATA

TYPE:	Payload
OWNER:	CIS
LAUNCH DATE:	30.50 Mar 1993
DRY MASS (KG):	3000
MAIN BODY:	Cylinder; 1.3 m diameter by 17 m long
MAJOR APPENDAGES:	Solar panels
ATTITUDE CONTROL:	Active, 3-axis
ENERGY SOURCES:	On-board propellants, explosive charge (?)

EVENT DATA

DATE:	1 Dec 1994	LOCATION:	6.5 S, 243.0 E
TIME:	1111 GMT	ASSESSED CAUSE:	Unknown
ALTITUDE:	275 km		

PRE-EVENT ELEMENTS

EPOCH:	94335.21831221	MEAN ANOMALY:	119.6648
RIGHT ASCENSION:	124.7826	MEAN MOTION:	16.06466469
INCLINATION:	65.0063	MEAN MOTION DOT/2:	.00787680
ECCENTRICITY:	.0069696	MEAN MOTION DOT DOT/6:	.0000095760
ARG. OF PERIGEE:	239.7651	BSTAR:	.00073936

DEBRIS CLOUD DATA

MAXIMUM ΔP:	Unknown
MAXIMUM ΔI:	Unknown

COMMENTS

Cosmos 2238 was a member of the Cosmos 699-type and the first such spacecraft to breakup for 7 years. As many as 26 debris were detected; however, additional debris probably decayed before they were observed. Cosmos is the 17th fragmentation of a Cosmos 699 class payload.

Insufficient data to construct a Gabbard diagram.

SATELLITE DATA

TYPE:	Payload
OWNER:	CIS
LAUNCH DATE:	27.44 Apr 1993
DRY MASS (KG):	5700
MAIN BODY:	Cone-cylinder; 2.7 m diameter by 6.3 m length
MAJOR APPENDAGES:	Solar panels
ATTITUDE CONTROL:	Active, 3-axis
ENERGY SOURCES:	On-board propellants, explosive charge

EVENT DATA

DATE:	27 Apr 1993	LOCATION:	61N, 81E (asc)
TIME:	1044 GMT	ASSESSED CAUSE:	Deliberate
ALTITUDE:	200 km		

PRE-EVENT ELEMENTS *

EPOCH:	93119.28633059	MEAN ANOMALY:	283.6524	
RIGHT ASCENSION:	51.8515	MEAN MOTION:	16.26199828	
INCLINATION:	70.3602	MEAN MOTION DOT/2:	.02823100	ECCENTRICITY:
.0032877	MEAN MOTION DOT DOT/6:		.000019668	
ARG. OF PERIGEE:	76.8057	BSTAR:	.00077017	

* Note: Element Set 1 not generated until ~2 days after the event.

DEBRIS CLOUD DATA

MAXIMUM ΔP:	7.1 min
MAXIMUM ΔI:	0.4 deg

COMMENTS

Although this event was originally thought to be a fragmentation of the Soyuz final stage rocket body, it was actually the payload that fragmented. This event occurred near orbital insertion into the operational orbit. It is unclear whether the payload was attached at the time of the event. The payload malfunctioned and self-destructed. There were 25 initial element sets available after launch. NAVSPASUR reported tracking approximately 27 objects on 30 April 1993, and detected as many as 20 more unknowns. The final official piece count associated with this event was 172 objects. Due to the very low altitude, most objects decayed from this cloud within 2 weeks of launch. No cataloged element sets were released until almost 2 days after the event.

REFERENCE DOCUMENTS

The Fragmentation of the Cosmos 2237 & 2243 Rocket Bodies, D. J. Nauer, TBE Technical Report CS93-LKD-016, Teledyne Brown Engineering, Colorado Springs, 15 June 1993.

History of Soviet/Russian Satellite Fragmentations-A Joint U.S.-Russian Investigation, N. L. Johnson et al, Kaman Sciences Corporation, October 1995.

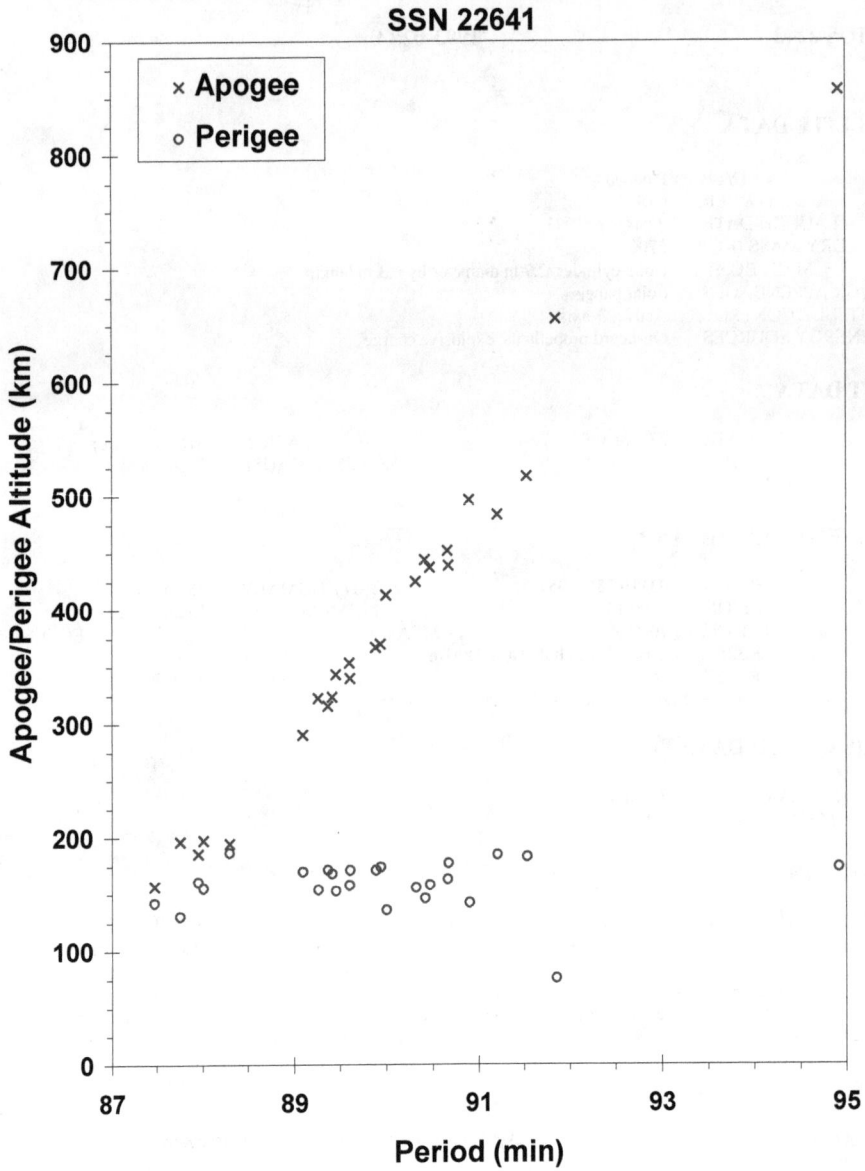

SSN 22641

Naval Space Surveillance System generated 25 initial element sets on the Cosmos 2243 fragmentation that are plotted on the Gabbard diagram above.

COSMOS 2259   1993-045A   22716

SATELLITE DATA

TYPE:	Payload
OWNER:	CIS
LAUNCH DATE:	14.69 Jul 1993
DRY MASS (KG):	5700
MAIN BODY:	Cone-cylinder; 2.7 m diameter by 6.3 m length
MAJOR APPENDAGES:	Solar panels
ATTITUDE CONTROL:	Active, 3-axis
ENERGY SOURCES:	On-board propellants, explosive charge

EVENT DATA

DATE:	25 Jul 1993	LOCATION:	Unknown
TIME:	Unknown	ASSESSED CAUSE:	Deliberate
ALTITUDE:	Unknown		

PRE-EVENT ELEMENTS

EPOCH:	93205.96411483	MEAN ANOMALY:	292.3177
RIGHT ASCENSION:	134.4696	MEAN MOTION:	16.09525981
INCLINATION:	67.1310	MEAN MOTION DOT/2:	0.00638090
ECCENTRICITY:	0.0113387	MEAN MOTION DOT DOT/6:	0.000023099
ARG. OF PERIGEE:	68.9805	BSTAR:	0.00025239

DEBRIS CLOUD DATA

MAXIMUM ΔP:	Unknown
MAXIMUM ΔI:	Unknown

COMMENTS

Spacecraft was destroyed after a malfunction prevented reentry and landing in the CIS. Event identified by Russian officials during investigation cited below.

REFERENCE DOCUMENT

History of Soviet/Russian Satellite Fragmentations-A Joint U.S.-Russian Investigation, N. L. Johnson et al, Kaman Sciences Corporation, October 1995.

Insufficient data to construct a Gabbard diagram.

COSMOS 2262   1993-057A   22789

SATELLITE DATA

TYPE:	Payload
OWNER:	CIS
LAUNCH DATE:	7.56 Sep 1993
DRY MASS (KG):	6000
MAIN BODY:	Cylinder; 2.4 m diameter by 7 m length
MAJOR APPENDAGES:	Solar panels
ATTITUDE CONTROL:	Active, 3-axis
ENERGY SOURCES:	On-board propellants, explosive charge

EVENT DATA

DATE:	18 Dec 1993	LOCATION:	65N, 107E (dsc)
TIME:	0711 GMT	ASSESSED CAUSE:	Deliberate
ALTITUDE:	195 km		

PRE-EVENT ELEMENTS

EPOCH:	93352.09835999	MEAN ANOMALY:	294.6647		
RIGHT ASCENSION:	209.9170	MEAN MOTION:	16.17608693		
INCLINATION:	64.8761	MEAN MOTION DOT/2:	.00554324	ECCENTRICITY:	
.0065884	MEAN MOTION DOT DOT/6:		.0		
ARG. OF PERIGEE:	66.1310	BSTAR:	.00022099		

DEBRIS CLOUD DATA

MAXIMUM ΔP:	8.7 min *
MAXIMUM ΔI:	0.8 deg *

* Based on uncataloged debris data

COMMENTS

Spacecraft was destroyed with a planned detonation. Fifth fragmentation of this sub-type (Cosmos 2031 subclass). Early elements on 43 objects (including the parent) were collected; at least 179 objects were reported by the NAVSPOC for early passes through the NAVSPASUR fence.

REFERENCE DOCUMENTS

The Fragmentation of Cosmos 2262, Technical Report CS94-LKD-006, Teledyne Brown Engineering, Colorado Springs, Colorado, 31 December 1993.

History of Soviet/Russian Satellite Fragmentations-A Joint U.S.-Russian Investigation, N. L. Johnson et al, Kaman Sciences Corporation, October 1995.

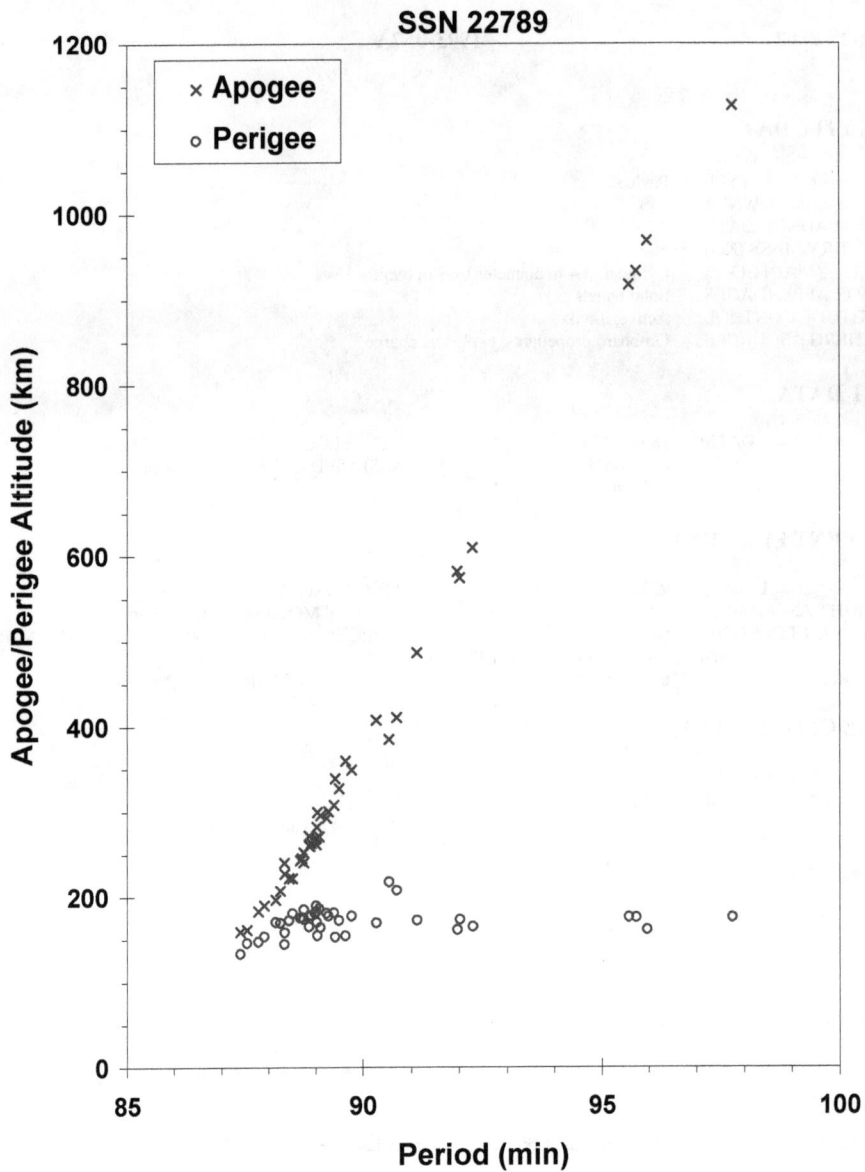

SSN 22789

x Apogee
o Perigee

Apogee/Perigee Altitude (km)

Period (min)

Cosmos 2262 debris cloud based upon 43 early element sets provided by the NAVSPOC.

GORIZONT 29 ULLAGE MOTOR   1993-072E   22925

SATELLITE DATA

TYPE:	Proton Block DM SOZ Ullage Motor
OWNER:	CIS
LAUNCH DATE:	18.58 Nov 1993
DRY MASS (KG):	~55 kg
MAIN BODY:	~0.6 m by 0.6 m by 1.0 m
MAJOR APPENDAGES:	None
ATTITUDE CONTROL:	None
ENERGY SOURCES:	On-board propellants?

EVENT DATA

DATE:	6-7 Sep 2000	LOCATION:	Unknown
TIME:	between 1918–0253 GMT	ASSESSED CAUSE:	Propulsion
ALTITUDE:	Unknown		

PRE-EVENT ELEMENTS

EPOCH:	00250.18110680	MEAN ANOMALY:	305.0033
RIGHT ASCENSION:	135.7916	MEAN MOTION:	6.55809618
INCLINATION:	46.7439	MEAN MOTION DOT/2:	.00601672
ECCENTRICITY:	.4592082	MEAN MOTION DOT DOT/6:	.00000031378
ARG. OF PERIGEE:	109.1361	BSTAR:	.00059159

DEBRIS CLOUD DATA

MAXIMUM ΔP:	Unknown
MAXIMUM ΔI:	Unknown

COMMENTS

This is the 22nd breakup event for an object of this class, and the first of the year 2000. The breakups are assessed to be caused by residual propellants. Russian officials have been aware of the problem since 1992 and have made design changes, although the date of full implementation is unknown. The environmental consequence of the breakup will be short-lived; the object is in catastrophic decay from a geosynchronous transfer orbit. Latest estimate of the breakup time is between 1918 GMT, 6 September and 0253 GMT, 7 September.

REFERENCE DOCUMENT

"September Breakup is 22nd in Series", The Orbital Debris Quarterly News, NASA JSC, October 2000. Available online at http://www.orbitaldebris.jsc nasa.gov/newsletter/pdfs/ODQNv5i4.pdf.

Insufficient data to construct a Gabbard diagram.

CLEMENTINE R/B 1994-004B 22974

SATELLITE DATA

TYPE:	Titan II Second Stage
OWNER:	US
LAUNCH DATE:	25.69 Jan 1994
DRY MASS (KG):	2860
MAIN BODY:	Cylinder
MAJOR APPENDAGES:	None
ATTITUDE CONTROL:	None at time of the event.
ENERGY SOURCES:	On-board propellants

EVENT DATA

DATE:	7 Feb 1994	LOCATION:	59S, 126W (dsc)
TIME:	1719 GMT	ASSESSED CAUSE:	Propulsion
ALTITUDE:	260 km		

PRE-EVENT ELEMENTS

EPOCH:	94038.24510489	MEAN ANOMALY:	208.0182
RIGHT ASCENSION:	47.9208	MEAN MOTION:	16.13665058
INCLINATION:	66.9945	MEAN MOTION DOT/2:	.01050211
ECCENTRICITY:	.0027030	MEAN MOTION DOT DOT/6:	.0000059221
ARG. OF PERIGEE:	152.2460	BSTAR:	.00081413

DEBRIS CLOUD DATA

MAXIMUM ΔP:	5.6 min *
MAXIMUM ΔI:	0.6 deg *

* Based on uncataloged debris data

COMMENTS

First Titan II Second Stage to violently fragment. NAVSPOC reported observing a maximum of 364 objects in the early debris cloud, and the NAVSPOC released 45 element sets. Engineering analysis by the manufacturer (Martin Marietta) indicates no known failure mechanism, although unspent on-board propellants were present.

REFERENCE DOCUMENT

The Fragmentation of the Clementine Rocket Body, TBE Technical Report CS94-LKD-010, Teledyne Brown Engineering, Colorado Springs, Colorado, 31 March 1994.

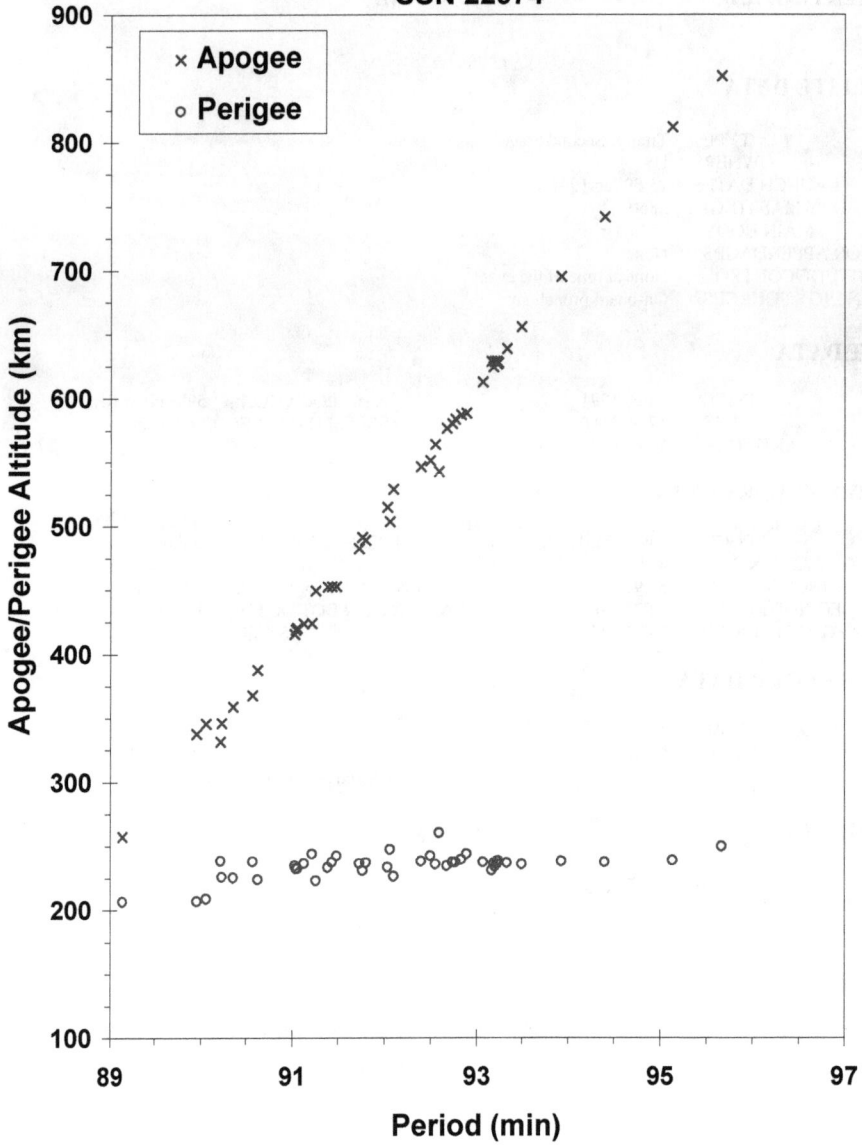

Gabbard diagram of 45 NAVSPOC element sets.

SATELLITE DATA

TYPE:	Pegasus HAPS
OWNER:	USA
LAUNCH DATE:	19.71 May 1994
DRY MASS (KG):	97
MAIN BODY:	Cylinder; 0.97 m diameter by 0.93 m length
MAJOR APPENDAGES:	None
ATTITUDE CONTROL:	None at time of the event.
ENERGY SOURCES:	On-board propellants and pressurants

EVENT DATA

DATE:	3 Jun 1996	LOCATION:	67 S, 56 E (asc)
TIME:	1518 GMT	ASSESSED CAUSE:	Propulsion
ALTITUDE:	625 km		

PRE-EVENT ELEMENTS

EPOCH:	96155.10100506	MEAN ANOMALY:	108.3711
RIGHT ASCENSION:	197.8565	MEAN MOTION:	14.56780581
INCLINATION:	81.9749	MEAN MOTION DOT/2:	0.00000158
ECCENTRICITY:	0.0165742	MEAN MOTION DOT DOT/6:	0
ARG. OF PERIGEE:	249.9583	BSTAR:	0.000025815

DEBRIS CLOUD DATA

MAXIMUM ΔP:	27.9 min
MAXIMUM ΔI:	2.4 deg

COMMENTS

The Pegasus HAPS vehicle was employed for only the second time. It failed to place its payload into the assigned circular orbit and had an estimated 5-8 kg of residual propellant plus propellant for attitude control on board. The fragmentation of the small, 2-year-old upper stage produced a record number of more than 750 tracked debris. This is about an order of magnitude more than can be expected for the small dry mass of the R/B of 97 kg. Observations suggest that the debris are physically small with a high radar reflectivity. Investigations suggest that a regulator failure led to overpressurization of the propellant tank that in turn ruptured.

REFERENCE DOCUMENT

"Major Satellite Breakup in June", N. Johnson, <u>Orbital Debris Quarterly News</u>, NASA JSC, September 1996, p. 2 and 11. Available online at http://www.orbitaldebris.jsc.nasa.gov/newsletter/pdfs/ODQNV1i2.pdf.

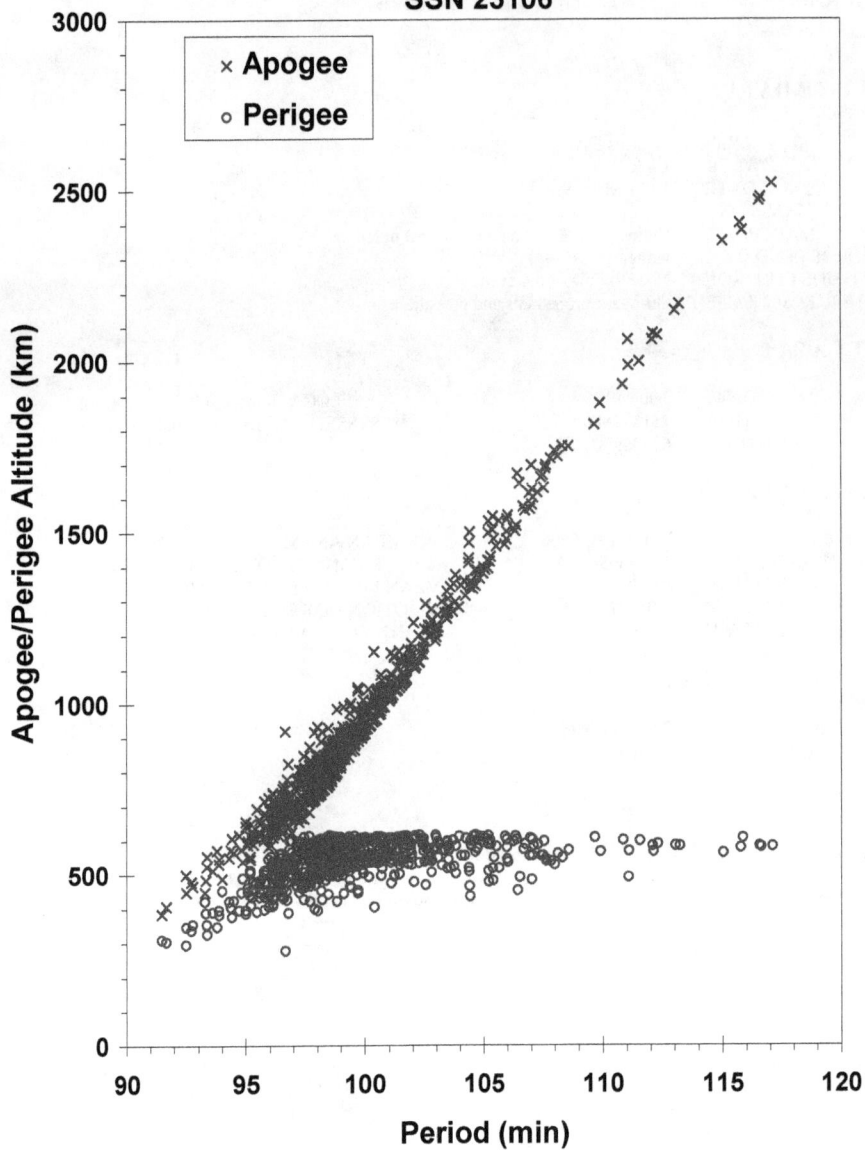

SSN 23106

Step II R/B debris cloud of 713 fragments as of August 29, 1996 as
reconstructed from the US SSN database.

SATELLITE DATA

TYPE:	Mission Related Debris
OWNER:	CIS
LAUNCH DATE:	6.99+ Jul 1994
DRY MASS (KG):	55
MAIN BODY:	Ellipsoid; 0.6 m diameter by 1 m length
MAJOR APPENDAGES:	None
ATTITUDE CONTROL:	None
ENERGY SOURCES:	On-board propellants

EVENT DATA

DATE:	Approx. 21 Oct 1995	LOCATION:	Unknown
TIME:	Unknown	ASSESSED CAUSE:	Propulsion
ALTITUDE:	Unknown		

PRE-EVENT ELEMENTS

EPOCH:	95293.99530492	MEAN ANOMALY:	2.33725319
RIGHT ASCENSION:	157.0951	MEAN MOTION:	321.8211
INCLINATION:	47.0485	MEAN MOTION DOT/2:	.00002472
ECCENTRICITY:	.7223127	MEAN MOTION DOT DOT/6:	.0000
ARG. OF PERIGEE:	127.9520	BSTAR:	.0010694

DEBRIS CLOUD DATA

MAXIMUM ΔP:	Unknown
MAXIMUM ΔI:	Unknown

COMMENTS

Parent satellite was one of two small engine units that are routinely released after the first burn of the Proton fourth stage. The nature of these objects was identified by Dr. Boris V. Chernlatiev, Deputy Constructor for the Energiya NPO, in October 1992. The cause of this fragmentation is assumed to be related to the residual hypergolic propellants on board and failure of the membrane separating the fuel and oxidizer. NAVSPASUR observed 114 objects that were associated with this breakup.

REFERENCE DOCUMENTS

The Fragmentation of Proton Debris, D. J. Nauer, TBE Technical Report CS93-LKD-004, Teledyne Brown Engineering, Colorado Springs, 31 December 1992.

Analysis of Fragmentations From December 1992 - February 1993, TBE Technical Report CS93-LKD-010, Teledyne Brown Engineering, Colorado Springs, 30 March 1993.

History of Soviet/Russian Satellite Fragmentations-A Joint U.S.-Russian Investigation, N. L. Johnson et al, Kaman Sciences Corporation, October 1995.

"Identification and Resolution of an Orbital Debris Problem with the Proton Launch Vehicle", B. V. Cherniatiev et al, Proceedings of the First European Conference on Space Debris, April 1993.

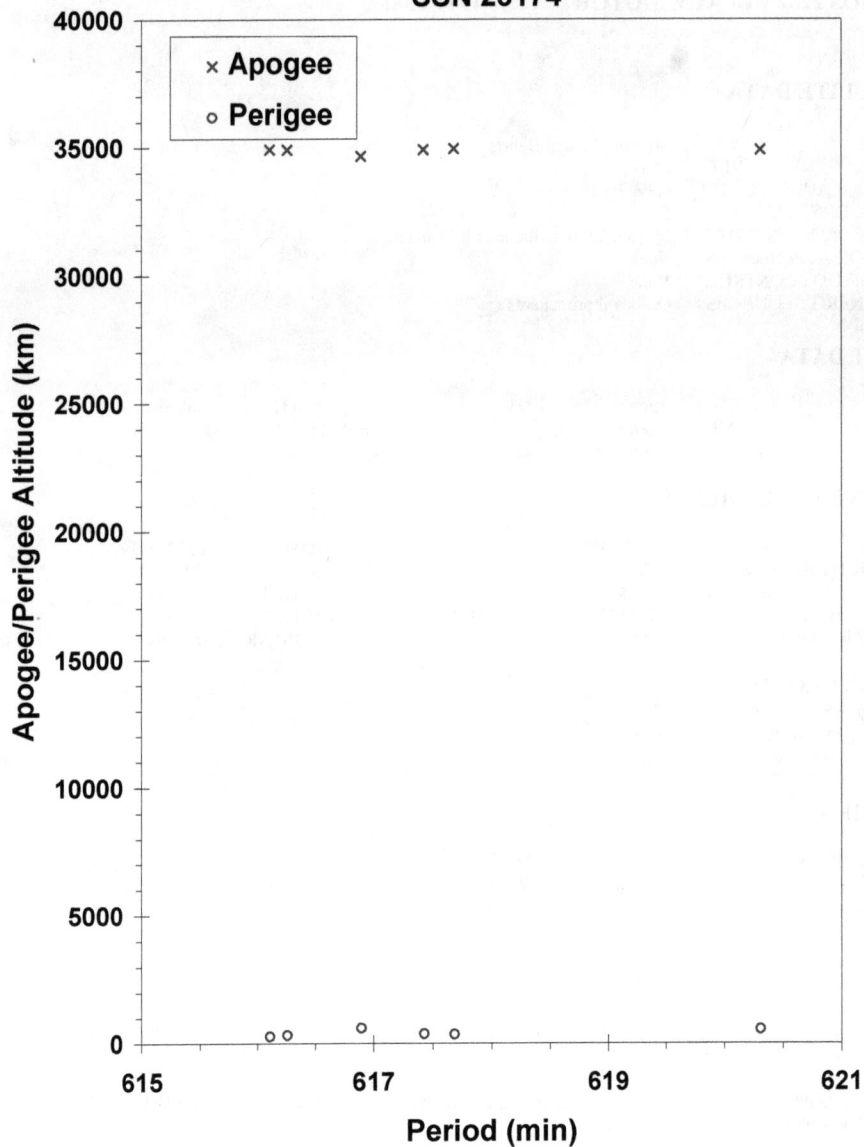

SSN 23174

Cosmos 2282 ullage motor debris cloud of 6 fragments assembled by NAVSPOC.

SATELLITE DATA

TYPE:	Mission Related Debris
OWNER:	CIS
LAUNCH DATE:	31.60 Oct 1994
DRY MASS (KG):	55
MAIN BODY:	Ellipsoid; 0.6 m diameter by 1 m length
MAJOR APPENDAGES:	None
ATTITUDE CONTROL:	None
ENERGY SOURCES:	On-board propellants

EVENT DATA

DATE:	Prior to 0547, 11 May 1995	LOCATION:	Unknown
TIME:	Unknown	ASSESSED CAUSE:	Propulsion
ALTITUDE:	Unknown		

PRE-EVENT ELEMENTS

EPOCH:	95130.00087914	MEAN ANOMALY:	317.6470
RIGHT ASCENSION:	200.4799	MEAN MOTION:	11.93599702
INCLINATION:	46.9113	MEAN MOTION DOT/2:	.99999999
ECCENTRICITY:	0.2007574	MEAN MOTION DOT DOT/6:	34693-4
ARG. OF PERIGEE:	63.6124	BSTAR:	.021116

DEBRIS CLOUD DATA

MAXIMUM ΔP:	Unknown
MAXIMUM ΔI:	Unknown

COMMENTS

Parent satellite was one of two small engine units that are routinely released after the first burn of the Proton fourth stage. The nature of these objects was identified by Dr. Boris V. Cherniatiev, Deputy Constructor for the Energiya NPO and Mr. Nicholas Johnson of Kaman Sciences, in October, 1992. The cause of this fragmentation appears to be related to the residual hypergolic propellants on board and failure of the membrane separating the fuel and oxidizer. NAVSPASUR observed up to 13 objects that were associated with this breakup on 11 May 95. This was the eleventh in a series of fragmentations of this object type.

REFERENCE DOCUMENTS

The Fragmentation of Proton Debris, D. J. Nauer, TBE Technical Report CS93-LKD-004, Teledyne Brown Engineering, Colorado Springs, 31 December 1992.

Analysis of Fragmentations From December 1992 - February 1993, TBE Technical Report CS93-LKD-010, Teledyne Brown Engineering, Colorado Springs, 30 March 1993.

Identification and Resolution of an Orbital Debris Problem with the Proton Launch Vehicle, B. V. Cherniatiev, et al, First European Conference on Space Debris, 5-7 April 1993.

History of Soviet/Russian Satellite Fragmentations-A Joint U.S.-Russian Investigation, N. L. Johnson et al, Kaman Sciences Corporation, October 1995.

Insufficient data to construct a Gabbard diagram.

SATELLITE DATA

TYPE:	Rokot Third Stage
OWNER:	CIS
LAUNCH DATE:	26.13 Dec 1994
DRY MASS (KG):	1000
MAIN BODY:	Cylinder; 2.4 m diameter by 2.8 m length
MAJOR APPENDAGES:	None
ATTITUDE CONTROL:	Unknown
ENERGY SOURCES:	On-board propellants

EVENT DATA

DATE:	26 Dec 1994	LOCATION:	51.6S, 53W (asc)
TIME:	0627 GMT	ASSESSED CAUSE:	Unknown
ALTITUDE:	1880 km		

PRE-EVENT ELEMENTS

EPOCH:	94361.79150546	MEAN ANOMALY:	66.1014
RIGHT ASCENSION:	172.1572	MEAN MOTION:	11.27113018
INCLINATION:	64.8297	MEAN MOTION DOT/2:	-.00000043
ECCENTRICITY:	0.0188748	MEAN MOTION DOT DOT/6:	00000-0
ARG. OF PERIGEE:	292.0126	BSTAR:	00000+0

DEBRIS CLOUD DATA

MAXIMUM ΔP:	4.5 min
MAXIMUM ΔI:	0.2 deg

COMMENTS

Parent satellite was the Rokot third stage. The Rokot is an SS-19 ICBM based vehicle with a new third stage referred to as Breaz. All three stages are fueled with UDMH/N204. NAVSPASUR observed 34 objects that were associated with this breakup.

SSN 23440

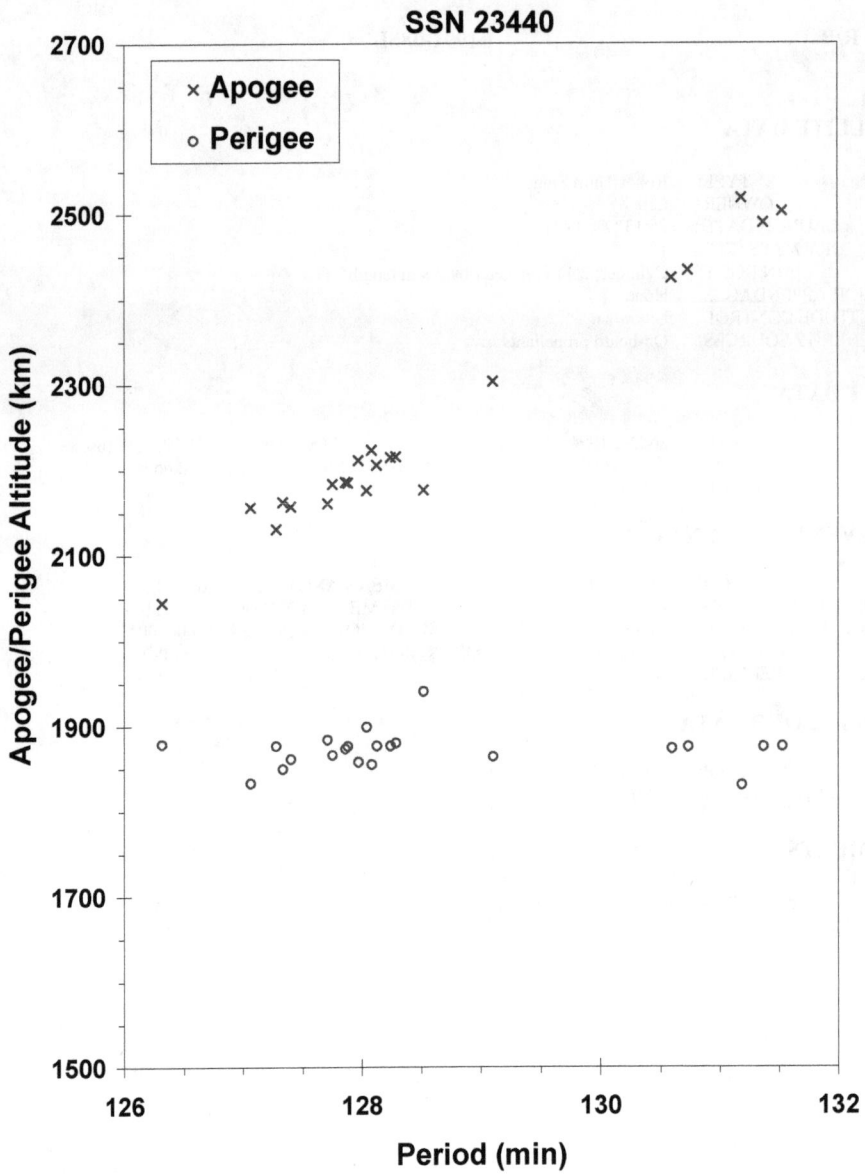

Gabbard diagram for RS-15 R/B debris cloud of 23 fragments
as reconstructed from the US SSN database.

COSMOS 2313   1995-028A   23596

SATELLITE DATA

TYPE:	Payload
OWNER:	CIS
LAUNCH DATE:	8.20 Jun 1995
DRY MASS (KG):	3000
MAIN BODY:	Cylinder; 1.3 m diameter by 17 m length
MAJOR APPENDAGES:	Solar arrays
ATTITUDE CONTROL:	Active, 3-axis
ENERGY SOURCES:	On-board propellants, explosive charge?

EVENT DATA

DATE:	26 June 1997	LOCATION:	44 N, 173 E (asc.)
TIME:	0257 GMT	ASSESSED CAUSE:	Unknown
ALTITUDE:	285 km		

PRE-EVENT ELEMENTS

EPOCH:	97176.10173599	MEAN ANOMALY:	124.6445
RIGHT ASCENSION:	342.0749	MEAN MOTION:	16.02369895
INCLINATION:	65.0221	MEAN MOTION DOT/2:	0.00306537
ECCENTRICITY:	0.0084335	MEAN MOTION DOT DOT/6:	0.0000069339
ARG. OF PERIGEE:	234.6794	BSTAR:	0.00033322

DEBRIS CLOUD DATA

MAXIMUM ΔP:	Unknown
MAXIMUM ΔI:	Unknown

COMMENTS

Cosmos 2313 was the second spacecraft of its type to breakup since November 1987. Prior to the current event 17 spacecraft of this class (Cosmos 699) have experienced breakups in low Earth orbit. In the 1980's procedures were introduced to deplete remaining propellants at the end of mission, reducing orbital lifetime at the same time. Cosmos 2313 performed such a maneuver during 22-23 April 1997 and was close to reentry at the time of the event. Earlier spacecraft breakups resulted in up to 150 or more trackable debris. The cause of the event may well not be propellant related, but by reducing the orbital lifetime recent vehicles have decayed before the trigger mechanism could activate. At least 90 debris were detected after this event.

REFERENCE DOCUMENT

"Three Satellite Breakups During May-June", The Orbital Debris Quarterly News, NASA JSC, July 1997, p. 2. Available online at http://www.orbitaldebris.jsc nasa.gov/newsletter/pdfs/ODQNv2i3.pdf.

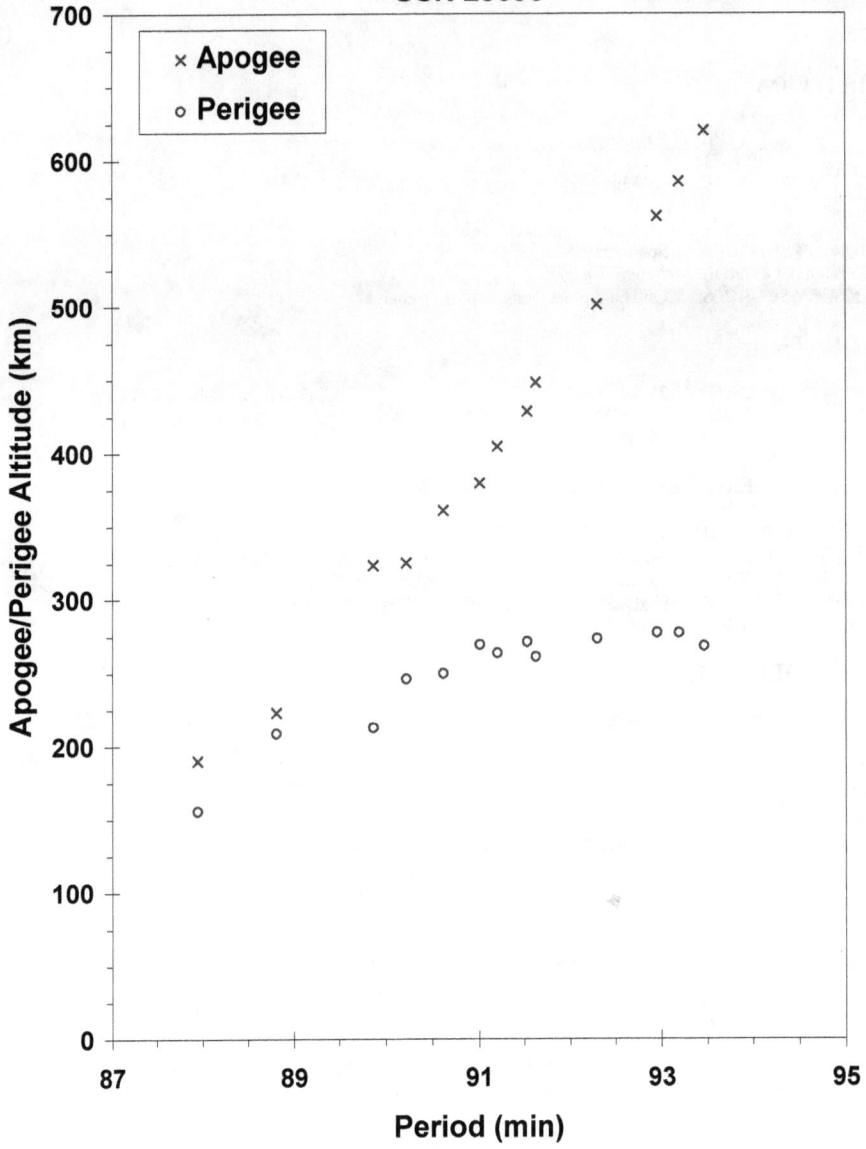

Cosmos 2313 debris cloud of 13 fragments 1 day to 2 weeks after the event as reconstructed from the US SSN database.

CERISE   **1995-033B**   **23606**

SATELLITE DATA

TYPE:	Payload
OWNER:	France
LAUNCH DATE:	7.68 Jul 1995
DRY MASS (KG):	50
MAIN BODY:	Box; 0.6 m by 0.3 m by 0.3 m
MAJOR APPENDAGES:	6 m long gravity-gradient boom; solar panels
ATTITUDE CONTROL:	Gravity-gradient stabilized
ENERGY SOURCES:	Battery

EVENT DATA

DATE:	24 Jul 1996	LOCATION:	38 S, 60 E (asc)
TIME:	0948 GMT	ASSESSED CAUSE:	Collision
ALTITUDE:	685 km		

PRE-EVENT ELEMENTS

EPOCH:	96205.39273562	MEAN ANOMALY:	292.8048
RIGHT ASCENSION:	141.7519	MEAN MOTION:	14.67264268
INCLINATION:	98.1025	MEAN MOTION DOT/2:	0.00000083
ECCENTRICITY:	0.0008991	MEAN MOTION DOT DOT/6:	0
ARG. OF PERIGEE:	67.4104	BSTAR:	0.000023247

DEBRIS CLOUD DATA

MAXIMUM ΔP:	N/A
MAXIMUM ΔI:	N/A

COMMENTS

The incident marked the first time that two objects in the U.S. satellite catalog are known to have accidentally run into one another. The CERISE spacecraft (Satellite Number 23606, International Designator 1995-033B) is a microsatellite of British design. The other participant in the encounter was Satellite Number 18208 (International Designator 1986-019RF), which was generated in November 1986, when ESA's SPOT 1 rocket body broke up into nearly 500 tracked debris. The orbit of this fragmentation debris at the time of the collision was 660 km by 680 km at an inclination of 98.45 degrees. The collision, which occurred with a relative velocity of 14.8 km/s, produced only a single piece of debris large enough to be tracked, i.e., the upper portion of the gravity-gradient boom. Analysis of the manufacturer of the spacecraft bus, Surrey Satellite Technology Ctd. at the University of Surrey, United Kingdom suggested that the 6 m, gravity-gradient boom had been severed at 3.1-3.2 meter from its base.

Using USAF Space Command's COMBO (Computation of Miss Between Orbits) program, a close approach of less than 1 km between Satellite 23606 and Satellite 18208 was determined by NASA JSC to have taken place at 0948 GMT on 24 July over the southern Indian Ocean. Naval Space Operations Center (NAVSPOC) at Dahlgren, Virginia, replicated the NASA findings and, using direct observational data and special perturbation theory, was able to refine the miss distance uncertainty to within 137 m. In addition, NAVSPOC identified a minor perturbation in the orbit of Satellite 18208 that occurred about the time of the event.

REFERENCE DOCUMENTS

"First Natural Collision of Cataloged Earth Satellites", N. Johnson, The Orbital Debris Quarterly News, NASA JSC, September 1996, p. 1. Available online at http://www.orbitaldebris.jsc nasa.gov/newsletter/pdfs/ODQNV1i2.pdf.

"Collision of CERISE with Space Debris", F. Alby et al, Proceedings of the Second European Conference on Space Debris, SP-393, p. 589-596.

"First 'Confirmed' Natural Collision Between Two Cataloged Satellites", T. Payne, Proceedings of the Second European Conference on Space Debris, SP-393, p. 597-600.

"Predicting Conjunctions with Trackable Space Debris: Some Recent Experiences", E. L. Jenkins and P. W. Schumacher, Jr., AAS 97-014, 20th Annual AAS Guidance and Control Conference, February 1997.

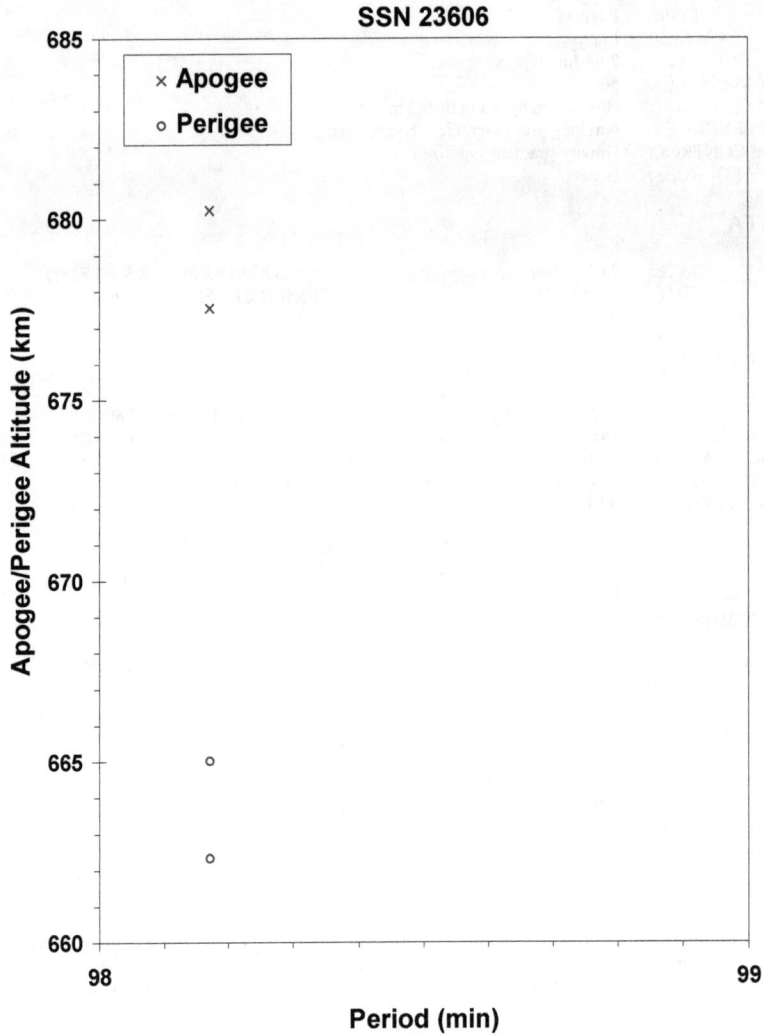

CERISE debris cloud of 2 fragments 4 days after the event as reconstructed from the US SSN database.

COSMOS 2316-2318 ULLAGE MOTOR   1995-037K   23631

SATELLITE DATA

TYPE:	Proton Block DM SOZ Ullage Motor
OWNER:	CIS
LAUNCH DATE:	24.66 Jul 1995
DRY MASS (KG):	~55 kg
MAIN BODY:	~0.6 m by 0.6 m by 1.0 m
MAJOR APPENDAGES:	None
ATTITUDE CONTROL:	None
ENERGY SOURCES:	On-board propellants?

EVENT DATA

DATE:	21 Nov 2000	LOCATION:	Unknown
TIME:	Unknown	ASSESSED CAUSE:	Propulsion
ALTITUDE:	Unknown		

PRE-EVENT ELEMENTS

EPOCH:	00324.99357911	MEAN ANOMALY:	90.3648
RIGHT ASCENSION:	200.0539	MEAN MOTION:	4.50149006
INCLINATION:	64.4375	MEAN MOTION DOT/2:	.00164632
ECCENTRICITY:	.5787543	MEAN MOTION DOT DOT/6:	.00000030156
ARG. OF PERIGEE:	213.7574	BSTAR:	.00048999

DEBRIS CLOUD DATA

MAXIMUM ΔP:	Unknown
MAXIMUM ΔI:	Unknown

COMMENTS

This is the 4th breakup of 2000 and the 23rd breakup of a Proton SOZ motor (see Orbital Debris Quarterly Newsletter V, Issue 4, p. 2.) This object is associated with the 24 July 1995 launch (1995-037) of the Cosmos 2316-2318 satellites. These members of the GLONASS series are equivalent to GPS/Navstar satellites and reside in middle Earth orbit. This object was one of two pieces left in the transfer orbit and is assessed to be one of the SOZ ullage/orientation motor units. As of November 21, 2000, this object had been on orbit 5 years and 121 days.

REFERENCE DOCUMENT

"SOZ Ullage Motor Breakup", The Orbital Debris Quarterly News, NASA JSC, January 2001. Available online at http://www.orbitaldebris.jsc.nasa.gov/newsletter/pdfs/odqnv6i1.pdf.

Insufficient data to construct a Gabbard diagram.

RADUGA 33 R/B   1996-010D   23797

SATELLITE DATA

TYPE:	Proton Block DM Fourth Stage
OWNER:	CIS
LAUNCH DATE:	19.36 Feb 1996
DRY MASS (KG):	3400 (?)
MAIN BODY:	Cylinder; 3.7 m diameter by 6.3 m length
MAJOR APPENDAGES:	None
ATTITUDE CONTROL:	Active, 3-axis
ENERGY SOURCES:	On-board propellants, pressurants, and batteries

EVENT DATA

DATE:	19 Feb 1996	LOCATION:	0.2 N, 88.8 E (dsc)
TIME:	14.59 GMT	ASSESSED CAUSE:	Propulsion
ALTITUDE:	36511 km		

POST-EVENT ELEMENTS

EPOCH:	96058.46760248	MEAN ANOMALY:	359.9314
RIGHT ASCENSION:	280.4138	MEAN MOTION:	2.23172282
INCLINATION:	48.7	MEAN MOTION DOT/2:	.0002158
ECCENTRICITY:	.7321111	MEAN MOTION DOT DOT/6:	.0000
ARG. OF PERIGEE:	1.7779	BSTAR:	.00068491

DEBRIS CLOUD DATA

MAXIMUM ΔP:	Unknown
MAXIMUM ΔI:	Unknown

COMMENTS

The first burn of this stage was successful and indicates that 23797 was in a transfer orbit. However, prior to the first pass through the NAVSPOC fence, 23797 fragmented. Twenty (20) pieces were observed during this first pass. During a subsequent pass, 196 pieces were observed that were associated with the upperstage. Stage apparently broke up after main engine restart for GEO apogee maneuver.

REFERENCE DOCUMENT

"Satellite Fragmentations in 1996", N. Johnson, The Orbital Debris Quarterly News, NASA JSC, January 1997, p. 1. Available online at http://www.orbitaldebris.jsc.nasa.gov/newsletter/pdfs/ODQNv2i1.pdf.

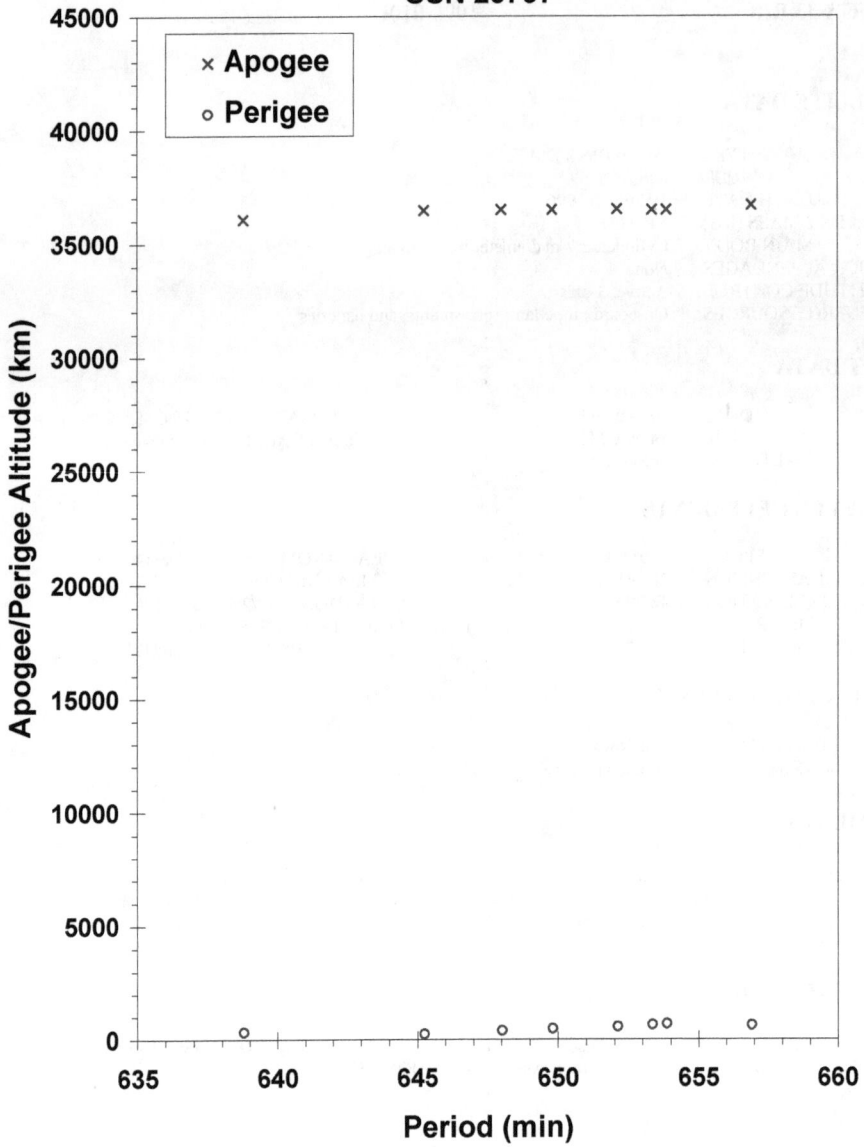

SSN 23797

Apogee/Perigee Altitude (km) vs **Period (min)**

Legend:
- × Apogee
- ○ Perigee

Gabbard diagram from Raduga 33 from NAVSPOC elements.

GORIZONT 32 ULLAGE MOTOR   1996-034F   23887

SATELLITE DATA

TYPE:	Mission Related Debris
OWNER:	CIS
LAUNCH DATE:	25.09 May 1996
DRY MASS (KG):	~55
MAIN BODY:	Ellipsoid; 0.6 m by 1 m
MAJOR APPENDAGES:	None
ATTITUDE CONTROL:	Unknown
ENERGY SOURCES:	On-board Propellants

EVENT DATA

DATE:	~13 Dec 1999	LOCATION:	Unknown
TIME:	Unknown	ASSESSED CAUSE:	Propulsion
ALTITUDE:	Unknown		

PRE-EVENT ELEMENTS

EPOCH:	99347.02294368	MEAN ANOMALY:	314.8549
RIGHT ASCENSION:	194.3249	MEAN MOTION:	9.75630550
INCLINATION:	46.4558	MEAN MOTION DOT/2:	.00969995
ECCENTRICITY:	.2950283	MEAN MOTION DOT DOT/6:	.0000015400
ARG. OF PERIGEE:	75.9037	BSTAR:	.00055450

CATALOGED DEBRIS CLOUD DATA

MAXIMUM ΔP:	Unknown
MAXIMUM ΔI:	Unknown

COMMENTS

This is 21st breakup event for an object of this class, and the third in 1999. The breakups are believed to be caused by residual propellants. Russian officials have been aware of the problem since 1992 and have made design changes, although the date of full implementation is unknown. This is the most recently launched object to breakup (age = ~3.5 years). The environmental consequence of the breakup was short-lived; the object was in catastrophic decay from a geosynchronous transfer orbit.

REFERENCE DOCUMENTS

The Fragmentation of Proton Debris, D. J. Nauer, TBE Technical Report CS93-LKD-004, Teledyne Brown Engineering, Colorado Springs, 31 December 1992.

Analysis of Fragmentations From December 1992 - February 1993, TBE Technical Report CS93-LKD-010, Teledyne Brown Engineering, Colorado Springs, 30 March 1993.

Insufficient data to construct a Gabbard diagram.

COSMOS 2343   1997-024A   24805

SATELLITE DATA

TYPE:	Payload
OWNER:	CIS
LAUNCH DATE:	15.51 May 1997
DRY MASS (KG):	6000
MAIN BODY:	Cylinder; 2.4 m diameter by 7 m length
MAJOR APPENDAGES:	Solar arrays
ATTITUDE CONTROL:	Active, 3-axis
ENERGY SOURCES:	On-board propellants, explosive charge

EVENT DATA

DATE:	16 Sep 1997	LOCATION:	58.2 N, 157.5 E (asc.)
TIME:	2208 GMT	ASSESSED CAUSE:	Deliberate
ALTITUDE:	230 km		

PRE-EVENT ELEMENTS

EPOCH:	97258.16080604	MEAN ANOMALY:	247.0345
RIGHT ASCENSION:	1.1478	MEAN MOTION:	16.06645410
INCLINATION:	64.8485	MEAN MOTION DOT/2:	0.00206295
ECCENTRICITY:	0.0048612	MEAN MOTION DOT DOT/6:	0.000026376
ARG. OF PERIGEE:	113.5945	BSTAR:	0.00022999

DEBRIS CLOUD DATA

MAXIMUM ΔP:	7.3 min
MAXIMUM ΔI:	0.9 deg

COMMENTS

Cosmos 2343 was the sixth of the Cosmos 2031 class of spacecraft that debuted in 1989 but was not flown since 1993. In all five previous missions (1989-1993), the spacecraft was deliberately exploded at the end of mission. Previous missions of this type include Cosmos 2031, Cosmos 2101, Cosmos 2163, Cosmos 2225, and Cosmos 2262. All such events have occurred over Eastern Russia. This event, as with three of the previous events, occurred over the Kamchatka Peninsula. Highest previous piece count for large debris for this class of vehicle was 180, although more were probably created. Due to the low altitude of the breakup, the debris were short-lived.

REFERENCE DOCUMENT

"International LEO Spacecraft Breakup in September", N. Johnson, The Orbital Debris Quarterly News, NASA JSC, October 1997, p. 2. Available online at http://www.orbitaldebris.jsc.nasa.gov/newsletter/pdfs/ODQNv2i4.pdf.

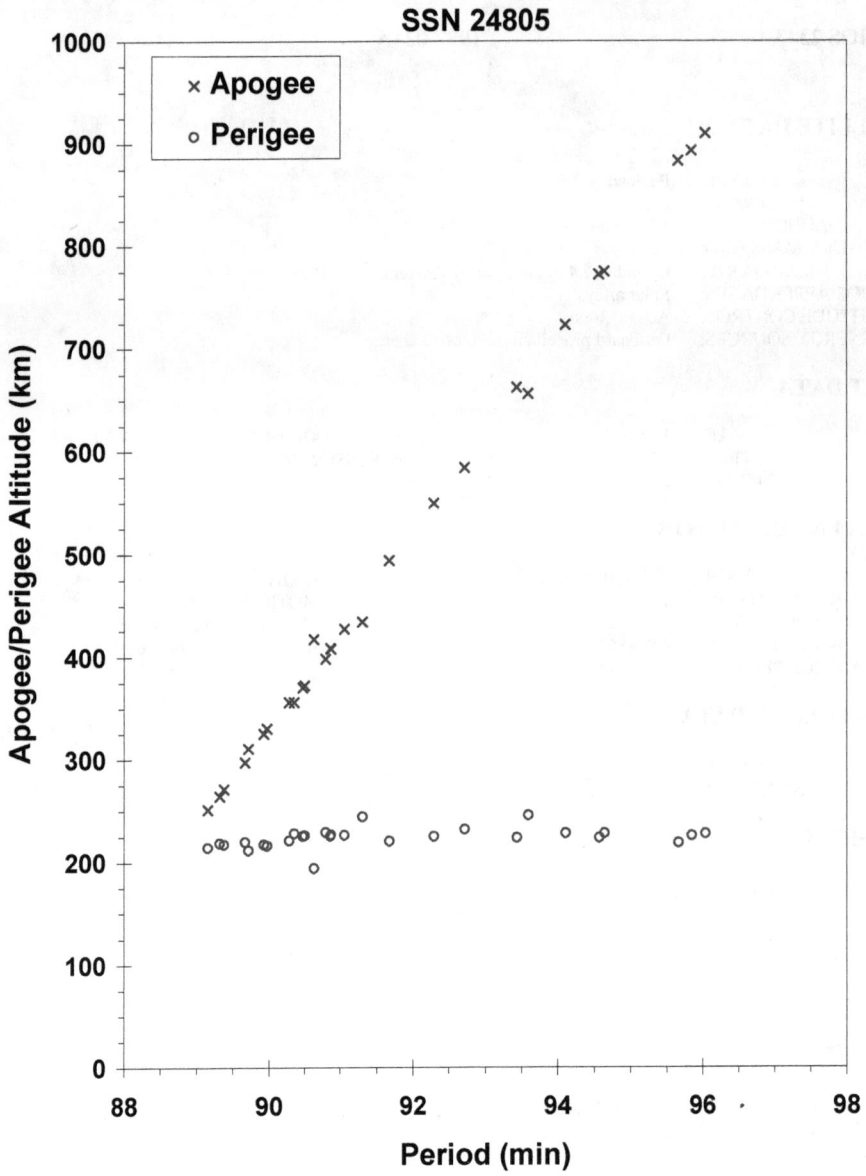

SSN 24805

Cosmos 2343 debris cloud of 28 fragments as reconstructed from the US SSN database.

KUPON ULLAGE MOTOR   1997-070F   **25054**

SATELLITE DATA

TYPE:	Mission Related Debris
OWNER:	CIS
LAUNCH DATE:	12.71 Nov 1997
DRY MASS (KG):	55
MAIN BODY:	Ellipsoid; 0.6 m diameter by 1 m length
MAJOR APPENDAGES:	None
ATTITUDE CONTROL:	None
ENERGY SOURCES:	On-board propellant

EVENT DATA

DATE:	14 Feb 2007	LOCATION:	Unknown
TIME:	Unknown	ASSESSED CAUSE:	Propulsion
ALTITUDE:	Unknown		

PRE-EVENT ELEMENTS

EPOCH:	07044.95683864	MEAN ANOMALY:	35.6246
RIGHT ASCENSION:	14.2124	MEAN MOTION:	5.48131368
INCLINATION:	46.649	MEAN MOTION DOT/2:	0.00001445
ECCENTRICITY:	.5113669	MEAN MOTION DOT DOT/6:	0.0
ARG. OF PERIGEE:	267.6957	BSTAR:	0.00013146

DEBRIS CLOUD DATA

MAXIMUM ΔP:	Unknown
MAXIMUM ΔI:	Unknown

COMMENTS

The object was in a decaying geosynchronous transfer orbit; this event marks the 36th known breakup of a Proton Block DM SOZ ullage motor since 1984. By accident, an observer in Finland captured about 20 of the debris with two CCD cameras less than 24 hours after the event. The debris would have been too faint to be seen with the naked eye. An estimated 60 debris were detected by the US SSN.

REFERENCE DOCUMENTS

History of Soviet/Russian Satellite Fragmentations-A Joint U.S.-Russian Investigation, N. L. Johnson et al, Kaman Sciences Corporation, October 1995.

"Four Satellite Breakups in February Add to Debris Population", The Orbital Debris Quarterly News, NASA JSC, April 2007. Available online at http://www.orbitaldebris.jsc nasa.gov/newsletter/pdfs/ODQNv11i2.pdf.

Insufficient data to construct a Gabbard diagram.

COSMOS 2347   1997-079A   25088

SATELLITE DATA

TYPE:	Payload
OWNER:	CIS
LAUNCH DATE:	9.30 Dec 1997
DRY MASS (KG):	3000
MAIN BODY:	1.3 m diameter by 17 m length, plus solar arrays
MAJOR APPENDAGES:	Solar arrays
ATTITUDE CONTROL:	Active, 3-axis
ENERGY SOURCES:	On-board propellants, explosive charge?

EVENT DATA

DATE:	22 Nov 1999	LOCATION:	31.6N, 4.3E
TIME:	0440 GMT	ASSESSED CAUSE:	Unknown
ALTITUDE:	370 km		

PRE-EVENT ELEMENTS

EPOCH:	99325.85267585	MEAN ANOMALY:	85.1293
RIGHT ASCENSION:	332.8746	MEAN MOTION:	15.83563975
INCLINATION:	65.0115	MEAN MOTION DOT/2:	.00295116
ECCENTRICITY:	.0134056	MEAN MOTION DOT DOT/6:	.000036131
ARG. OF PERIGEE:	273.4567	BSTAR:	.00065869

DEBRIS CLOUD DATA

MAXIMUM ΔP:	2.834 min
MAXIMUM ΔI:	0.22 deg

COMMENTS

Cosmos 2347 was the 19th spacecraft of this type known to have experienced a major fragmentation. Such events were common prior to 1988, but only three breakups have occurred during the past 12 years: Cosmos 2347, Cosmos 2238 (1 Dec 1994), and Cosmos 2313 (26 June 1997). In this case Cosmos 2347 had performed a standard end-of-mission maneuver on 19 November 1999, a little more than 2 days before the breakup. Extensive analyses of these events have been conducted, although the cause is still unknown in the open literature.

A second breakup of Cosmos 2347 was discovered on 10 December when the spacecraft's orbit had decayed to 175 km by 250 km. Three dozen new debris were detected after the second event, but the very low altitude made it difficult to assess accurately the number of large debris. Prior spacecraft (especially Cosmos 1220, 1260, and 1306) also experienced multiple fragmentations.

REFERENCE DOCUMENT

"Satellite Breakups Increase in Last Quarter of 1999", The Orbital Debris Quarterly News, NASA JSC, January 2000. Available online at http://www.orbitaldebris.jsc nasa.gov/newsletter/pdfs/ODQNv5i1.pdf.

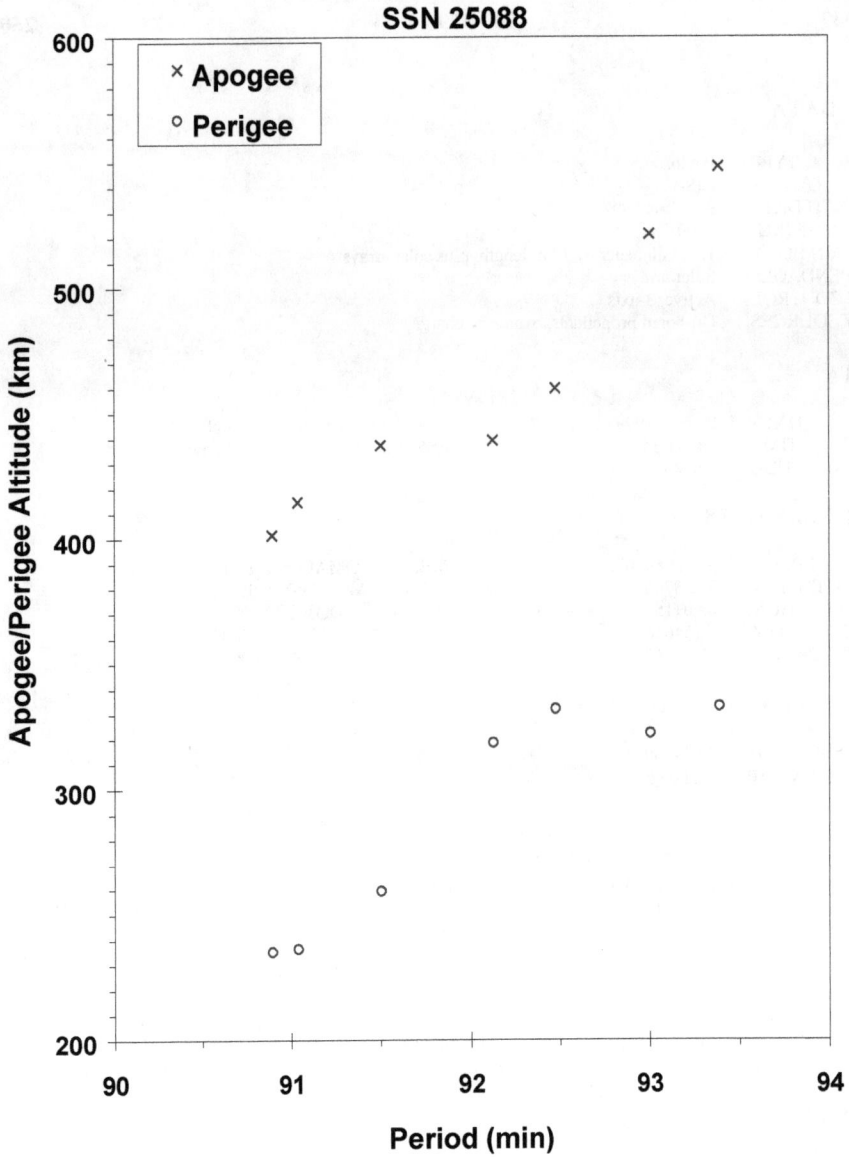

Cosmos 2347 debris cloud of 7 fragments within 1 day of the event as reconstructed from the US SSN database.

SATELLITE DATA

TYPE:	Proton Block DM Fourth Stage
OWNER:	CIS
LAUNCH DATE:	24.97 Dec 1997
DRY MASS (KG):	3400 (?)
MAIN BODY:	Cylinder; 3.7 m diameter by 6.3 m length
MAJOR APPENDAGES:	None
ATTITUDE CONTROL:	Active, 3-axis
ENERGY SOURCES:	On-board propellants

EVENT DATA

DATE:	25 Dec 1997	LOCATION:	0.3 S, 91.2 E (dsc)
TIME:	0550 GMT	ASSESSED CAUSE:	Propulsion
ALTITUDE:	35995 km		

POST-EVENT ELEMENTS

EPOCH:	97359.90803672	MEAN ANOMALY:	359.8589
RIGHT ASCENSION:	92.0594	MEAN MOTION:	2.26580509
INCLINATION:	51.4499	MEAN MOTION DOT/2:	-0.00000507
ECCENTRICITY:	0.7304004	MEAN MOTION DOT DOT/6:	0
ARG. OF PERIGEE:	1.0181	BSTAR:	0

DEBRIS CLOUD DATA

MAXIMUM ΔP:	Unknown
MAXIMUM ΔI:	Unknown

COMMENTS

The commercial Asiasat 3 spacecraft was launched by a Proton launch vehicle at 2319 GMT, 24 December 1997. The fourth stage completed its first burn successfully about 0035 GMT, 25 December, placing the R/B-S/C combination into a GTO. When the vehicle reached its first apogee, the main engine was restarted but shutdown within 1 second, apparently due to a catastrophic failure. The US Space Surveillance Network detected less than 10 objects, and by 9 January only 1-2 were still being observed. The fragmentation is similar to the breakup of the Raduga 33 upper stage on 19 Feb 1996. In that case, nearly 200 debris were detected by the SSN.

REFERENCE DOCUMENT

"Recent Satellite Fragmentation Investigations", N. Johnson, The Orbital Debris Quarterly News, January 1998, p. 3. Available online at http://www.orbitaldebris.jsc.nasa.gov/newsletter/pdfs/ODQNv3i1.pdf.

Insufficient data to construct a Gabbard diagram.

COMETS R/B   1998-011B   25176

SATELLITE DATA

TYPE:	H-II Second Stage
OWNER:	Japan
LAUNCH DATE:	21.33 Feb 1998
DRY MASS (KG):	3000
MAIN BODY:	Spheroid + cylinder + cone; 4 m diameter by 10.1 m length
MAJOR APPENDAGES:	None
ATTITUDE CONTROL:	Active, 3-axis
ENERGY SOURCES:	On-board propellants

EVENT DATA

DATE:	21 Feb 1998	LOCATION:	Unknown
TIME:	Unknown	ASSESSED CAUSE:	Propulsion
ALTITUDE:	Unknown		

POST-EVENT ELEMENTS

EPOCH:	98054.59975400	MEAN ANOMALY:	162.0601
RIGHT ASCENSION:	294.3031	MEAN MOTION:	13.51967368
INCLINATION:	30.0458	MEAN MOTION DOT/2:	0.0002873
ECCENTRICITY:	0.1097485	MEAN MOTION DOT DOT/6:	-0.000003104
ARG. OF PERIGEE:	194.5714	BSTAR:	0.00029603

DEBRIS CLOUD DATA

MAXIMUM ΔP:	Unknown
MAXIMUM ΔI:	Unknown

COMMENTS

According to a NASDA report, a welding failure caused the LE-5A engine shutdown that stranded the Japanese COMETS satellite (1998-011A) in an elliptical orbit. The failure occurred 47 seconds into an apogee-raising maneuver. The report determined from telemetry data that hot combustion gases managed to penetrated special welding, called brazing, between nickel alloy cooling tubes in the lowest part of the combustion chamber near the top of the engine's nozzle skirt. Burning through the tubes, combustion gases quickly caused a fire, which triggered the engine shutdown. The report concludes the accident was caused by a manufacturing flaw and not a fundamental design problem. At least three dozen debris were detected by optical sensors in Hawaii.

REFERENCE DOCUMENT

"The Upper Stage Breakups in One Week Top February Debris Activity", The Orbital Debris Quarterly News, NASA JSC, April 1998, p. 1. Available online at http://www.orbitaldebris.jsc.nasa.gov/newsletter/pdfs/ODQNv3i2.pdf.

Insufficient data to construct a Gabbard diagram.

FENGYUN 1C   **1999-025A**   **25730**

SATELLITE DATA

TYPE:	Payload
OWNER:	PRC
LAUNCH DATE:	10.06 May 1999
DRY MASS (KG):	950
MAIN BODY:	Box; 1.5 m by 1.5 m by 1.5 m
MAJOR APPENDAGES:	Solar Panels, 1.5 m by 4 m
ATTITUDE CONTROL:	Active, 3-axis
ENERGY SOURCES:	On-board propellants

EVENT DATA

DATE:	11 Jan 2007	LOCATION:	35N, 100E (asc)
TIME:	2226 GMT	ASSESSED CAUSE:	Deliberate
ALTITUDE:	860 km		

PRE-EVENT ELEMENTS

EPOCH:	07011.90621003	MEAN ANOMALY:	94.0215
RIGHT ASCENSION:	1.7411	MEAN MOTION:	14.11820274
INCLINATION:	98.6464	MEAN MOTION DOT/2:	0.00000180
ECCENTRICITY:	.0013513	MEAN MOTION DOT DOT/6:	0.0
ARG. OF PERIGEE:	266.0357	BSTAR:	0.00012153

DEBRIS CLOUD DATA

MAXIMUM ΔP:	33.4 min
MAXIMUM ΔI:	5.0 deg

COMMENTS

The debris cloud created by this anti-satellite test represents the worst contamination of low Earth orbit in history. More than half the identified debris were thrown into orbits exceeding a mean altitude of 850 km, meaning that much of the 10 cm and larger debris will be in orbit for decades or centuries.

REFERENCE DOCUMENT

"Chinese Anti-satellite Test Creates Most Severe Orbital Debris Cloud in History", The Orbital Debris Quarterly News, NASA JSC, April 2007.
Available online at http://www.orbitaldebris.jsc nasa.gov/newsletter/pdfs/ODQNv11i2.pdf.

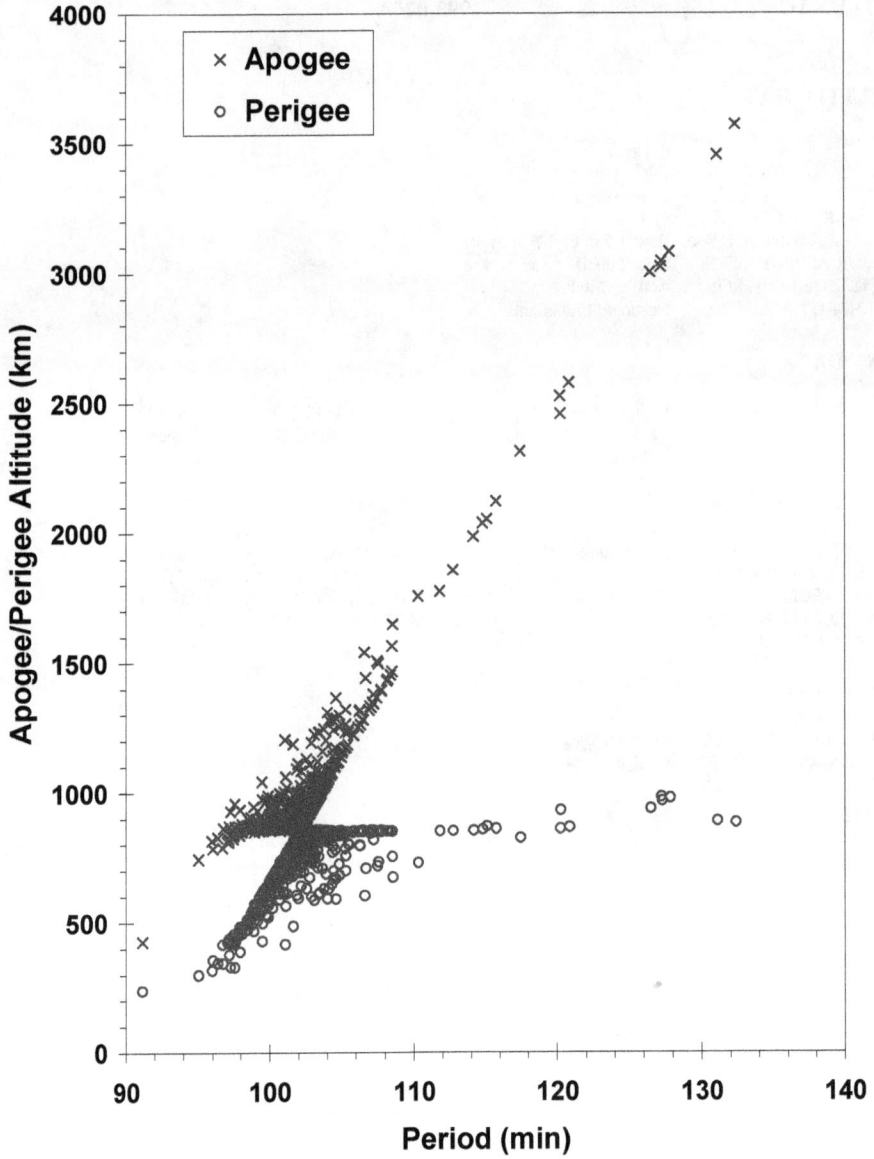

Fengyun 1C debris cloud remnant of 2000+ cataloged fragments 6 months after the event as reconstructed from the US SSN database.

CBERS 1   1999-057A   25940

SATELLITE DATA

TYPE:	Payload
OWNER:	PRC/Brazil
LAUNCH DATE:	14.14 Oct 1999
DRY MASS (KG):	1450
MAIN BODY:	Box: 1.8 m by 2.0 m by 2.2 m
MAJOR APPENDAGES:	6.3 m x 2.6 m Solar Panels
ATTITUDE CONTROL:	None at time of event
ENERGY SOURCES:	On-board propellant?

EVENT DATA

DATE:	18 Feb 2007	LOCATION:	35N, 128E (asc)
TIME:	1107 GMT	ASSESSED CAUSE:	Unknown
ALTITUDE:	780 km		

PRE-EVENT ELEMENTS

EPOCH:	07049.17726620	MEAN ANOMALY:	250.9413
RIGHT ASCENSION:	88.9135	MEAN MOTION:	14.34483847
INCLINATION:	98.2175	MEAN MOTION DOT/2:	-0.00000185
ECCENTRICITY:	.0007520	MEAN MOTION DOT DOT/6:	0.0
ARG. OF PERIGEE:	109.2997	BSTAR:	-0.000051172

DEBRIS CLOUD DATA

MAXIMUM ΔP:	3.4 min
MAXIMUM ΔI:	0.3 deg

COMMENTS

The spacecraft exceeded its expected lifetime and had been moved to a retirement orbit in August 2003. It is unclear whether or not it was entirely passivated. A total of 60 debris had been officially cataloged by late 2007. The rocket body associated with this payload (1999-057C, 25942) experienced an unrelated, but significant fragmentation six months after launch, creating over 300 pieces of trackable debris, of which more than half remain in orbit today.

REFERENCE DOCUMENT

"Four Satellite Breakups in February Add to Debris Population", The Orbital Debris Quarterly News, NASA JSC, April 2007. Available online at http://www.orbitaldebris.jsc nasa.gov/newsletter/pdfs/ODQNv11i2.pdf.

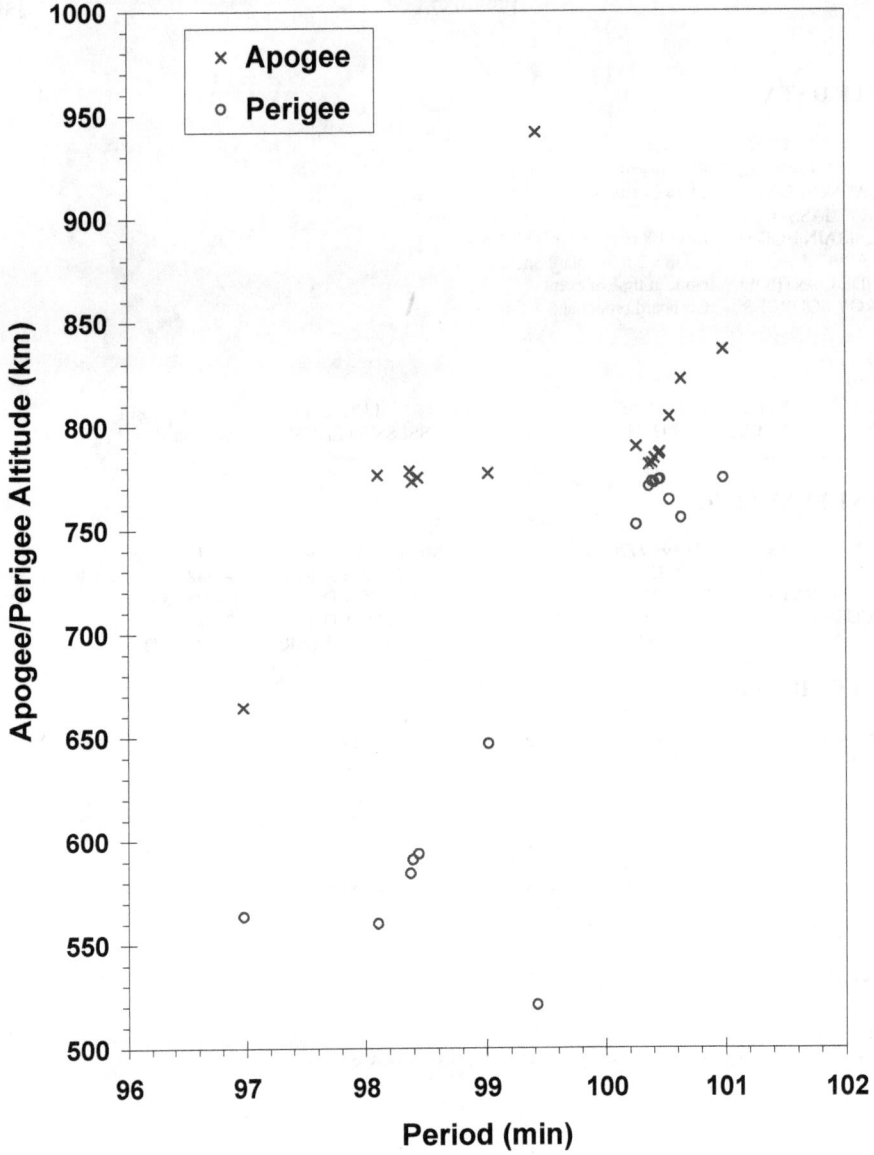

SSN 25940

CBERS 1 debris cloud of 16 cataloged fragments 2 weeks after the event as reconstructed from the US SSN database.

CBERS-1/SACI-1 R/B   **1999-057C**   **25942**

SATELLITE DATA

TYPE:	Long March 4 third stage
OWNER:	PRC
LAUNCH DATE:	14.14 Oct 1999
DRY MASS (KG):	1000
MAIN BODY:	Cylinder nozzle; 2.9 m diameter by ~5 m length
MAJOR APPENDAGES:	None
ATTITUDE CONTROL:	None
ENERGY SOURCES:	On-board propellants

EVENT DATA

DATE:	11 Mar 2000	LOCATION:	51.2S, 48.5W
TIME:	1304 UTC	ASSESSED CAUSE:	Propulsion
ALTITUDE:	741 km		

PRE-EVENT ELEMENTS

EPOCH:	00069.14898026	MEAN ANOMALY:	43.0989
RIGHT ASCENSION:	145.5131	MEAN MOTION:	14.46866365
INCLINATION:	98.5373	MEAN MOTION DOT/2:	.00001603
ECCENTRICITY:	.0012467	MEAN MOTION DOT DOT/6:	.0
ARG. OF PERIGEE:	316.9224	BSTAR:	.00045410

DEBRIS CLOUD DATA

MAXIMUM ΔP:	10.985 min
MAXIMUM ΔI:	0.99 deg

COMMENTS

This is the second Long March 4 to breakup in only four missions. The first breakup (flight 2) occurred on 4 Oct 1990, 1 month after launch. Long March 4 missions did not resume until 1999, when two more were flown. This breakup involved the second 1999 mission (flight 4) and occurred 5 months after launch. This event has created more trackable debris than the 1990 breakup, with more than 300 pieces tracked by the SSN. Chinese officials were aware of the international concern following the 1990 breakup and had pledged to adopt countermeasures before the 1999 missions. Passivation of this vehicle was attempted.

REFERENCE DOCUMENT

"Analyzing the Cause of LM-4 (A)'s Upper Stage's Disintegration and the Countermeasures", W. X. Zang and S. Y. Liao, 5th International Conference of Pacific Basin Societies, 6-9 Jun 1993, Shanghai.

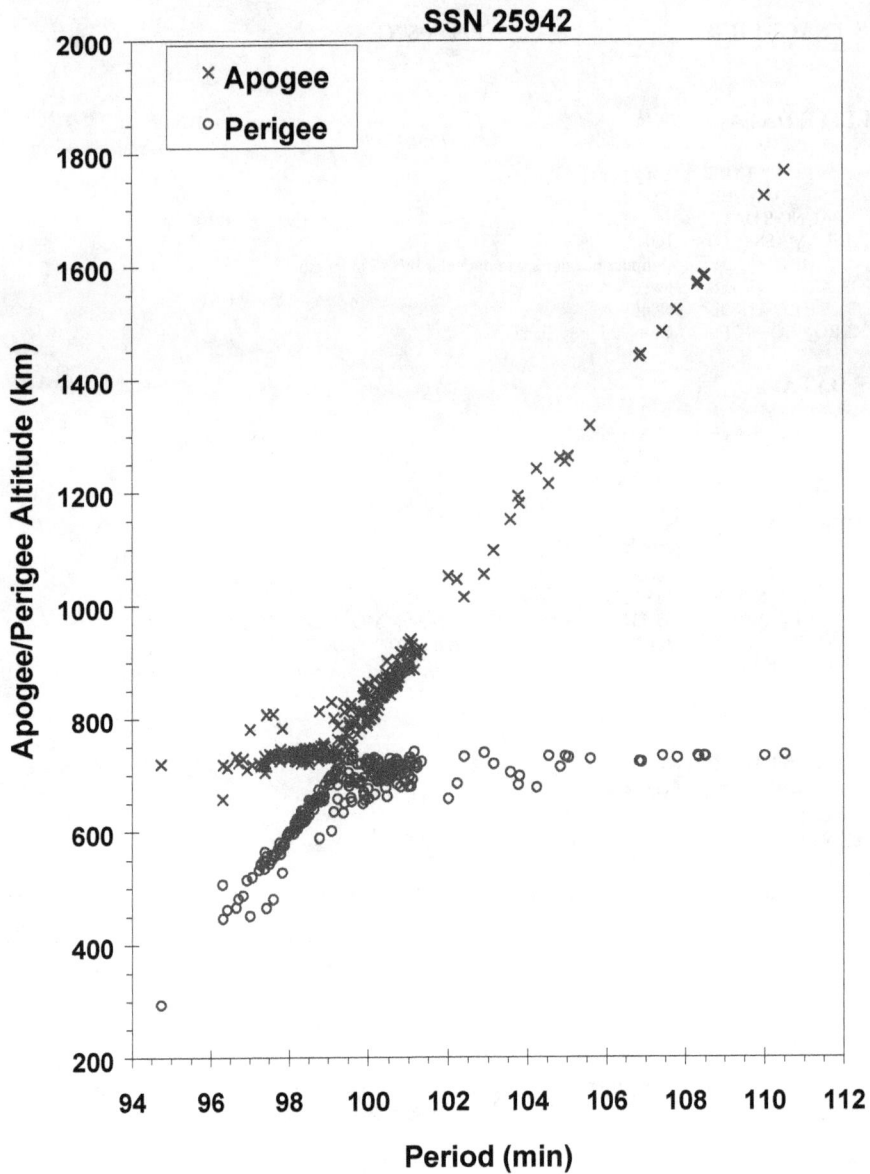

CBERS-1 / SACI-1 long March 4 third stage debris cloud of 280 fragments within 1 day of the event as reconstructed from the US SSN database.

COSMOS 2367	1999-072A	26040

SATELLITE DATA

TYPE:	Payload
OWNER:	CIS
LAUNCH DATE:	26.33 Dec 1999
DRY MASS (KG):	~3000
MAIN BODY:	Cylinder; 1.3 m diameter by 17 m length, plus solar arrays
MAJOR APPENDAGES:	Solar Arrays
ATTITUDE CONTROL:	Active 3-axis
ENERGY SOURCES:	On-board propellant, explosive charge

EVENT DATA

DATE:	21 Nov 2001	LOCATION:	38.3 S, 17.7 W
TIME:	1414Z	ASSESSED CAUSE:	Unknown
ALTITUDE:	410 km		

PRE-EVENT ELEMENTS

EPOCH:	03325.57054648	MEAN ANOMALY:	199.8631
RIGHT ASCENSION:	55.0233	MEAN MOTION:	15.51939724
INCLINATION:	65.0021	MEAN MOTION DOT/2:	.00131711
ECCENTRICITY:	.0008788	MEAN MOTION DOT DOT/6:	.0
ARG. OF PERIGEE:	257.3641	BSTAR:	.0021441

DEBRIS CLOUD DATA

MAXIMUM ΔP:	10.62 min*
MAXIMUM ΔI:	1.28 deg*

* Based on uncataloged debris data

COMMENTS

Cosmos 2367 was the 20th spacecraft of this type (Cosmos 699 class) known to have experienced a major fragmentation. The previous spacecraft in this series was Cosmos 2347, which experienced two fragmentations, one each in Nov and Dec of 1999. Cosmos 2367 was still in its operational orbit at the time of the event. Over 100 pieces were detected by the SSN 1 week after the breakup. Based upon other observations, the actual number of pieces probably exceeded 300. Although some debris were thrown into orbits with apogees above 1000 km, in general the debris were short-lived.

REFERENCE DOCUMENT

"Two Major Satellite Breakups Near End of 2001", The Orbital Debris Quarterly News, NASA JSC, January 2002. Available online at http://www.orbitaldebris.jsc.nasa.gov/newsletter/pdfs/ODQNv7i1.pdf.

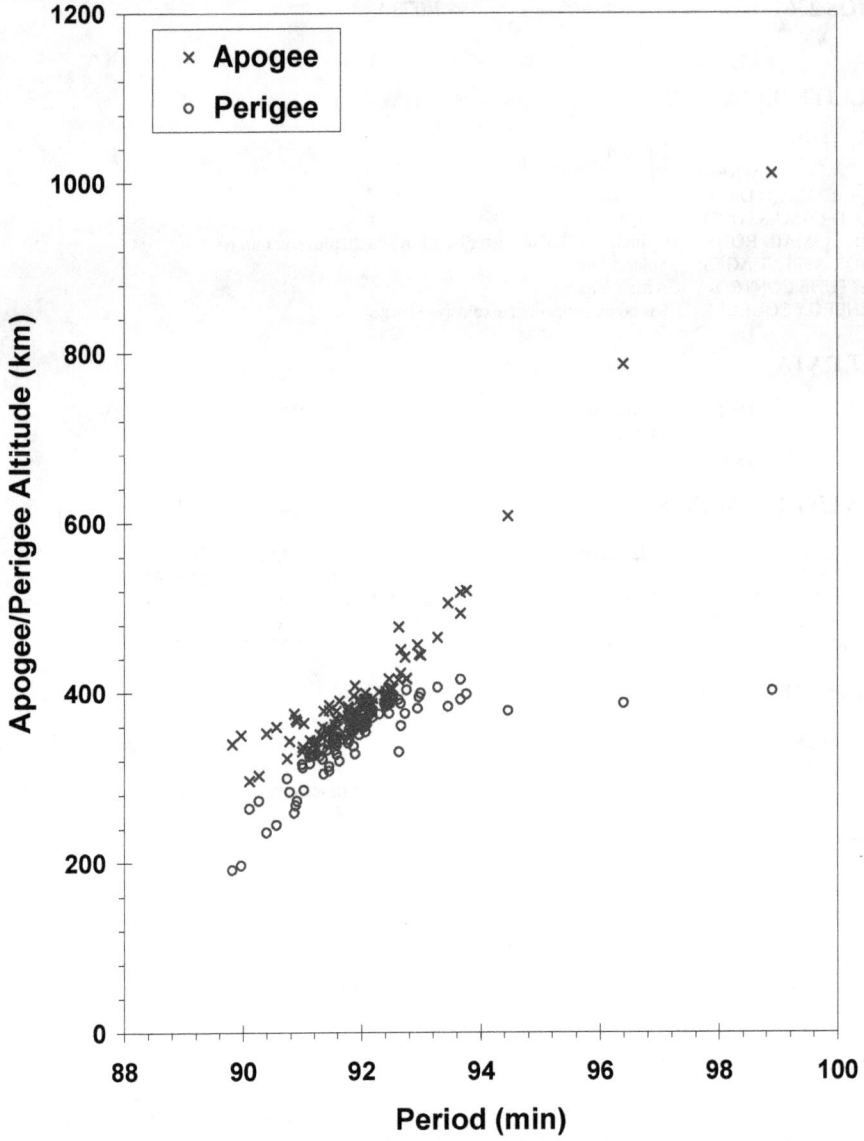

SSN 26040

Cosmos 2367 debris cloud of 103 fragments 1 week after the breakup as reconstructed from the US SSN database.

COSMOS 2371 ULLAGE MOTOR   2000-036E   26398

SATELLITE DATA

TYPE:	Mission Related Debris
OWNER:	CIS
LAUNCH DATE:	4.99 Jul 2000
DRY MASS (KG):	55
MAIN BODY:	Ellipsoid; 0.6 m diameter by 1 m length
MAJOR APPENDAGES:	None
ATTITUDE CONTROL:	None
ENERGY SOURCES:	On-board propellant

EVENT DATA

DATE:	~1 Sep 2006	LOCATION:	Unknown
TIME:	Unknown	ASSESSED CAUSE:	Propulsion
ALTITUDE:	Unknown		

PRE-EVENT ELEMENTS

EPOCH:	06244.59746638	MEAN ANOMALY:	38.1280
RIGHT ASCENSION:	18.3906	MEAN MOTION:	3.86574836
INCLINATION:	46.8834	MEAN MOTION DOT/2:	0.00005467
ECCENTRICITY:	.6151900	MEAN MOTION DOT DOT/6:	0.0
ARG. OF PERIGEE:	248.4110	BSTAR:	0.00034737

DEBRIS CLOUD DATA

MAXIMUM ΔP:	98.8 min
MAXIMUM ΔI:	0.2 deg

COMMENTS

The object was in a decaying geosynchronous transfer orbit; this event marks the 35th known breakup of a Proton Block DM SOZ ullage motor since 1984. Only a handful of debris was detected from this event.

REFERENCE DOCUMENTS

History of Soviet/Russian Satellite Fragmentations-A Joint U.S.-Russian Investigation, N. L. Johnson et al, Kaman Sciences Corporation, October 1995.

"Three More Satellites Involved in Fragmentations", The Orbital Debris Quarterly News, NASA JSC, October 2006. Available online at http://www.orbitaldebris.jsc nasa.gov/newsletter/pdfs/ODQNv10i4.pdf.

SSN 26398

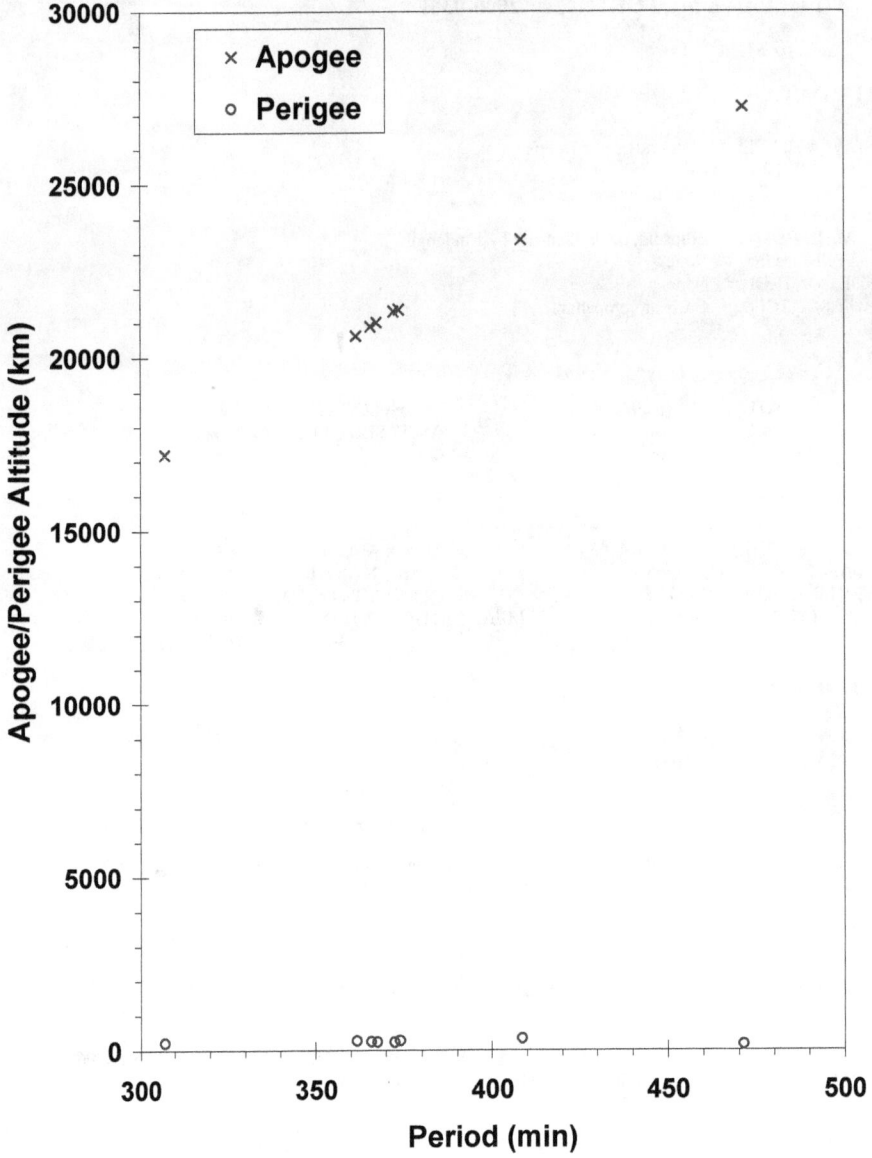

Cosmos 2371 SOZ Motor debris cloud of 7 cataloged fragments plus the parent a day after the event. The debris pieces were never cataloged.

TES R/B   **2001-049D**   **26960**

SATELLITE DATA

TYPE:	PSLV Final Stage
OWNER:	India
LAUNCH DATE:	22.20 Oct 2001
DRY MASS (KG):	~900
MAIN BODY:	Cylinder; 2.8 m diameter by 2.6 m length
MAJOR APPENDAGES:	None
ATTITUDE CONTROL:	None
ENERGY SOURCES:	On-board hypergolic propellants

EVENT DATA

DATE:	19 Dec 2001	LOCATION:	25 S, 340 E
TIME:	~1140Z	ASSESSED CAUSE:	Propulsion
ALTITUDE:	670 km		

PRE-EVENT ELEMENTS

EPOCH:	01352.90695581	MEAN ANOMALY:	316.4909
RIGHT ASCENSION:	65.6004	MEAN MOTION:	14.85657962
INCLINATION:	97.9010	MEAN MOTION DOT/2:	-.00000443
ECCENTRICITY:	.0088752	MEAN MOTION DOT DOT/6:	.0
ARG. OF PERIGEE:	44.3375	BSTAR:	-.000041058

DEBRIS CLOUD DATA

MAXIMUM ΔP:	9.86 min*
MAXIMUM ΔI:	3.06 deg*

* Based on uncataloged debris data

COMMENTS

This is the first known breakup associated with the PSLV fourth stage. While 332 fragments were initially detected by the SSN, 326 debris were cataloged. The vehicle employed hypergolic propellants that were not passivated after payload delivery. Some of the debris could remain in orbit for several years or longer.

REFERENCE DOCUMENT

"Two Major Satellite Breakups Near End of 2001", The Orbital Debris Quarterly News, NASA JSC, January 2002. Available online at http://www.orbitaldebris.jsc.nasa.gov/newsletter/pdfs/ODQNv7i1.pdf.

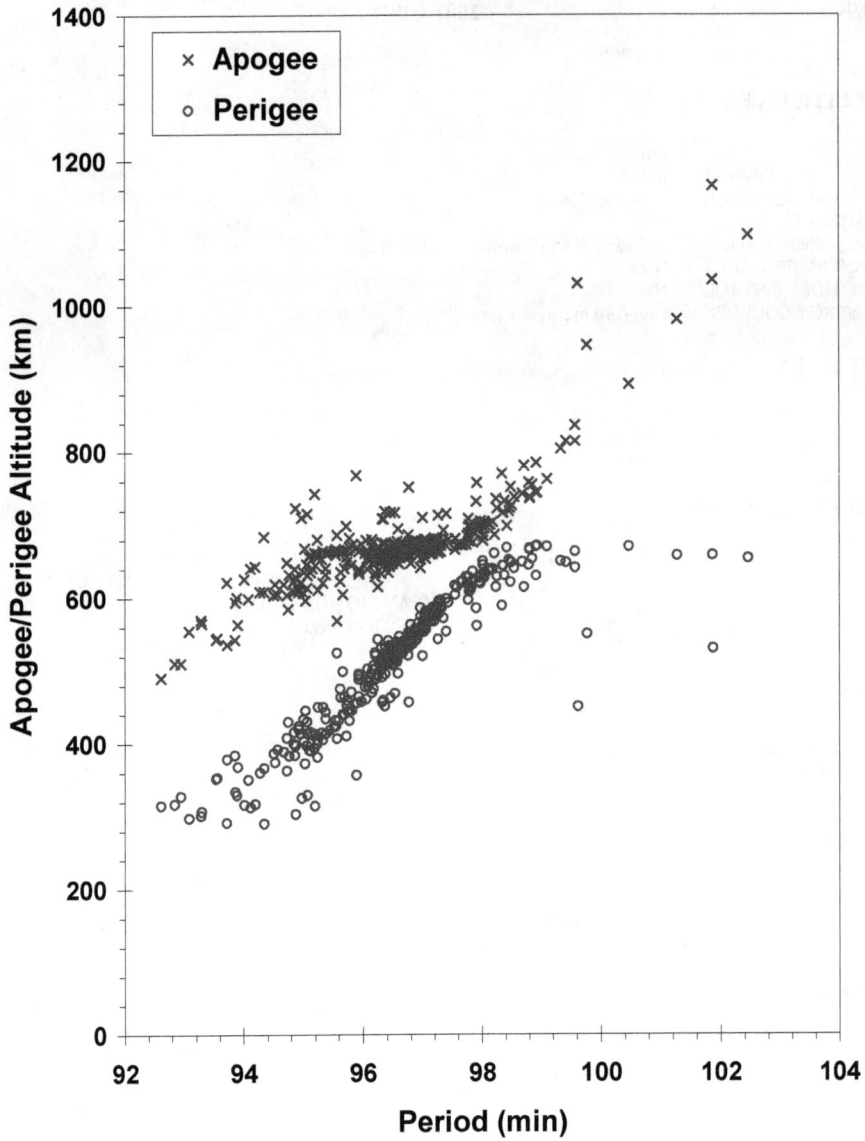

SSN 26960

TES R/B debris cloud of 332 fragments 1 weeks after the breakup as reconstructed from the US SSN database.

COSMOS 2383   2001-057A   27053

SATELLITE DATA

TYPE:	Payload
OWNER:	CIS
LAUNCH DATE:	21.17 Dec 2001
DRY MASS (KG):	3000
MAIN BODY:	Cylinder; 1.3 m diameter by 17 m length, plus solar arrays
MAJOR APPENDAGES:	Solar Arrays
ATTITUDE CONTROL:	Active 3-axis
ENERGY SOURCES:	On-board propellant; explosive charge

EVENT DATA

DATE:	28 Feb 2004	LOCATION:	26S, 100E (dsc)
TIME:	~1930 GMT	ASSESSED CAUSE:	Unknown
ALTITUDE:	265 km		

PRE-EVENT ELEMENTS

EPOCH:	04059.20843210	MEAN ANOMALY:	95.7196
RIGHT ASCENSION:	358.8049	MEAN MOTION:	15.87352021
INCLINATION:	64.9763	MEAN MOTION DOT/2:	0.00416036
ECCENTRICITY:	.0131275	MEAN MOTION DOT DOT/6:	0.000069430
ARG. OF PERIGEE:	262.9022	BSTAR:	0.00074756

DEBRIS CLOUD DATA

MAXIMUM ΔP:	3.4 min
MAXIMUM ΔI:	0.3 deg

COMMENTS

Cosmos 2383 was the 21st spacecraft of this type (Cosmos 699 class) known to have experienced a major fragmentation. The previous spacecraft in this series was Cosmos 2367, which experienced a fragmentation at an altitude just above the International Space Station in 2001. This event also produced debris crossing the ISS orbit altitude. Approximately 50 debris were detected by the US SSN. Fortunately, all the cataloged debris from this fragmentation were short-lived. Of the 48 spacecraft of this type, 21 have experienced fragmentation events.

REFERENCE DOCUMENT

"Fragmentation of Cosmos 2383", The Orbital Debris Quarterly News, NASA JSC, April 2004.
Available online at http://www.orbitaldebris.jsc nasa.gov/newsletter/pdfs/ODQNv8i2.pdf.

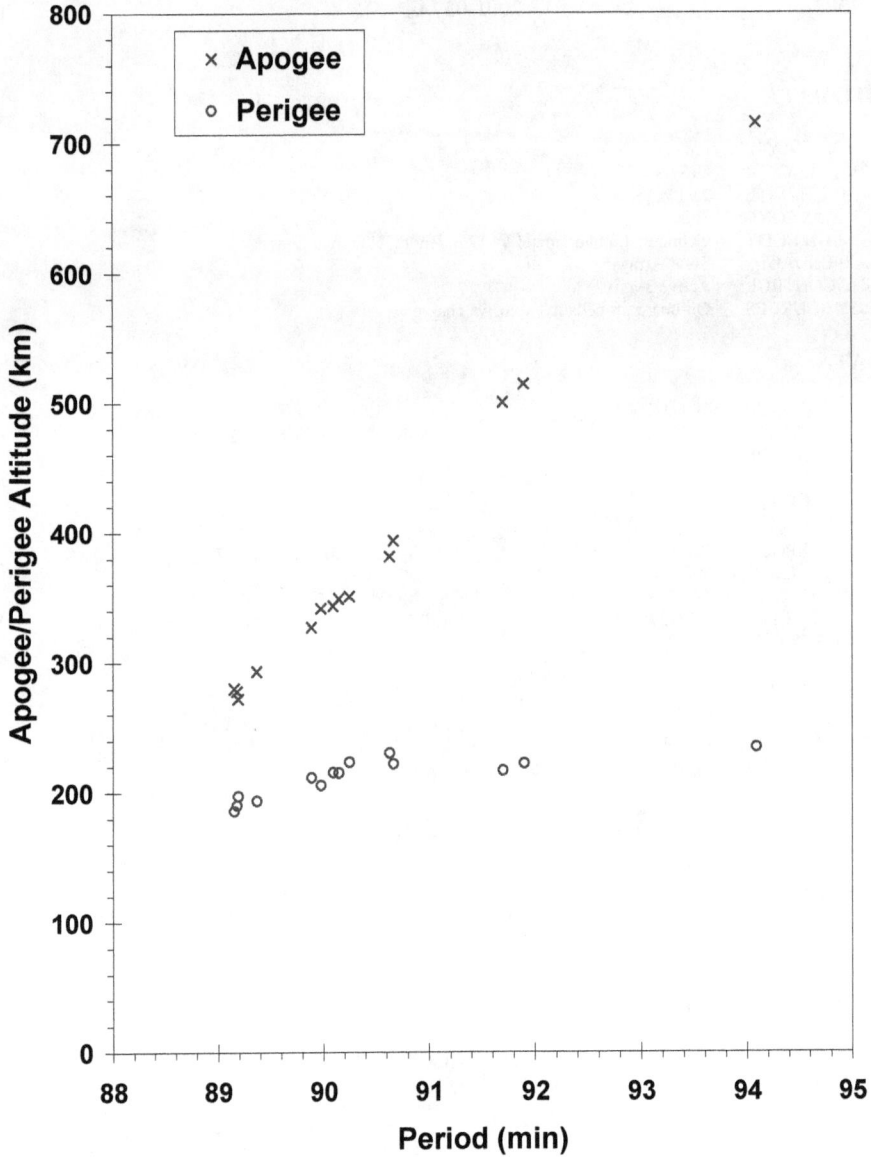

SSN 27053

Cosmos 2383 debris cloud of 14 cataloged fragments a few days after the event as reconstructed from the US SSN database.

COSMOS 2392 ULLAGE MOTOR   2002-037E   27474

SATELLITE DATA

TYPE:	Mission Related Debris
OWNER:	CIS
LAUNCH DATE:	25.63 Jul 2002
DRY MASS (KG):	55
MAIN BODY:	Ellipsoid; 0.6 m diameter by 1 m length
MAJOR APPENDAGES:	None
ATTITUDE CONTROL:	None
ENERGY SOURCES:	On-board propellant

EVENT DATA

DATE:	1 Jun 2005	LOCATION:	Unknown
TIME:	Unknown	ASSESSED CAUSE:	Propulsion
ALTITUDE:	Unknown		

PRE-EVENT ELEMENTS

EPOCH:	05151.71140009	MEAN ANOMALY:	222.2107
RIGHT ASCENSION:	143.5118	MEAN MOTION:	15.06786995
INCLINATION:	63.6569	MEAN MOTION DOT/2:	0.00075759
ECCENTRICITY:	.0418207	MEAN MOTION DOT DOT/6:	0.000007275
ARG. OF PERIGEE:	140.9987	BSTAR:	0.00057187

DEBRIS CLOUD DATA

MAXIMUM ΔP:	5.2 min
MAXIMUM ΔI:	1.0 deg

COMMENTS

This event marks the 33rd known breakup of a Proton Block DM SOZ ullage motor since 1984. The motor experienced a second event a month later about 29 June. About 40 new debris were seen after each event.

REFERENCE DOCUMENT

History of Soviet/Russian Satellite Fragmentations-A Joint U.S.-Russian Investigation, N. L. Johnson et al, Kaman Sciences Corporation, October 1995.

"Recent Satellite Breakups", The Orbital Debris Quarterly News, NASA JSC, July 2005. Available online at http://www.orbitaldebris.jsc nasa.gov/newsletter/pdfs/ODQNv9i3.pdf.

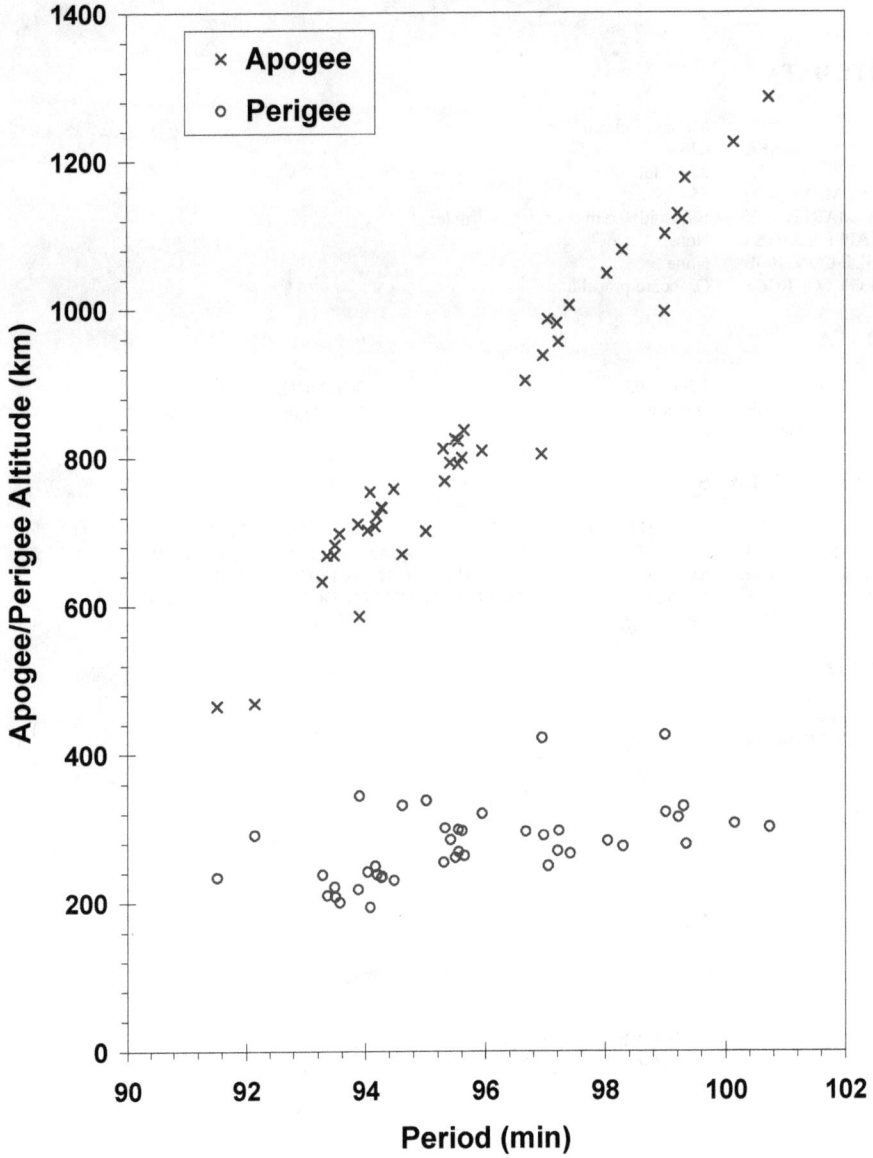

SOZ Motor debris cloud in July 2005 after the second breakup event.

COSMOS 2392 ULLAGE MOTOR   2002-037F   27475

SATELLITE DATA

TYPE:	Mission Related Debris
OWNER:	CIS
LAUNCH DATE:	25.63 Jul 2002
DRY MASS (KG):	55
MAIN BODY:	Ellipsoid; 0.6 m diameter by 1 m length
MAJOR APPENDAGES:	None
ATTITUDE CONTROL:	None
ENERGY SOURCES:	On-board propellant

EVENT DATA

DATE:	29 Oct 2004	LOCATION:	Unknown
TIME:	Unknown	ASSESSED CAUSE:	Propulsion
ALTITUDE:	Unknown		

PRE-EVENT ELEMENTS

EPOCH:	04302.83463691	MEAN ANOMALY:	199.8896
RIGHT ASCENSION:	56.1075	MEAN MOTION:	15.09294101
INCLINATION:	63.6401	MEAN MOTION DOT/2:	0.00146672
ECCENTRICITY:	.0436823	MEAN MOTION DOT DOT/6:	0.000010335
ARG. OF PERIGEE:	161.8395	BSTAR:	0.00073391

DEBRIS CLOUD DATA

MAXIMUM ΔP:	Unknown
MAXIMUM ΔI:	Unknown

COMMENTS

This event marks the 31st known breakup of a Proton Block DM SOZ ullage motor since 1984. More than 60 debris were detected by the Naval electronic fence. No debris were in orbit long enough to be cataloged.

REFERENCE DOCUMENTS

History of Soviet/Russian Satellite Fragmentations-A Joint U.S.-Russian Investigation, N. L. Johnson et al, Kaman Sciences Corporation, October 1995.

"Recent Satellite Breakups", The Orbital Debris Quarterly News, NASA JSC, January 2005.
Available online at http://www.orbitaldebris.jsc nasa.gov/newsletter/pdfs/ODQNv9i1.pdf.

Insufficient data to construct a Gabbard diagram.

COSMOS 2399   2003-035A   27856

SATELLITE DATA

TYPE:	Payload
OWNER:	CIS
LAUNCH DATE:	12.60 Aug 2003
DRY MASS (KG):	~6000
MAIN BODY:	Cylinder; 2.4 m diameter by 7 m length
MAJOR APPENDAGES:	Solar Arrays
ATTITUDE CONTROL:	Active, 3-axis
ENERGY SOURCES:	On-board propellants, explosive charge

EVENT DATA

DATE:	9 Dec 2003	LOCATION:	64.8 N, 135.4 E
TIME:	0129Z	ASSESSED CAUSE:	Deliberate
ALTITUDE:	189.33 km		

PRE-EVENT ELEMENTS

EPOCH:	03342.92270571	MEAN ANOMALY:	296.9639
RIGHT ASCENSION:	136.8172	MEAN MOTION:	16.22926227
INCLINATION:	64.9062	MEAN MOTION DOT/2:	.01025110
ECCENTRICITY:	.0055948	MEAN MOTION DOT DOT/6:	.0000073532
ARG. OF PERIGEE:	63.7269	BSTAR:	.00028689

DEBRIS CLOUD DATA

MAXIMUM ΔP:	7.34 min*
MAXIMUM ΔI:	0.08 deg*

                                        * Based on uncataloged debris data

COMMENTS

Cosmos 2399 was the seventh of the Cosmos 2031 class of spacecraft that debuted in 1989 but was not flown since 1997. In all six previous missions (1989-1997), the spacecraft was deliberately exploded at the end of mission. Previous missions of this type include Cosmos 2031, Cosmos 2101, Cosmos 2163, Cosmos 2225, Cosmos 2262, and Cosmos 2343. All such events have occurred over Eastern Russia. Highest previous piece count for large debris for this class of vehicle was 180, although more were probably created. Approximately 22 debris were detected by the SSN. Due to the low altitude of the breakup, the debris were short-lived.

REFERENCE DOCUMENT

"Satellite Fragmentations in 2003", The Orbital Debris Quarterly News, NASA JSC, January 2004. Available online at http://www.orbitaldebris.jsc nasa.gov/newsletter/pdfs/ODQNv8i1.pdf.

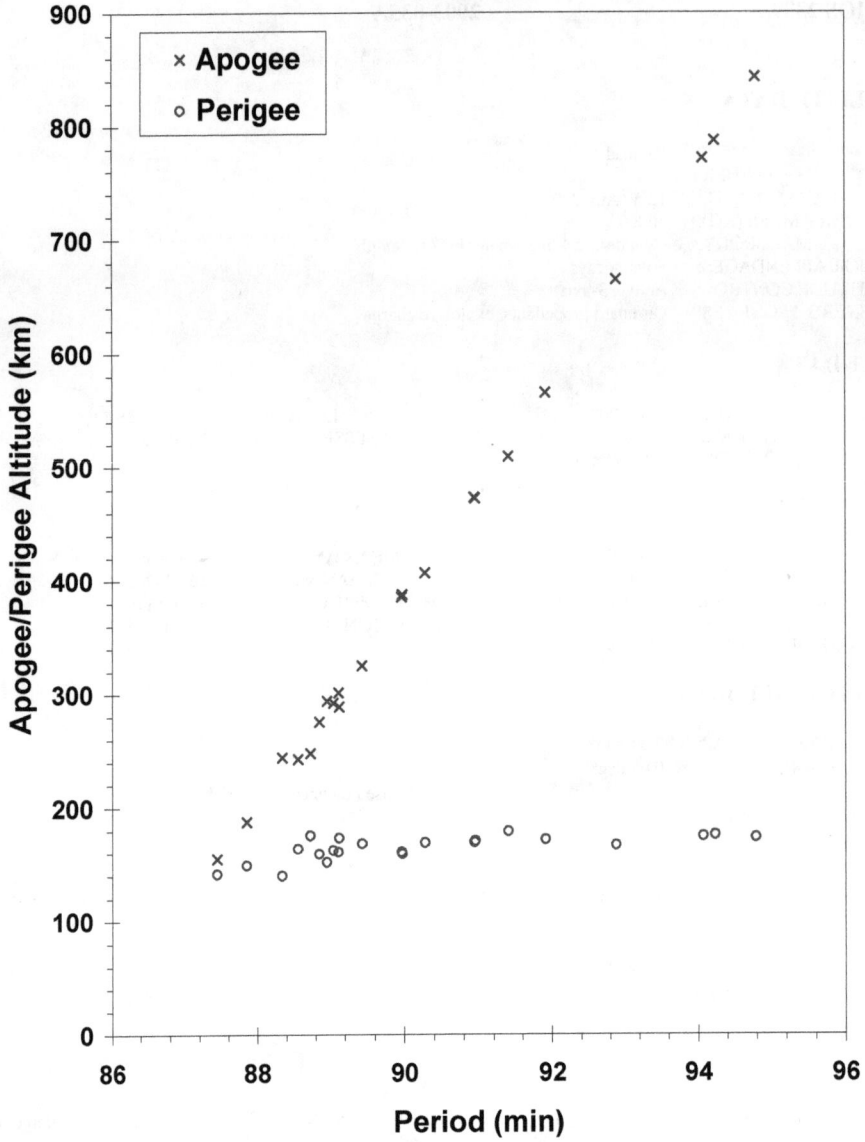

Cosmos 2399 debris cloud of 22 fragments 1 day after the breakup as reconstructed from the US SSN database.

ALOS-1 R/B   2006-002B   28932

SATELLITE DATA

TYPE:	Rocket Body
OWNER:	Japan
LAUNCH DATE:	24.06 Jan 2006
DRY MASS (KG):	~3000
MAIN BODY:	Cylinder; 4.0 m diameter by 10.6 m length
MAJOR APPENDAGES:	None
ATTITUDE CONTROL:	None at time of event
ENERGY SOURCES:	On-board propellant

FIRST EVENT DATA

DATE:	8 Aug 2006	LOCATION:	5N, 131E (asc)
TIME:	1407 GMT	ASSESSED CAUSE:	Unknown
ALTITUDE:	685 km		

SECOND EVENT DATA

DATE:	27 Aug 2006	LOCATION:	76S, 135E (asc)
TIME:	1618 GMT	ASSESSED CAUSE:	Unknown
ALTITUDE:	695 km		

PRE-EVENT ELEMENTS

EPOCH:	06220.18259253	MEAN ANOMALY:	147.5075
RIGHT ASCENSION:	300.3416	MEAN MOTION:	14.8204486
INCLINATION:	98.1944	MEAN MOTION DOT/2:	0.00000143
ECCENTRICITY:	.0106899	MEAN MOTION DOT DOT/6:	0.0
ARG. OF PERIGEE:	211.9623	BSTAR:	0.000023737

DEBRIS CLOUD DATA

MAXIMUM ΔP:	1.0 min
MAXIMUM ΔI:	0.0 deg

COMMENTS

The first event shed four pieces of debris, the second event shed more than 15 pieces. The parent experienced little if any change in orbit and the debris from both events were ejected with relatively low delta velocity from the parent, making a propulsion explosion unlikely for the cause of breakup. All the debris appeared to have high area-to-mass ratios, causing the ejected debris to decay within six months of the events. The parent body remains in orbit as of August 2007. Another H-IIA second stage (2006-037B) experienced two minor fragmentation events a few months later. Possible source of the debris is light-weight insulation material.

REFERENCE DOCUMENTS

"Three More Satellites Involved in Fragmentations", The Orbital Debris Quarterly News, NASA JSC, October 2006. Available online at http://www.orbitaldebris.jsc.nasa.gov/newsletter/pdfs/ODQNv10i4.pdf.

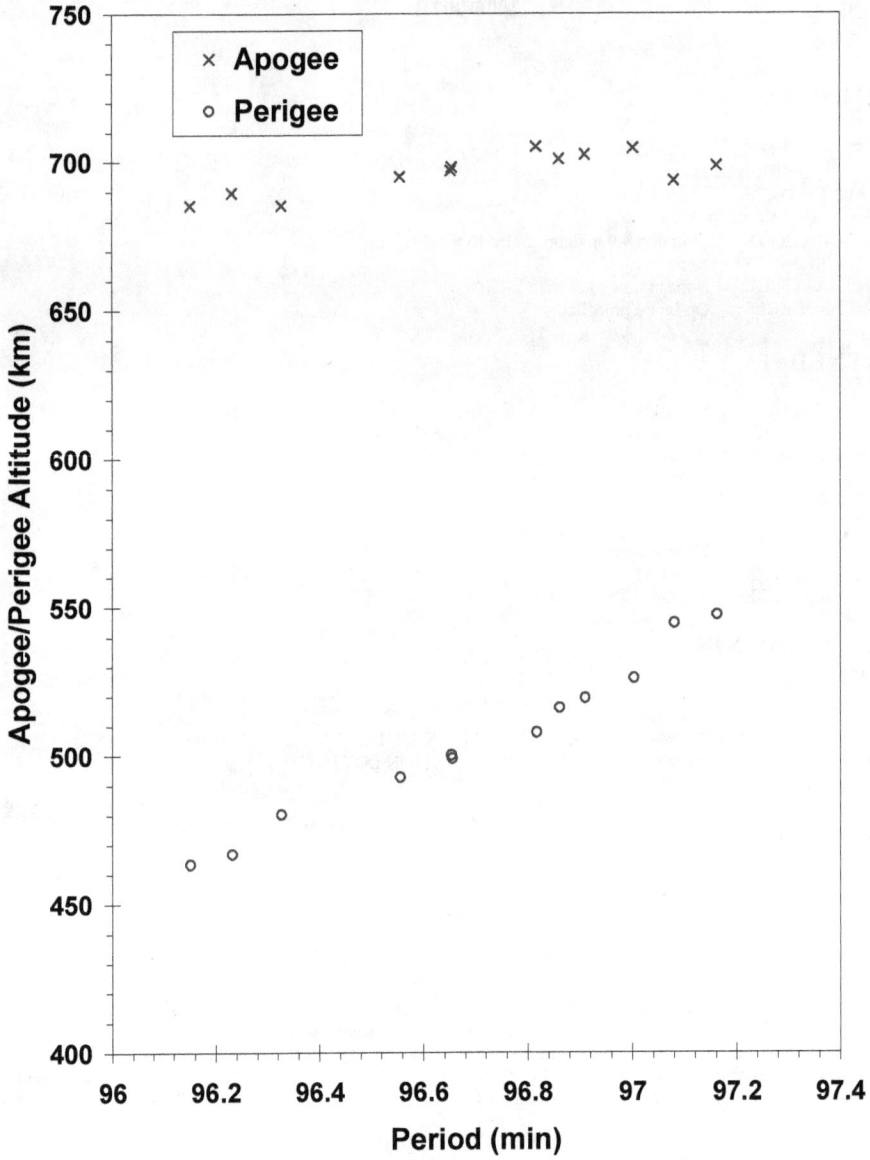

The ALOS R/B debris cloud from the second event, a few days after the event. The parent body is the piece with the highest perigee, at the right of the graph.

ARABSAT 4 BRIZ-M R/B   **2006-006B**   **28944**

SATELLITE DATA

TYPE:	Rocket Body
OWNER:	CIS
LAUNCH DATE:	28.84 Feb 2006
DRY MASS (KG):	2600
MAIN BODY:	Cylinder; 2.5 m diameter by 2.6 m length
MAJOR APPENDAGES:	None
ATTITUDE CONTROL:	None
ENERGY SOURCES:	On-board propellants

EVENT DATA

DATE:	19 Feb 2007	LOCATION:	31S, 135E (asc)
TIME:	1721 GMT	ASSESSED CAUSE:	Propulsion
ALTITUDE:	7,640 km		

PRE-EVENT ELEMENTS

EPOCH:	07050.57138199	MEAN ANOMALY:	134.5177
RIGHT ASCENSION:	213.0506	MEAN MOTION:	5.25304422
INCLINATION:	51.4995	MEAN MOTION DOT/2:	0.00000185
ECCENTRICITY:	.5083307	MEAN MOTION DOT DOT/6:	0.0
ARG. OF PERIGEE:	197.8403	BSTAR:	0.00029198

DEBRIS CLOUD DATA

MAXIMUM ΔP:	48.6 min
MAXIMUM ΔI:	2.6 deg

COMMENTS

This upper stage malfunctioned, stranding itself and its payload in an eccentric orbit. The cause of the breakup is assessed to be related to the ~8 metric tons of unused propellants. The breakup occurred over Southern Australia, and was captured by three amateur astronomers, which showed expansion of a faint cloud around the exploded fourth stage. Over a thousand objects 10 cm or larger were estimated by the US SSN.

REFERENCE DOCUMENT

"Two More Incidents Add to Growing Space Debris", Space News, February 26, 2007.

"Four Satellite Breakups in February Add to Debris Population", The Orbital Debris Quarterly News, NASA JSC, April 2007. Available online at http://www.orbitaldebris.jsc.nasa.gov/newsletter/pdfs/ODQNv11i2.pdf.

407

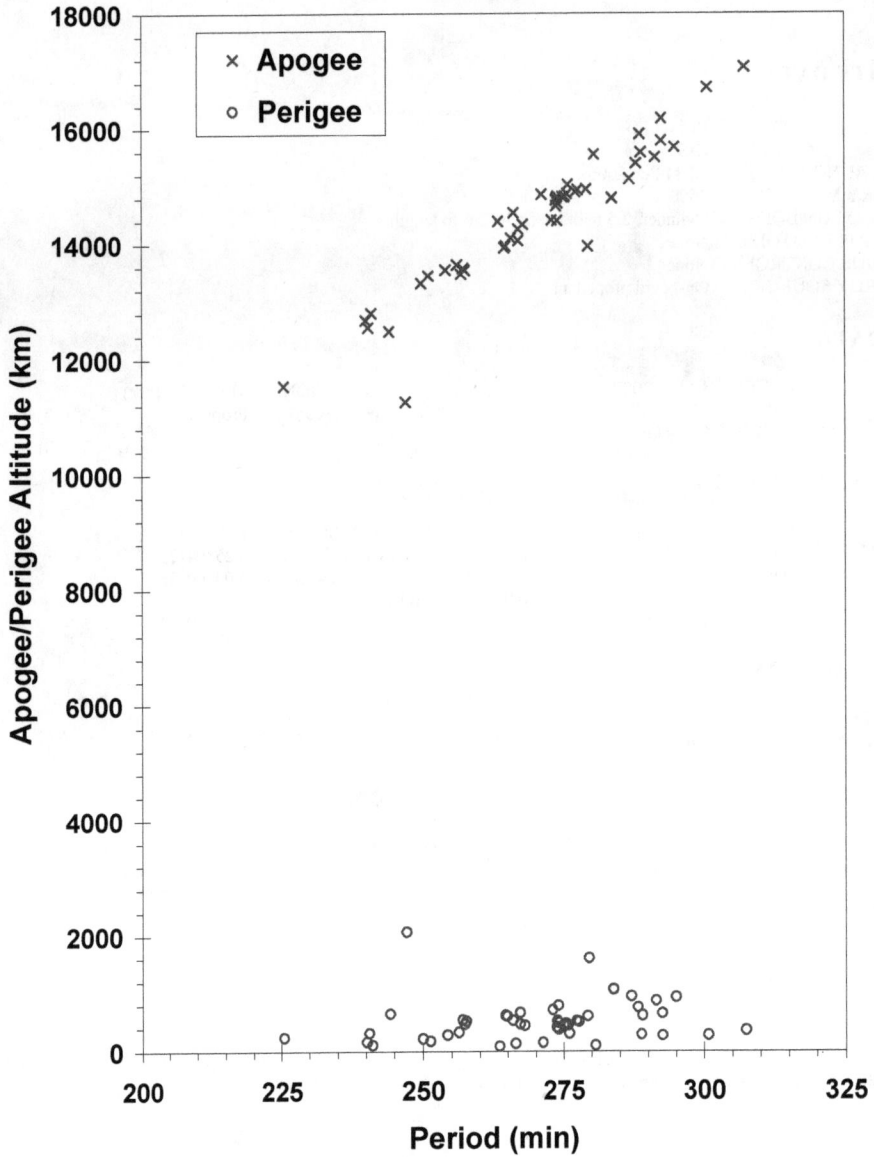

SSN 28944

The Briz-M R/B debris cloud of uncataloged fragments 10 days after the event.

SATELLITE DATA

TYPE:	Rocket Body
OWNER:	Japan
LAUNCH DATE:	11.19 Sep 2006
DRY MASS (KG):	~3000
MAIN BODY:	Cylinder; 4.0 m diameter by 10.6 m length
MAJOR APPENDAGES:	None
ATTITUDE CONTROL:	None at time of event
ENERGY SOURCES:	On-board propellant

EVENT DATA

DATE:	28 Dec 2006	LOCATION:	58S, 131E (asc)
TIME:	1729 GMT	ASSESSED CAUSE:	Unknown
ALTITUDE:	470 km		

PRE-EVENT ELEMENTS

EPOCH:	06361.50058695	MEAN ANOMALY:	347.5081
RIGHT ASCENSION:	117.7926	MEAN MOTION:	15.35084918
INCLINATION:	97.2357	MEAN MOTION DOT/2:	0.00002838
ECCENTRICITY:	.0043549	MEAN MOTION DOT DOT/6:	0.0
ARG. OF PERIGEE:	12.7250	BSTAR:	0.000083212

DEBRIS CLOUD DATA

MAXIMUM ΔP:	2.6 min
MAXIMUM ΔI:	0.2 deg

COMMENTS

This was the second fragmentation of an H-IIA second stage in 2006 (see 2006-002B). No debris were cataloged from the first event, but at least 20 new objects were detected. However, the rocket body experienced a second fragmentation event on 25 July 2007, releasing at least 15 new debris.

REFERENCE DOCUMENT

"Significant Increase in Satellite Breakups During 2006", The Orbital Debris Quarterly News, NASA JSC, January 2007. Available online at http://www.orbitaldebris.jsc nasa.gov/newsletter/pdfs/ODQNv11i1.pdf.

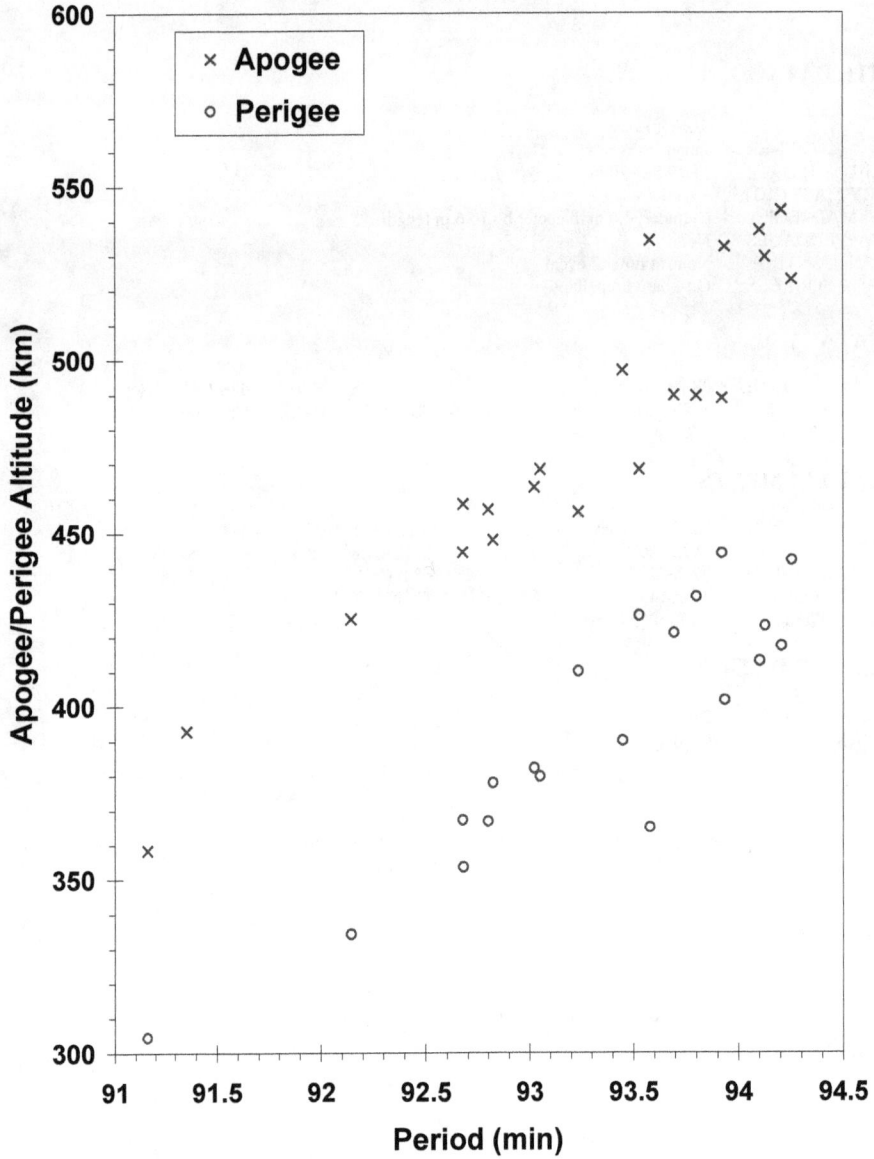

SSN 29394

H-IIA debris cloud 2 weeks after the first event. The debris were never cataloged.

410

COSMOS 2423   **2006-039A**   **29402**

SATELLITE DATA

TYPE:	Payload
OWNER:	CIS
LAUNCH DATE:	14.57 Sep 2006
DRY MASS (KG):	~6000
MAIN BODY:	Cylinder; 2.4 m diameter by 7 m length
MAJOR APPENDAGES:	Solar Arrays
ATTITUDE CONTROL:	Active, 3-axis
ENERGY SOURCES:	On-board propellants, explosive charge

EVENT DATA

DATE:	17 Nov 2006	LOCATION:	Unknown
TIME:	~1800 GMT	ASSESSED CAUSE:	Deliberate
ALTITUDE:	210 km		

PRE-EVENT ELEMENTS

EPOCH:	06321.75318443	MEAN ANOMALY:	8.8408
RIGHT ASCENSION:	358.1498	MEAN MOTION:	16.11335386
INCLINATION:	64.8900	MEAN MOTION DOT/2:	0.00257180
ECCENTRICITY:	.0061777	MEAN MOTION DOT DOT/6:	0.0000071463
ARG. OF PERIGEE:	106.6782	BSTAR:	0.00018281

DEBRIS CLOUD DATA

MAXIMUM ΔP:	5.8 min
MAXIMUM ΔI:	0.5 deg

COMMENTS

Cosmos 2423 was the eighth of the Cosmos 2031 class of spacecraft, which debuted in 1989. In all seven previous missions, the spacecraft was deliberately exploded at the end of mission. Cosmos 2423 had the shortest lifetime of any of this class of spacecraft, two months instead of a usual four months. Because these spacecraft are deliberately exploded at a low altitude, the lifetime of the trackable debris cloud is usually measured in days.

REFERENCE DOCUMENTS

"Significant Increase in Satellite Breakups During 2006", The Orbital Debris Quarterly News, NASA JSC, January 2007. Available online at http://www.orbitaldebris.jsc.nasa.gov/newsletter/pdfs/ODQNv11i1.pdf.

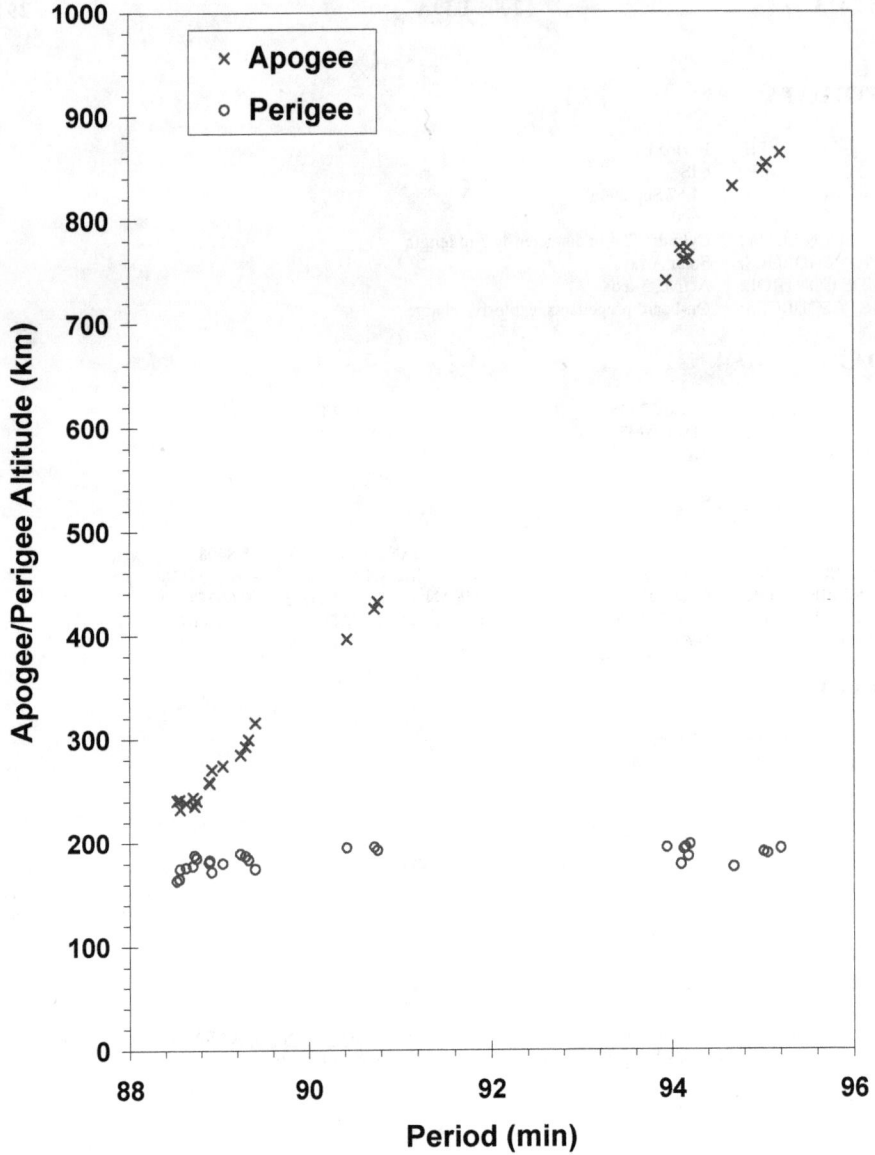

Cosmos 2423 debris cloud of 28 fragments a few days after the event as reconstructed from the US SSN database.

DMSP 5D-3 F17 R/B   **2006-050B**   **29523**

SATELLITE DATA

TYPE:	Rocket Body
OWNER:	US
LAUNCH DATE:	4.58 Nov 2006
DRY MASS (KG):	2850
MAIN BODY:	Cylinder; 4.0 m diameter by 12 m length
MAJOR APPENDAGES:	None
ATTITUDE CONTROL:	None at time of event
ENERGY SOURCES:	On-board propellant

EVENT DATA

DATE:	4 Nov 2006	LOCATION:	Unknown
TIME:	Unknown	ASSESSED CAUSE:	Unknown
ALTITUDE:	Unknown		

PRE-EVENT ELEMENTS

EPOCH:	06308.62553010	MEAN ANOMALY:	119.1776
RIGHT ASCENSION:	307.4245	MEAN MOTION:	14.13676442
INCLINATION:	98.7885	MEAN MOTION DOT/2:	-0.00000044
ECCENTRICITY:	.0022197	MEAN MOTION DOT DOT/6:	0.0
ARG. OF PERIGEE:	240.7178	BSTAR:	0.0

DEBRIS CLOUD DATA

MAXIMUM ΔP:	9.5 min
MAXIMUM ΔI:	0.3 deg

COMMENTS

This was the first major event associated with a Delta 4 second stage. Debris appeared to separate from the parent body in a retrograde direction soon after orbit insertion. Over 60 pieces were eventually cataloged from this event. The rocket body did not see any performance degradation and was reentered directly after payload delivery. The cause of the debris release is under investigation to ensure any countermeasures can be implemented for future Delta 4 missions.

REFERENCE DOCUMENTS

"Significant Increase in Satellite Breakups During 2006", The Orbital Debris Quarterly News, NASA JSC, January 2007. Available online at http://www.orbitaldebris.jsc.nasa.gov/newsletter/pdfs/ODQNv11i1.pdf.

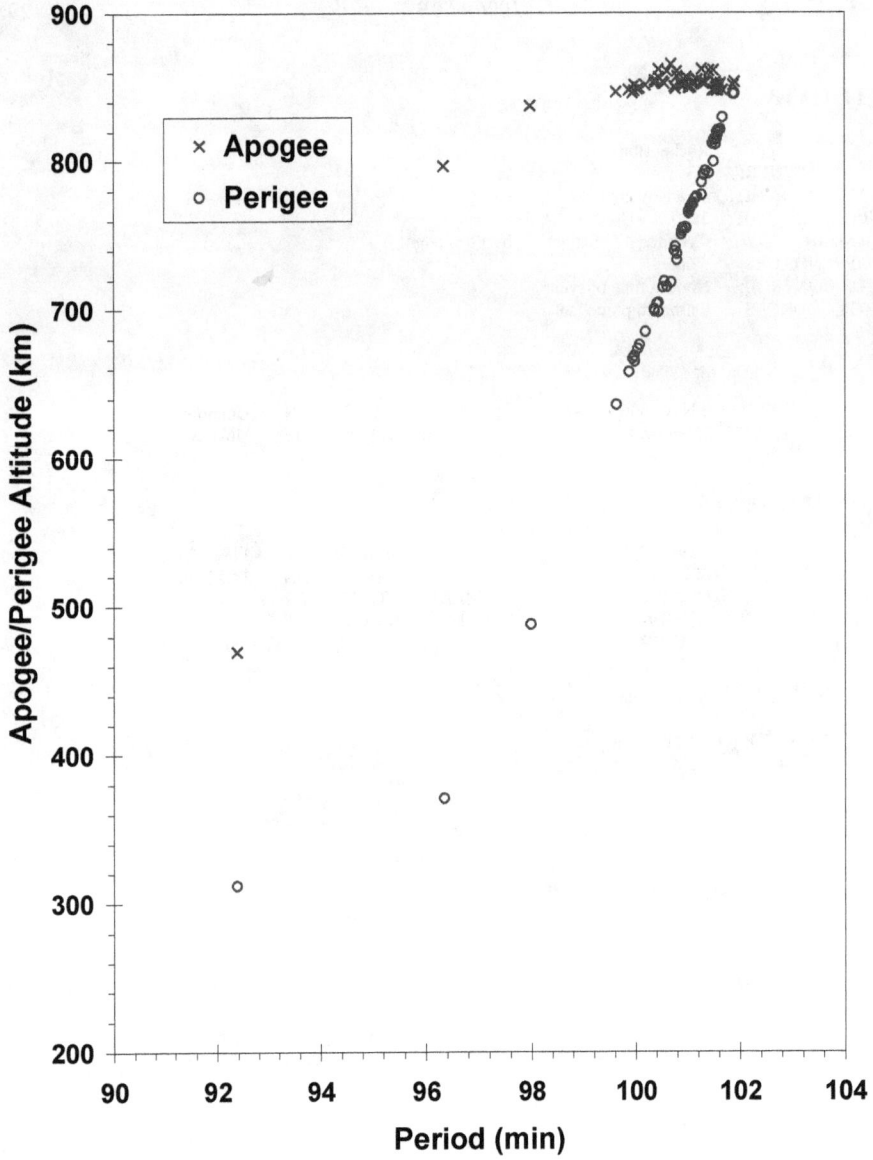

Delta 4 debris cloud of 62 fragments less than two weeks after the event.

BEIDOU 2A   **2007-003A**   **30323**

SATELLITE DATA

TYPE:	Paylaod
OWNER:	China
LAUNCH DATE:	02.69 Feb 2007
DRY MASS (KG):	~1100
MAIN BODY:	Unknown
MAJOR APPENDAGES:	Unknown
ATTITUDE CONTROL:	Unknown at time of event
ENERGY SOURCES:	On-board propellant

EVENT DATA

DATE:	02 Feb 2007	LOCATION:	Unknown
TIME:	Unknown	ASSESSED CAUSE:	Unknown
ALTITUDE:	Unknown		

PRE-EVENT ELEMENTS

EPOCH:	07033.44400535	MEAN ANOMALY:	182.1815
RIGHT ASCENSION:	9.6534	MEAN MOTION:	1.91691569
INCLINATION:	24.9838	MEAN MOTION DOT/2:	-0.00000543
ECCENTRICITY:	.0043549	MEAN MOTION DOT DOT/6:	0.0
ARG. OF PERIGEE:	179.6150	BSTAR:	0.0

DEBRIS CLOUD DATA

MAXIMUM ΔP:	5.5 min
MAXIMUM ΔI:	0.2 deg

COMMENTS

The spacecraft experienced problems soon after insertion into a geosynchronous transfer orbit. In March 2007, the spacecraft was recovered and maneuvered into GEO. As many as 100 debris were detected by the US SSN.

REFERENCE DOCUMENT

"Four Satellite Breakups in February Add to Debris Population", The Orbital Debris Quarterly News, NASA JSC, January 2007. Available online at http://www.orbitaldebris.jsc nasa.gov/newsletter/pdfs/ODQNv11i2.pdf.

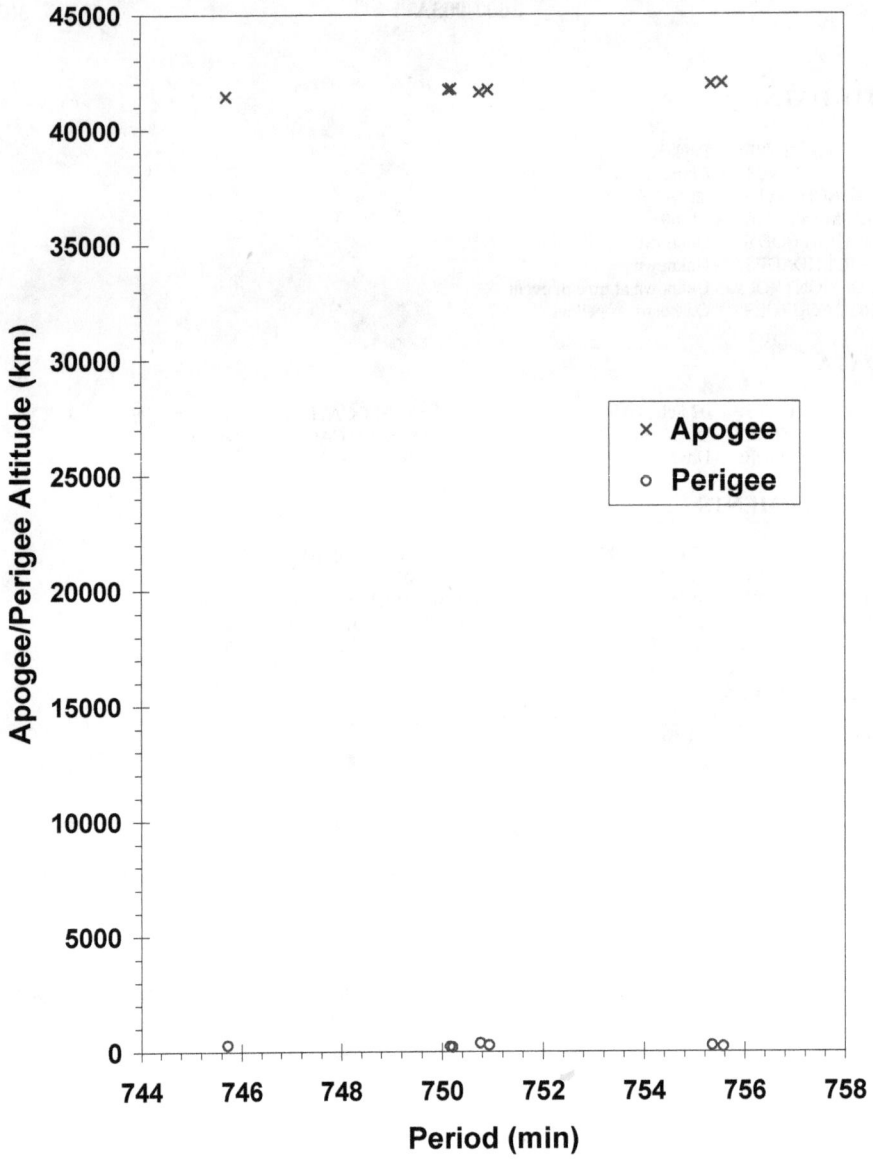

Identified Beidou debris cloud 2 weeks after the event as reconstructed from the US SSN database.

3.0   SATELLITE ANOMALOUS EVENTS

This section describes anomalous events identified throughout the years of orbital debris research associated with this report. No exhaustive search for anomalous events has yet been conducted, although the following compilation should represent the most significant events noted thus far.

3.1   Background and Status

As defined in the introduction of this report, an **anomalous event** is the unplanned separation, usually at low velocity, of one or more detectable objects from a satellite that remains essentially intact. The assessment that the configuration of the parent satellite has not changed significantly is to a degree subjective and is often based on indirect parameters and not on detailed imagery.

Anomalous events can be caused by material deterioration of items such as thermal blankets, protective shields, or solar panels and by impacts of small debris, either natural or man-made. Other satellite deteriorations, e.g., paint debonding, are known to take place, but are undetectable with the sensors of the US SSN. Interestingly, 29 of the 51 satellites in this section are US, 18 are CIS, 2 are French, and 1 each from Japan and Canada. Of the 51 satellites, 34 are payloads and 17 are rocket bodies. These events are summarized in Tables 3.1 and 3.2.

Because of the usually low velocity of debris ejection and the potential delay in detecting debris liberated in small numbers, the accuracy of the calculated time of separation is often degraded. Hence, only the month and year of each event are provided, although in some cases the time of the event has been narrowed to a shorter interval. In many cases, there were more than one anomalous event per parent object, the month and year of the additional events are provided when possible. As in the previous section, orbital altitudes are cited to the nearest 5 km based on a mean Earth radius and on the last element set prior to the assessed event date.

Anomalous event debris often exhibit unusually high decay rates, which are indicative of high area-to-mass ratios. This feature, coupled with the normal small size of the debris, hinders official tracking and cataloging. Consequently, some debris are observed but are lost or decay before being assigned a permanent catalog number. The numbers of cataloged debris listed in this section are only from the anomalous events and do <u>not</u> include normal mission related debris identified with the particular launch nor the parent itself.

Historically, anomalous events have often been confused with satellite breakups and have not been the subject of separate, extensive analyses. The list of events in this section is known to be incomplete. Several other satellites have been tentatively tagged as sources of anomalous events. Moreover, preliminary satellite catalog surveys suggest that additional anomalous events have occurred but remain unrecognized as such. Table 3.2 suggests a potential correlation of anomalous events with high solar activity. This section will be updated as future studies warrant.

For additional information on anomalous events, see "Environmentally induced Debris Sources", N. L. Johnson, Second World Space Congress, October 2002.

TABLE 3.1 HISTORY OF SATELLITE ANOMALOUS EVENTS BY LAUNCH DATE

NAME	INTERNATIONAL DESIGNATOR	CATALOG NUMBER	LAUNCH DATE	FIRST EVENT DATE	KNOWN EVENTS	CATALOGED DEBRIS	IN-ORBIT DEBRIS	APOGEE (KM)	PERIGEE (KM)	INCLINATION (DEG)
VANGUARD 3	1959-007A	20	18-Sep-59	14-Feb-06	1	2	1	3310	510	33.4
TRANSIT 5B-2	1963-049B	704	5-Dec-63	9/10-Jan-98	1	1	1	1110	1060	90.1
OPS 4412 (TRANSIT 9)	1964-026A	801	4-Jun-64	Dec-80	4	4	0	930	845	90.5
COSMOS 44 R/B	1964-053B	877	28-Aug-64	Nov-90	1	1	1	775	655	65.1
OPS 4988 (GREB 6)	1965-016A	1271	9-Mar-65	Nov-80	1	1	1	935	900	70.1
OPS 4682 (SNAPSHOT)	1965-027A	1314	3-Apr-65	1-Nov-79	7	56	54	1320	1270	90.3
OPS 8480 (TRANSIT 5B-6)	1965-048A	1420	24-Jun-65	Aug-80	4	12	4	1135	1025	89.9
FR-1 R/B	1965-101B	1815	6-Dec-65	21-Mar-03	1	1	1	660	655	75.8
OPS 1509 (TRANSIT 10)	1965-109A	1864	22-Dec-65	30-Nov-96	2	2	1	1065	895	89.1
OPS 1593 (TRANSIT 11)	1966-005A	1952	28-Jan-66	Apr-80	4	7	1	1205	855	89.8
OPS 1117 (TRANSIT 12)	1966-024A	2119	26-Mar-66	Jul-81	1	3	0	1115	890	89.9
NIMBUS 2	1966-040A	2173	15-May-66	Nov-97	>5	38	38	1175	1095	100.4
OPS 0856 (MIDAS 11)	1966-077A	2403	19-Aug-66	Mar-91	4	25	25	3710	3660	89.7
OPS 0100 (TRANSIT 15)	1967-034A	2754	14-Apr-67	Sep-92	1	7	4	1065	1035	90.1
OPS 7218 (TRANSIT 16)	1967-048A	2807	18-May-67	Feb-95	1	2	1	1090	1060	89.6
OPS 4947 (TRANSIT 17)	1967-092A	2965	25-Sep-67	Apr-81	4	7	0	1110	1035	89.3
COSMOS 206 R/B	1968-019B	3151	14-Mar-68	Nov-90	1	0	0	515	450	81.2
ISIS 1	1969-009A	3669	30-Jan-69	24-May-07	1	1	1	3455	580	88.5
TRANSIT 19	1970-067A	4507	27-Aug-70	7-Mar-98	1	1	0	1205	945	90.0
METEOR 1-7 R/B	1971-003B	4850	20-Jan-71	Jun-87	1	1	0	665	535	81.2
METEOR 1-12 R/B	1972-049B	6080	30-Jun-72	Sep-89	1	1	0	935	860	81.2
COSMOS 539	1972-102A	6319	21-Dec-72	21-Apr-02	1	1	0	1380	1340	74.0
GEOS 3 R/B	1975-027B	7735	9-Apr-75	Mar-78	1	3	2	845	835	115.0
KYOKKOH 1 (EXOS-A)	1978-014A	10664	4-Feb-78	Jan-88	2	2	0	4220	760	65.0
SEASAT	1978-064A	10967	27-Jun-78	Jul-83	>12	15	1	780	780	108.0
COSMOS 1043	1978-094A	11055	10-Oct-78	Feb-93	1	1	0	435	435	81.2
TIROS-N	1978-096A	11060	13-Oct-78	Sep-87	2	5	0	855	835	99.0
NIMBUS 7 R/B	1978-098B	11081	24-Oct-78	May-81	2	1	0	955	935	99.3
NOAA 6	1979-057A	11416	27-Jun-79	Sep-92	2	1	0	810	795	98.7
METEOR 2-7	1981-043A	12456	14-May-81	4-Mar-04	1	8	8	895	825	81.3
METEOR 2-7 R/B	1981-043B	12457	14-May-81	Oct-96	1	1	1	920	825	81.3
MOLNIYA 3-16 R/B	1981-054E	12519	9-Jun-81	Jul-98	1	0	0	33415	85	62.0
NOAA 7	1981-059A	12553	23-Jun-81	26-Jul-93	2	6	1	835	830	98.9
OSCAR 30	1985-066A	15935	3-Aug-85	Nov-86	2	2	2	1255	1000	89.9
COSMOS 1689 R/B	1985-090B	16111	3-Oct-85	5-May-02	1	1	1	565	510	97.7

TABLE 3.1 HISTORY OF SATELLITE ANOMALOUS EVENTS BY LAUNCH DATE (CONT'D)JSC 63927

NAME	INTERNATIONAL DESIGNATOR	CATALOG NUMBER	LAUNCH DATE	FIRST EVENT DATE	KNOWN EVENTS	CATALOGED DEBRIS	IN-ORBIT DEBRIS	APOGEE (KM)	PERIGEE (KM)	INCLINATION (DEG)
OSCAR 30	1985-066A	15935	3-Aug-85	Nov-86	2	1	1	1255	1000	89.9
COSMOS 1823	1987-020A	17535	20-Feb-87	Apr-May-97	3	3	3	1525	1480	73.6
METEOR 2-17	1988-005A	18820	30-Jan-88	21-Jun-05	1	20	20	960	930	82.5
COSMOS 1939 R/B	1988-032B	19046	20-Apr-88	30-Jul-96	2	2	2	655	585	97.6
COBE	1989-089A	20322	18-Nov-89	Mar-93	12	78	2	885	870	99.0
NADEZHDA 2 R/B	1990-017B	20509	27-Feb-90	22-Jun-05	1	1	1	1015	950	83.0
HST	1990-037B	20580	24-Apr-90	5-Aug-03	1	1	0	575	570	28.5
OKEAN 3	1991-039A	21397	4-Jun-91	12-Oct-98	1	1	1	665	620	82.5
SARA	1991-050E	21578	17-Jul-91	22-Aug-03	3	4	3	730	730	98.1
ERS-1 R/B	1991-050F	21610	17-Jul-91	1-Apr-01	1	1	0	770	770	98.2
EKA 1 (START 1)	1993-014A	22561	25-Mar-93	4-Mar-98	1	2	2	970	685	75.8
START 1 R/B	1993-014B	22562	25-Mar-93	Late-02	Multiple	42	41	920	680	75.8
COSMOS 2297 R/B	1994-077B	23405	24-Nov-94	Jun-98	2?	1	0	845	845	71.0
KOREASAT 1 R/B	1995-041B	23640	5-Aug-95	6-Dec-95	1	1	0	1375	935	26.7
RADARSAT R/B	1995-059B	23711	4-Nov-95	30-Jan-96	1	2	0	1495	935	100.6
FUSE	1999-035A	25791	24-Jun-99	6-Jun-04	1	8	2	760	745	25.0
IKONOS 2	1999-051A	25919	24-Sep-99	19-Mar-01	1	1	0	680	680	98.2
				TOTALS		388	229			

418

TABLE 3.2 HISTORY OF SATELLITE ANOMALOUS EVENTS BY EVENT DATE

NAME	INTERNATIONAL DESIGNATOR	CATALOG NUMBER	LAUNCH DATE	FIRST EVENT DATE	KNOWN EVENTS	CATALOGED DEBRIS	IN-ORBIT DEBRIS	APOGEE (KM)	PERIGEE (KM)	INCLINATION (DEG)
GEOS 3 R/B	1975-027B	7735	9-Apr-75	Mar-78	1	3	1	845	835	115.0
OPS 4682 (SNAPSHOT)	1965-027A	1314	3-Apr-65	1-Nov-79	7	56	54	1320	1270	90.3
OPS 1593 (TRANSIT 11)	1966-005A	1952	28-Jan-66	Apr-80	4	7	1	1205	855	89.8
OPS 8480 (TRANSIT 5B-6)	1965-048A	1420	24-Jun-65	Aug-80	4	12	4	1135	1025	89.9
OPS 4988 (GREB 6)	1965-016A	1271	9-Mar-65	Nov-80	1	1	1	935	900	70.1
OPS 4412 (TRANSIT 9)	1964-026A	801	4-Jun-64	Dec-80	4	4	0	930	845	90.5
OPS 4947 (TRANSIT 17)	1967-092A	2965	25-Sep-67	Apr-81	4	7	0	1110	1035	89.3
NIMBUS 7 R/B	1978-098B	11081	24-Oct-78	May-81	2	1	0	955	935	99.3
OPS 1117 (TRANSIT 12)	1966-024A	2119	26-Mar-66	Jul-81	1	3	0	1115	890	89.9
SEASAT	1978-064A	10967	27-Jun-78	Jul-83	>12	15	1	780	780	108.0
OSCAR 30	1985-066A	15935	3-Aug-85	Nov-86	2	1	1	1255	1000	89.9
METEOR 1-7 R/B	1971-003B	4850	20-Jan-71	Jun-87	1	1	0	665	535	81.2
TIROS-N	1978-096A	11060	13-Oct-78	Sep-87	2	5	0	855	835	99.0
KYOKKOH 1 (EXOS-A)	1978-014A	10664	4-Feb-78	Jan-88	2	2	0	4220	760	65.0
METEOR 1-12 R/B	1972-049B	6080	30-Jun-72	Sep-89	1	1	1	935	860	81.2
COSMOS 44 R/B	1964-053B	877	28-Aug-64	Nov-90	1	1	1	775	655	65.1
COSMOS 206 R/B	1968-019B	3151	14-Mar-68	Nov-90	1	0	0	515	450	81.2
OPS 0856 (MIDAS 11)	1966-077A	2403	19-Aug-66	Mar-91	4	25	25	3710	3660	89.7
OPS 0100 (TRANSIT 15)	1967-034A	2754	14-Apr-67	Sep-92	1	7	4	1065	1035	90.1
NOAA 6	1979-057A	11416	27-Jun-79	Sep-92	2	1	0	810	795	98.7
COSMOS 1043	1978-094A	11055	10-Oct-78	Feb-93	1	1	0	435	435	81.2
COBE	1989-089A	20322	18-Nov-89	Mar-93	12	78	2	885	870	99.0
NOAA 7	1981-059A	12553	23-Jun-81	26-Jul-93	2	6	1	835	830	98.9
OPS 7218 (TRANSIT 16)	1967-048A	2807	18-May-67	Feb-95	1	2	1	1090	1060	89.6
KOREASAT 1 R/B	1995-041B	23640	5-Aug-95	6-Dec-95	1	1	0	1375	935	26.7
RADARSAT R/B	1995-059B	23711	4-Nov-95	30-Jan-96	1	2	0	1495	935	100.6
COSMOS 1939 R/B	1988-032B	19046	20-Apr-88	30-Jul-96	2	2	2	655	585	97.6
METEOR 2-7 R/B	1981-043B	12457	14-May-81	Oct-96	1	1	1	920	825	81.3
OPS 1509 (TRANSIT 10)	1965-109A	1864	22-Dec-65	30-Nov-96	2	2	1	1065	895	89.1
COSMOS 1823	1987-020A	17535	20-Feb-87	Apr-May-97	3	3	3	1525	1480	73.6
NIMBUS 2	1966-040A	2173	15-May-66	Nov-97	>5	38	38	1175	1095	100.4
TRANSIT 5B-2	1963-049B	704	5-Dec-63	9/10-Jan-98	1	1	1	1110	1060	90.1
EKA 1 (START 1)	1993-014A	22561	25-Mar-93	4-Mar-98	1	2	2	970	685	75.8
TRANSIT 19	1970-067A	4507	27-Aug-70	7-Mar-98	1	1	0	1205	945	90.0
COSMOS 2297 R/B	1994-077B	23405	24-Nov-94	Jun-98	2?	1	0	845	845	71.0

TABLE 3.2 HISTORY OF SATELLITE ANOMALOUS EVENTS BY EVENT DATE (CONT'D)

NAME	INTERNATIONAL DESIGNATOR	CATALOG NUMBER	LAUNCH DATE	FIRST EVENT DATE	KNOWN EVENTS	CATALOGED DEBRIS	IN-ORBIT DEBRIS	APOGEE (KM)	PERIGEE (KM)	INCLINATION (DEG)
MOLNIYA 3-16 R/B	1981-054E	12519	9-Jun-81	Jul-98	1	0	0	33415	85	62.0
OKEAN 3	1991-039A	21397	4-Jun-91	12-Oct-98	1	1	1	665	620	82.5
IKONOS 2	1999-051A	25919	24-Sep-99	19-Mar-01	1	1	0	680	680	98.2
ERS-1 R/B	1991-050F	21610	17-Jul-91	1-Apr-01	1	1	0	770	770	98.2
COSMOS 539	1972-102A	6319	21-Dec-72	21-Apr-02	1	1	0	1380	1340	74.0
COSMOS 1689 R/B	1985-090B	16111	3-Oct-85	5-May-02	1	2	2	565	510	97.7
START 1 R/B	1993-014B	22562	25-Mar-93	Late-02	Multiple	42	41	920	680	75.8
FR-1 R/B	1965-101B	1815	6-Dec-65	21-Mar-03	1	1	1	660	655	75.8
HST	1990-037B	20580	24-Apr-90	5-Aug-03	1	1	0	575	570	28.5
SARA	1991-050E	21578	17-Jul-91	22-Aug-03	3	4	3	730	730	98.1
METEOR 2-7	1981-043A	12456	14-May-81	4-Mar-04	1	8	8	895	825	81.3
FUSE	1999-035A	25791	24-Jun-99	6-Jun-04	1	8	2	760	745	25.0
METEOR 2-17	1988-005A	18820	30-Jan-88	21-Jun-05	1	20	20	960	930	82.5
NADEZHDA 2 R/B	1990-017B	20509	27-Feb-90	22-Jun-05	1	1	1	1015	950	83.0
VANGUARD 3	1959-007A	20	18-Sep-59	14-Feb-06	1	2	2	3310	510	33.4
ISIS 1	1969-009A	3669	30-Jan-69	24-May-07	1	1	1	3455	580	88.5
				TOTALS		388	229			

420

3.2   *Identified Satellite Anomalous Events*

Much like section 2.2 above, this section identifies particulars for the limited number of anomalous events thus far cataloged. There is no Gabbard Diagram included with these events, and each page often refers to multiple events. The first known date of the first anomalous event is categorized for each satellite. Where possible the best estimate of the cause and potential failure are noted in the comments section.

VANGUARD 3 **1959-007A** **20**

SATELLITE DATA

TYPE:	Payload
OWNER:	US
LAUNCH DATE:	18 September 1959
DRY MASS (KG):	45
MAIN BODY:	50.8 cm sphere with third stage attached
MAJOR APPENDAGES:	66 cm boom(s)
ATTITUDE CONTROL:	spin stabilized

EVENT DATA

KNOWN EVENTS:	1
FIRST DATE:	14 February 2006

APOGEE	PERIGEE	PERIOD	INCLINATION
3310	510	125.14 min	33.4 deg

COMMENTS

At the time of the event, Vanguard 3 was the 5th oldest object in orbit. Two objects released, although the second object was not cataloged until May 2007.

REFERENCE DOCUMENT

"First Satellite Breakups of 2006", The Orbital Debris Quarterly News, NASA JSC, July 2006.
Available online at http://www.orbitaldebris.jsc.nasa.gov/newsletter/pdfs/ODQNv10i3.pdf.

"Detection of Debris from Chinese ASAT Test Increases; One Minor Fragmentation Event in Second Quarter of 2007", The Orbital Debris Quarterly News, NASA JSC, July 2007.
Available online at http://www.orbitaldebris.jsc.nasa.gov/newsletter/pdfs/ODQNv11i3.pdf.

TRANSIT 5B-2   **1963-049B**   **704**

SATELLITE DATA

TYPE:	Payload
OWNER:	US
LAUNCH DATE:	5 December 1963
DRY MASS (KG):	75
MAIN BODY:	Octagon; 0.46 m diameter by 0.5 m length
MAJOR APPENDAGES:	Boom
ATTITUDE CONTROL:	None at the time of event

EVENT DATA

KNOWN EVENTS:	1
FIRST DATE:	9-10 January 1998

APOGEE	PERIGEE	PERIOD	INCLINATION
1110 km	1060 km	106.98 min	90.1 deg

COMMENTS

One of several Transit-class spacecraft involved in anomalous events. Spacecraft may have experienced earlier anomalous events in 1960's and 1970's. Only one object associated with January 1998 event.

OPS 4412 (TRANSIT 9)   1964-026A   **801**

SATELLITE DATA

TYPE:	Payload
OWNER:	US
LAUNCH DATE:	4 June 1964
DRY MASS (KG):	60
MAIN BODY:	Octagonal cylinder; 0.5 m by 0.4 m
MAJOR APPENDAGES:	4 solar panels; gravity-gradient boom
ATTITUDE CONTROL:	Gravity-gradient boom

EVENT DATA

KNOWN EVENTS:	4
FIRST DATE:	December 1980

APOGEE	PERIGEE	PERIOD	INCLINATION
930 km	845 km	102.7 min	90.5 deg

COMMENTS

Second event observed July 1982. The third event occurred in May 1994. Fourth event date not determined but also close to May 1994. First fragment decayed rapidly; the second decayed more slowly. Two latest pieces not cataloged as of publication date. One of several known Transits involved in anomalous events.

COSMOS 44 R/B   **1964-053B**   **877**

SATELLITE DATA

TYPE:	Vostok Final Stage
OWNER:	CIS
LAUNCH DATE:	28 August 1964
DRY MASS (KG):	1440
MAIN BODY:	Cylinder; 2.6 m diameter by 3.8 m length
MAJOR APPENDAGES:	None
ATTITUDE CONTROL:	None at time of event

EVENT DATA

KNOWN EVENTS:	1
FIRST DATE:	Late-1990

APOGEE	PERIGEE	PERIOD	INCLINATION
775 km	655 km	99.1 min	65.1 deg

COMMENTS

Cosmos 44 was the first prototype spacecraft of the Meteor 1 program. This is one of several Vostok final stages associated with this old program to shed a piece of debris since 1987.

OPS 4988 (GREB 6)   **1965-016A**   **1271**

SATELLITE DATA

TYPE:	Payload
OWNER:	US
LAUNCH DATE:	9 May 1965
DRY MASS (KG):	40
MAIN BODY:	Sphere
MAJOR APPENDAGES:	Unknown
ATTITUDE CONTROL:	Unknown

EVENT DATA

KNOWN EVENTS:	1
FIRST DATE:	November 1980

APOGEE	PERIGEE	PERIOD	INCLINATION
935 km	900 km	103.4 min	70.1 deg

COMMENTS

No other events observed.

OPS 4682 (SNAPSHOT)   **1965-027A**   **1314**

SATELLITE DATA

TYPE:	Payload (attached to Agena D upper stage)
OWNER:	US
LAUNCH DATE:	3 April 1965
DRY MASS (KG):	2500 (approx.)
MAIN BODY:	Cylinder-cone; 1.5 m by 11.6 m
MAJOR APPENDAGES:	None
ATTITUDE CONTROL:	None at time of event

EVENT DATA

KNOWN EVENTS:	7
FIRST DATE:	November 1979

APOGEE	PERIGEE	PERIOD	INCLINATION
1320 km	1270 km	111.5 min	90.3 deg

COMMENTS

Six additional events observed: Dec 1980, Aug 1981, Mar 1983, Aug 1983, Nov 1983, and Jan 1985. Decay rates of all debris are nominal for this altitude. One debris was administratively decayed in February 1989.

REFERENCE DOCUMENTS

Investigation of Certain Anomalies Associated with Object 1314, A US Nuclear Powered Satellite, G. T. DeVere, Technical Memorandum 85-S-001, Headquarters NORAD/ADCOM, DCS/Plans, March 1985 (Appendix TM-85-001A, Secret).

Environmentally-Induced Debris Sources, N. L. Johnson, NASA Lyndon B. Johnson Space Center, Second World Space Congress, 2002.

OPS 8480 (TRANSIT 5B-6)   **1965-048A**   **1420**

SATELLITE DATA

TYPE:	Payload
OWNER:	US
LAUNCH DATE:	24 June 1965
DRY MASS (KG):	60
MAIN BODY:	Octagonal cylinder; 0.5 m by 0.4 m
MAJOR APPENDAGES:	4 solar panels; gravity-gradient boom
ATTITUDE CONTROL:	Gravity-gradient

EVENT DATA

KNOWN EVENTS:	Multiple
FIRST DATE:	August 1980

APOGEE	PERIGEE	PERIOD	INCLINATION
1135 km	1025 km	106.8 min	89.9 deg

COMMENTS

Three additional events observed: one 2 days after the initial event, one in June 1981, and the most recent in late 1999. All debris appear very small. One of several known Transits involved in anomalous events.

REFERENCE DOCUMENT

Environmentally-Induced Debris Sources, N. L. Johnson, NASA Lyndon B. Johnson Space Center, Second World Space Congress, 2002.

FR-1 R/B   **1965-101B**   **1815**

SATELLITE DATA

TYPE:	Rocket Body
OWNER:	US
LAUNCH DATE:	6 December 1965
DRY MASS (KG):	26
MAIN BODY:	Cylinder, 0.64 m diameter by 2.53 m length
MAJOR APPENDAGES:	None
ATTITUDE CONTROL:	None at the time of the event

EVENT DATA

KNOWN EVENTS:	1
FIRST DATE:	21 March 2003

APOGEE	PERIGEE	PERIOD	INCLINATION
660 km	655 km	97.89 min	75.8 deg

COMMENTS

There was only one piece cataloged from the relatively small Scout R/B stage.

OPS 1509 (TRANSIT 10)   **1965-109A**   **1864**

SATELLITE DATA

TYPE:	Payload
OWNER:	US
LAUNCH DATE:	22 December 1965
DRY MASS (KG):	60
MAIN BODY:	Octagon; 0.5 m diameter by 0.4 m length
MAJOR APPENDAGES:	4 vanes
ATTITUDE CONTROL:	None at time of event

EVENT DATA

KNOWN EVENTS:	2
FIRST DATE:	30 November 1996

APOGEE	PERIGEE	PERIOD	INCLINATION
1065 km	895 km	104.66 min	89.1 deg

COMMENTS

One of several Transit-class spacecraft involved in anomalous events. Two debris objects (one cataloged and one not cataloged) were being tracked in 1997.

OPS 1593 (TRANSIT 11)   **1966-005A**   **1952**

SATELLITE DATA

TYPE:	Payload
OWNER:	US
LAUNCH DATE:	28 January 1966
DRY MASS (KG):	60
MAIN BODY:	Octagonal cylinder; 0.5 m by 0.4 m
MAJOR APPENDAGES:	4 solar panels; gravity-gradient boom
ATTITUDE CONTROL:	Gravity-gradient

EVENT DATA

KNOWN EVENTS:	3
FIRST DATE:	April 1980

APOGEE	PERIGEE	PERIOD	INCLINATION
1205 km	855 km	105.8 min	89.8 deg

COMMENTS

Two additional events observed: Sep 1980 and Jul 1983. Last event may have originated with a piece of debris from earlier event. One of several known Transits involved in anomalous events.

OPS 1117 (TRANSIT 12)   **1966-024A**   **2119**

SATELLITE DATA

TYPE:	Payload
OWNER:	US
LAUNCH DATE:	26 March 1966
DRY MASS (KG):	60
MAIN BODY:	Octagonal cylinder; 0.5 m by 0.4 m
MAJOR APPENDAGES:	4 solar panels; gravity-gradient boom
ATTITUDE CONTROL:	Gravity-gradient

EVENT DATA

KNOWN EVENTS:	1
FIRST DATE:	July 1981

APOGEE	PERIGEE	PERIOD	INCLINATION
1115 km	890 km	105.1 min	89.9 deg

COMMENTS

No other events observed. One of several known Transits involved in anomalous events.

NIMBUS 2   **1966-040A**   **2173**

SATELLITE DATA

TYPE:	Payload
OWNER:	US
LAUNCH DATE:	15 May 1966
DRY MASS (KG):	414
MAIN BODY:	Conical skeleton; 1.45 m diameter by 3.0 m length
MAJOR APPENDAGES:	2 Paddles
ATTITUDE CONTROL:	None at time of event

EVENT DATA

KNOWN EVENTS:	Multiple
FIRST DATE:	November 1997

APOGEE	PERIGEE	PERIOD	INCLINATION
1175 km	1095 km	108.03 min	100.4 deg

COMMENTS

A single piece of debris was detected on 16 November 1997. Separation may have occurred about 1 November. Numerous debris were released from the late 1990's to 2001. See cited reference below.

REFERENCE DOCUMENT

Environmentally-Induced Debris Sources, N. L. Johnson, NASA Lyndon B. Johnson Space Center, Second World Space Congress, 2002.

OPS 0856 (MIDAS 11)   **1966-077A**   **2403**

SATELLITE DATA

TYPE:	Agena D Stage
OWNER:	US
LAUNCH DATE:	19 August 1966
DRY MASS (KG):	600
MAIN BODY:	Cylinder; 1.5 m diameter by 8 m length
MAJOR APPENDAGES:	None
ATTITUDE CONTROL:	None at time of events

EVENT DATA

KNOWN EVENTS:	4
FIRST DATE:	March 1991

APOGEE	PERIGEE	PERIOD	INCLINATION
3710 km	3660 km	167.5 min	89.7 deg

COMMENTS

Second, third, and fourth events observed on 16 June 1992, 23 June 1992, and 1 November 1995 respectively. Additional events may have occurred.

OPS 0100 (TRANSIT 15)   **1967-034A**   **2754**

SATELLITE DATA

TYPE:	Payload
OWNER:	US
LAUNCH DATE:	14 April 1967
DRY MASS (KG):	60
MAIN BODY:	Octagonal cylinder; 0.5 m by 0.4 m
MAJOR APPENDAGES:	4 solar panels; gravity-gradient boom
ATTITUDE CONTROL:	Gravity-gradient

EVENT DATA

KNOWN EVENTS:	1
FIRST DATE:	September 1992

APOGEE	PERIGEE	PERIOD	INCLINATION
1065 km	1035 km	106.2	90.1

COMMENTS

Event most likely happened around 28 September 1992 based on element data near the event time and analysis using the COMBO algorithm in the SATRAK Astrodynamics Toolkit. One of several Transit-class satellites involved in anomalous events.

REFERENCE DOCUMENT

Environmentally-Induced Debris Sources, N. L. Johnson, NASA Lyndon B. Johnson Space Center, Second World Space Congress, 2002.

OPS 7218 (TRANSIT 16)   **1967-048A**   **2807**

SATELLITE DATA

TYPE:	Payload
OWNER:	US
LAUNCH DATE:	18 May 1967
DRY MASS (KG):	60
MAIN BODY:	Octagonal cylinder; 0.5 m by 0.4 m
MAJOR APPENDAGES:	4 solar panels; gravity-gradient boom
ATTITUDE CONTROL:	Gravity-gradient boom

EVENT DATA

KNOWN EVENTS:	1
DATE:	11/12 February 1995

APOGEE	PERIGEE	PERIOD	INCLINATION
1090 km	1060 km	106.12 min	89.6 deg

COMMENTS

One piece of debris liberated. One of several Transit-class satellites involved in anomalous events.

OPS 4947 (TRANSIT 17)   **1967-092A**   **2965**

SATELLITE DATA

TYPE:	Payload
OWNER:	US
LAUNCH DATE:	25 September 1967
DRY MASS (KG):	60
MAIN BODY:	Octagonal cylinder; 0.5 m by 0.4 m
MAJOR APPENDAGES:	4 solar panels; gravity-gradient boom
ATTITUDE CONTROL:	Gravity-gradient

EVENT DATA

KNOWN EVENTS:	4
FIRST DATE:	April 1981

APOGEE	PERIGEE	PERIOD	INCLINATION
1110 km	1035 km	106.7 min	89.3 deg

COMMENTS

Second event observed in August 1986. One of several known Transits involved in anomalous events.

COSMOS 206 R/B 1968-019B **3151**

SATELLITE DATA

TYPE:	Vostok Final Stage
OWNER:	CIS
LAUNCH DATE:	14 March 1968
DRY MASS (KG):	1440
MAIN BODY:	Cylinder; 2.6 m diameter by 3.8 m length
MAJOR APPENDAGES:	None
ATTITUDE CONTROL:	None at the time of event

EVENT DATA

KNOWN EVENTS:	1
FIRST DATE:	Late-1990

APOGEE	PERIGEE	PERIOD	INCLINATION
515 km	450 km	94.3 min	81.2 deg

COMMENTS

Cosmos 206 was a prototype spacecraft of the Meteor 1 program. This is one of several Vostok final stages to shed a piece of debris since 1987. One piece of debris was released, although never officially cataloged.

ISIS 1   **1969-009A**   **3669**

SATELLITE DATA

TYPE:	Payload
OWNER:	Canada
LAUNCH DATE:	30 January 1969
DRY MASS (KG):	240
MAIN BODY:	Cylinder; 1.27 m diameter x 1.07 m length
MAJOR APPENDAGES:	Several antennae
ATTITUDE CONTROL:	None at the time of event

EVENT DATA

KNOWN EVENTS:	1
FIRST DATE:	24.82 May 2007

APOGEE	PERIGEE	PERIOD	INCLINATION
3455 km	580 km	127.57 min	88.5 deg

COMMENTS

One piece of debris cataloged. Altitude at the time of breakup was approximately 2940 km.

REFERENCE DOCUMENT

"Two Minor Satellite Fragmentations Identified in the Third Quarter", The Orbital Debris Quarterly News, NASA JSC, October 2007. Available online at: http://www.orbitaldebris.jsc nasa.gov/newsletter/pdfs/ODQNv11i4.pdf

TRANSIT 19   **1970-067A**   **4507**

SATELLITE DATA

TYPE:	Payload
OWNER:	US
LAUNCH DATE:	27 August 1970
DRY MASS (KG):	60
MAIN BODY:	Octagon; 0.5 m diameter by 0.4 m length
MAJOR APPENDAGES:	4 solar panels; gravity-gradient boom
ATTITUDE CONTROL:	None at time of event

EVENT DATA

KNOWN EVENTS:	1
FIRST DATE:	7 March 1998

APOGEE	PERIGEE	PERIOD	INCLINATION
1205 km	945 km	106.75 min	90.0 deg

COMMENTS

One of several Transit-class satellites involved in anomalous events.

METEOR 1-7 R/B   **1971-003B**   **4850**

SATELLITE DATA

TYPE:	Vostok Final Stage
OWNER:	CIS
LAUNCH DATE:	20 January 1971
DRY MASS (KG):	1440
MAIN BODY:	Cylinder; 2.6 m diameter by 3.8 m length
MAJOR APPENDAGES:	None
ATTITUDE CONTROL:	None at time of the event.

EVENT DATA

KNOWN EVENTS:	1
FIRST DATE:	June 1987

APOGEE	PERIGEE	PERIOD	INCLINATION
665 km	535 km	96.7 min	81.2 deg

COMMENTS

No other events observed. One of several Vostok final stages to be involved in anomalous events.

METEOR 1-12 R/B   **1972-049B**   **6080**

SATELLITE DATA

TYPE:	Vostok Final Stage
OWNER:	CIS
LAUNCH DATE:	30 June 1972
DRY MASS (KG):	1440
MAIN BODY:	Cylinder; 2.6 m diameter by 3.8 m length
MAJOR APPENDAGES:	None
ATTITUDE CONTROL:	None at time of the event.

EVENT DATA

KNOWN EVENTS:	1
FIRST DATE:	September 1989

APOGEE	PERIGEE	PERIOD	INCLINATION
935 km	860 km	102.9 min	81.2 deg

COMMENTS

No other events observed. One of several Vostok final stages involved in anomalous events.

COSMOS 539 1972-102A **6319**

SATELLITE DATA

TYPE:	Payload
OWNER:	CIS
LAUNCH DATE:	21 December 1972
DRY MASS (KG):	600
MAIN BODY:	Unknown
MAJOR APPENDAGES:	Unknown
ATTITUDE CONTROL:	None at the time of the event

EVENT DATA

KNOWN EVENTS:	1
FIRST DATE:	April 2002

APOGEE	PERIGEE	PERIOD	INCLINATION
1380 km	1340 km	112.9 min	74.0 deg

COMMENTS

One piece of debris cataloged. It was concluded that because of the deduced debris ejecta velocity, Cosmos 539 was apparently struck by a small meteoroid or man-made object.

REFERENCE DOCUMENTS

Environmentally-Induced Debris Sources, N. L. Johnson, NASA Lyndon B. Johnson Space Center, Second World Space Congress, 2002.

"A New Collision in Space?", The Orbital Debris Quarterly News, NASA JSC, July 2002. Available online at http://www.orbitaldebris.jsc.nasa.gov/newsletter/pdfs/ODQNv7i3.pdf

GEOS 3 R/B   **1975-027B**   **7735**

SATELLITE DATA

TYPE:	Delta Second Stage (2410)
OWNER:	US
LAUNCH DATE:	9 April 1975
DRY MASS (KG):	900
MAIN BODY:	Cylinder-nozzle; 2.4 m diameter by 8 m length
MAJOR APPENDAGES:	None
ATTITUDE CONTROL:	None at time of the event.

EVENT DATA

KNOWN EVENTS:	1
FIRST DATE:	March 1978

APOGEE	PERIGEE	PERIOD	INCLINATION
845 km	835 km	101.7 min	115.0 deg

COMMENTS

Only one event noted with three fragments cataloged 12 March 1978. There was repeated mis-tagging of objects during 1978 among rocket body and debris. One fragment lost in 1978 and administratively decayed in 1983. This event may be related to series of major Delta second stage breakups.

KYOKKOH 1 (EXOS-A)   1978-014A   **10664**

SATELLITE DATA

TYPE:	Payload
OWNER:	Japan
LAUNCH DATE:	4 February 1978
DRY MASS (KG):	103
MAIN BODY:	Octagonal cylinder; 0.95 m by 0.8 m
MAJOR APPENDAGES:	3 small booms
ATTITUDE CONTROL:	Unknown

EVENT DATA

KNOWN EVENTS: 2

FIRST DATE: January 1988

APOGEE	PERIGEE	PERIOD	INCLINATION
4219 km	760 km	134.0 min	65.0 deg

COMMENTS

First event (object 18816) may have occurred much earlier than the January 1988 date. Catalog actions taken at the end of 1988. The second event occurred in June 1992. Object 22008 led a short life, decaying on 2 August 1992.

SEASAT   **1978-064A**   **10967**

SATELLITE DATA

TYPE:	Payload (attached to Agena R/B)
OWNER:	US
LAUNCH DATE:	27 June 1978
DRY MASS (KG):	2300
MAIN BODY:	Cylinder; 1.5 m diameter by 21 m length
MAJOR APPENDAGES:	2 solar panels; 1 antenna panel; miscellaneous booms
ATTITUDE CONTROL:	None at time of event

EVENT DATA

KNOWN EVENTS:	Multiple
FIRST DATE:	July 1983

APOGEE	PERIGEE	PERIOD	INCLINATION
780 km	780 km	100.5 min	108.0 deg

COMMENTS

On average, one piece of debris is released per year, but sometimes in groups. Most debris experience very rapid decay for this altitude. Last known release was in 2007.

REFERENCE DOCUMENTS

"Environmentally-Induced Debris Sources," N.L. Johnson. Journal of Advances in Space Research, Vol. 34, Issue 5, 2004, pp. 993-999.

"Detection of Debris from Chinese ASAT Test Increases; One Minor Fragmentation Event in Second Quarter of 2007", The Orbital Debris Quarterly News, NASA JSC, July 2007.
Available online at http://www.orbitaldebris.jsc nasa.gov/newsletter/pdfs/ODQNv11i3.pdf.

COSMOS 1043   **1978-094A**   **11055**

SATELLITE DATA

TYPE:	Payload
OWNER:	USSR
LAUNCH DATE:	10 October 1978
DRY MASS (KG):	2200 (est.)
MAIN BODY:	Cylinder; dimensions ~1.5 m diameter by 5 m length
MAJOR APPENDAGES:	Solar panels; payload panels; gravity-gradient boom
ATTITUDE CONTROL:	None at time of the event.

EVENT DATA

KNOWN EVENTS:	1
FIRST DATE:	February 1993

APOGEE	**PERIGEE**	**PERIOD**	**INCLINATION**
437 km	435 km	94.9 min	81.2 deg

COMMENTS

No other satellite of this type has experienced an anomalous event. The piece was cataloged on 28 Feb 93 and decayed on 11 Mar 93. Given prior cataloging practices, other spacecraft could have experienced similar events that went unrecorded.

TIROS N   **1978-096A**   **11060**

SATELLITE DATA

TYPE:	Payload
OWNER:	US
LAUNCH DATE:	13 October 1978
DRY MASS (KG):	725
MAIN BODY:	Cylinder; 1.9 m diameter by 3.7 m length
MAJOR APPENDAGES:	1 solar panel
ATTITUDE CONTROL:	None at time of the event

EVENT DATA

KNOWN EVENTS:	2
FIRST DATE:	September 1987

APOGEE	PERIGEE	PERIOD	INCLINATION
855 km	835 km	101.9 min	99.0 deg

COMMENTS

Both fragments from the first event decayed rapidly during winter of 1988-89. A second event associated with 1978-096A occurred on 23 Feb 96 liberating 1 piece.

NIMBUS 7 R/B   **1978-098B**   **11081**

SATELLITE DATA

TYPE:	Delta Second Stage (2910)
OWNER:	US
LAUNCH DATE:	24 October 1978
DRY MASS (KG):	900
MAIN BODY:	Cylinder-nozzle; 2.4 m diameter by 8 m length
MAJOR APPENDAGES:	None
ATTITUDE CONTROL:	None at time of the event.

EVENT DATA

KNOWN EVENTS:	2
FIRST DATE:	May 1981

APOGEE	PERIGEE	PERIOD	INCLINATION
955 km	935 km	104.0 min	99.3 deg

COMMENTS

Second anomalous event apparently occurred about January 1987. A more prolific event in December 1981 is tentatively categorized as a satellite breakup (see Section 2). The cataloged debris section above refers only to the new fragment observed after the second anomalous event and does not include the Delta second stage that is accounted for in the tables of Section 2. These events may be related to the series of major Delta second stage breakups.

NOAA 6   1979-057A   11416

SATELLITE DATA

TYPE:	Payload
OWNER:	US
LAUNCH DATE:	27 June 1979
DRY MASS (KG):	723
MAIN BODY:	Rectangular; 3.71 m by 1.88 m
MAJOR APPENDAGES:	Solar panels
ATTITUDE CONTROL:	3-axis reaction control

EVENT DATA

KNOWN EVENTS: 2

FIRST DATE: September 1992

APOGEE	PERIGEE	PERIOD	INCLINATION
810 km	795 km	100.8 min	98.68 deg

COMMENTS

One piece of debris cataloged from the first event. A second event took place in June 1995 with one piece of debris liberated, but none cataloged.

REFERENCE DOCUMENT

Environmentally-Induced Debris Sources, N. L. Johnson, NASA Lyndon B. Johnson Space Center, Second World Space Congress, 2002.

METEOR 2-7   **1981-043A**   **12456**

SATELLITE DATA

TYPE:	Payload
OWNER:	CIS
LAUNCH DATE:	14 May 1981
DRY MASS (KG):	2750
MAIN BODY:	Cylinder
MAJOR APPENDAGES:	Large Solar Arrays
ATTITUDE CONTROL:	None at the time of event

EVENT DATA

KNOWN EVENTS:	1
FIRST DATE:	March 2004

APOGEE	PERIGEE	PERIOD	INCLINATION
895 km	825 km	102.15 min	81.3 deg

COMMENTS

Eight pieces of debris cataloged, may have been from two individual events, the origination date of the pieces is not conclusive. The rocket body associated with the launch of this spacecraft experienced an anomalous event over seven years earlier. The events are unrelated.

METEOR 2-7 R/B **1981-043B** **12457**

SATELLITE DATA

TYPE:	Vostok Final Stage
OWNER:	CIS
LAUNCH DATE:	14 May 1981
DRY MASS (KG):	1440
MAIN BODY:	Cylinder; 2.6 m diameter by 3.8 m length
MAJOR APPENDAGES:	None
ATTITUDE CONTROL:	None at time of the event.

EVENT DATA

KNOWN EVENTS:	1
FIRST DATE:	October 1996

APOGEE	PERIGEE	PERIOD	INCLINATION
920 km	825 km	102.41 min	81.3 deg

COMMENTS

One of several Vostok upper stages involved in anomalous events.

MOLNIYA 3-16 R/B   **1981-054E**   **12519**

SATELLITE DATA

TYPE:	Molniya Final Stage
OWNER:	CIS
LAUNCH DATE:	9 Jun 1981
DRY MASS (KG):	1100
MAIN BODY:	Cylinder; 2.7 m diameter by 3 m length
MAJOR APPENDAGES:	None
ATTITUDE CONTROL:	None

EVENT DATA

KNOWN EVENTS:	1
FIRST DATE:	July 1998

APOGEE	PERIGEE	PERIOD	INCLINATION
33415 km	85 km	583.42 min	62.0 deg

COMMENTS

No debris was cataloged from this event.

NOAA 7   1981-059A   12553

SATELLITE DATA

TYPE:	Payload
OWNER:	US
LAUNCH DATE:	23 June 1981
DRY MASS (KG):	723
MAIN BODY:	Rectangular; 3.71 m by 1.88 m
MAJOR APPENDAGES:	Solar panels
ATTITUDE CONTROL:	None at the time of event

EVENT DATA

KNOWN EVENTS:	2
FIRST DATE:	26.5 July 1993

APOGEE	PERIGEE	PERIOD	INCLINATION
835 km	830 km	101.6 min	98.9 deg

COMMENTS

Two objects were detected by the NAVSPOC and subsequently entered in the US SSN catalog. A piece separation analysis by the NAVSPOC identified the precise time these objects separated from the parent. It is unclear whether this event is a small breakup or whether other explanations such as spacecraft degradation could explain this separation. Unless other evidence is uncovered, this event will be classified as an anomalous event. The NOAA 7 payload was inactive for 3 years prior to this event. In 1997 a second, more curious event occurred. Three new debris appeared simultaneously with a discrete decrease in the orbital period of NOAA 7 of approximately 1 second.

REFERENCE DOCUMENT

Environmentally-Induced Debris Sources, N. L. Johnson, NASA Lyndon B. Johnson Space Center, Second World Space Congress, 2002.

OSCAR 30   **1985-066A**   **15935**

SATELLITE DATA

TYPE:	Payload
OWNER:	US
LAUNCH DATE:	3 August 1985
DRY MASS (KG):	60
MAIN BODY:	Octagonal cylinder; 0.5 m by 0.4 m
MAJOR APPENDAGES:	4 solar panels; gravity-gradient boom
ATTITUDE CONTROL:	Gravity-gradient boom

EVENT DATA

KNOWN EVENTS:	1
FIRST DATE:	27 December 1991

APOGEE	PERIGEE	PERIOD	INCLINATION
1255 km	1000 km	107.8 min	89.9 deg

COMMENTS

Other debris pieces are associated with this dual payload launch. The most recent event identified (SCC 21878) apparently originated from Oscar 30 on 27 December 1991, when a portion of the gravity-gradient boom broke off. One of several Transit-class satellites involved in anomalous events.

REFERENCE DOCUMENT

"The Transit System," L. Lee Pryor, AIAA Paper 92-1708, Applied Physics Laboratory, 1992.

COSMOS 1689 R/B **1985-090B** **16111**

SATELLITE DATA

TYPE:	Vostok Final Stage
OWNER:	CIS
LAUNCH DATE:	3 October 1985
DRY MASS (KG):	1440
MAIN BODY:	Cylinder; 2.6 m diameter by 3.8 m length
MAJOR APPENDAGES:	None
ATTITUDE CONTROL:	None at the time of the event.

EVENT DATA

KNOWN EVENTS:	1
FIRST DATE:	May 2002

APOGEE	PERIGEE	PERIOD	INCLINATION
565 km	510 km	95.4 min	97.7 deg

COMMENTS

One of several Vostok stages involved in anomalous events.

COSMOS 1823   **1987-020A**   **17535**

SATELLITE DATA

TYPE:	Payload
OWNER:	CIS
LAUNCH DATE:	20.20 Feb 1987
DRY MASS (KG):	1500
MAIN BODY:	Cylinder; 2.4 m diameter by 4 m length
MAJOR APPENDAGES:	Gravity-gradient boom; 10 small solar panels
ATTITUDE CONTROL:	Gravity gradient

EVENT DATA

KNOWN EVENTS:	3
FIRST DATE:	Apr-May 1997

APOGEE	PERIGEE	PERIOD	INCLINATION
1525 km	1480 km	116.0 min	73.6 deg

COMMENTS

Cosmos 1823 appears to have experienced three separate anomalous events, two in 1997 and one in 1999. Because Cosmos 1823 suffered a serious fragmentation in December 1987, the anomalous debris pieces may have been loosely attached to the spacecraft, then separated after continued exposure to the space environment or change in attitude of the spacecraft remnant.

REFERENCE DOCUMENTS

History of Soviet/Russian Satellite Fragmentations-A Joint U.S.-Russian Investigation, N. L. Johnson et al, Kaman Sciences Corporation, October 1995.

Environmentally-Induced Debris Sources, N. L. Johnson, NASA Lyndon B. Johnson Space Center, Second World Space Congress, 2002.

METEOR 2-17   **1988-005A**   **18820**

SATELLITE DATA

TYPE:	Payload
OWNER:	CIS
LAUNCH DATE:	30 January 1988
DRY MASS (KG):	2750
MAIN BODY:	Cylinder
MAJOR APPENDAGES:	Large Solar Arrays
ATTITUDE CONTROL:	None at the time of event

EVENT DATA

KNOWN EVENTS:	1
FIRST DATE:	21 June 2005

APOGEE	PERIGEE	PERIOD	INCLINATION
960	930	103.96 min	82.5 deg

COMMENTS

Twenty debris were cataloged during the period March 2006 – May 2007. Debris found in both slightly retrograde and posigrade orbits. Debris do not exhibit high area-to-mass ratios. Cause of event likely similar to Meteor 2-7.

COSMOS 1939 R/B   **1988-032B**   **19046**

SATELLITE DATA

TYPE:	Vostok Final Stage
OWNER:	CIS
LAUNCH DATE:	20 April 1988
DRY MASS (KG):	1440
MAIN BODY:	Cylinder; 2.6 m diameter by 3.8 m length
MAJOR APPENDAGES:	None
ATTITUDE CONTROL:	None at the time of the event.

EVENT DATA

KNOWN EVENTS:	2
FIRST DATE:	30 July 1996

APOGEE	PERIGEE	PERIOD	INCLINATION
655 km	585 km	97.14 min	97.6 deg

COMMENTS

One of several Vostok final stages involved in anomalous events.

COBE   **1989-089A**   **20322**

SATELLITE DATA

TYPE:	Payload
OWNER:	US
LAUNCH DATE:	18.61 November 1989
DRY MASS (KG):	2265
MAIN BODY:	Cylinder; with protective shield, 4.0 m diameter by 5.8 m length
MAJOR APPENDAGES:	3 - 8.6 m solar arrays
ATTITUDE CONTROL:	Spin stabilized; gyroscopes

EVENT DATA

KNOWN EVENTS:	Multiple
FIRST DATE:	January 1993

APOGEE	PERIGEE	PERIOD	INCLINATION
885 km	870 km	102.5 min	99.0 deg

COMMENTS

At least 12 separate event dates have been calculated by the NAVSPOC, and other events are certain to have occurred. Through December 1993 the satellite remained active, and the cause of the separations could be determined. No degradation of satellite performance was reported by the satellite operators.

REFERENCE DOCUMENT

Environmentally-Induced Debris Sources, N. L. Johnson, NASA Lyndon B. Johnson Space Center, Second World Space Congress, 2002.

NADEZHDA 2 R/B   **1990-017B**   **20509**

SATELLITE DATA

TYPE:	Rocket Body
OWNER:	CIS
LAUNCH DATE:	27 February 1990
DRY MASS (KG):	1434
MAIN BODY:	Cylinder; 2.4 m diameter x 6.0 m length
MAJOR APPENDAGES:	None
ATTITUDE CONTROL:	None at the time of the event

EVENT DATA

KNOWN EVENTS:	1
FIRST DATE:	22 June 2005

APOGEE	PERIGEE	PERIOD	INCLINATION
1015	950	104.65 min	83.0 deg

COMMENTS

A piece was detected in a more eccentric and higher inclination orbit, indicating that this event may have been a collision with a small, uncataloged object or meteoroid.

REFERENCE DOCUMENT

"Recent Satellite Breakups", The Orbital Debris Quarterly News, NASA JSC, July 2005.
Available online at http://www.orbitaldebris.jsc nasa.gov/newsletter/pdfs/ODQNv9i3.pdf.

HST   **1990-037B**   **20580**

SATELLITE DATA

TYPE:	Payload
OWNER:	US
LAUNCH DATE:	24 April 1990
DRY MASS (KG):	10863
MAIN BODY:	Cylinder
MAJOR APPENDAGES:	Two Solar Array Panels
ATTITUDE CONTROL:	CMG controlled

EVENT DATA

KNOWN EVENTS:	1
FIRST DATE:	August 2003

APOGEE	PERIGEE	PERIOD	INCLINATION
575 km	570 km	96.1 min	28.5 deg

COMMENTS

The Hubble Space Telescope debris decayed rapidly after the event.

OKEAN 3   **1991-039A**   **21397**

SATELLITE DATA

TYPE:	Payload
OWNER:	CIS
LAUNCH DATE:	4 June 1991
DRY MASS (KG):	1922
MAIN BODY:	Cylinder; 1.4-0.8 m diameter by 3.5 m length
MAJOR APPENDAGES:	Solar arrays, payload trays, radar antenna
ATTITUDE CONTROL:	Gravity-gradient

EVENT DATA

KNOWN EVENTS:	1
DATE:	12 October 1998

APOGEE	PERIGEE	PERIOD	INCLINATION
665 km	620 km	97.5 min	82.5 deg

COMMENTS

First event for this type object. No other events observed.

REFERENCE DOCUMENT

Environmentally-Induced Debris Sources, N. L. Johnson, NASA Lyndon B. Johnson Space Center, Second World Space Congress, 2002.

SARA   **1991-050E**   **21578**

SATELLITE DATA

TYPE:	Payload
OWNER:	France
LAUNCH DATE:	17 July 1991
DRY MASS (KG):	26
MAIN BODY:	Cube; 360 mm per side
MAJOR APPENDAGES:	Several deployable 5 m long antennae
ATTITUDE CONTROL:	None at the time of events

EVENT DATA

KNOWN EVENTS:	3
FIRST DATE:	August 2003

APOGEE	PERIGEE	PERIOD	INCLINATION
730 km	730 km	99.4 min	98.1 deg

COMMENTS

This French "microsat" was no longer active at the time of the events. Some objects may be a piece broken off from any of the long antennae. Follow on events occurred on 17 April 2005 and 15 October 2006. Four debris total have been cataloged from the parent object.

ERS-1 R/B   **1991-050F**   **21610**

SATELLITE DATA

TYPE:	Ariane 40 Rocket Body
OWNER:	France
LAUNCH DATE:	17 July 1991
DRY MASS (KG):	1720
MAIN BODY:	Cylinder; 2.6 m diameter by 10 m length
MAJOR APPENDAGES:	None
ATTITUDE CONTROL:	None

EVENT DATA

KNOWN EVENTS: 1

FIRST DATE: April 2001

APOGEE	PERIGEE	PERIOD	INCLINATION
770 km	770 km	100.2 min	98.2 deg

COMMENTS

One piece of debris cataloged. Parent object was in a sun-synchronous orbit at the time of the event. First occurrence of an anomalous event with an Ariane R/B.

REFERENCE DOCUMENT

Environmentally-Induced Debris Sources, N. L. Johnson, NASA Lyndon B. Johnson Space Center, Second World Space Congress, 2002.

EKA 1 (START 1)   **1993-014A**   **22561**

SATELLITE DATA

TYPE:	Payload
OWNER:	CIS
LAUNCH DATE:	25 March 1993
DRY MASS (KG):	260
MAIN BODY:	Two cylinders; < 1.5 m diameter
MAJOR APPENDAGES:	Solar panels; gravity-gradient boom
ATTITUDE CONTROL:	Gravity-gradient

EVENT DATA

KNOWN EVENTS:	Multiple
FIRST DATE:	4 March 1998

APOGEE	PERIGEE	PERIOD	INCLINATION
970 km	685 km	101.43 min	75.8 deg

COMMENTS

EKA 1 is a test payload prior to launches of small communications satellites. First orbital launch of Start-1 booster derived from SS-20/SS-25 missiles.

START 1 R/B   **1993-014B**   **22562**

SATELLITE DATA

TYPE:	Rocket Body
OWNER:	CIS
LAUNCH DATE:	25 March 1993
DRY MASS (KG):	200
MAIN BODY:	Cylinder; 1.4 m diameter x 2.5 m length
MAJOR APPENDAGES:	None
ATTITUDE CONTROL:	None at the time of the event

EVENT DATA

KNOWN EVENTS:	Multiple
FIRST DATE:	Late 2002

APOGEE	PERIGEE	PERIOD	INCLINATION
920 km	680 km	100.85 min	75.8 deg

COMMENTS

Dozens of pieces have been cataloged from this rocket body, starting in late 2002 through 2005. Exact time and date of the event(s) is unknown. The payload associated with this rocket body, START-1, also experienced an anomalous event in March of 1998, the events are unrelated.

COSMOS 2297 R/B   **1994-077B**   **23405**

SATELLITE DATA

TYPE:	Zenit Second Stage
OWNER:	CIS
LAUNCH DATE:	24 November 1994
DRY MASS (KG):	8300
MAIN BODY:	Cylinder; 3.9 m diameter by 12 m length
MAJOR APPENDAGES:	None
ATTITUDE CONTROL:	None

EVENT DATA

KNOWN EVENTS:	2?
FIRST DATE:	June 1998

APOGEE	PERIGEE	PERIOD	INCLINATION
845 km	845 km	101.82 min	71.0 deg

COMMENTS

One piece of debris was cataloged.

REFERENCE DOCUMENTS

History of Soviet/Russian Satellite Fragmentations-A Joint U.S.-Russian Investigation, N. L. Johnson et al, Kaman Sciences Corporation, October 1995.

Environmentally-Induced Debris Sources, N. L. Johnson, NASA Lyndon B. Johnson Space Center, Second World Space Congress, 2002.

KOREASAT 1 R/B 1995-041B **23640**

SATELLITE DATA

TYPE:	Delta Second Stage
OWNER:	US
LAUNCH DATE:	5 August 1995
DRY MASS (KG):	900
MAIN BODY:	Cylinder; 2.4 m diameter by 8 m length
MAJOR APPENDAGES:	None
ATTITUDE CONTROL:	None at time of the event.

EVENT DATA

KNOWN EVENTS:	1
DATE:	6 December 1995

APOGEE	PERIGEE	PERIOD	INCLINATION
1375 km	935 km	108.5 min	26.7 deg

COMMENTS

One piece was liberated.

RADARSAT R/B   **1995-059B**   **23711**

SATELLITE DATA

TYPE:	Delta Second Stage
OWNER:	US
LAUNCH DATE:	5 November 1995
DRY MASS (KG):	900
MAIN BODY:	Cylinder; 2.4 m diameter by 8 m length
MAJOR APPENDAGES:	None
ATTITUDE CONTROL:	None

EVENT DATA

KNOWN EVENTS:	1
DATE:	30 January 1996

APOGEE	PERIGEE	PERIOD	INCLINATION
1495 km	935 km	109.7 min	100.6 deg

COMMENTS

One piece was liberated.

FUSE   **1999-035A**   **25791**

SATELLITE DATA

TYPE:	Payload
OWNER:	US
LAUNCH DATE:	24 June 1999
DRY MASS (KG):	1360
MAIN BODY:	Box; 1.3 m by 0.9 m by 0.9 m
MAJOR APPENDAGES:	Two 3.5 m^2 solar Arrays
ATTITUDE CONTROL:	Three-axis stabilized

EVENT DATA

KNOWN EVENTS:	1
FIRST DATE:	6 June 2004

APOGEE	PERIGEE	PERIOD	INCLINATION
760 km	745 km	99.90 min	25.0 deg

COMMENTS

Eight pieces of debris were cataloged from this event. An additional piece was detected but never cataloged. The event might have been coincidental with a "safe mode" entry around 5 June 2004, which cause the closing and reopening of several sensor doors. Five of the cataloged debris had reentered within 8 months indicating higher than normal area-to-mass ratios. The event did not affect vehicle operations or performance.

REFERENCE DOCUMENT

"FUSE Satellite Releases Unexpected Debris", The Orbital Debris Quarterly News, NASA JSC, July 2004. Available online at http://www.orbitaldebris.jsc nasa.gov/newsletter/pdfs/ODQNv8i3.pdf.

IKONOS 2   **1999-051A**   **25919**

SATELLITE DATA

TYPE:	Payload
OWNER:	US
LAUNCH DATE:	24.76 September 1999
DRY MASS (KG):	735
MAIN BODY:	Box; 1.8 m by 1.8 m by 1.6 m
MAJOR APPENDAGES:	3 solar panels
ATTITUDE CONTROL:	3 axis stabilization

EVENT DATA

KNOWN EVENTS: 1

DATE: 19 March 2001

APOGEE	PERIGEE	PERIOD	INCLINATION
680 km	678 km	98.3 min	98.2 deg

COMMENTS

One piece was liberated. A very high ballistic coefficient resulted in the anomalous debris object reentering on 11 April 2001.

4.0   OTHER SATELLITES ASSOCIATED WITH FRAGMENTATIONS

4.1   Aerodynamic Associations with Fragmentation Events

A change from earlier editions of the <u>History of On-Orbit Satellite Fragmentations</u> was to remove fragmentation events associated with aerodynamic effects at the time of reentry from Chapter 2 and into this section. Twenty-four such events have occurred between 1994 and August 2007. Because of the orbit elements of the parent object at the time of fragmentation, only three of these events showed any cataloged debris and all parent objects reentered within 1 year of the event (most reentered within a few days). It is understood that only a fraction of these fragmentations can be detected, because of the short remaining life of the parent and debris created. These events have no impact to the mid or long term debris environment, and therefore it was deemed more appropriate to separate these from the fragmentations in Chapter 2. The parent object for these aerodynamic events shall not be considered "fragmentation debris" when discussing object categorization. As mentioned, three of these events (1964-006D, 1978-083A, and 1980-028A) produced cataloged debris other than the parent, and these three debris objects represent the difference between the decayed fragmentation debris count in Table 1.3.2 and the decayed fragmentation debris count in Tables 2.1 and 2.2.

The following missions, listed by international designator in Table 4.1-1 and by event date in 4.1-2, have been determined to been solely related to aerodynamic effects at the time of reentry, and therefore did not contribute to the orbital environment.

TABLE 4.1-1 HISTORY OF SATELLITE AERODYNAMIC EVENTS BY LAUNCH DATE

NAME	INTERNATIONAL DESIGNATOR	CATALOG NUMBER	LAUNCH DATE	EVENT DATE	REENTRY DATE	DEBRIS CATALOGED	APOGEE (KM)	PERIGEE (KM)	INCLINATION (DEG)	COMMENT
ELEKTRON 1/2 R/B	1964-006D	751	30-Jan-64	13-Feb-98	15-Feb-98	2	56315	90	56.2	VOSTOK FINAL STAGE
COSMOS 41 R/B	1964-049E	898	22-Aug-64	4-Apr-04	7-May-04	1	~35750	~115	64.5	MOLNIYA FINAL STAGE
COSMOS 41 DEB	1964-049F	13091	22-Aug-64	30-Dec-02	31-Dec-02	1	1200	85	64.4	
COSMOS 1030	1978-083A	11015	6-Sep-78	14-Aug-04	17-Aug-04	2	~4560	~95	61.9	
COSMOS 1172	1980-028A	11758	12-Apr-80	23-Dec-97	26-Dec-97	2	5125	75	61.8	
MOLNIYA 3-16	1981-054A	12512	9-Jun-81	5-Feb-98	10-Feb-98	1	7670	85	62.1	
MOLNIYA 3-16 R/B	1981-054E	12519	9-Jun-81	1-Jul-98	30-Apr-99	1	33415	85	62.0	MOLNIYA FINAL STAGE
MOLNIYA 3-19	1982-083A	13432	27-Aug-82	13-Jan-02	13-Jan-02	1	2075	95	62.3	
COSMOS 1658	1985-045A	15808	11-Jun-85	12-Nov-05	12-Nov-05	1	1730	80	62.1	
MOLNIYA 3-26	1985-091A	16112	3-Oct-85	21-Feb-01	22-Feb-01	1	5690	80	62.6	
MOLNIYA 1-66 R/B	1985-103D	16223	28-Oct-85	13-Jan-03	13-Jan-03	1	~1600	~120	62.4	MOLNIYA FINAL STAGE
COSMOS 1701	1985-105A	16235	9-Nov-85	29-Apr-01	11-May-01	1	25570	85	62.9	
COSMOS 1849	1987-048A	18083	4-Jun-87	27-Jan-03	4-Feb-03	1	7450	95	62.1	
COSMOS 1966	1988-076A	19445	30-Aug-88	~02-Nov-05	10-Nov-05	1	11535	90	62.9	
MOLNIYA 3-35	1989-043A	20052	8-Jun-89	14-Dec-01	14-Dec-01	1	595	65	61.9	
MOLNIYA 3-36	1989-094A	20338	28-Nov-89	19-May-00	20-May-00	1	7145	75	63.6	
MOLNIYA 3-36 R/B	1989-094B	20339	28-Nov-89	28-Jun-00	4-Jul-00	1	1530	80	63.7	MOLNIYA FINAL STAGE
MOLNIYA 1-77	1990-039A	20583	26-Apr-90	24-Feb-05	25-Feb-05	1	1710	75	62.0	
MOLNIYA 3-38 R/B	1990-052D	20649	13-Jun-90	Sep-06	13-Sep-06	1	37710	130	62.4	MOLNIYA FINAL STAGE
MOLNIYA 1-82	1991-053A	21630	1-Aug-91	8-Oct-04	9-Oct-04	1	1510	75	61.7	
MOLNIYA 1-83 R/B	1992-011D	21900	4-Mar-92	26-Sep-06	26-Sep-06	1	1090	70	62.0	MOLNIYA FINAL STAGE
MOLNIYA 3-44	1993-025A	22633	21-Apr-93	25-Jan-04	25-Jan-04	1	~1000	~90	63.4	
ETS-VI R/B	1994-056B	23231	28-Aug-94	31-Mar-95	2-Apr-95	1	4840	100	28.6	H-II SECOND STAGE
HELLAS SAT-2 R/B	2003-020B	27812	13-May-03	11-Dec-04	12-Dec-04	1	10300	90	17.5	ATLAS V
					TOTAL	27				

TABLE 4.1-2 HISTORY OF SATELLITE AERODYNAMIC EVENTS BY EVENT DATE

NAME	INTERNATIONAL DESIGNATOR	CATALOG NUMBER	LAUNCH DATE	EVENT DATE	REENTRY DATE	DEBRIS CATALOGED	APOGEE (KM)	PERIGEE (KM)	INCLINATION (DEG)	COMMENT
ETS-VI R/B	1994-056B	23231	28-Aug-94	31-Mar-95	2-Apr-95	1	4840	100	28.6	H-II SECOND STAGE
COSMOS 1172	1980-028A	11758	12-Apr-80	23-Dec-97	26-Dec-97	2	5125	75	61.8	
MOLNIYA 3-16	1981-054A	12512	9-Jun-81	5-Feb-98	10-Feb-98	1	7670	85	62.1	
ELEKTRON 1/2 R/B	1964-006D	751	30-Jan-64	13-Feb-98	15-Feb-98	2	56315	90	56.2	VOSTOK FINAL STAGE
MOLNIYA 3-16 R/B	1981-054E	12519	9-Jun-81	1-Jul-98	30-Apr-99	1	33415	85	62.0	MOLNIYA FINAL STAGE
MOLNIYA 3-36	1989-094A	20338	28-Nov-89	19-May-00	20-May-00	1	7145	75	63.6	
MOLNIYA 3-36 R/B	1989-094B	20339	28-Nov-89	28-Jun-00	4-Jul-00	1	1530	80	63.7	MOLNIYA FINAL STAGE
MOLNIYA 3-26	1985-091A	16112	3-Oct-85	21-Feb-01	22-Feb-01	1	5690	80	62.6	
COSMOS 1701	1985-105A	16235	9-Nov-85	29-Apr-01	11-May-01	1	25570	85	62.9	
MOLNIYA 3-35	1989-043A	20052	8-Jun-89	14-Dec-01	14-Dec-01	1	595	65	61.9	
MOLNIYA 3-19	1982-083A	13432	27-Aug-82	13-Jan-02	13-Jan-02	1	2075	95	62.3	
COSMOS 41 DEB	1964-049F	13091	22-Aug-64	30-Dec-02	31-Dec-02	1	1200	85	64.4	
MOLNIYA 1-66 R/B	1985-103D	16223	28-Oct-85	13-Jan-03	13-Jan-03	1	~1600	~120	62.4	MOLNIYA FINAL STAGE
COSMOS 1849	1987-048A	18083	4-Jun-87	27-Jan-03	4-Feb-03	1	7450	95	62.1	
MOLNIYA 3-44	1993-025A	22633	21-Apr-93	25-Jan-04	25-Jan-04	1	~1000	~90	63.4	
COSMOS 41 R/B	1964-049E	898	22-Aug-64	4-Apr-04	7-May-04	1	~35750	~115	64.5	MOLNIYA FINAL STAGE
COSMOS 1030	1978-083A	11015	6-Sep-78	14-Aug-04	17-Aug-04	2	~4560	~95	61.9	
MOLNIYA 1-82	1991-053A	21630	1-Aug-91	8-Oct-04	9-Oct-04	1	1510	75	61.7	
HELLAS SAT-2 R/B	2003-020B	27812	13-May-03	11-Dec-04	12-Dec-04	1	10300	90	17.5	ATLAS V
MOLNIYA 1-77	1990-039A	20583	26-Apr-90	24-Feb-05	25-Feb-05	1	1710	75	62.0	
COSMOS 1966	1988-076A	19445	30-Aug-88	-02-Nov-05	10-Nov-05	1	11535	90	62.9	
COSMOS 1658	1985-045A	15808	11-Jun-85	12-Nov-05	12-Nov-05	1	1730	80	62.1	
MOLNIYA 3-38 R/B	1990-052D	20649	13-Jun-90	Sep-06	13-Sep-06	1	37710	130	62.4	MOLNIYA FINAL STAGE
MOLNIYA 1-83 R/B	1992-011D	21900	4-Mar-92	26-Sep-06	26-Sep-06	1	1090	70	62.0	MOLNIYA FINAL STAGE
					TOTAL	27				

4.2 Spurious Associations with Fragmentation Events

Satellite fragmentation lists compiled by other organizations, in particular by the National Security Council and AFSSS, were carefully reviewed during the preparation of the fourth edition of the History of On-Orbit Satellite Fragmentations. However, due to the frequent exchange of information within the small orbital debris and space operations community and the long period during which satellite fragmentation lists have been maintained, no current list is completely independent from all others.

These reviews also revealed the need to define better the terms "satellite breakup" and "anomalous event" as discussed in Section 1.0. Many "breakup" lists have historically included entries related to normal launch and mission activities, which resulted in numbers of debris in excess of the handful usually observed on these occasions. Some researchers have been misled by tracking difficulties and cataloging procedures, which may cause late cataloging or misidentification of debris, superficially giving the appearance of fragmentations. A higher than average number of debris alone is not sufficient to assume a satellite fragmentation. Such pitfalls can generally be avoided by conducting analyses with complete satellite element set data rather than the limited orbital data available in the U.S. Satellite Catalog.

The following space missions, listed by international designator, have been examined in detail and have failed to qualify as either satellite breakup or anomalous event as set forth in Section 1.0. The source of debris associated with nearly all of these flights is of a mission related nature. Underlined items indicate the alleged source of the debris.

Table 4.2: Spurious Association with Fragmentations by Launch Date

INT'L Des.	S/C COMMON NAME	R/B	TOTAL DEBRIS	DEBRIS ON-ORBIT
1963-014	FTV 1169	Agena B spacecraft	147	50
1965-073	C 86-90	Cosmos 3	5	5
1965-112	C 103	Cosmos 3	13	0
1967-001	INTELSAT 2-F2	Delta 1 R/B (2): FW-4	17	1
1967-011	Diademe 1	Diamant	13	0
1967-024	C 149	Cosmos 2	16	0
1967-086	C 176	Cosmos 2	9	0
1968-117	C 261	Cosmos 2	22	0
1969-021	C 269	Cosmos 3	21	0
1970-005	C 320	Cosmos 2	5	0
1970-033	C 334	Cosmos 2	3	0
1970-065	C 359	Molniya	3	0
1972-078	C 523	Cosmos 2	10	0
1973-027	Skylab 1	Saturn V	22	0

1973-075	C 601	Cosmos 2	12	0
1974-074	C 686	Cosmos 2	18	0
1974-104	Salyut 4	Proton	17	0
1976-012	C 801	Cosmos 2	15	0
1976-037	C 816	Cosmos 3	23	0
1976-057	Salyut 5	Proton	8	0
1976-124	C 885	Cosmos 3	17	0
1977-042	C 913	Cosmos 3	20	0
1977-097	Salyut 6	Proton	104	0
1977-111	C 965	Cosmos 3	25	0
1978-043	C 1004	Soyuz	5	0
1978-120	C 1065	Cosmos 3	6	0
1979-008	C 1074	Soyuz	5	0
1979-063	C 1112	Cosmos 3	24	0
1980-047	C 1186	Cosmos 3	25	0
1980-067	C 1204	Cosmos 3	22	0
1980-083	C 1215	Cosmos 3	2	0
1981-093	SJ-2/-2A/-2B	CZ-2B	6	0
1981-097	C 1311	Cosmos 3	24	0
1982-006	OPS 2849	Titan 3B Agena	4	3
1982-007	C 1335	Cosmos 3	22	0
1982-033	Salyut 7	Proton	197	0
1982-034	C 1351	Cosmos 3	24	0
1982-076	C 1397	Cosmos 3	22	0
1983-034	C 1453	Cosmos 3	22	0
1983-049	C 1465	Cosmos 3	8	0
1983-091	C 1494	Cosmos 3	25	0
1983-101	C 1501	Cosmos 3	24	0
1984-008	STTW-T1	CZ-3	2	0
1984-104	C 1601	Cosmos 3	28	0
1985-021	GEOSAT	Atlas 41E (OIS R/B)	4	3
1985-050	C 1662	Cosmos 3	27	0
1985-075	C 1677	Tsyklon	2	0
1985-097	C 1697	Zenit	4	4
1986-017	Mir	Proton	323	0

1986-024	**C 1736**	**Tsyklon**	28	1
1986-030	C 1741	Cosmos 3	6	6
1986-052	C 1763	**Cosmos 3**	3	3
1986-067	**C 1776**	Cosmos 3	28	0
1986-101	C 1809	**Tsyklon**	9	9
1988-019	**C 1932**	Tsyklon	3	2
1988-065	C 1960	Cosmos 3	28	0
1988-067	FSW-1 2	**CZ-2C**	5	0
1988-113	**C 1985**	Tsyklon	36	0
1989-012	C 2002	**Cosmos 3**	10	0
1989-100	**C 2053**	Tsyklon	37	0
1990-012	C 2059	**Cosmos 3**	10	0
1990-038	C 2075	Cosmos 3	14	0
1990-104	C 2106	Tsyklon	28	0
1995-008	C 2306	Cosmos 3	23	0

For more information on these events, see History of On-orbit Satellite Fragmentations, 4th Ed., Jan. 1990; the Interagency Group (Space) Report on Orbital Debris, 1989; and Soviet Space Programs, 1976-80, Part 3, May 1985.

5.0 SATELLITES NOT ASSOCIATED WITH BREAKUPS

The table below identifies specific SSN numbers of objects, which possess the same International Designator year and number but are not associated with the indicated event. For example, 1961-015C was an Ablestar rocket body, which broke up. The mission deployed two objects (Transit 4A and Solrad 3/Injun 1) that were not associated with the rocket body explosion. Those two objects are not counted in the 1961-015 totals, although they definitely are associated with the 1961-015 international designator.

Occasionally it is not obvious whether an object should be included in a fragmentation event. In those cases historical research and historical Satellite Catalogs usually reveal whether an object should be included in the count. The list below represents the best summary of excluded objects. The parent object is always considered a fragment. Aerodynamic breakups are included in this list if they produced cataloged fragmentation other than the parent object.

The list below is formatted as follows: The international designator and number of excluded debris in parenthesis are followed by the SSN numbers, which are not debris. A blank line separates years.

Int'l Designator

1961-015 (2) -   116 117

1962-057 (0) -

1963-047 (0) -

1964-006 (27) -   746 748 750 14427 14428 15786 16544 16545 16546 16547 16548 18589 18686 19010 19173
19990 19991 19992 19993 19994 19995 19996 19997 19998 20101 20224 21621
1964-070 (1) -   920

1965-012 (1) -   1095
1965-020 (3) -   1267 1268 1269
1965-082 (1) -   1624
1965-088 (23)   1707 1708 1740 1741 1784 1785 1786 1787 1788 1789 1790 1791 1792 1793 1794 1795 1796
1797 1798 1799 1800 1801 1802

1966-012 (2) -   2012 2014
1966-046 (3) -   2186 2189 2190
1966-056 (3) -   2255 2256 2511
1966-059 (1) -   2291
1966-088 (0) –
1966-101 (0) –

1968-003 (1) -   3096
1968-025 (1) -   3170
1968-081 (5) -   3428 3429 3430 3431 5999
1968-090 (0) -
1968-091 (1) -   3505
1968-097 (0) -

1969-029 (1) -   3835

1969-064 (1) -   4051
1969-082 (10) - 4111 4132 4166 4168 4237 4247 4256 4257 4259 4295

1970-025 (2) -   4362 4363
1970-089 (1) -   4597
1970-091 (0) -

1971-015 (1) -   4965
1971-106 (4) -   5650 5664 5665 5672

1972-058 (1) -   6126

1973-017 (1) -   6398
1973-021 (2) -   6434 6436
1973-086 (1) -   6920

1974-015 (1) -   7218
1974-089 (3) -   7529 7530 7531
1974-103 (1) -   7588

1975-004 (1) -   7615
1975-052 (2) -   7924 7965
1975-080 (1) -   8192
1975-102 (1) -   8417

1976-063 (1) -   8933
1976-067 (2) -   9013 9016
1976-072 (1) -   9048
1976-077 (1) -   9057
1976-105 (3) -   9496 9497 9506
1976-120 (2) -   9604 9605
1976-123 (4) -   9623 9624 9639 9640
1976-126 (3) -   9643 9644 9645

1977-027 (3) -   9912 9913 9921
1977-047 (3) -   10060 10066 10089
1977-065 (3) -   10143 10145 10156
1977-068 (3) -   10151 10152 10167
1977-092 (6) -   10366 10367 10368 10408 10484 11571
1977-121 (1) -   10532

1978-026 (2) -   10702 10703
1978-083 (3) -   11016 11017 11076
1978-098 (2) -   11080 18605
1978-100 (4) -   11084 11085 11086 11177

1979-017 (3) -   11279 11291 11322
1979-033 (2) -   11334 11367
1979-058 (3) -   11418 11423 11555
1979-077 (3) -   11512 11513 11550
1979-104 (3) -   11645 24754 25098

1980-021 (1) -   11730
1980-028 (4) -   11759 11760 11761 11762
1980-030 (1) -   11766
1980-057 (3) -   11872 11873 11888

```
1980-085 (3) -    12033 12034 12035
1980-089 (1) -    12055

1981-016 (4) -    12304 12305 12306 12311
1981-028 (1) -    12365
1981-031 (3) -    12377 12378 12384
1981-053 (1) -    12508
1981-058 (3) -    12548 12549 12561
1981-071 (3) -    12629 12630 12680
1981-072 (1) -    12632
1981-088 (5) -    12818 12819 12820 12821 12822
1981-089 (1) -    12829
1981-108 (3) -    12934 12935 12940

1982-038 (1) -    13151
1982-055 (2) -    13260 13261
1982-088 (1) -    13509
1982-115 (4) -    13685 13686 13692 13693

1983-020 (3) -    13901 13903 20413
1983-022 (2) -    13924 14477
1983-038 (6) -    14036 14037 14038 14041 14042 14043
1983-044 (1) -    14065
1983-070 (3) -    14183 14184 14191
1983-075 (5) -    14208 14209 14229 14631 14928
1983-127 (7) -    14590 14591 14592 14593 14594 14595 14607

1984-011 (6) -    14681 14688 14689 14692 14695 14696
1984-083 (1) -    15168
1984-106 (6) -    15333 15334 15335 15336 15337 17358
1984-114 (2) -    15385 15386

1985-030 (1) -    15654
1985-037 (7) -    15697 15698 15699 15700 15701 15702 15715
1985-039 (1) -    15735
1985-042 (5) -    15755 15770 15771 15772 15774
1985-082 (1) -    16055
1985-094 (6) -    16138 16140 16141 16142 16143 16144
1985-108 (1) -    16262
1985-118 (10) -   16396 16397 16398 16399 16403 16404 16405 16406 16407 16445
1985-121 (5) -    16434 16435 16436 16437 16438

1986-019 (3) -    16613 16614 16616
1986-059 (1) -    16896
1986-069 (0) -

1987-004 (1) -    7298
1987-020 (4) -    17536 26111 26601 26982
1987-059 (2) -    18185 18186
1987-062 (1) -    18215
1987-068 (1) -    18312
1987-078 (3) -    18350 18351 18353
1987-079 (6) -    18355 18356 18357 18358 18359 18360
1987-108 (1) -    18714
1987-109 (5) -    18715 18716 18717 18718 18722
```

1988-007 (1) -   18824
1988-023 (1) -   18986
1988-040 (1) -   19121
1988-085 (6) -   19501 19502 19503 19504 19505 21751
1988-109 (3) -   19687 19688 19690

1989-001 (6) -   19749 19750 19751 19752 19753 19754
1989-004 (5) -   19765 19766 19767 19768 19776
1989-006 (1) -   19772
1989-039 (7) -   20024 20025 20026 20027 20028 20044 20082
1989-052 (5) -   20107 20108 20109 20110 20115
1989-054 (1) -   20125
1989-056 (2) -   20137 20138
1989-089 (79) - 20322 20324 20328 22625 22683 22695 22747 22748 22749 22750 22751 22752 22753 22754
                 22755 22756 22757 22758 22759 22760 22761 22762 22763 22764 22765 22766 22767 22768
                 22769 22770 22771 22772 22773 22774 22775 22776 22820 22852 22853 22854 22855 22856
                 22857 22858 22972 23053 23054 23055 23056 23057 23058 23059 23060 23061 23062 23063
                 23064 23065 23066 23067 23068 23069 23070 23071 23072 23073 23074 23075 23076 23077
                 23078 23079 23080 23081 23082 23083 23084 23085 23086
1989-100 (38) - 20389 20397 20398 20408 20467 20468 20515 20522 20531 20532 20637 20640 20802 20803
                 20821 20822 20823 20911 21020 21021 21022 21023 21042 21043 21064 21205 21206 21207
                 21537 21540 21767 21768 21769 21770 21771 21772 21773 21774
1989-101 (6) -   20391 20392 20393 20394 20400 21648

1990-045 (6) -   20619 20620 20621 20622 20623 20630
1990-081 (7) -   20788 20789 20790 20792 20793 20797 20798
1990-087 (1) -   20829
1990-102 (5) -   20953 20954 20955 20958 21046
1990-105 (1) -   20978
1990-110 (6) -   21006 21007 21008 21009 21010 21011

1991-003 (3) -   21055 21056 21058
1991-009 (8) -   21100 21101 21102 21103 21104 21105 21106 21107
1991-010 (5) -   21111 21112 21113 21122 21129
1991-015 (4) -   21139 21140 21142 21904
1991-025 (6) -   21216 21217 21218 21219 21220 21221
1991-068 (6) -   21728 21729 21730 21731 21732 21733
1991-071 (1) -   21742
1991-075 (1) -   21765
1991-082 (4)   21800 21801 21825 21836

1992-021 (3) -   21939 21940 21942
1992-041 (8) -   22027 22028 22033 27484 27485 27486 27487 27675
1992-047 (6) -   22056 22057 22058 22059 22060 22061
1992-082 (5) -   22245 22246 22247 22248 22249
1992-088 (5) -   22269 22270 22271 22272 22273
1992-091 (1) -   22281
1992-093 (5) -   22284 22290 22291 22292 22293

1993-016 (3) -   22565 22575 22576
1993-018 (1) -   22586
1993-028 (1) -   22642
1993-045 (1) -   22717
1993-057 (2) -   22790 22953
1993-072 (5) -   22907 22908 22909 22910 22926

1994-004 (2) -	22973 22987
1994-029 (1) -	23105
1994-038 (5) -	23168 23169 23170 23171 23172
1994-069 (5) -	23327 23328 23329 23330 23339
1994-085 (1) -	23439
1995-028 (1) -	23597
1995-033 (3) -	23605 23607 23608
1995-037 (9) -	23620 23621 23622 23623 23624 23625 23626 23627 23630
1996-010 (4) -	23794 23795 23796 23824
1996-034 (5) -	23880 23881 23882 23883 23886
1997-024 (1) -	24806
1997-070 (5) -	25045 25046 25047 25048 25053
1997-079 (1) -	25089
1997-086 (3) -	25126 25127 25128
1998-011 (1) -	25175
1999-025 (3) -	25731 25732 25733
1999-057 (2) -	25940 25941
1999-072 (1) -	26041
2000-036 (5) -	26394 26395 26396 26397 26399
2001-049 (3) -	26957 26958 26959
2001-057 (1) -	27054
2002-037 (6) -	27470 27471 27472 27473 27476 27494
2003-035 (6) -	27857 28084 28085 28086 28087 28088
2006-002 (1) -	28931
2006-006 (1) -	28943
2006-037 (4) -	29393 29395 29396 29493
2006-039 (2) -	29397 29403
2006-050 (5) -	29522 29524 29525 29600 29637
2007-003 (1) -	30324

6.0 SATELLITES NOT ASSOCIATED WITH ANOMALOUS EVENTS

The table below identifies specific SSN numbers of objects, which possess the same International Designator year and number but are not associated with the indicated anomalous event. The list below represents the best summary of excluded objects. Parent object is not considered a fragment.

The list below is formatted as follows: The international designator and number of excluded debris in parenthesis are followed by the SSN numbers that are not debris.

Int'l Designator

1959-007 (1) -   20

1963-049 (12) -   703 704 705 706 715 753 2432 2620 2930 4586 6182 6283

1964-026 (5) -   801 805 806 809 2986
1964-053 (2) -   876 877

1965-016 (9) -   1208 1244 1245 1271 1272 1291 1292 1293 1310
1965-027 (3) -   1314 1315 1316
1965-048 (4) -   1420 1425 1428 1435
1965-101 (4) -   1814 1815 1934 1935
1965-109 (5) -   1864 1865 2086 2226 2353

1966-005 (6) -   1952 1953 2140 2141 2889 2989
1966-024 (2) -   2119 2120
1966-040 (2) -   2173 2174
1966-077 (3) -   2403 2411 2412

1967-034 (4) -   2754 2755 2777 2778
1967-048 (4) -   2807 2811 17723 19222
1967-092 (4) -   2965 2967 2994 3122

1968-019 (2) -   3150 3151

1969-009 (2) -   3669 3670

1970-067 (5) -   4507 4515 5036 5447 6372

1971-003 (2) -   4849 4850

1972-049 (2) -   6079 6080
1972-102 (2) -   6319 6320

1975-027 (2) -   7734 7735

1978-014 (6) -   10664 10665 12329 12330 12331 12406
1978-064 (1) -   10967
1978-094 (2) -   11055 11056
1978-096 (3) -   1060 11061 11062
1978-098 (2) -   11080 11081

1979-057 (3) -   11416 11419 11634

1981-043 (3) -   12456 12457 15769
1981-054 (5) -   12512 12513 12514 12515 12519
1981-059 (3) -   12553 12559 12560

1985-066 (6) -   15935 15936 15938 15950 15951 16020
1985-090 (2) -   16110 16111

1987-020 (2) -   17535 17536 (there are over 100 pieces of fragmentation as well)

1988-005 (2) -   18820 18821
1988-032 (2) -   19045 19046

1989-089 (26) -   20322 20323 29683 29684 29685 29686 29687 29688 29689 29690 29691 29693 29694 29695
                  29696 29697 29698 29699 29700 29701 29702 29703 29704 29705 29706 29707

1990-017 (2) -   20508 20509
1990-037 (3) -   20579 20580 22920

1991-039 (3) -   21397 21398 21842
1991-050 (6) -   21574 21575 21576 21577 21578 21610

1993-014 (5) -   22561 22562 22567 22568 22599

1994-077 (11) -   23404 23405 23406 23407 23408 23409 23410 23417 23418 23419 27760

1995-041 (3) -   23639 23640 23641
1995-059 (2) -   23710 23711

1999-035 (2) -   25791 25792
1999-051 (3) -   25919 25920 25921

www.ingramcontent.com/pod-product-compliance
Lightning Source LLC
Chambersburg PA
CBHW080122220326
41598CB00032B/4920